Computational Maps
in the Visual Cortex

Risto Miikkulainen
James A. Bednar
Yoonsuck Choe
Joseph Sirosh

Computational Maps in the Visual Cortex

With 177 Figures, 47 in Full Color

 Springer

Risto Miikkulainen
Department of Computer Sciences
   and Institute for Neuroscience
The University of Texas at Austin
Austin, TX 78712-0233
USA
http://www.cs.utexas.edu/users/risto

James A. Bednar
School of Informatics
The University of Edinburgh
5 Forrest Hill
Edinburgh, EH1 2QL UK
http://homepages.inf.ed.ac.uk/jbednar

Yoonsuck Choe
Department of Computer Science
Texas A&M University
College Station, TX 77843-3112
USA
http://faculty.cs.tamu.edu/choe

Joseph Sirosh
Fair, Isaac & Company, Inc.
San Diego, CA 92129
USA
http://nn.cs.utexas.edu/keyword?sirosh

Cover illustration: The image includes parts of figures 5.29 and 13.22*b* appearing in the text.

ISBN-10: 0-387-22024-0   Printed on acid-free paper.
ISBN-13: 978-0387-22024-6

Printed in China.   (EVB)

9 8 7 6 5 4 3 2 1

springeronline.com

To our families

Foreword

Biological structures can be seen as collections of special devices coordinated by a matrix of organization. Devices are difficult to evolve and are meticulously conserved through the eons. Organization is a fluid medium capable of rapid adaptation. The brain carries organizational fluidity to the extreme. In its context, typical devices are ion channels, transmitters and receptors, signaling pathways, whole individual neurons or specific circuit patterns. The border line between what is to be called device and what a feat of organization is flowing, given that in time organized sub-systems solidify into devices. In spite of the neurosciences' traditional concentration on devices, their aiming point on the horizon must be to understand the principles by which the nervous system ties vast arrays of internal and external variables into one coherent purposeful functional whole — to understand the brain's mechanism of organization.

For that purpose a crucial methodology is in silico experimentation. Computer simulation is a convenient tool for testing functional ideas, a sharp weapon for distinguishing those that work from those that don't. To be sure, many alternatives can only be decided by direct experiment on the substrate, not by modeling. However, if a functional idea can be debunked as flawed once tried in silico it would be a waste to make it the subject of a decade of experimentation or discussion.

The venture of understanding the function and organization of the visual system illustrates this danger. Without much exaggeration it can be said that none of the academically formulated functional ideas could be shown to work on just any visual input. There is at present growing awareness that that is not due to lack of ingenuity but rather to a matter of principle: given the tremendous variability of the visual environment, no simple, intellectually coherent device can work in all situations. Object contours cannot be found solely by local contrast detection, the obvious direct mechanism, but only by coordination with other subsystems. The ambiguity plaguing the subsystems individually can be reduced only by global coordination between them. Thus, without understanding the phenomenon of organization we will not understand vision.

There is an even stronger reason to study organization. When trying to model brain function in silico, we have the tendency to first understand and solve the specific problem at hand in our own head and then create specific circuits and devices accordingly. This approach has long dominated the venture of artificial intelligence, and certainly also the field of computer vision. However, what may in the brain act like a fixed device may be an artifact of standardized experimental conditions and may in reality be the result of spontaneous organization. The devices (algorithms) in our computers are created by a separate process, in the mind of programmers. For the brain, there is no independent programmer (and evolution should not lightly be invoked as such). For the brain, there is no clear-cut separation between generation and execution of "algorithms." The interdigitated processes of evolution, ontogenesis, learning, brain state organization and, in the case of man at least, culture and education, are autonomously organizing the brain's functionality. The work of science will only have been done once we understand the principles of organization that not only coordinate subsystems but also create them. Only these principles are fixed, what they produce may to a large extent be due to accidents and circumstances.

This book is highly relevant to the goal of understanding organization. It summarizes and integrates an important body of work, accumulated over decades, aimed at describing and understanding the organization of the vertebrate visual system. Maps and columnar structures are a dominant theme of cortical organization. Due to an important wealth of experimental work on the substrate and in silico the mechanisms by which these structures are organized seem now before our eyes. The riddle of how less than 10^9 bits of genetic information are able to determine the arrangement of 10^{14} synaptic connections in ontogenesis is resolved by the demonstration that a relatively simple, genetically determined and controlled repertoire of cellular behavior is sufficient to understand the ontogenesis of regular connection patterns. The fundamental motivation behind hundreds of experimental studies of the ontogenesis of retinotopic connection patterns and also a sizable part of the work on cortical maps (on which this book concentrates) is the hope to elucidate the general mechanisms behind the development of the brain's wiring patterns. This work has led to very clear-cut conclusions painting a convincing and coherent picture. There is a regrettable reluctance of neurobiology to broadcast such conclusions as the message of fundamental importance that they constitute, so that there is a mission still to be accomplished here. This book is an important step in that direction. It employs the tool of computer simulation to show the validity of the principles that have emerged, to teach them, to develop them further and prepare them for application to novel cases.

Physics has found an ultimate receptacle and means of transmission of its results, in the form of mathematical descriptions and paradigmatic experiments. In distinction, biology still has to find the mode of knowledge formulation with which to capture the essence of the tremendous wealth of detailed results it has produced and is producing at a prodigious rate, a mode of formulation that makes it possible to close chapters and transmit conclusions to next generations of biologists. Theoretical biology is routinely applying mathematics to what I am calling here devices, but these individual mathematical formulations do not add up to a coherent canon, are rather as disparate as the devices to which they apply. There is, however, definite hope that a mathematical framework can be found for the phenomenon of organization. It has often been remarked that physics is deliberately studying the simple and that biology by force is concerned with the complex. But then, what is irreducibly complex? Seen under the right perspective even complex matters may come under the sway of relatively simple conceptual frameworks. Where this is not possible there can be no science and art must reign. No doubt, there are domains of irreducible complexity, but I doubt that the mechanisms of organization form one. Meticulous study of paradigmatic cases is necessary to penetrate that domain, and the study of vision at the cortical level, the focus of a tremendous body of scientific work, is sure to play a central role here.

The eternal discussion of nature vs. nurture, of prenatal vs. postnatal organization, has taken a very interesting turn in the context of cortical map formation. As will be discussed in these pages, neither side can possibly win. The methods that life has chosen here give the intriguing feeling that they contain a message of great importance for organization in general, if only we found the right perspective. It all gives the impression that evolution, far from having labored to develop and genetically encode specific devices for specific purposes, is just lightly playing its usual games, that just new tunes are played on a long-existing piano, the behavioral repertoire of living cells. Ocularity stripes evidently are not a tremendously clever and hard-won trick of evolution to exploit some complex vision problem, but turn out to naturally result from the collision of two retinotopic mappings trying to carve out common territory. This message is forcefully brought home by the famous experiment of Constantine-Paton and Law, in which this situation was artificially created in a frog, promptly resulting in ocularity stripes on the tectum for the first time in the evolution of that frog.

All that organization is about is the coordination of subsystems under a purpose. It is interesting to see how the conclusions propagated in this book perfectly illustrate and concretize that general theme. The function of the primary cortices is not constructed in isolation, with afferents to be plugged in later, like a fully constructed computer to which peripherals are connected, but structuring the cortices is more of an exercise in adaptation to the periphery and to other subsystems. Purpose of a specific kind may be brought in by the prenatal simulation (within the retina, or in the pontine region, if the PGO hypothesis advanced here is correct) of biologically significant stimuli. Here, evolution has to labor and make it clear to the new-born human baby, for instance, that the face of the mother is a most interesting and important stimulus. But evolution does so in a parsimonious fashion, laying down a mere schema of the face, which together with filter properties of the immature visual system and simple behavioral patterns of the mother suffice to identify examples as soon as the eyes are open. A possibly very general principle of learning may lie here. In order to extract essential structure from the environment in learning, it is first necessary to identify and separate from the background what is biologically significant. The general principle to identify significant patterns might be based on schematic descriptions of significant structures in the learning brain and its ability to map them into the environment, schemas being defined by evolution (or as the result of previous learning). When a pattern has been recognized, it is separated from the background. The brain thus avoids being swamped by masses of irrelevant information. A likely candidate mechanism for this separation is synchrony coding discussed here in the chapters on perceptual grouping.

It is my impression that the time is ripe for a major attack on the general problem of organization. Molecular biology and information technology are both hitting a serious complexity barrier. This can only be overcome by a shift of attention from the details of large systems to their organizing principles. Science can only conquer this domain with the help of insight gained on paradigmatic cases. The organization of visual cortex in perinatal ontogenesis may prove decisive in this role.

Bochum,
July 2004

Christoph von der Malsburg
Institut für Neuroinformatik, Ruhr-Universität Bochum;
Departments of Computer Science and Neurobiology,
University of Southern California

Preface

For several decades, the visual cortex has been the source of new theories and ideas about how the brain processes information. The visual cortex is easily accessible through a number of recording and imaging techniques and allows mapping high-level behavior relatively directly to neural mechanisms. It has also been the focal point in the emerging field of computational neuroscience. Several key ideas, such as input-driven self-organization, representing information on topographic maps, and temporal coding, originate from the mechanisms observed in the visual cortex. Understanding the computations in the visual cortex is therefore an important step toward a general computational brain theory.

Although computational theories of the visual cortex have existed for about 30 years, it has been difficult to test these theories experimentally and computationally. In the last 10 years or so the situation has finally started to change, for two reasons. First, it has become technically possible to measure how the visual cortex develops in response to external input, and how visual functions depend on low-level cortical mechanisms. Second, the available computational power has increased by several orders of magnitude. This technological confluence makes it possible for the first time to constrain and test precise computational models about how the visual cortex develops and functions, and why it has the organization it does. Computational models have gradually become an integral part of neuroscience theory.

The research in this area is far from unified. Several models exist to explain phenomena such as how ocular dominance and orientation preferences develop, how visual illusions and aftereffects arise, and how binding and segmentation take place, but it is not possible to see how they could function together in the visual cortex. Also, much of the research involves reimplementing ideas that have been around for several decades. There is no common overview of the field, nor is there a software framework on which future research could be based. This book is intended to fill these gaps: It presents a comprehensive, unified computational theory of the visual cortex as a laterally connected self-organizing map, it puts the theory in the context of past and current research in the area, and it is accompanied by a major software tool, *Topographica*, for modeling computational maps in the cortex in general.

For more than a decade, our research group at the University of Texas at Austin has worked on computational modeling of the visual cortex. Our perspective is to focus not only on the map-like structure of the cortex, but also take into account the dynamical processes that take place through lateral interaction and synchronization. It turns out that many developmental and functional phenomena depend on such processes, giving the model a unique explanatory power. This level of explanation is highly appropriate for understanding many visual processing phenomena; it is also a level where the theories are verifiable, leading to many predictions and proposals for future biological experiments. The book demonstrates how a number of phenomena follow from these principles, including columnar map organization and patchy connectivity, recovery from retinal and cortical injury, psychophysical phenomena such as tilt aftereffects and contour integration, and newborn preference for faces. Computational models are used to gain a precise understanding of existing data, and to make specific predictions for future experimental and theoretical research.

Our aim is to use the theory as a launching point to promote further research in this area. The principles of the models are described in detail, as are the techniques that make them work in practice, including parameter settings and scaling to different sizes and purposes. Most significantly, the book is accompanied by software, animations and demonstrations freely available on the Internet through http://topographica.org. *Topographica* is a general software tool for simulating cortical maps that allows neuroscientists to put together sophisticated computational experiments of their own design. As examples, the site contains specific models and demos described in this book. In this way, the book and the software are designed to complement each other, serving as a practical and a theoretical foundation for future research in computational neuroscience. Such a contribution, we believe, will significantly facilitate research in this area in the future.

The LISSOM project and the development of *Topographica* have benefited from the suggestions and contributions of many researchers, in fact too many to be listed here. We would especially like to thank Bill Geisler, Teuvo Kohonen, and Christoph von der Malsburg for substantial contributions of both ideas and critique over the years. Les Cohen, Larry Cormack, Joydeep Ghosh, Ben Kuipers, Bruce McCormick, Ray Mooney, Bruce Porter, Eyal Seidemann, Peter Stone, Chris Williams, and David Willshaw provided inspiration and guidance as doctoral committee members and as colleagues. Many research ideas were refined in discussions with Mike Arbib, Tony Bell, David Brainard, Dan Butts, Cara Cashon, Dmitri Chklovskii, Gary Cottrell, Jack Cowan, Michael Crair, Yang Dan, Peter Dayan, Scania de Schonen, Eizaburo Doi, Dawei Dong, Shimon Edelman, Steven Eglen, James Elder, Jeff Elman, Jerry Feldman, David Field, Peter Fox, Uli Frauenfelder, Nigel Goddard, Geoff Goodhill, Anatoli Gorchetchnikov, Steve Grossberg, Seung Kee Han, Seong-Whan Lee, Mike Hasselmo, Robert Hecht-Nielsen, Mike Hines, Geoff Hinton, David Horn, Fred Howell, Patrik Hoyer, Aapo Hyvärinen, Risto Ilmoniemi, Masumi Ishikawa, Naoum Issa, Mark Johnson, George Kalarickal, Pentti Kanerva, Sami Kaski, Krista Lagus, Pat Langley, Daniel Lee, Soo-Young Lee, Christian Lehmann, Ping Li, Jyh-Charn Liu, Xiuwen Liu, Jay McClelland, Brian MacWhinney, Gary Marcus, Denis Mareschal, Vinod Menon, Ken Miller, Klaus Obermayer, Erkki Oja, Bruno Olshausen, Remus Osan, Larry Parsons, Jim Reggia, Pamela Reinagel, Helge Ritter, Adrian Roberts, Eytan Ruppin, Terry Sejnowski, Lokendra Shastri, Harel Shouval, Hava Siegelmann, Michael Stryker, Mriganka Sur, John Taylor, Simon Thorpe, Dave Touretzky, David van Essen, Rufin VanRullen, Thomas Wachtler, DeLiang Wang, Mike Weliky, and Len White. Several former and current members of the University of Texas Neural Networks Research Group contributed to the design and implementation of the models and experiments, including Gautam Agarwal, Justine Blackmore, Judah De Paula, Igor Farkas, Andrea Haessly, Stefanie Jegelka, Amol Kelkar, Jeff Provost, Joe Reisinger, Yaron Silberman, Yiu Fai Sit, Tal Tversky, and Vinod Valsalam.

The research was supported in part by the National Institute of Mental Health (under Human Brain Project grant 1R01-MH66991 through Steven Koslow and Michael Hirsch), the National Science Foundation (under grants EIA-0303609, IIS-9811478, and IRI-9309273 through Darleen Fisher, Larry Reeker, and Su-Shing Chen, as well as by supercomputer grants IRI-94000P and IRI-930005P), and the College of Natural Sciences, the University of Texas at Austin (under a Dean's Research Fellowship).

Austin, Edinburgh, College Station, San Diego,   *Risto Miikkulainen*
November 2004   *James A. Bednar*
                                                                      Yoonsuck Choe
                                                                      Joseph Sirosh

Contents

Part I FOUNDATIONS

Part III CONSTRUCTING VISUAL FUNCTION

Part V EVALUATION AND FUTURE DIRECTIONS

Appendices

List of Figures

List of Tables

FOUNDATIONS

1

Introduction

How can a system as complex as the human visual system be constructed? How can it be specified genetically, still allowing it to adapt to the environment? How can it perform complicated functions such as recognizing faces and identifying coherent objects immediately and automatically?

This book aims at developing a computational theory of the visual cortex to answer these questions. While these questions have been open for quite some time, and much experimental work remains to be done to answer them conclusively, computational models serve an important role in this process: They provide a formal description of the principles and processes that are going on in biology. It is possible to use the models in lieu of biology, to test ideas that are difficult to establish experimentally, and to direct experiment to areas that are not understood well. Once verified, computational models provide a precise theory of the system.

The computational theory is expressed in detail in LISSOM, a laterally connected self-organizing map model of the visual cortex. LISSOM models the structure, development, and function of the visual cortex at the level of maps and their connections. The theory is based on three computational principles: Cortical columns constitute the basic computational unit, the units continuously adapt to visual and internal input, and the units synchronize and desynchronize their activity. Simulated experiments with LISSOM demonstrate how a wide variety of phenomena follow from these principles, including columnar map organization and patchy connectivity, recovery from retinal and cortical injury, psychophysical phenomena such as tilt aftereffects and contour integration, and newborn preference for faces. The model is used to gain a precise understanding of existing data, and to make specific predictions for future experimental and theoretical research.

The LISSOM model therefore suggests specific, computational answers to the above questions: (1) The cortical structures are constructed through input-driven self-organization; (2) the self-organization is driven both by external visual inputs and by genetically determined internal inputs; and (3) perceptual grouping takes place automatically through synchronization of neuronal activity, mediated by self-organized lateral connections. In this chapter, these three hypotheses are motivated in detail and the approach to verifying them computationally is outlined, providing a roadmap for the rest of the book.

1.1 Input-Driven Self-Organization

Current computing systems lag far behind humans and animals at many important information-processing tasks. One potential reason is that brains have far greater complexity, e.g. 10^{11} neurons and 10^{14} synapses compared with 10^8 transistors (Burger and Goodman 1997; Kandel, Schwartz, and Jessell 2000; Shepherd 2003). Designing specific blueprints for systems with 10^{14} components is beyond human engineering for the foreseeable future. How does nature manage to do it? One clue is that the genome has fewer than 10^5 genes in total, which means that any encoding scheme for the connections must be extremely compact (Lander et al. 2001; Venter et al. 2001). The first main hypothesis to be tested in this book is that instead of being specified directly genetically, *the structure in the visual cortex is constructed by input-driven self-organization.* Let us review the motivation for this idea in more detail.

The structure of the mammalian early visual areas is now well understood. Nerve fibers from the retina project to an intermediate region called the lateral geniculate nucleus (LGN), from which the fibers project to the primary visual cortex (V1). The Nobel prize winning studies of Hubel and Wiesel (1959, 1965, 1974) showed that neurons in the primary visual cortex are responsive to particular features in the input, such as a line of a particular orientation at a particular location in the visual field. Together, the locations on the retina to which a neuron responds are called the receptive field of the neuron. Neurons in a vertical column in the cortex have similar receptive fields and feature preferences. Vertical groups of neurons with the same orientation preference are called orientation columns, and vertical groups with the same eye preference are called ocular dominance columns; such groups may also be selective for direction of movement, spatial frequency, and color. This organization is shown schematically in Figure 1.1. The feature preferences gradually vary across the surface of the cortex in characteristic spatial patterns called cortical feature maps.

Many researchers have argued that such maps develop through self-organization of input connections from the thalamus and are shaped by visual experience (Shatz 1992). A number of classic experiments by Hubel, Wiesel and other researchers showed that altering the visual environment can drastically change the organization of input connections, ocular dominance columns, and orientation columns (Hubel and Wiesel 1962, 1974; Hubel, Wiesel, and LeVay 1977). The animal is most susceptible during a critical period of early life, typically a few weeks. For example, if a kitten is raised with both eyes sutured shut, its cortex will be abnormally organized, without ocular dominance and orientation columns. If the eyes are opened only after a critical period of a few weeks, the animal will be blind for life, even though the eyes and the LGN are perfectly normal. Similarly, if kittens are raised in environments containing only vertical or horizontal contours, their ability to see other orientations suffers significantly. In the cortex, most cells develop preferences for these

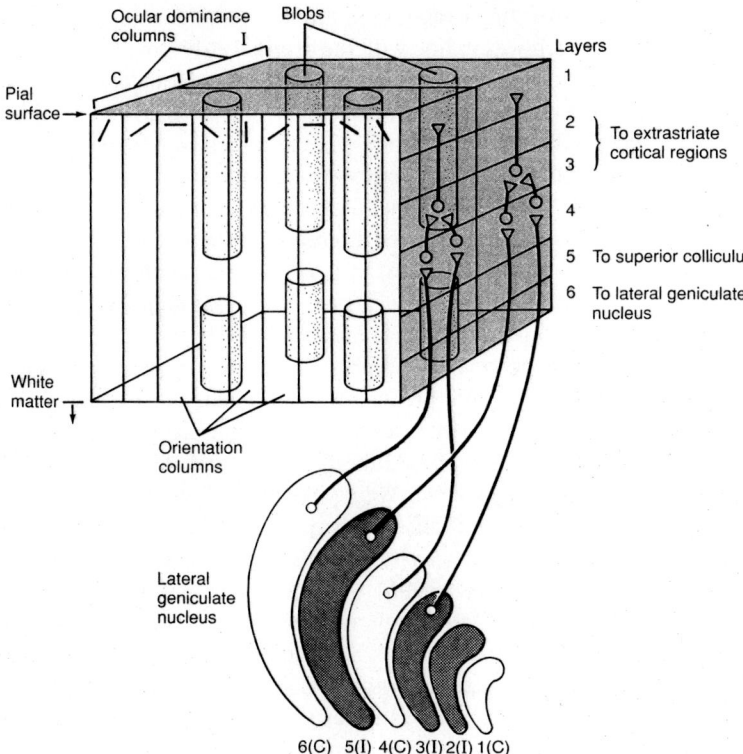

Fig. 1.1. Columnar organization of the primary visual cortex. This classic diagram illustrates an example patch of V1, responsive to one retinotopic location in the visual field. This patch includes an ocular dominance column for each eye, and a set of orientation columns within each ocular dominance column. Orientation preference changes along the length of the area shown, and ocular dominance along its width. Vertically, the receptive field properties are the same. Structures such as blobs, which analyze color, are scattered throughout the columns. Reprinted with permission from Kandel et al. (1991), copyright 1991 by McGraw-Hill.

particular orientations, and do not respond well to the other orientations (Blakemore and Cooper 1970; Blakemore and van Sluyters 1975; Hirsch and Spinelli 1970; Sengpiel, Stawinski, and Bonhoeffer 1999). Such experiments indicate that visual inputs are crucial for normal cortical organization, and suggest that the cortex tunes itself to the distribution of visual inputs.

How do such environmentally tuned feature preferences develop, and how do they become organized across the cortex? Since the 1970s, computational models have been used to demonstrate that both the preferences and their organization can result from a statistical learning algorithm that performs a nonlinear approximation of the distribution of visual inputs. The experiments in this book follow this tradition. An important, novel part of our theory is that lateral connections between columns self-organize to establish the competition and cooperation necessary for this process.

The earlier theories of the visual cortex did not include a significant role for the lateral connections, which was in line with the original experimental results. Altering the visual environment of the young animal changes the organization of its afferents; lateral connections were assumed to be necessary only to provide a stable environment for the afferent adaptation, and they were assumed to be isotropic, as they are in the retina. In the adult, the visual cortex was thought to be a collection of filters for visual input, and the properties of the filters (such as orientation preference) were thought to be defined by the patterns of afferent synapses. Possible lateral interactions between cells across the cortex were generally not taken into account, partly for simplicity, and partly because there did not exist sufficient neurobiological data to form well-defined theories about these interactions.

Over the last decade, however, a number of exciting results about lateral intracortical connectivity and dynamic processes in the visual cortex have emerged: (1) Lateral connections primarily connect areas with similar properties, such as neurons with the same orientation preference (Gilbert, Hirsch, and Wiesel 1990; Gilbert and Wiesel 1989; Löwel and Singer 1992; Weliky, Kandler, Fitzpatrick, and Katz 1995). (2) The lateral connections are initially uniform, but they become patchy during early development as a result of neural activity (Callaway and Katz 1990, 1991; Löwel and Singer 1992; Ruthazer and Stryker 1996). (3) Lateral connections develop at approximately the same time as orientation columns and ocular dominance columns form (Burkhalter, Bernardo, and Charles 1993; Katz and Callaway 1992). (4) By integrating information over large portions of the cortex, these connections appear to assist in the grouping of simple features such as edges into perceptual objects (Singer, Gray, Engel, König, Artola, and Bröcher 1990; von der Malsburg and Singer 1988). (5) The visual cortex is not static after maturation, but can adapt rapidly (in minutes) to retinal lesions and similar changes in the visual input. Several researchers have hypothesized that lateral connections play an important role in this adaptability (Gilbert and Wiesel 1992; Kapadia, Gilbert, and Westheimer 1994; Pettet and Gilbert 1992).

The new understanding of cortical development and function thus differs drastically from the old. It now appears that the adult visual cortex is a continuously adapting recurrent structure in a dynamic equilibrium, capable of rapid changes in response to altered visual environments. The lateral connections develop cooperatively and simultaneously with the thalamocortical afferents, and visual experience dynamically changes the lateral interactions throughout life.

In this book, a unified, dynamic computational model of such mechanisms in the visual cortex is developed. A single self-organizing process determines how both afferent and lateral connections develop in early life. This same process also continuously adapts the adult cortical structure during visual processing and may play an important role in perception. The model therefore provides strong computational support for the idea that cortical structure develops based on input-driven self-organization.

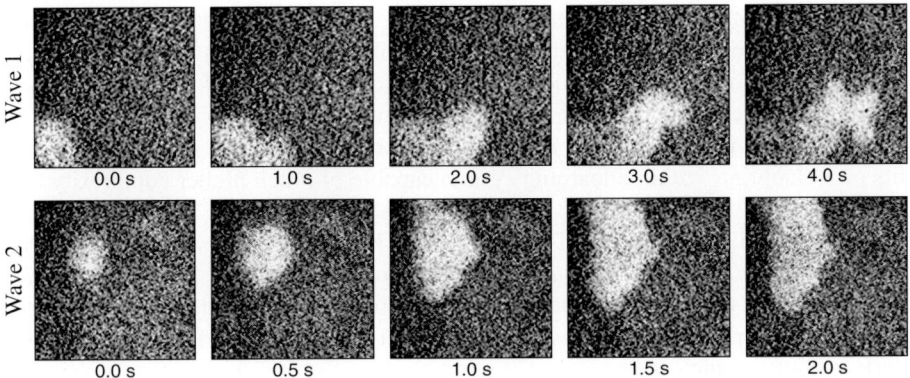

Fig. 1.2. Spontaneous activity in the retina. Each of the frames shows calcium concentration imaging of approximately 1 mm^2 of newborn ferret retina; the plots are a measure of how active the retinal cells are. Light gray indicates areas of increased activity. This activity is spontaneous (internally generated), because the photoreceptors have not yet developed at this time. From left to right, the frames on the top row form a 4-second sequence showing the start and expansion of a wave of activity. The bottom row shows a similar wave 30 seconds later. Later chapters will show that this type of correlated activity can explain how orientation selectivity develops before eye opening. Reprinted with permission from Feller et al. (1996), copyright 1996 by the American Association for the Advancement of Science; gray scale reversed.

1.2 Constructing Visual Function

The experiments with LISSOM will show that the self-organizing algorithm is powerful enough to construct structure from visual inputs starting from an initially uniform, unorganized state. However, there are two problems with this result: (1) Self-organization takes time, and the animal would not be able to act on visual input until the process is almost complete. (2) The self-organized structure depends critically on the specific input patterns available: if the visual environment is variable, the organism may not develop predictably, and what the learning algorithm discovers may not be the information most relevant to the organism.

In contrast, visual development in nature is highly stable, and the visual cortex of most animals is partially organized already at birth (or eye-opening). Such robustness could be achieved with a specific, fixed genetic blueprint, but (as was discussed above) there is not enough information available in the genome to represent it.

Recent experimental findings in neuroscience suggest that nature may have found a clever way to utilize self-organization to achieve the same result. Developing sensory systems are now known to be spontaneously active even before birth, i.e. before they could be learning from the environment (see O'Donovan 1999; Wong 1999 for reviews; Figure 1.2). This spontaneous, internal activity may actually guide the process of cortical development, acting as genetically specified training patterns for a learning algorithm (Constantine-Paton, Cline, and Debski 1990; Hirsch 1985; Jouvet 1998; Katz and Shatz 1996; Marks, Shaffery, Oksenberg, Speciale, and Roffwarg 1995; Roffwarg, Muzio, and Dement 1966; Shatz 1990, 1996; Sur and Leamey 2001). For a biological species, being able to control the training patterns can guarantee that each organism has a rudimentary level of performance from the start. Such training would also ensure that initial development does not depend on the details of the external environment. Thus, internally generated patterns can preserve the benefits of a blueprint, within a learning system capable of much higher complexity and performance.

The second main hypothesis tested in this book is that *the input-driven self-organization is based on internally generated patterns as well as external visual inputs.* Internal patterns drive the initial development, and the external environment completes the process. The result is a compact specification of a complex high-performance product.

This idea will be implemented in LISSOM, and illustrated on two visual capabilities where both genetic and environmental influences play a strong role: orientation processing and face detection. At birth, newborns can already discriminate between two orientations (Slater and Johnson 1998; Slater, Morison, and Somers 1988), and animals have neurons and brain regions selective for particular orientations even before their eyes open (Chapman and Stryker 1993; Crair, Gillespie, and Stryker 1998; Gödecke, Kim, Bonhoeffer, and Singer 1997). Yet, as reviewed above, orientation processing circuitry in these same areas can also be strongly affected by visual experience (Blakemore and van Sluyters 1975; Sengpiel et al. 1999). Internally generated patterns make it easier to build an effective orientation map from later environmental input, and they are crucial for explaining the experimental data. Similarly, newborns already prefer facelike patterns soon after birth, but face-processing ability takes months or years of experience to develop fully (Goren, Sarty, and Wu 1975; Johnson and Morton 1991; see de Haan 2001 for a review). Pattern generators can be used to specify such species-specific structure: If the visual system model is trained with simple three-dot patterns before birth, the newborn system prefers facelike schematics the same way human infants do, and gradually learns to recognize real faces through similar developmental phases.

These results suggest that self-organization driven by both internal and external inputs can be used to build complex, plastic, robust structures that would be too complex to determine directly genetically, and too fragile to learn from external inputs. Pattern generation is ubiquitous in nature, and could also be utilized in engineering of complex artificial systems in general.

1.3 Perceptual Grouping

In addition to understanding how the observed structures in the visual cortex emerge, it is important to understand what role they play in visual processing. Because LISSOM is a functional computational model, it can be tested in simulated neurobiological and psychophysical experiments. It is therefore ideal for testing hypotheses about the functional phenomena that arise from the self-organized structures.

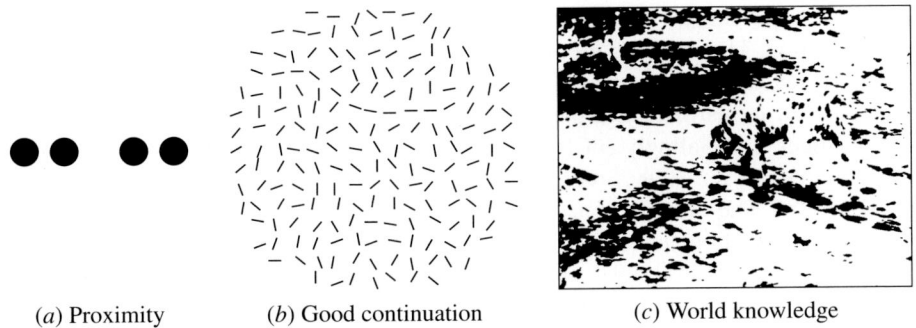

(*a*) Proximity   (*b*) Good continuation   (*c*) World knowledge

Fig. 1.3. Perceptual grouping tasks. Perceptual grouping is the process of identifying constituents in the visual scene that together form a coherent object. Perceptual grouping can take place at many different levels, from the very low level (*a*), to the very high level (*c*). (*a*) Grouping by proximity. The two black disks that are close to each other appear to form a unit. Thus, two groups are perceived: one on the left and another on the right. (*b*) Grouping by good continuation. In the random background of oriented edges (or contour elements), it is easy to notice the long, continuous sequence of edges that runs horizontally from the top-left of the circular area toward the right and slightly down. The task of detecting such contours is known as contour integration. (*c*) Grouping requiring world knowledge. In this seemingly unintelligible image lurks a Dalmatian dog sniffing on the pavement (a photograph by R. C. James; the dog is in the top right of the image, facing left). Without world knowledge, e.g. experience with dogs, leaves, etc., it would be impossible to group together the dots that form the Dalmatian.

Perhaps the most significant such function is perceptual grouping, or the process of identifying the constituents in the visual scene that together form a coherent object (Grossberg, Mingolla, and Ross 1997; Watt and Phillips 2000; Zucker 1995). The complexity of such tasks varies widely, and they can take place at various levels of the visual processing hierarchy (Figure 1.3). Different grouping principles are utilized at the different levels, including those based on spatial, temporal, and chromatic relationships (Geisler and Super 2000). At the level of orientation maps, perceptual grouping is manifested in contour integration, and a large body of neurobiological and psychophysical data is available to constrain, validate, and test the models. In this book, the LISSOM model will be used to test the hypothesis that *contour integration is an automatic function of the orientation map in the visual cortex, based on synchronized neuronal activity mediated by self-organized lateral connections.*

A typical visual input for the contour integration task is shown in Figure 1.3*b*. The input consists of a series of short oriented edge segments (contour elements) aligned along a continuous path, embedded in a background of randomly oriented contour elements. The task is to identify the longest continuous contour in this scene. Contour integration is an appropriate problem for computational analysis because the relationships between constituents of the image are neither too simple to be interesting (as in Figure 1.3*a* where the distance between the centers of the disks is the only grouping criteria), nor too complex to be represented (as in Figure 1.3*c* where complex world knowledge is required).

Most importantly, contour integration is believed to occur relatively early in the visual system. The response properties and connection patterns found in the primary visual cortex have exactly the right properties for explaining contour integration performance in terms of neural mechanisms (Field, Hayes, and Hess 1993; Geisler, Perry, Super, and Gallogly 2001; Li 1998; McIlhagga and Mullen 1996; Pettet, McKee, and Grzywacz 1998; Stettler, Das, Bennett, and Gilbert 2002; Yen and Finkel 1997, 1998). The lateral connections run along collinear or cocircular paths, and these areas are often activated together (Bosking, Zhang, Schofield, and Fitzpatrick 1997; Dalva and Katz 1994; Gilbert 1992; Katz and Callaway 1992; Löwel and Singer 1992; McGuire, Gilbert, Rivlin, and Wiesel 1991; Weliky et al. 1995). As discussed above, there is strong evidence that these structures are self-organized, driven by neural input (Blakemore and Cooper 1970; Blakemore and van Sluyters 1975; Hirsch and Spinelli 1970; Hubel and Wiesel 1962, 1974; Hubel et al. 1977; Ruthazer and Stryker 1996; White, Coppola, and Fitzpatrick 2001). Such specific patterns of connectivity are well suited for forming a consistent, coherent activation in response of a continuous contour.

One major question is how coherent percepts are represented in the cortex. The task consists of two parts: binding, i.e. grouping together separate constituent representations in the visual scene into a coherent object, and segmentation, i.e. segregating such coherently bound representations into different objects. With static activity, it is hard to represent binding and segmentation in a constantly changing sensory environment (von der Malsburg 1981, 1986a). Several researchers have proposed that temporal coding through synchronization, spike timing, phase differences, or other temporal information, could solve the problem (Eckhorn, Reitboeck, Arndt, and Dicke 1990; Horn and Opher 1998; Kammen, Holmes, and Koch 1989; Reitboeck, Stoecker, and Hahn 1993; Terman and Wang 1995; von der Malsburg 1986b; Wang 1995). Indeed, experiments with cats have shown that presentation of coherent objects gives rise to synchronized firing of neurons in the visual cortex, and presenting separate objects causes no synchronization (Eckhorn, Bauer, Jordan, Kruse, Munk, and Reitboeck 1988; Gray, Konig, Engel, and Singer 1989; Gray and Singer 1987; Singer 1993). Such coherent firing of neurons may be a possible representation for grouping.

In this book, the mechanisms of self-organized lateral connections and synchronization between groups of spiking neurons are brought together into an integrated developmental and functional model of the visual cortex. The results support the hypothesis that much of contour integration is performed in V1, based on these mechanisms. The work also suggests that similar mechanisms could be in use at higher levels, providing insights into perceptual grouping in general.

1.4 Approach

The above three hypotheses will be tested in a computational framework called LISSOM, or laterally interconnected synergetically self-organizing map. LISSOM is a computational map model of the visual cortex developed in our laboratory over the

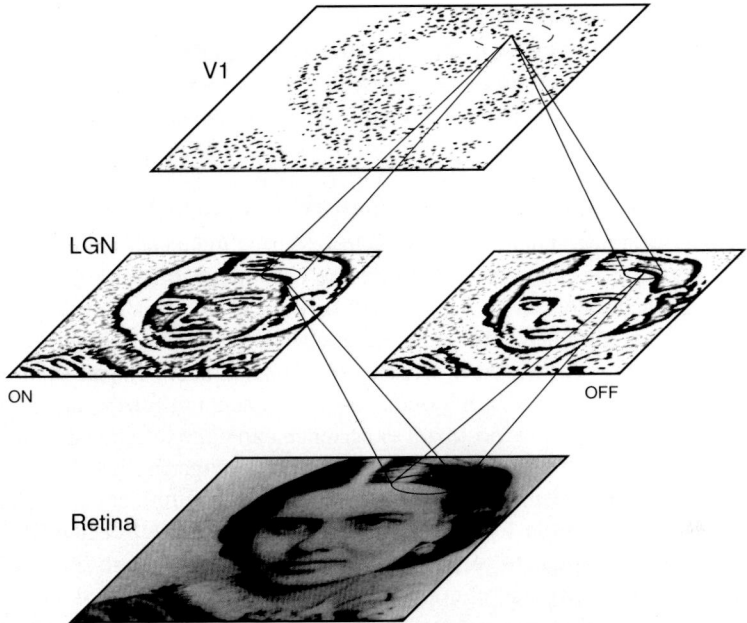

Fig. 1.4. Basic LISSOM model of the primary visual cortex. The core of the model consists of a two-dimensional array of computational units representing columns in V1. These units receive input from the retinal receptors through the ON/OFF channels of the LGN, and from other columns in V1 through lateral connections. The solid circles and lines delineate the receptive fields of two sample units in the LGN and one in V1, and the dashed circle in V1 outlines the lateral connections of the V1 unit. The LGN and V1 activation in response to a sample input on the retina is displayed in gray-scale coding from white to black (low to high). The V1 responses are patchy because each neuron is selective for a particular combination of image features (Figure 1.1), and only certain combinations exist in the image. This basic LISSOM model will be used in Part II to understand input-driven self-organization, cortical plasticity, and functional effects of adapting lateral connections. In Part III, the model is further extended with subcortical and higher level areas to study prenatal and postnatal development, and in Part IV, with binding and segmentation circuitry in V1 to model perceptual grouping.

past 10 years, building on about 30 years of map modeling research in the literature. LISSOM's core is a two-dimensional array of computational units corresponding to columns in V1, which receive inputs from the retina through the ON/OFF channels of the LGN and from other columns in V1 through lateral connections (Figure 1.4). The units learn through Hebbian adaptation, and compete with other units in a self-organizing map structure (Hebb 1949; Kohonen 2001; von der Malsburg 1973). The hypotheses are tested by analyzing the behavior of this model through simulated neurobiological and psychophysical experiments.

The input-driven self-organization hypothesis is tested in four ways: (1) In a number of specific experiments where each individual feature of visual inputs to the cortex, such as topographic order, eye dominance, orientation, and direction of movement, is learned and represented in the cortex; (2) in a combined simulation where a large cortical model self-organizes to represent all these features simultaneously; (3) in an adult-plasticity experiment where the cortex repairs itself after retinal or cortical damage; and (4) in a functional experiment where visual aftereffects are shown to arise from these same mechanisms in the normal adult system.

The pattern generation hypothesis will be evaluated by building and testing HLISSOM, a hierarchical model that includes both subcortical and higher visual areas. The goal is to understand how internal and external inputs affect the organization and function of the visual cortex. Because the orientation processing circuitry has been mapped out in detail in animals, it will be used as a verifiable test case for the pattern generation approach. The same techniques will then be applied to face processing, where they will be used as a basis for a unified theory for the phenomenon. The goal is to demonstrate how internal activity can account for the newborn structure in each system, and how postnatal experience can complete this developmental process. In each case, the model is first validated by comparing it with existing experimental results, and then used to derive predictions for future experiments.

The contour integration hypothesis will be studied in the PGLISSOM model, where LISSOM is extended to perform perceptual grouping through spiking neurons and long-range excitatory lateral connections. Grouping is measured as the degree of synchrony among neural populations, and such synchrony is established through the lateral connections. This model shows how the statistical structure in the visual environment determines the structure of the visual cortex, which in turn determines its grouping performance. The model therefore provides a computational account of the possible neural mechanisms of contour integration.

In addition to providing computational support for the above three hypotheses, the LISSOM framework constitutes a general computational theory of representation and learning in the visual cortex. The learning mechanisms extract correlations in the input that allow representing visual information efficiently in a sparse, redundancy-reduced code. Such representations are separable and generalizable, and serve as an effective foundation for later stages of visual processing, such as pattern recognition. These computational principles are abstractions of what the cortex is doing, but they are also general principles that could be used in constructing artificial systems.

The LISSOM approach is intended to serve as a starting point for future explorations in computational understanding of the visual cortex. The models described in this book are freely available on the Internet under the *Topographica* project (http://topographica.org). In this project, a general simulator for computational modeling of cortical maps is being developed, intended to support further research in this general area. We believe that the current confluence of experimental data on cortical maps and such newly available computational tools will lead to major progress in understanding how the brain processes visual information.

1.5 Guide for the Reader

The book is divided into five parts. First, the biological background is reviewed for the core constituents of LISSOM, i.e. for self-organization, lateral connections, genetic vs. environmentally driven development, and temporal coding. The computational foundations of LISSOM, such as the neuron models, synchronization, learning, and self-organizing maps, are also discussed. However, the specific biological and psychophysical evidence and prior modeling work for each individual experiment is reviewed in the individual chapters throughout the book.

Part II focuses on mechanisms of input-driven self-organization. The basic architecture of the LISSOM computational map model of V1 is presented, and demonstrated to develop a map organization and patchy lateral connections based on regularities in the visual input. The same self-organization processes are shown to account for plasticity of the adult cortex, and give rise to psychophysical phenomena such as the tilt aftereffect.

Part III demonstrates how genetic and environmental influences can be combined in input-driven self-organization. The LISSOM model of V1 is first expanded outward into a multi-level model containing subcortical areas and higher visual maps, capable of processing both natural images and internally generated input. This model demonstrates a synergy of nature and nurture in developing orientation preferences, and allows gaining insight into high-level phenomena such as infant face processing.

Perceptual grouping is studied in Part IV. To gain insight into this process, the LISSOM model is extended inward to include spiking units and separate excitatory and inhibitory components in cortical columns. The resulting temporal coding and self-organization processes are demonstrated in detail, and shown to work together. The model is shown to account for low-level perceptual grouping phenomena such as contour integration performance under varying conditions, integration of illusory contours, and differences in grouping performance across the different areas of the cortex.

In Part V, laterally connected self-organizing maps are shown to result in efficient visual representations well suited for higher level processing and for practical applications. Techniques are developed for scaling the approach to very large maps, including possibly the entire visual cortex. The assumptions and predictions of LISSOM are reviewed and evaluated in terms of biological research results and opportunities. Connections are made to related and complementary work in cortical modeling and cognitive science, and future directions are outlined.

2

Biological Background

In later chapters, computational simulations are presented that describe how the human visual system develops and functions. In order to make such simulations a useful tool for understanding natural systems, they are based on detailed anatomical, neurophysiological, and psychological evidence. In this chapter, the organization of the visual system in humans and higher animals is reviewed, and biological evidence is discussed for structures and processes that are important for later chapters, such as lateral connections, externally and internally driven development, and temporal coding. Computational principles for modeling these phenomena are reviewed in the next chapter. Biological evidence for each specific phenomenon modeled will be reviewed in each chapter separately, and the general biological foundations of the model are evaluated in Chapter 16.

2.1 Visual System Organization

The adult visual system has been studied experimentally in a number of mammalian species, including human, monkey, cat, ferret, and tree shrew. For a variety of reasons, many of the important results have been measured in only one or a subset of these species, but they are generally expected to apply to the others as well. This book focuses on the human visual system, but also relies on data from these animals where human data are not available.

Figure 2.1 shows a diagram of the main feedforward pathways in the human visual system (see e.g. Daw 1995; Kandel et al. 2000; Wandell 1995 for reviews). Other mammalian species have a similar organization. During visual perception, light entering the eye is detected by the retina, an array of photoreceptors and related cells on the inside of the rear surface of the eye. The cells in the retina encode the light levels at a given location as patterns of electrical activity in neurons called ganglion cells. This activity is called visually evoked activity. Retinal ganglion cells are densest in a central region called the fovea, corresponding to the center of gaze; they are much less dense in the periphery. Output from the ganglion cells travels through neural connections to the lateral geniculate nucleus of the thalamus, or LGN, at the base

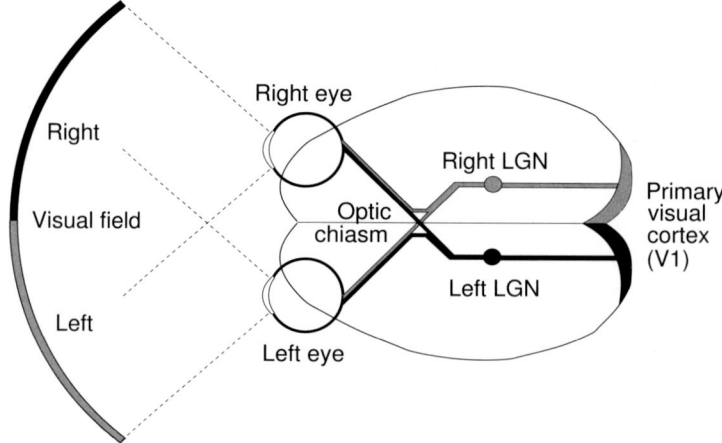

Fig. 2.1. Human visual pathways (top view). Visual information travels in separate pathways for each half of the visual field. For example, light entering the eye from the right hemifield reaches the left half of the retina, on the rear surface of each eye. The right hemifield inputs from each eye join at the optic chiasm, and travel to the LGN of the left thalamus, then to primary visual cortex, or area V1, of the left hemisphere. Signals from each eye are kept segregated into different neural layers in the LGN, and are combined in V1. There are also smaller pathways from the optic chiasm and LGN to other subcortical structures, such as the superior colliculus and pulvinar (not shown).

of each side of the brain. From the LGN, the signals continue to the primary visual cortex, or V1 (also called striate cortex and area 17) at the rear of the brain. V1 is the first cortical site of visual processing; the previous areas are termed subcortical. The output from V1 goes on to many different higher cortical areas, including areas that underlie object and face processing (see e.g. Merigan and Maunsell 1993; Van Essen, Anderson, and Felleman 1992 for reviews). Much smaller pathways also go from the optic nerve and LGN to subcortical structures such as the superior colliculus and pulvinar. In humans these subcortical pathways are involved primarily in eye movements and attention (LaBerge 1995; LaBerge and Buchsbaum 1990; Wallace, McHaffie, and Stein 1997). The LISSOM model focuses on V1 and the structures to which it connects, as reviewed below.

2.1.1 Early Visual Processing

At the photoreceptor level, the representation of the visual field is much like an image, but significant processing of this information occurs in the subsequent subcortical and early cortical stages (see e.g. Daw 1995; Kandel et al. 2000 for reviews).

First, retinal ganglion cells perform a type of edge detection on the input, responding most strongly to borders between bright and dark areas. Figure 2.2*a,b* illustrates the two main types of such neurons, ON-center and OFF-center. An ON-center retinal ganglion cell responds most strongly to a spot of light surrounded by

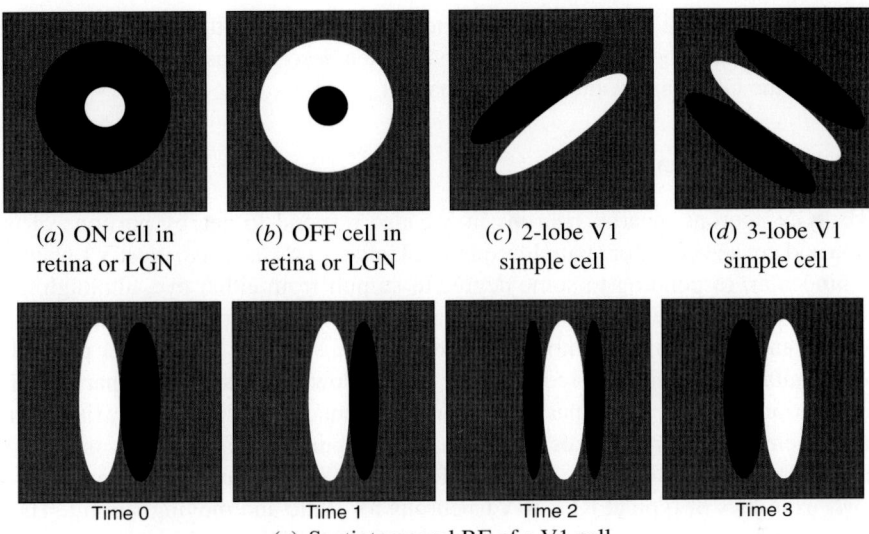

| (a) ON cell in retina or LGN | (b) OFF cell in retina or LGN | (c) 2-lobe V1 simple cell | (d) 3-lobe V1 simple cell |

| Time 0 | Time 1 | Time 2 | Time 3 |

(e) Spatiotemporal RF of a V1 cell

Fig. 2.2. Receptive field types in retina, LGN and V1. Each diagram shows a receptive field on the retina for one neuron. Areas of the retina where light spots excite this neuron are plotted in white (ON areas), areas where dark spots excite it are plotted in black (OFF areas), and areas with little effect are plotted in medium gray. The size of the RFs varies, but they all have the same basic shape and they are all spatially localized, i.e. their ON and OFF areas cover a small specific portion of the retina. (a) ON cells are found in the retina and LGN, and prefer light areas surrounded by dark. (b) OFF cells have the opposite preferences, responding most strongly to a dark area surrounded by light. RFs for both ON and OFF cells are isotropic, i.e. have no preferred orientation. Starting in V1, most cells in primates have orientation-selective RFs instead. The V1 RFs can be classified into a few basic spatial types, of which the two most common are shown above: (c) A two-lobe arrangement, favoring a $45°$ edge with dark in the upper left and light in the lower right, and (d) a three-lobe pattern, favoring a $135°$ white line against a dark background. Both types of RF are often represented with Gabor functions (Daugman 1980; Jones and Palmer 1987). RFs of all orientations are found in V1, but those representing the cardinal axes (horizontal and vertical) are more common. Many neurons are also sensitive for the direction of movement of these patterns, i.e. their RFs are spatiotemporal. For such a neuron, successive snapshots of the spatial RF at different times are shown in (e); together they form a spatiotemporal RF selective for a vertical light bar moving to the right. A model for the ON and OFF cells will be introduced in Chapter 4 and for the simple and spatiotemporal V1 cells in Chapter 5.

dark, located in a region of the retina called its receptive field, or RF. An OFF-center ganglion cell instead prefers a dark area surrounded by light. The size of the preferred spot determines the spatial frequency preference of the neuron; neurons preferring large spots have a low preferred spatial frequency, and vice versa.

Neurons in the LGN have properties similar to retinal ganglion cells, and are also arranged retinotopically, so that nearby LGN cells respond to nearby portions of the retina. The ON-center cells in the retina connect to the ON cells in the LGN, and the OFF cells in the retina connect to the OFF cells in the LGN. Because of this independence, the ON and OFF cells are often described as separate processing channels: the ON channel and the OFF channel.

2.1.2 Primary Visual Cortex

Like LGN neurons, nearby neurons in V1 also respond to nearby portions of the retina and are selective for spatial frequency. Unlike LGN neurons, most V1 neurons are binocular, responding to some degree to stimuli from either eye, although they usually prefer one eye or the other. They are also selective for the orientation of the stimulus and its direction of movement. In addition, some V1 cells prefer particular color combinations (such as red/green or blue/yellow borders), and disparity (relative positions on the two retinas). V1 neurons respond most strongly to stimuli that match their feature preferences, although they respond to approximate matches as well (Hubel and Wiesel 1962, 1968; see Ringach 2004 for a review). Figure 2.2*c–e* shows examples of typical RFs of V1 neurons for static and moving stimuli. These neurons are simple cells, i.e. neurons whose ON and OFF regions are located at specific areas of the retinal field. Other neurons (complex cells) respond to the same configuration of light and dark over a range of positions. LISSOM models the simple cells only, which are thought to be the first in V1 to show orientation selectivity.

V1, like the other parts of the cortex, is composed of a two-dimensional, slightly folded sheet of neurons and other cells. If flattened, human V1 would cover an area of nearly four square inches. It contains at least 150 million neurons, each making hundreds or thousands of specific connections with other neurons in the cortex and in subcortical areas like the LGN (Wandell 1995). The neurons are arranged in six layers with different anatomical characteristics (using Brodmann's scheme for numbering laminations in human V1, as described by Henry 1989; Figure 2.6). Input from the thalamus goes through afferent connections to V1, typically terminating in layer 4 (Casagrande and Norton 1989; Henry 1989). Neurons in the other layers form local connections within V1 or connect to higher visual processing areas. For instance, many neurons in layers 2 and 3 have long-range lateral connections to the surrounding neurons in V1 (Gilbert et al. 1990; Gilbert and Wiesel 1983; Hirsch and Gilbert 1991). There are also extensive feedback connections from higher areas (Van Essen et al. 1992). Lateral connections play a central role in the LISSOM model, and will be discussed in detail in Section 2.2.

At a given location on the cortical sheet, the neurons in a vertical section through the cortex respond most strongly to the same eye of origin, stimulus orientation, spatial frequency, and direction of movement. It is customary to refer to such a section as a column (Gilbert and Wiesel 1989). The LISSOM model will treat each column as a single unit, thus representing the cortex as a purely two-dimensional surface. This model is a useful approximation because it greatly simplifies the analysis while retaining the basic functional features of the cortex.

Nearby columns generally have similar, but not identical, preferences; slightly more distant columns have more dissimilar preferences. Preferences repeat at regular intervals (approximately 1–2 mm) in every direction, which ensures that each

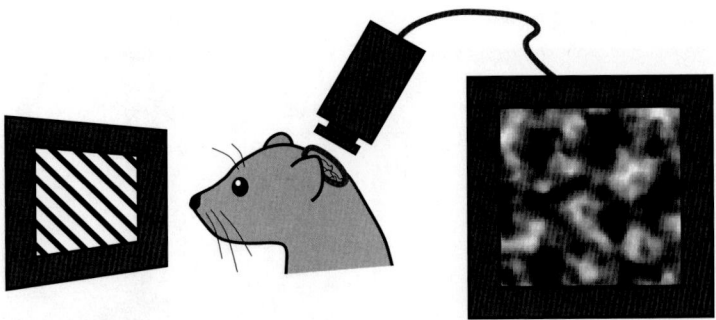

Fig. 2.3. Measuring cortical maps. Optical imaging techniques allow neuronal preferences to be measured for large numbers of neurons at once (Blasdel and Salama 1986). In such experiments, part of the skull of a laboratory animal is removed by surgery, exposing the surface of the visual cortex. Visual patterns are then presented to the eyes, and a video camera records either light absorbed by the cortex or light given off by voltage-sensitive fluorescent chemicals that have been applied to it. Depending on the neural activity, there will be small differences in the emitted or reflected light, and these differences can be amplified by repeated presentations and averaging. The results are an indirect measure of the average two-dimensional pattern of neural activity resulting from a particular stimulus. Measurements can then be compared between different stimulus conditions, e.g. different orientations, determining which stimulus is most effective at activating each small patch of neurons. Figure 2.4 and later figures in this chapter will show maps of orientation preference computed using these techniques. Adapted from Weliky et al. (1995).

type of preference is represented across the retina. This arrangement of preferences forms a smoothly varying map for each dimension. For example, stimulus orientation is represented across the cortex in an orientation map of the retinal input (Blasdel 1992a; Blasdel and Salama 1986; Grinvald, Lieke, Frostig, and Hildesheim 1994; Ts'o, Frostig, Lieke, and Grinvald 1990). Figure 2.3 shows how such maps can be measured experimentally in animals, and Figure 2.4 displays an example orientation map from monkey cortex. In an orientation map, each location on the retina is mapped to a region on the map, with each possible orientation at that retinal location represented by different but nearby orientation-selective cells. Other mammalian species have largely similar orientation maps, although they differ in details (Müller, Stetter, Hubener, Sengpiel, Bonhoeffer, Gödecke, Chapman, Löwel, and Obermayer 2000; Rao, Toth, and Sur 1997).

Other stimulus features are represented in a similar fashion as maps, including those for direction of motion and ocular dominance (left or right eye preference; Blasdel 1992a; Crowley and Katz 2000; Löwel 1994; Obermayer and Blasdel 1993; Shatz and Stryker 1978; Shmuel and Grinvald 1996; Weliky, Bosking, and Fitzpatrick 1996). These maps are overlaid so that a hierarchical representation of the input features emerges (Figure 2.5). The primary organization in the hierarchy is retinotopy. Neurons that respond to the same location are divided into those that respond primarily to the left eye and those that respond primarily to the right eye. Each

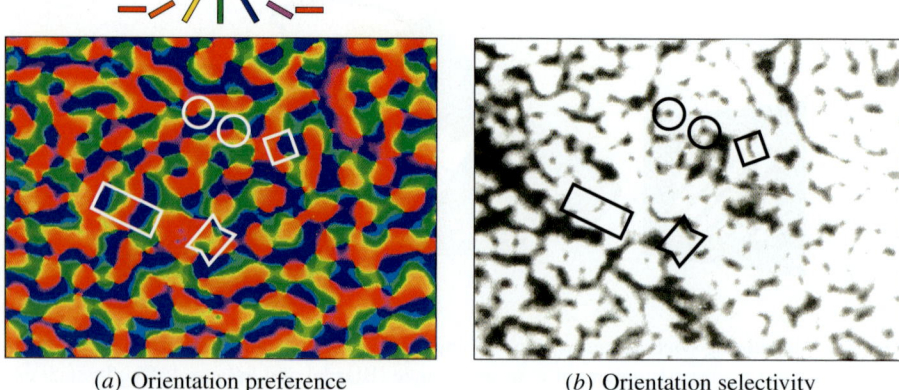

(*a*) Orientation preference   (*b*) Orientation selectivity

Fig. 2.4. Orientation map in the macaque. (*a*) Orientation preference and (*b*) orientation selectivity maps in a 7.5 mm × 5.5 mm area of adult macaque monkey V1, measured by optical imaging techniques. Each neuron in (*a*) is colored according to the orientation it prefers, using the color key on top. Nearby neurons in the map generally prefer similar orientations, forming groups of the same color called iso-orientation patches. Other qualitative features are also found. Linear zones are straight lines along which the orientations change continuously, like a rainbow; a linear zone is marked with a long white rectangle. Pinwheels are points around which orientations change continuously. They often occur in matched pairs: such a pair is circled in white. At saddle points a long patch of one orientation is nearly bisected by another; one saddle point is marked with a bowtie. Fractures are sharp transitions from one orientation to a very different one; a fracture between red and blue (without purple in between) is marked with a white square. Orientation selectivity measures how closely the input must match the neuron's preferred orientation for it to respond. As shown in (*b*), neurons at pinwheel centers and fractures tend to be less selective (dark areas) in the optical imaging response, whereas iso-orientation patches, linear zones and saddle points tend to be more selective (light areas). Reprinted with permission from Blasdel (1992b), copyright 1992 by the Society for Neuroscience; annotations added and brightness increased.

group is further divided into areas that respond to particular orientations. In turn, each orientation-selective patch is often further subdivided into two patches, each preferring opposite directions of motion (Shmuel and Grinvald 1996; Weliky et al. 1996). Other stimulus features (such as spatial frequency and color) are represented as well, but are not as well organized at the large scale (Issa, Trepel, and Stryker 2001; Landisman and Ts'o 2002b). Simulations with LISSOM will show how the hierarchical map-like organization arises automatically from input-driven self-organization, and how it constitutes an efficient way to represent visual information.

2.1.3 Face and Object Processing

Beyond V1 in primates are dozens of extrastriate visual areas that can be arranged into a rough hierarchy (Van Essen et al. 1992). The relative locations of the areas in this hierarchy are largely consistent across individuals of the same species. Non-primate species have fewer higher areas, and in at least one mammal (the least shrew,

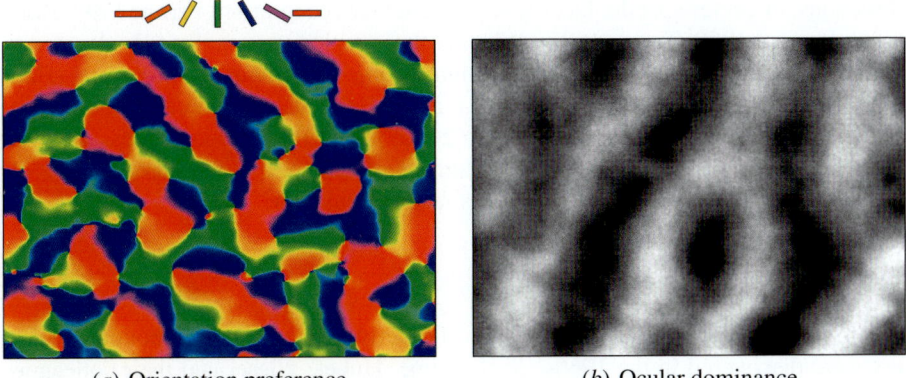

(*a*) Orientation preference   (*b*) Ocular dominance

Fig. 2.5. Hierarchical organization of feature preferences in the macaque. The images illustrate orientation and ocular dominance patches in a 4 mm × 3 mm area of the cortical surface in the macaque monkey, measured through optical imaging. (*a*) The cells are colored according to their orientation preference as in Figure 2.4*a*. (*b*) The same cells are colored in gray scale from white to black according to how strongly they prefer input from the left vs. the right eye. Each neuron is sensitive to a combination of feature values, in this case a line of a particular orientation in the left or the right eye at a particular location on the visual field. These maps are shown superimposed in Figure 5.3, revealing more fine-grained interactions between the maps. Plot (*a*) reprinted with permission from Blasdel (1992b) and plot (*b*) from Blasdel (1992a), copyright 1992 by the Society for Neuroscience.

a tiny rodent-like creature) V1 is the only visual area (Catania, Lyon, Mock, and Kaas 1999). Although the higher levels have not been studied as thoroughly as V1, the basic circuitry within each region is thought to be largely similar to V1. Even so, the functional properties differ greatly, in part because their connections with other regions are different. For instance, neurons in higher areas tend to have larger retinal receptive fields, respond to stimuli at a greater range of positions, and process more complex visual features (Ghose and Ts'o 1997; Haxby, Horwitz, Ungerleider, Maisog, Pietrini, and Grady 1994; Kandel et al. 2000; Rolls 2000; Wang, Tanaka, and Tanifuji 1996). In particular, extrastriate cortical regions that respond preferentially to faces have been found both in adult monkeys (using single-neuron studies and optical imaging; Gross, Rocha-Miranda, and Bender 1972; Hasselmo, Rolls, and Baylis 1989; Rolls 1992; Rolls, Baylis, Hasselmo, and Nalwa 1989; Wang et al. 1996) and adult humans (using functional magnetic resonance imaging, or fMRI; Halgren, Dale, Sereno, Tootell, Marinkovic, and Rosen 1999; Kanwisher, McDermott, and Chun 1997; Puce, Allison, Gore, and McCarthy 1995). Any such cell or region that responds stronger to faces than to other similar stimuli is called face selective.

The face-selective areas receive visual input via V1. They are loosely segregated into different regions that process faces in different ways. For instance, some areas perform face detection, i.e. respond unspecifically to many facelike stimuli (de Gelder and Rouw 2000, 2001). Others selectively respond to facial expressions, gaze directions, or prefer specific faces (i.e. perform face recognition; Perrett 1992; Rolls 1992; Sergent 1989; Treves 1997). Whether these regions are exclusively devoted to face processing, or also process other common objects, remains controversial (Hanson, Matsuka, and Haxby 2004; Haxby, Gobbini, Furey, Ishai, Schouten, and Pietrini 2001; Kanwisher 2000; Tarr and Gauthier 2000). LISSOM will model areas involved in face detection (and not face recognition or other types of face processing), although these areas do not have to process faces exclusively.

2.1.4 Input-Driven Self-Organization

The first hints of how these complicated yet orderly structures come about in the cortex were discovered in the 1960s. At that time, Hubel, Wiesel and their colleagues conducted a number of experiments where they showed that altering the visual environment drastically changes the organization of the visual cortex (Hubel and Wiesel 1962, 1974; Hubel et al. 1977). For example, if a kitten's vision is impaired by suturing the eyes shut, the visual cortex becomes disorganized, lacking orientation selectivity and ocular dominance patches. Such an effect is most dramatic during the critical period, typically within a few weeks after birth: If the eyes are kept shut until after the critical period, the animal actually becomes blind. If the animal (e.g. a ferret) is reared in the dark instead of suturing the eyes shut, the visual system becomes similarly impaired, although to a lesser extent (White et al. 2001), suggesting that abnormal visual stimulation through the closed eyelids is more harmful than receiving none at all. These results show how important normal visual stimuli are during the critical period to ensure that the visual system develops normally.

Development has been shown to depend on input in several more specific experiments as well. For example, kittens can be raised in an environment with only vertical or horizontal features, and as a result, they are unable to respond well to other orientations (Blakemore and Cooper 1970; Blakemore and van Sluyters 1975; Hirsch and Spinelli 1970). Similar results have been reported for ocular dominance in ferrets: If one eye is sutured shut during the critical period, the animal loses the ability to respond to inputs from that eye as an adult (Issa, Trachtenberg, Chapman, Zahs, and Stryker 1999). Further, the auditory cortex has been shown to become sensitive to visual inputs when the projections from the retina are surgically connected to it (Sharma, Angelucci, and Sur 2000; Sur, Garraghty, and Roe 1988).

These experimental results convincingly demonstrate that the connections in the cortex are shaped by environmental input. Part II of the book focuses on understanding the mechanisms underlying this process, showing that input-driven self-organization is able to construct the observed structures even from an initially uniform, unordered starting point, based on suitable input. However, how much of the organization is indeed due to environmentally driven self-organization and how much is genetically determined is open to a considerable debate (as will be reviewed in Section 2.3). A solution to this question is proposed in Part III, showing how genetically specified self-organization followed by environmentally driven self-organization can account for many of the observed phenomena in visual development.

2.2 Lateral Connections

As was discussed in Section 1.1, the modern understanding of the visual cortex as a continuously adapting dynamic system has caused us to reconsider the role of lateral connections in cortical development and function. Lateral interactions seem to play a much larger role than previously believed, a role that we are only now beginning to understand. Because complex recurrent systems are difficult to study experimentally, computational models are crucial in developing a detailed theory about lateral connections in the cortex.

LISSOM is the first computational theory specifically designed for this purpose. It allows self-organization and analysis of lateral connections to take place in a functioning visual cortex model. The biological foundations of the LISSOM approach are discussed below, followed by a review of current ideas about the role of lateral connections in the cortex (for more details, see e.g. Sirosh, Miikkulainen, and Choe 1996b).

2.2.1 Organization

Long-range lateral connections form a dense, highly patterned network within the cortex. Each connection extends over several millimeters and gives rise to clusters of axon endings at regular intervals (Figure 2.6; Fisken, Garey, and Powell 1975; Gilbert and Wiesel 1979; Schwark and Jones 1989). In the primary visual cortex these connections can be 6–8 mm long, i.e. cover a substantial percentage of the V1 area. They are reciprocal, i.e. if area A connects to B, then B connects back to A. Long-range connections are found in layers 2, 3, 5, and 6; they are longest in layers 2 and 3. The lateral connection patterns in the different layers are aligned, and the dendritic arbor of pyramidal cells in layer 3 matches the axonal clusters (Burkhalter and Bernardo 1989; Gilbert and Wiesel 1989; Katz and Callaway 1992; Livingstone and Hubel 1984b; Luhmann, Martínez Millán, and Singer 1986; Lund, Yoshioka, and Levitt 1993; Rockland 1985; Rockland, Lund, and Humphrey 1982; see Douglas and Martin 2004 for a review).

About 80% of the long-range connections synapse on excitatory pyramidal cells, while the remaining 20% synapse on inhibitory interneurons (Gilbert et al. 1990; McGuire et al. 1991). Imaging studies and other measurements indicate a substantial amount of long-range inhibition in the cortex, more than predicted by the above 80–20 distribution; moreover, at high contrasts the net effect is strongly inhibitory (Section 16.1.4; Grinvald et al. 1994; Hata, Tsumoto, Sato, Hagihara, and Tamura 1993; Hirsch and Gilbert 1991; Weliky et al. 1995). One fundamental assumption of the LISSOM model is that the lateral excitatory and inhibitory connections serve different roles in the visual cortex; both kinds of connections are therefore included in the LISSOM models in this book.

Long-range lateral connections are clustered in patches whose distribution corresponds closely to the organization of receptive fields in the sensory map, especially orientation. The connections of a given neuron target neurons in other areas that have similar orientation preferences, aligned along the preferred orientation of the neuron

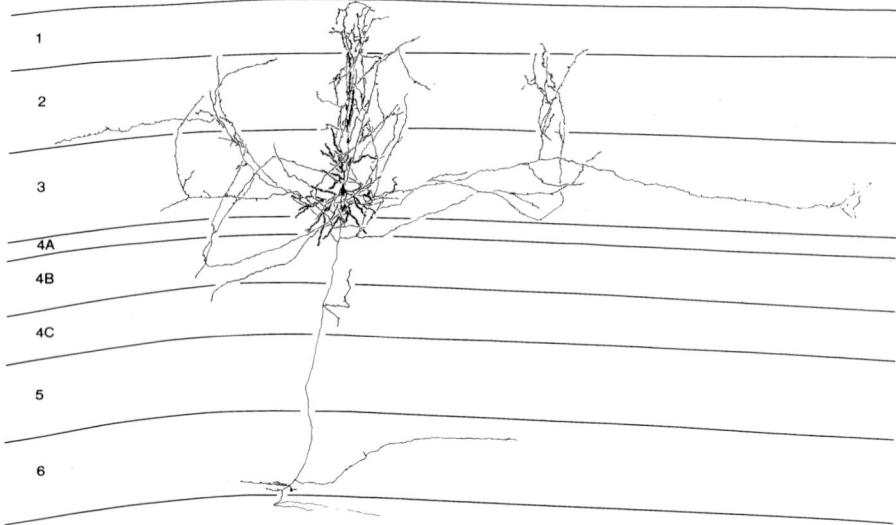

Fig. 2.6. Long-range lateral connections in the macaque. Lateral connections, also sometimes called horizontal or intrinsic connections, run parallel to the cortical surface. In the visual cortex they extend over several millimeters and sprout branches at intervals. The branches form a local cluster of connections to other cells in the region, as shown for this layer 3 pyramidal cell in the macaque visual cortex (injected with horseradish peroxidase: The dendrites are shown with thick lines and axon collaterals with thin lines, and the horizontal scale is approximately 2.3mm). Such clusters occur only in regions with similar functional properties as the parent cell. Reprinted with permission from Gilbert et al. (1990; adapted from McGuire et al. 1991), copyright 1990 by Cold Spring Harbor Laboratory Press.

(Figure 2.7; Bosking et al. 1997; Fitzpatrick, Schofield, and Strote 1994; Gilbert et al. 1990; Gilbert and Wiesel 1989; Malach, Amir, Harel, and Grinvald 1993; Schmidt, Kim, Singer, Bonhoeffer, and Löwel 1997; Sincich and Blasdel 2001; Weliky et al. 1995). In the immediate vicinity of each neuron, the connection patterns are relatively unspecific, but over larger distances they closely follow the orientation preferences. To a lesser degree, the patterns are also shaped by other perceptual features such as ocular dominance and spatial frequency (Bauman and Bonds 1991; De Valois and Tootell 1983; Löwel 1994; Löwel and Singer 1992; Vidyasagar and Mueller 1994).

For computational efficiency, most prior models of self-organization represented the lateral connections as a simple isotropic function. Later chapters will demonstrate that specific connection patterns are important for several developmental and functional phenomena, including self-organization, efficient representations, certain visual illusions, and perceptual grouping. For this reason, LISSOM will specifically simulate the development of patchy lateral connectivity.

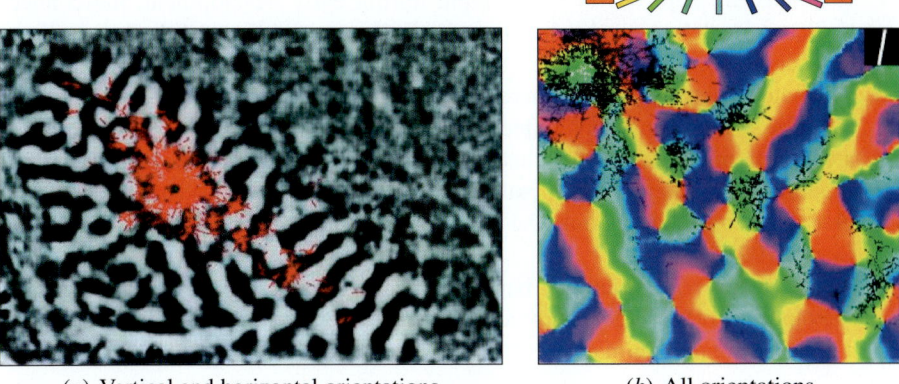

(a) Vertical and horizontal orientations   (b) All orientations

Fig. 2.7. Lateral connections in the tree shrew orientation map. (a) The vertical and horizontal orientation preferences in a 8 mm × 5 mm section of V1 in the adult tree shrew, measured using optical imaging. The areas responding to vertical stimuli are plotted in black and those responding to horizontal stimuli in white. Vertical in the visual field (90°) corresponds to a diagonal line at 135° in this plot. The small green dot in the middle marks the site where a patch of vertical-selective neurons were injected with a tracer chemical. The neurons to which that chemical propagated through lateral connections are colored red. Short-range lateral connections target all orientations equally, but long-range connections go to neurons that have similar orientation preferences and are extended along the orientation preference of the source neurons. (b) The same information plotted on a 2.5 mm × 2 mm section of the full orientation map to the right and below the injection site. The injected neurons are colored greenish cyan (80°), and connect to other neurons with similar preferences. Measurements in monkeys show similar patchiness, but in monkey the connections do not usually extend as far along the orientation axis of the neuron (Sincich and Blasdel 2001). These results, theoretical analysis, and computational models suggest that the lateral connections play a significant role in orientation processing (Bednar and Miikkulainen 2000b; Gilbert 1998; Sirosh 1995). Reprinted with permission from Bosking et al. (1997), copyright 1997 by the Society for Neuroscience.

2.2.2 Development

Lateral connectivity patterns have been found to form gradually during early development. Before eye opening, lateral connections grow exuberantly and to long distances in a short period (Callaway and Katz 1990). The connections are then pruned into well-defined clusters (Callaway and Katz 1990, 1991; Dalva and Katz 1994; Gilbert 1992; Katz and Callaway 1992; Löwel and Singer 1992; Luhmann et al. 1986). What process drives such pruning? Enormous amounts of genetic information would be required to specify each connection and each synaptic weight of the neurons in a cortical map. Instead, lateral connections develop in an activity-dependent manner. Several observations support this view:

1. When activity in ferret V1 is silenced using tetrodotoxin during early development, lateral connections remain broad and unspecific, and do not become patchy (Ruthazer and Stryker 1996).

2. If kittens are deprived of visual input during early development, the connections are much less patchy than normal (Callaway and Katz 1991; Ruthazer and Stryker 1996).
3. The patchy patterns can be altered by changing the input to the developing cortex. The resulting patterns reflect correlations in the input. For example, when a kitten is made strabismic, thereby removing correlations between the visual inputs in the two eyes, the lateral connections in the primary visual cortex organize differently, linking only the regions responding to the same eye (Löwel and Singer 1992).
4. In the mouse somatosensory barrel cortex, sensory deprivation (by sectioning the input nerve) results in shorter and sparser lateral connections compared with a normally reared animal (McCasland, Bernardo, Probst, and Woolsey 1992).

These observations suggest that lateral connections, like afferent connections, develop based on correlations in the input. The development of these different types of connections may actually be strongly related. Lateral connection patterns form approximately at the same time as the afferent connections organize into topographic maps (Burkhalter et al. 1993; Dalva and Katz 1994; Katz and Callaway 1992). Although each individual lateral connection is weak, their total effect on neural activity can be substantial (Gilbert et al. 1990), and they can thereby affect how the afferent connections develop. Changes in afferent connections then change the activity patterns in the cortex, which in turn influences the organization of lateral connections. In this manner, the two sets of connections develop synergetically, eventually evolving to a state of equilibrium in the adult animal. This principle is formalized and tested in detail in the LISSOM model.

2.2.3 Computational and Functional Hypotheses

Given the above observations, several possible functions have been proposed for the long-range lateral connections in the cortex. The list below is by no means complete, but it represents several of the views currently debated, including those put forward in later chapters of this book.

Modulating and Controlling Cortical Responses

1. Lateral connections may amplify weak stimuli and suppress strong stimuli, thus normalizing cortical activity (Somers, Toth, Todorov, Rao, Kim, Nelson, Siapas, and Sur 1996; Stemmler, Usher, and Niebur 1995).
2. They may modulate responses to achieve sharp orientation tuning and hyperacuity (Edelman 1996; Sabatini 1996; Somers et al. 1996).
3. They may combine responses to establish rotational and scaling invariance (Edelman 1996; Wiskott and von der Malsburg 1996).
4. They may mediate competition and synchronization over large distances of cortex (Taylor and Alavi 1996; Usher, Stemmler, and Niebur 1996; Wang 1996).
5. They may selectively enhance and suppress responses to implement attention and control (Taylor and Alavi 1996).

Representing and Associating Information

1. Lateral connections may store information that allows decorrelating visual input and filtering out known statistical redundancies in the cortical representations (Barlow and Földiák 1989; Dong 1996; Ghahramani and Hinton 1998; Sirosh, Miikkulainen, and Bednar 1996a).
2. They may help establish direction selectivity and motion sensitivity (Ernst, Pawelzik, Sahar-Pikielny, and Tsodyks 2001; Marshall 1990).
3. They alone may be responsible for orientation selectivity in the cortex (Adorján, Levitt, Lund, and Obermayer 1999; Ernst et al. 2001).
4. They may store information for feature binding and grouping, such as Gestalt rules (Choe and Miikkulainen 1997; Edelman 1996; Polat, Norcia, and Sagi 1996; Prodöhl, Würtz, and von der Malsburg 2003; Singer et al. 1990; von der Malsburg and Singer 1988; Wang 1996).
5. They may associate representations at different sensory cortices, serving as a foundation for multi-modal integration (Choe 2002; Lewis and Van Essen 2000; Shipp, Blanton, and Zeki 1998).

Mediating Development, Plasticity, and Learning

1. Lateral interactions may play a crucial role in the development of cortical columns, such as those representing orientation, ocular dominance, spatial frequency, and direction selectivity (Bednar and Miikkulainen 2003b; Dong 1996; Edelman 1996; Sirosh et al. 1996a).
2. They may mediate reorganization of the cortex in response to drastic changes in the input environment (such as retinal lesions and input deprivation; Gilbert and Wiesel 1992; Kapadia et al. 1994; Pettet and Gilbert 1992; Sirosh et al. 1996a).
3. They may mediate the perceptual learning processes observed as early as the primary visual cortex by encoding local associations (Dong 1996; Edelman 1996; Usher et al. 1996).
4. They may form the substrate for encoding memories as attractors in the cortical network (Miikkulainen 1992; Taylor and Alavi 1996).
5. Shared lateral connections may explain how similar orientation maps can develop for both eyes, even if the eyes are alternately sutured shut so that they never experience similar input (Kim and Bonhoeffer 1994; Shouval, Goldberg, Jones, Beckerman, and Cooper 2000).

Mediating Visual Phenomena

1. Lateral connections may mediate visual comparisons, such as those necessary for object recognition, figure-ground discrimination, and segmentation (Edelman 1996; Marshall and Alley 1996; Somers et al. 1996; Sporns, Tononi, and Edelman 1991; Wang 1996; Wiskott and von der Malsburg 1996).

2. They may mediate perceptual filling in, such as compensating for blind spots, perceptual completion and illusory contours (Choe 2001; Finkel and Edelman 1989; Grossberg and Mingolla 1985; Li 1998, 1999; Somers et al. 1996; Usher et al. 1996).

3. They may be responsible for visual illusions, such as the tilt illusion, brightness-contrast illusion, and Poggendorf illusion, which involve interactions between neighboring feature detectors (Bednar and Miikkulainen 2000b; Usher et al. 1996; Yu and Choe 2004; Yu, Yamauchi, and Choe 2004).

4. Adaptation of lateral connections may be responsible for temporary, pattern-specific visual aftereffects, due to increasing lateral inhibition between activated neurons (Barlow 1990; Bednar 1997; Bednar and Miikkulainen 2000b).

5. Lateral connections between different ocular dominance areas and disparity-selective neurons may contribute to binocular fusion, depth perception and stereo vision (Cormack and Riddle 1996; Löwel 1994; Löwel and Singer 1992; Petrov 2002).

The LISSOM model is based on the idea that lateral connections are crucial for the computations that take place in the visual cortex. In Part II of the book, inhibitory long-range lateral connections are shown to play a central role in self-organization. The LISSOM visual cortex is in a dynamic equilibrium, constantly adapting to both external and internal input. As a result, the observed structures of feature preferences develop, as well as patchy lateral connectivity between them (Chapter 5). The mechanisms are also active in the adult, implementing repair after retinal or cortical damage (Chapter 6), and resulting in psychophysical phenomena such as visual illusions and aftereffects (Chapter 7). The experimental data specific to these phenomena will be reviewed in the beginning of those chapters. Part IV will focus on excitatory lateral connections, showing how they can mediate binding and segmentation in a spiking-neuron model of the visual cortex. Part III, however, will focus on how environmentally and internally directed self-organization can implement a synergy of nature and nurture in development. The theoretical and biological foundations for this idea are reviewed next.

2.3 Genetic Versus Environmental Factors in Development

The LISSOM model will demonstrate how input-driven self-organization can account for the afferent and lateral connection structures in the visual cortex. As was discussed in Chapter 1, the most obvious source for such inputs is the visual environment during early life. However, the visual cortex already has a significant amount of structure before the visual experience begins, i.e. at birth or at eye opening. Such structure must be at least partially determined genetically. Why is it useful to include both genetic and environmental influences in constructing the visual cortex, and how can a developmental process combine them? These issues will be discussed in this section, providing the motivation for the computational studies of prenatal and post-natal development in Part III.

2.3.1 Bias/Variance Tradeoff

Why did evolution result in a developmental process that utilizes both genetic and environmental information, as opposed to a pure hardwiring or a pure tabula rasa learning process? This issue can be understood in terms of the well-known bias/variance tradeoff in machine learning (Geman, Bienenstock, and Doursat 1992; Utgoff and Mitchell 1982). Given a set of example inputs and outputs (the training set), a learning system needs to construct a mapping that produces correct outputs for new examples (the test set). There is often a very large number of possible mappings consistent with the training set, and they result in different outputs for the same test inputs. Which mapping will be selected is determined by the bias of the learner. The best results are obtained if the bias matches the problem and is strong (Haussler 1988). That way, the outputs for new examples are likely to be correct. Also, the same mapping is selected with different training sets and even with noisy training examples, i.e. the learner will have a low variance.

Unfortunately, it is not usually clear what the right bias is, and it is necessary to make the bias weaker. Which mapping will be selected will then depend more on the training data. As a result, the variance is increased: The selection of the mapping becomes unpredictable, determined based on which examples were included in the training set and the noise in those examples. Choosing an appropriate point in the bias/variance tradeoff therefore depends on how much is known about the problem in advance.

Biological systems face the same tradeoff: Neural structures can be determined genetically or learned from environmental inputs. A strong genetic bias is appropriate for organisms whose environment is predictable over many individual lifetimes, such as most invertebrates. For instance, the nematode worm *Caenorhabditis elegans* develops a nervous system with exactly 302 neurons in the same configuration in every individual (Sulston and Horvitz 1977). Such a strong bias allows the worm to function in its environment reliably and immediately.

However, the environment faced by mammals is much more complex and variable, and only the large-scale structures can be specified with a strong bias. The same sensory and motor areas appear in the same cortical locations in all individuals of the same species (Rakic 1988; Shatz 1996). These structures can still vary, but only in extreme cases such as prenatal injury (Goldman-Rakic 1980). The small-scale structures, on the other hand, are constructed primarily through interaction with the environment, and have a high variance. The number of neurons, their specific arrangements, and the patterns of connections differ between individuals of the same species (Shatz 1996).

The reason for the lower bias and higher variance in higher animals is that their environment is less predictable. If the individual were to be constructed with a strong bias, it would not be able to adapt to the different environments during its lifetime, and would perform poorly. On the other hand, learning is unreliable; if the right kind of input is not received at the right time, the individual may not develop a crucial skill (Blakemore and Cooper 1970; Hirsch and Spinelli 1970; Hubel and Wiesel 1974; Issa et al. 1999). Evolution has therefore determined a point in the bias/variance tradeoff that allows constructing a reliable but flexible system by combining genetic and environmental information.

How this idea can be utilized in constructing complex natural or artificial systems in general will be discussed in Sections 16.2.3 and 17.3.5. How it can be implemented specifically to construct the visual system of higher animals will be analyzed next.

2.3.2 Combining Genetic and Environmental Information

The large-scale structures of the brain, such as the pattern of the different brain areas, are constructed primarily through chemical gradients (Molnár, Higashi, and López-Bendito 2003; Rakic 1988; von der Malsburg and Willshaw 1977; Willshaw and von der Malsburg 1979). These gradients direct the growing connections to a general location on the cortical sheet. The gradients are largely unaffected by environmental stimuli, making the bias very strong. Incorporating environmental information into this process would be difficult, requiring a transduction mechanism between an environmental stimulus and the developmental hardware.

On the other hand, at the level of individual neurons and connections between small groups, sensory systems act as just such a transduction mechanism. In a sensory system, patterns in the environment are represented as patterns in neural activity, and these patterns in turn change how the orientation, ocular dominance, and similar map-level organizations in the cortex develop (as discussed above). At this level, the question becomes how *genetic* cues could be expressed. First, the system is structured to utilize information in input activity; second, the amount of information necessary to specify individual connections may be too large to store genetically.

The recent discovery of spontaneous activation provides an important clue: Much of the neural activity in developing sensory systems is not caused by the external environment, but generated internally in many cortical and subcortical sensory areas, such as the visual cortex, the retina, the auditory system, and the spinal cord (Feller et al. 1996; Lippe 1994; Meister, Wong, Baylor, and Shatz 1991; Peinado, Yuste, and Katz 1993; Wong, Meister, and Shatz 1993; Yuste, Nelson, Rubin, and Katz 1995; see O'Donovan 1999; Sengpiel and Kind 2002; Wong 1999 for reviews). This activity may express a genetic bias within a system that is designed to learn from the environment (Constantine-Paton et al. 1990; Maffei and Galli-Resta 1990; Marks et al. 1995; Roffwarg et al. 1966; Shatz 1990, 1996). The genetic information is represented in the same way at the neural level: as patterns of activity in the input seen by a brain area. The genome thus needs to specify only a pattern generator, a mechanism capable of producing visual-like patterns, rather than specifying individual connections.

The result is a genetic specification of potentially complex neural hardware. Such a specification is desirable in an evolutionary sense, because different functional architectures can be obtained by changing only a small part of the genome (Jouvet 1980). Random mutations in that portion of the genetic code would cause different patterns to be generated, which might lead to different cortical structures. Such a mechanism would facilitate evolutionary search, because it increases the chance that a chance mutation leads to a meaningful change.

The pattern generation hypothesis can potentially explain much of the experimental data on innate visual capabilities. The following two subsections present evidence that two specific types of internally generated activity patterns, retinal waves and ponto-geniculo-occipital waves, implement a genetic bias on visual cortex structures. These patterns will be crucial for the LISSOM model of how V1 and face-selective cortical areas are constructed, as will be discussed in detail in Part III.

2.3.3 Retinal Waves

In the developing retina of e.g. cats and ferrets, internally generated activity occurs as intermittent, local waves across groups of ganglion cells (Figure 1.2; Meister et al. 1991; Sirosh 1995; Wong et al. 1993). The waves begin before photoreceptors have developed (Maffei and Galli-Resta 1990), so they cannot result from visual input. Instead, they arise from spontaneous recurrent activity in networks of developing amacrine cells that provide input to the ganglion cells (Catsicas and Mobbs 1995; Feller et al. 1996; Shatz 1996). Like visual images, these waves are locally coherent in space and time (i.e. nearby ganglion cells are likely to be active continuously), and thus they could act as training input for the developing LGN and visual cortex (Shatz 1990).

Several experimenters have shown that interfering with the spontaneous activity can change how the visual system develops (Grubb, Rossi, Changeux, and Thompson 2003; McLaughlin, Torborg, Feller, and O'Leary 2003; Shatz 1990; Stellwagen and Shatz 2002). For instance, when the retinal waves are abolished, the inputs from the two eyes are no longer segregated in the LGN (Chapman 2000; Shatz 1996)). Similarly, when activity is silenced at the V1 level during early development, V1 neurons in mature animals are much less selective for orientation (Chapman and Stryker 1993). These results suggest that spontaneous activity is crucial for normal development of low-level vision.

Recent experiments have focused on whether spontaneous activity is merely permissive for development, perhaps by keeping newly formed connections alive until visual input occurs, or whether it is truly instructive, determining how the structures develop (Chapman, Gödecke, and Bonhoeffer 1999; Crair 1999; Katz and Shatz 1996; Miller, Erwin, and Kayser 1999; Penn and Shatz 1999; Sur, Angelucci, and Sharma 1999; Sur and Leamey 2001; Thompson 1997). For instance, Weliky and Katz (1997) artificially activated a large number of axons in the optic nerve of ferrets, thereby disrupting the pattern of spontaneous retinal activity. Even though this manipulation increased the total amount of activity, thereby making sure it was as permissive as before, V1 actually became less selective for orientation. Thus, spontaneous activity cannot only be permissive; it has at least some specific instructional role.

Similarly, pharmacologically increasing the number of retinal waves in one eye has been shown to prevent the LGN from developing normally (Stellwagen and Shatz 2002; cf. Crowley and Katz 2000). Yet, when waves are increased in *both* eyes, the

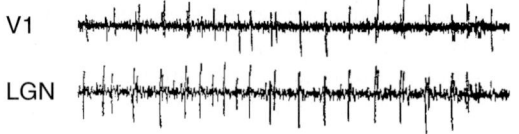

Fig. 2.8. Spontaneous activity in the cat PGO pathway. Each line shows a 65-second electrode recording from a cell in the indicated area during REM sleep in the cat. Spontaneous REM sleep activation in the pons of the brainstem is relayed to the LGN of the thalamus (bottom), to the primary visual cortex (top), and to many other regions in the cortex. It is not yet known what spatial patterns of visual cortex activation are associated with this temporal activity, or with other types of internally generated activity during sleep. However, such activity is largely genetically determined and could affect how the visual system develops. Reprinted with permission from Marks et al. (1995), copyright 1995 by Elsevier.

LGN develops normally, which again shows that the type of activity is important, not simply whether there is activity or not. However, what features of the activity are important are not known, because it has not yet been possible to manipulate the activity precisely. The LISSOM model will be used in Chapter 9 to study this issue computationally.

2.3.4 Ponto-Geniculo-Occipital Waves

Retinal waves are the best-studied source of spontaneous activity, because they are easily accessible to experimenters. However, other internally generated patterns may also be important for the development of the visual cortex. One example is the ponto-geniculo-occipital (PGO) waves that are the hallmark of rapid-eye-movement (REM) sleep in at least cats, ferrets, monkeys, and humans (see Steriade, Paré, Bouhassira, Deschênes, and Oakson 1989 for a review; Figure 2.8).

During and just before REM sleep, PGO waves originate in the brainstem and travel to the LGN, many areas of the visual cortex, and a variety of subcortical areas (see Callaway, Lydic, Baghdoyan, and Hobson 1987 for a review). In adults, PGO waves are strongly correlated with eye movements and with vivid visual imagery in dreams, suggesting that they activate the visual system as if they were visual inputs (Marks et al. 1995). Experimental studies also suggest that PGO waves are under genetic control: They elicit different activity patterns in different species (Datta 1997), and the eye movement patterns that are associated with PGO waves are more similar in identical twins than in unrelated age-matched subjects (Chouvet, Blois, Debilly, and Jouvet 1983). Thus, PGO waves are a possible source for genetically controlled training patterns for the visual system. But do they actually serve this role?

REM sleep has long been believed to be important for development, for two reasons (Roffwarg et al. 1966): Developing mammalian embryos spend a large percentage of their time in states that look much like adult REM sleep, and the duration of REM sleep is strongly correlated with how plastic the neural system is, both over development and across different species (also see the more recent review by Siegel 1999, as well as Jouvet 1980). Also consistent with Roffwarg et al.'s hypothesis, blocking REM sleep or the PGO waves alone has been found to *increase* the effect of visual experience during development (Marks et al. 1995; Oksenberg, Shaffery, Marks, Speciale, Mihailoff, and Roffwarg 1996; Pompeiano, Pompeiano, and Corvaja 1995). When the visual input to one eye of a normal kitten is blocked for a short time during a critical period, the cortical and LGN area devoted to signals from the other eye increases (Blakemore and van Sluyters 1975). When REM sleep (or just the PGO waves) is interrupted as well, the effect of blocking one eye's visual input is even stronger (Marks et al. 1995). This result suggests that REM sleep, and PGO waves in particular, limits or counteracts the effects of visual experience.

All of these characteristics suggest that PGO waves and other REM-sleep activity may be instructing development, like the retinal waves do (Jouvet 1980, 1998; Marks et al. 1995). However, due to limitations in experimental imaging equipment and techniques, it has not yet been possible to measure their two-dimensional spatial structure (Rector, Poe, Redgrave, and Harper 1997). Chapters 9 and 10 in Part III of this book will evaluate different candidates for internally generated activity, including retinal and PGO waves, and show what structure they would need to have to explain how maps and their connections develop in the visual cortex.

2.4 Temporal Coding

Part IV of the book will present a theory of perceptual grouping in the visual cortex, demonstrating that self-organized lateral excitatory connections play a crucial role in this process. The model assumes that binding and segmentation are based on temporal coding, i.e. timing of neuronal spiking events. In this section, experimental evidence for temporal coding will be reviewed. Computational models derived from these observations will be described and compared in the next chapter.

2.4.1 Binding Through Synchronization

Neurons are cells with the special property of being able to convey information in terms of electrical pulses, or spikes. Traditional neural network theories have hypothesized that the level of activation, or the spiking rate of neurons, forms the representation for perceptual events. However, as von der Malsburg (1981, 1986a,b) pointed out, such static representations suffer from the superposition catastrophe (Figure 2.9). This problem arises when distributed neural representations of two (or more) separate objects overlap: It is no longer clear which neuron represents which object (Figure 2.9*a*).

In contrast, if the representations for the individual objects alternate in time, binding and segmentation can occur naturally through temporal coding (Figure 2.9*b*). Von der Malsburg (1986b; 1987) hypothesized that perceptual grouping can be achieved in this way through synchronized and desynchronized firing of neurons. Temporal coding is therefore one way in which perceptual grouping can occur in the brain, but is there reason to believe that it does?

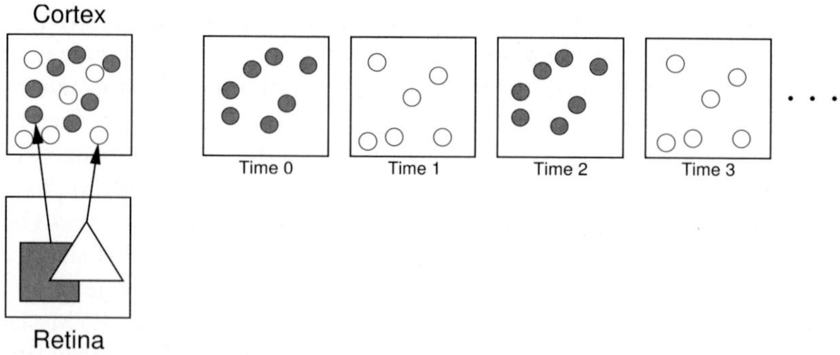

(a) Static representation   (b) Temporal coding

Fig. 2.9. Solving the superposition catastrophe through temporal coding. If firing rates of neurons alone are used to represent objects, multiple objects in the scene can result in confusion. (a) A square and a triangle are presented in the retina. In the cortex plot, the neurons responding to the square are colored gray, and those responding to the triangle white. When both populations of neurons are active at once, it is impossible to know which neuron is representing which object. This problem is known as the superposition catastrophe (von der Malsburg 1981, 1986a,b). One solution is temporal coding, where temporal information is used to separate the two populations. Neurons representing one object activate at one time step, and neurons representing the other object activate at the next time step, as shown in (b).

2.4.2 Experimental Evidence

Experimental studies have shown that coherent oscillations do indeed arise within populations of neurons. Such oscillations are usually observed as synchronized high-frequency waves near the 40 Hz γ-band (see Buzsáki and Draguhn 2004; Jefferys, Traub, and Whittington 1996 for reviews). To test whether such temporal representations are used in the visual system to present grouping, two approaches can be taken. One way is to present inputs to the visual system and measure the oscillations that result. The other is to alter the temporal properties in the input, preventing or enhancing synchronization, and measure the effect on perceptual performance.

Using the first approach, it has been possible to determine that activities of two populations with similar properties, such as the same orientation preference, do indeed synchronize when stimulated with a common input (Eckhorn et al. 1988; Gray et al. 1989; Gray and Singer 1987; Singer 1993). In one such study, electrical recordings were made on two sites in the cat visual cortex with non-overlapping receptive fields, while moving light bar(s) were swept across these receptive fields (Figure 2.10). When a single long bar was used as the input, the two populations representing distant sections of the long bar fired synchronously. However, when two short bars were swept in the same location as before but in the opposite direction of each other, the firing of the two populations was no longer synchronized. Interestingly, when two separate short bars were swept in the same direction, the two populations showed a weak but synchronized activity (Engel, König, Kreiter, and Singer 1991a; Engel, Kreiter, König, and Singer 1991b; Gray et al. 1989; Singer

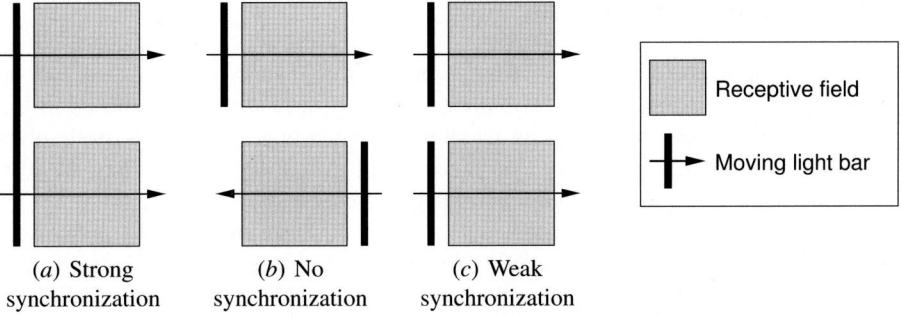

(a) Strong   (b) No   (c) Weak
synchronization   synchronization   synchronization

Fig. 2.10. Synchronization of one and two input objects in the cat. Moving bars of light were presented on two locations in the cat visual field where the receptive fields had no overlap, and the level of synchronization in the corresponding areas in the visual cortex were measured. (a) A single light bar moving across two receptive fields results in strong synchronization between the two neuronal populations. (b) Two separate bars moving in opposite directions result in no synchronization. (c) Two separate bars moving in the same direction result in weak synchronization. These results suggest that synchronization may indeed represent how likely the inputs are to belong to one and the same object. Adapted from Gray et al. 1989.

1993). These results suggest that synchronized firing of distant populations of neurons may represent the percept of a single coherent object, and desynchronized firing that of separate objects.

Another piece of evidence for synchronization-based grouping was obtained by manipulating the temporal properties in the visual input. Usher and Donnelly (1998) presented inputs where the object to be detected and the background were either flashed in synchrony (both object and background blink at the same time) or flashed asynchronously (object and background blink at different phases) over a period of time. The time-scale of the flashing was shorter than the integration time of the visual system so that such flashing could not be consciously perceived. Given such input, the subjects were asked to identify where the object appeared among one of four areas in the background. The percentage of correct responses turned out to be consistently higher when the object and background were flashing asynchronously. The percentage of correct responses increased as the phase difference between the flashing of object and background was increased. The explanation was that the timing of the inputs caused the temporal properties of neuronal firing to change and in turn caused the detection performance to differ. Flashing the object and background at different times would cause a slight phase shift between the neurons representing them, and such a shift helped distinguish the object from the background. Similar results have been reported by Fahle (1993), Lee and Blake (1999, 2001), Leonards and Singer (1998), Leonards, Singer, and Fahle (1996), Meyerson and Palmer (2004), Palmer (1999), and Wehrhahn and Westheimer (1993).

These results suggest that synchronization may indeed signal coherence in neural representations. The next issue is, how this synchronization is represented, i.e. what exactly is synchronized in the representation?

2.4.3 Modes of Synchronization

There are two ways in which synchrony can occur: (1) Individual neurons can be firing at the same time, and (2) population activity, i.e. number of neurons in the population firing per unit time, can oscillate in synchrony. Population oscillations are more general and include synchronized firing as a special case. They are also biologically the more likely candidate for several reasons.

Due to the stochastic nature of neuronal firing, it seems unlikely that individual neurons could synchronize their actual firing events. However, they could fire within a short time window so that the spikes are approximately aligned, and the whole group could exhibit synchrony (Lisman 1998; Menon 1990). Theoretical analysis also suggests that the oscillations found in the cortex result from a collective behavior of neurons. Such population oscillations are more robust and tolerant of random fluctuations (Menon 1990; Wilson and Cowan 1972).

In direct multi-electrode measurements, Eckhorn et al. (1988) discovered that synchrony in individual neurons is hard to find even when the number units firing and local field potential shows coherent oscillation, suggesting that population oscillation is the major mode of operation for binding of percepts. There is also indirect experimental evidence to support this hypothesis. When two almost simultaneous clicks are presented to a subject, they are initially heard as a single click, but as the interval between the clicks increases, the subject starts hearing two clicks instead of one. Interestingly, this transition from one click to two clicks occurs exactly at the frequency of population oscillations (Joliot, Ribary, and Llinás 1994). Apparently, the neuronal firing events within a single oscillation cycle are bound together even though the exact timing does not match, whereas the firings that occur in different cycles are perceived as separate.

For these reasons, most of the synchronizing models, including the model in this book, adopt the definition of synchrony in terms of population oscillations rather than that of individual neurons. Synchronization will be used in LISSOM to explain binding and segmentation phenomena, especially contour integration performance in humans. The detailed psychophysical evidence for these phenomena will be reviewed in Chapter 13.

2.5 Conclusion

Although the structure of the visual cortex has been well understood for several decades, the dynamic processes that develop this structure, maintain it, and represent visual information are still not well known. Lateral connections are believed to play a large role in all these processes, they may involve a synergy of nature and nurture through self-organization based on internal pattern generators, and achieve binding and segmentation through synchronization of activity.

Whereas such hypotheses are difficult to verify directly on biological systems, they can be implemented in computational models. Computational tests can lead to concrete predictions, and through further experiments, to a thorough understanding of the mechanisms underlying visual perception. Such an understanding is the main goal of this book. The computational principles and approaches on which it is based will be outlined in the next chapter.

3

Computational Foundations

As seen in the previous chapter, the visual system is a highly complex dynamical system, and it is difficult to integrate the scattered experimental results into a specific, coherent understanding of how the system is constructed and how it functions. A computational model provides a crucial tool for such integration: It constitutes a concrete implementation of the theory. Because all of its components must be implemented for the model to work, unstated assumptions must be made explicit. The model then shows what types of structure and behavior follow from those assumptions. The model can be tested just like animals or humans can, either to validate the theory or to provide predictions for future experimental tests.

This book introduces a comprehensive computational model of the visual cortex, built on findings from the past 30 years of research in computational neuroscience. These computational foundations are reviewed in this chapter, including the general models of neural computation, temporal coding, adaptation, and self-organizing maps.

3.1 Computational Units

A crucial issue in any computational model is what the appropriate level of abstraction is. Although in theory we could dissect and model each neuron at the smallest level of detail allowed by current technology (i.e. at the molecular level), in practice only a few measurements of the system parameters would be available at that level, and the resulting model would be largely underconstrained. Superfluous detail can also make it difficult to understand the model and to generate predictions based on it. Fortunately, such detailed simulations are often unnecessary for understanding high-level behavior, and more efficient abstractions can be used.

In this section, the models of computation in neurons and neuronal groups are reviewed at various levels of abstraction, evaluating which of their properties will be useful for understanding how the visual cortex develops and functions. Detailed models of the neuron are reviewed first, followed by gradually higher level abstractions (shown in Figure 3.1). As was discussed in Section 2.1.2, a cortical column is an appropriate computational unit for visual cortex models; the conclusion from this section is that integrate-and-fire and firing-rate models of the cortical column most efficiently capture the properties needed to understand their collective behavior.

3.1.1 Compartmental Models

Because neurons are generally believed to communicate through action potentials, or spikes, the most detailed models of the neuron focus on how spikes are generated and transmitted. The electrical currents that lead to spike generation are controlled by various ion channels in the fatty (lipid) membrane that encloses the cell body. The voltage across the membrane (i.e. the membrane potential) changes as these ions come in and go out of the neuron through the ion channel, and it is this voltage that determines whether the neuron generates a spike. Such dynamic change in state over time gives the neuron rich temporal dynamics that can be used to encode information.

Understanding the behavior of a neuron begins by modeling a small patch of the neuron membrane. Computational models of such patches are usually based on the Hodgkin–Huxley model of excitable membranes (Hodgkin and Huxley 1952). It consists of coupled differential equations for the membrane potential V and the fraction c_i of ion channels open for each channel type i (Gerstner and Kistler 2002; Rinzel and Ermentrout 1998):

$$
\begin{aligned}
C\frac{dV}{dt} &= -\sum_i I_i(c_i, V) + I(t), \\
\frac{dc_i}{dt} &= -\frac{c_i - c_{i,\infty}(V)}{\tau_i(V)},
\end{aligned}
\tag{3.1}
$$

where C is the membrane capacity, $I_i(c_i, V)$ is the current through ion channel i, and $I(t)$ is the externally applied input current. For a fixed membrane potential V, c_i approaches the steady-state level $c_{i,\infty}(V)$ with the time constant $\tau_i(V)$.

The Hodgkin–Huxley equation only describes an isolated patch of membrane, i.e. a single compartment. To model an entire neuron, it is first divided into major morphological sections corresponding to the axons, dendrites, and the cell body (Figure 3.1b). Each section is treated as an electrical conductor, usually represented as a cylinder or a cable (Rall 1962, 1977; Rall and Agmon-Snir 1998). Equations for the electrical behavior of a long cable can be solved analytically in simple cases, but for a realistic neuron model they need to be solved numerically. To do so, the cylinders are decomposed into smaller discrete compartments, each described using a membrane equation similar to Equation 3.1. The model for an entire neuron thus consists of a set of compartments, each with specific membrane properties and membrane voltage, all connected using electrical circuit theory. These models are simulated using software like NEURON (Hines and Carnevale 1997) or GENESIS (Bower and Beeman 1998) that are specifically designed to determine the appropriate compartments and solve the equations governing their electrical behavior (see e.g. Bower and Beeman 1998; Dayan and Abbott 2001; Lytton 2002 for reviews).

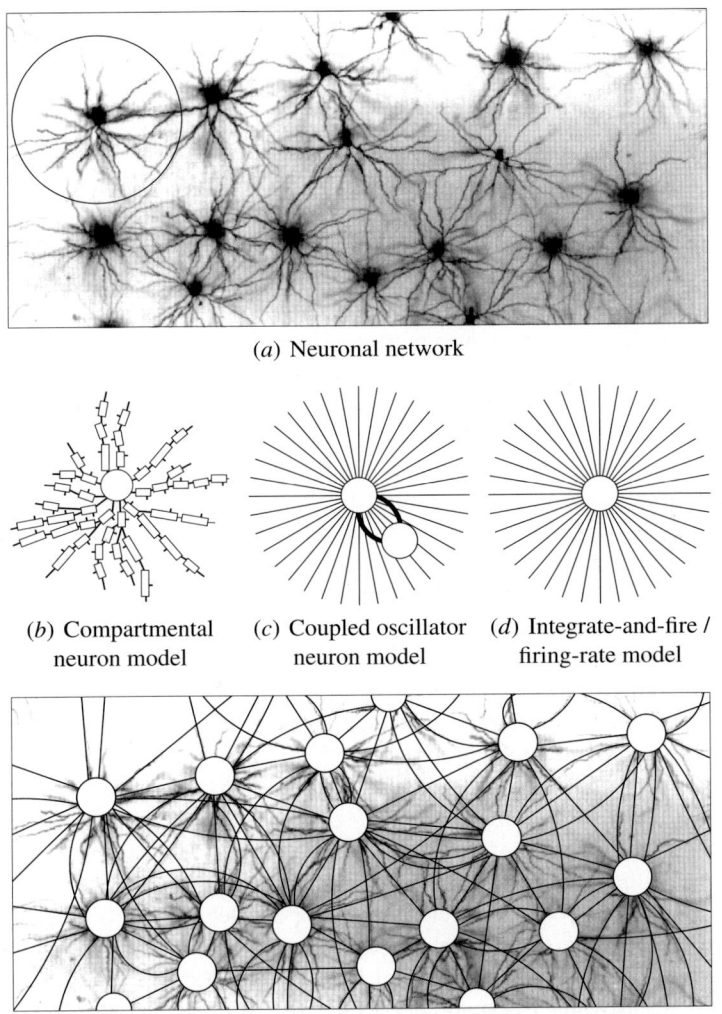

(a) Neuronal network

(b) Compartmental
neuron model

(c) Coupled oscillator
neuron model

(d) Integrate-and-fire /
firing-rate model

(e) Integrate-and-fire / firing-rate model of the network

Fig. 3.1. Computational abstractions of neurons and networks. Biological neurons can
be modeled at different levels of abstraction depending on the scale of the phenomena studied. (a) A microscopic image of pyramidal cells in a 1.4 mm × 0.7 mm area of layer III in
macaque temporo-occipital (TEO) area, injected individually with Lucifer Yellow (reprinted
with permission from Elston and Rosa 1998, copyright 1998 by Oxford University Press; circle added). Although this technique shows only a fraction of the neurons in a single horizontal
cross-section, it demonstrates the complex structure of individual neurons and their connectivity. (b) A detailed compartmental model of the top left neuron (circled). Each compartment
represents a small segment of the dendrite, and connections are established on the small dendritic spines, shown as line segments. (c) A coupled oscillator model of the neuron, consisting
of an excitatory and an inhibitory unit with recurrent coupling, and weighted connections with
other neurons in the network. (d) A model where a single variable describes the activation of
the neuron, corresponding to either the membrane potential (in the integrate-and-fire model),
or the average number of spikes per unit time (in the firing-rate model). (e) A high-level model
of a neuronal network. With the more abstract neurons, it is possible to simulate a number of
neurons and connections, allowing us to study phenomena at the level of networks and maps.

Compartmental models allow neurons to be represented in arbitrarily fine detail. They can be used effectively when experimental data are available to provide the parameters for the model, such as the size and shape of compartments and the distributions of ion channels (Doya, Selverston, and Rowat 1995). In such cases, they can be used to generate a very close fit to experimental data. For example, there are detailed models of cortical pyramidal neurons (Mainen and Sejnowski 1998) and cerebellar Purkinje cells (De Schutter and Bower 1994a,b).

However, large-scale cortical structures such as orientation maps are composed of millions of neurons, each making thousands of connections (Wandell 1995). Detailed data are available for only a very small sample of these neurons, and billions of parameters would have to be chosen arbitrarily for a compartmental model of such a map. The large number of components would make it difficult to understand its behavior, e.g. to determine which components are responsible for particular computations. Also, currently it is possible to simulate only a few neurons in such detail, due to limitations on computer memory and processing time.

The structures and phenomena studied in this book, i.e. cortical maps and perceptual behavior, depend crucially on having large numbers of neurons; on the other hand, they are not assumed to be sensitive to detailed membrane processes of individual neurons. Models of such phenomena must (and can) therefore use higher level abstractions of computational units. There are three major classes of such abstract models: coupled oscillators, integrate-and-fire neurons, and firing-rate neurons. These abstractions allow simulating large numbers of neurons and their connections, so that they theories about large-scale phenomena in the cortex can be tested. Each model will be described below in turn.

3.1.2 Coupled Oscillators

Coupled oscillator models focus on the temporal dynamics of pairs of neurons or neuron groups. The dynamics of each oscillator are determined by two variables x and y, representing the states of two coupled units, one of which is inhibitory, the other excitatory (Figure 3.1c; Horn and Opher 1998; Sabatini, Solari, and Secchi 2004; Terman and Wang 1995; von der Malsburg 1987; von der Malsburg and Buhmann 1992; Wang 1995, 1996; Wilson and Cowan 1972; some, like Chakravarthy and Ghosh 1996, use a single complex variable instead). The units are connected into a recursive loop where the excitatory unit activates the inhibitory unit, which in turn inhibits the excitatory unit. The activities of the units can be described with coupled differential equations that can be written in several different forms. In an example due to Terman and Wang (1995) and Wang (1999),

$$\frac{dx}{dt} = f(x) - y + z,$$
$$\frac{dy}{dt} = \epsilon[g(x) - y], \tag{3.2}$$

z is the input, ϵ is the coupling strength between the two units, and the functions $f(x)$ and $g(x)$ are chosen so that robust oscillation results. For example, with the cubical hyperbola $f(x) = 3x - x^3 + 2$, the height a and the slope b of sigmoid $g(x) = a[1 + \tanh(x/b)]$ can be tuned to obtain a robust limit cycle. When x rises in this system (initially due to external input z), $f(x)$, $g(x)$ and y increase. Once $f(x)$ starts to decrease, inhibition from y effectively turns x off. As a result, y also turns off, and the cycle repeats.

It is possible to interpret such an oscillator as a single neuron where the excitatory unit represents the membrane potential, and the inhibitory unit the change in potential resulting from ionic channel activation and deactivation (Wang 1999). However, more commonly, each of the units in the oscillator is interpreted as a pooled activity level of a population of neurons of the same cell type (pyramidal for the excitatory and stellate for the inhibitory unit), residing in the same cortical column (Menon 1991; Wang 1996). The oscillators can also be connected into a network, and based on the sign of the connection, their phases can become synchronized or desynchronized. Such coupled oscillator networks have been used in segmentation and binding tasks. For example, images such as aerial photographs or brain scans can be segmented into homogeneous regions (Liu and Wang 1999; von der Malsburg and Buhmann 1992), and speech can be segmented from background noise (Wang and Brown 1999). In each of these applications, desynchronization across oscillators represents segmentation and synchronization represents binding, establishing a temporal code.

One important advantage of coupled oscillator models is that they include only two variables, which makes them easier to analyze than compartmental models (FitzHugh 1961; Nagumo, Arimato, and Yoshizawa 1962). Unit activities can be represented in two-dimensional phase portraits, and behaviors such as limit-cycle oscillations identified. Even large-scale phenomena may sometimes be described theoretically (see Rinzel and Ermentrout 1998; Wang 1999 for reviews).

In summary, the coupled oscillator offers a description of the neuron at a higher level than the compartmental model does, allowing it to be analyzed more easily and used in applications. However, a further more efficient abstraction is still possible without losing the ability to perform temporal coding. Such a model is based on a single variable describing the membrane potential, as will be described in the next section.

3.1.3 Integrate-and-Fire Neurons

In the integrate-and-fire approach, a single variable corresponding to the membrane potential of a neuron is used to describe the state (Figure 3.1d). Such neurons accumulate the membrane potential from incoming signals, generate a spike when it exceeds a threshold, and reset the potential after each spike. A typical formulation of the general idea is

$$C\frac{dV}{dt} = I(t) - \frac{V}{R},\tag{3.3}$$

where V is the membrane potential, C its capacitance, R its resistance, and $I(t)$ is the input current (Lapicque 1907; see Gabbiani and Koch 1998 for a review). The effect of the incoming activity $I(t)$ is to build up the membrane potential over time. The $-V/R$, the leak term, retards the rise of the potential, and without further input, eventually returns it to the baseline level. Consequently, this model is also called the leaky integrate-and-fire neuron (Campbell, Wang, and Jayaprakash 1999; Nischwitz and Glünder 1995). When the membrane potential rises to the threshold level, the neuron spikes, and the potential is reset to the baseline. Such dynamics capture the aggregate behavior of the compartmental model well, and can be implemented efficiently computationally.

Several variations of the basic integrate-and-fire model have been proposed, and there are also formulations that unify many of them in a single framework (Gerstner 1998b; Hoppensteadt and Izhikevich 1997; Izhikevich 2003). A particularly efficient variation is the dynamic threshold model (Eckhorn et al. 1990; Reitboeck et al. 1993). The threshold is increased acutely after the neuron fires, and then decayed over time, simulating the refractory period of the neuron. Both the leaky synapse and the dynamic threshold are formulated using the same leaky-integration mechanism, implemented through convolution ($*$):

$$x(t) = X(t) * K(t), \qquad (3.4)$$

where $x(t)$ is the membrane potential or the threshold at time t and $X(t)$ is the impulse input representing a received or generated spike. The convolution kernel $K(t)$ is defined as

$$K(t) = \begin{cases} e^{-\lambda t} & \text{if } t \geq 0, \\ 0 & \text{otherwise,} \end{cases} \qquad (3.5)$$

where λ is the decay rate. A spike generates a single exponentially decaying potential over time, and multiple spikes generate a superposition of multiple decaying potentials.

The convolution can be calculated using the digital filter equation (Eckhorn et al. 1990) as

$$x(t) = X(t) + x(t-1) e^{-\lambda}, \qquad (3.6)$$

where t increases in discrete time steps. Any input from $X(t)$ causes a jump in $x(t)$, which then decays over time by the factor $e^{-\lambda}$. With this simple recursive equation, complicated neuron dynamics can be calculated efficiently. The temporal structure of the events is abstracted into a single variable, without storage or repeated calculations, which is ideal for large-scale simulations.

The integrate-and-fire model is efficient and theoretically well understood. Closed-form analytical solutions exist for simple cases, and even large networks can be analyzed theoretically (Gabbiani and Koch 1998; Gerstner and Kistler 2002; Meunier and Segev 2002). It has been used in several applications, including image segmentation of both static and moving objects, auditory analysis, motor control and reaching, range-image segmentation, sequence memory, and temporal pattern recognition (Campbell et al. 1999; Eckhorn et al. 1990; Glover, Hamilton, and Smith 2002; Hugh, Laubach, Nicolelis, and Henriquez 2002; Kuhlmann, Burkitt, Paolini, and Clark 2002; Rehn and Lansner 2004; Reitboeck et al. 1993; Sohn, Zhang, and Kaang 1999). It will also be used in Part IV of this book to understand how perceptual grouping occurs in the primary visual cortex.

3.1.4 Firing-Rate Neurons

The unit models reviewed so far can be used to understand the behavior of single neurons and the temporal coding that could take place in binding and segmentation. However, much of high-level behavior in the visual cortex (and elsewhere) does not require such detailed representations: The temporal behavior of the neurons is often not as important as their overall activity. The individual firing events can be abstracted into a general level of activation, or firing rate, and the activities of small groups of neurons can be aggregated into single computational units. For example, the force applied to a muscle and the firing rate of the muscle spindle are strongly correlated (Adrian 1926). Similarly, the firing rate of visual cortex neurons codes orientation and position of visual inputs (Hubel and Wiesel 1962, 1968). Focusing on firing rates alone leads to a much simpler and computationally tractable model.

The firing-rate model is again loosely based on the membrane potential of the neuron. This potential s is calculated as a sum of activities η_k of all neurons k that send their output to the neuron, multiplied by the connection weights w_k:

$$s = \sum_k \eta_k w_k. \tag{3.7}$$

Most models further abstract the membrane potential into a firing rate η, using a logistic (sigmoid) activation function σ:

$$\eta = \sigma(s) = 1/(1 + e^{-s}). \tag{3.8}$$

In this way, the activation (or firing rate) of the neuron is limited between 0 (i.e. minimum firing rate) and 1 (maximum rate), roughly modeling the activation function of real neurons. A piecewise linear approximation of σ can also be used in many cases, including the models in this book; it is faster to compute and results in qualitatively similar behavior.

Even though they are formulated at the single-neuron level, Equations 3.7 and 3.8 constitute a reasonable model for the response of small groups of neurons as well, such as cortical columns. In this interpretation, the amount of input stimulation (s) to the group is measured, and the total activation (or response) of the group is a logistic function of the input, limiting it between a minimum and a maximum value. Cortical column activation turns out to be a powerful abstraction for understanding the two-dimensional structure of the visual cortex, and will be used extensively in this book.

Firing-rate units can be used to simulate very large networks, and thereby even high-level behavior. Most neural network models in cognitive science and engineering, especially in natural language processing, reasoning, memory, speech recognition, and visual pattern recognition, are based on firing-rate units. They will be used in Parts II and III in this book to understand phenomena such as large-scale organization of the visual cortex, plasticity, visual illusions, and face detection.

3.2 Temporal Coding

Having reviewed the models for individual computational units, let us now turn to behavior in large groups of such units. Group behavior is an important issue for those models that generate spiking events, because they can form a temporal code through synchronization and desynchronization of firing events. The conditions under which synchronization occurs will be described in this section, as well as the computations and representations that are possible as a result.

In typical a implementation, spiking neurons are arranged into a two-dimensional topology with local excitatory connections and global inhibition (Terman and Wang 1995; von der Malsburg and Buhmann 1992; Wang 1995, 1996). Local excitation drives the phases of neighboring units closer to each other. The global inhibitor sums the activity of all excitatory units and inhibits them proportionally to this sum, thus implementing segmentation. The network is activated with external input, and after a while the peaks and valleys of the activity between different areas are compared: If they are synchronized, the areas are interpreted to represent components of a single coherent object. For example, Figure 3.2 illustrates how such a network binds the components of the cup together and those of the lamp together, and at the same time indicates that these are two separate objects.

Through analytical and computational studies it has been possible to character-ize the conditions under which such networks synchronize and desynchronize. Even though synchronization can be most naturally achieved by excitation, and desyn-chronization with inhibition, the process is quite complex and depends (among other factors) on whether the connections are excitatory or inhibitory, how long the axonal conduction delay is, and how much noise there is in the membrane potential.

First, even with the same types of connections (excitatory or inhibitory), either synchrony or desynchrony can be obtained with appropriate delays. More specifi-cally, there are four different combinations of connection and delay types: (1) Exci-tatory connections with no delay cause synchrony. Neurons that fire first cause the membrane potential of other neurons near threshold to reach the threshold quicker, thus decreasing the phase difference (Campbell et al. 1999; Gerstner and van Hem-men 1992; Han, Kim, and Kook 1998; Horn and Opher 1998; Kim 2004; Mirollo and Strogatz 1990; Terman and Wang 1995; Wang 1995, 1996). (2) Excitatory con-nections with sufficient delay can desynchronize. If the delay is long enough, it may take almost a full cycle for the excitatory contributions to get back to those neu-rons that fired first, causing them to fire even earlier and thus increasing the phase difference (Nischwitz and Glünder 1995). (3) Inhibitory connections without delay cause desynchrony. Their contribution keeps neurons near threshold from firing, in-creasing the phase difference (Han et al. 1998; Horn and Opher 1998; Nischwitz and Glünder 1995). (4) Inhibitory connections with appropriate delay synchronize. They delay firing of the other neuron until it coincides with the next spike (Horn and Opher 1998; Kim 2004; Kirillov and Woodward 1993; Lytton and Sejnowski 1991; Nischwitz and Glünder 1995; van Vreeswijk and Abbott 1994). These results show that both the connection type and the various temporal parameters involved in the neuron dynamics have a strong influence on synchronization behavior.

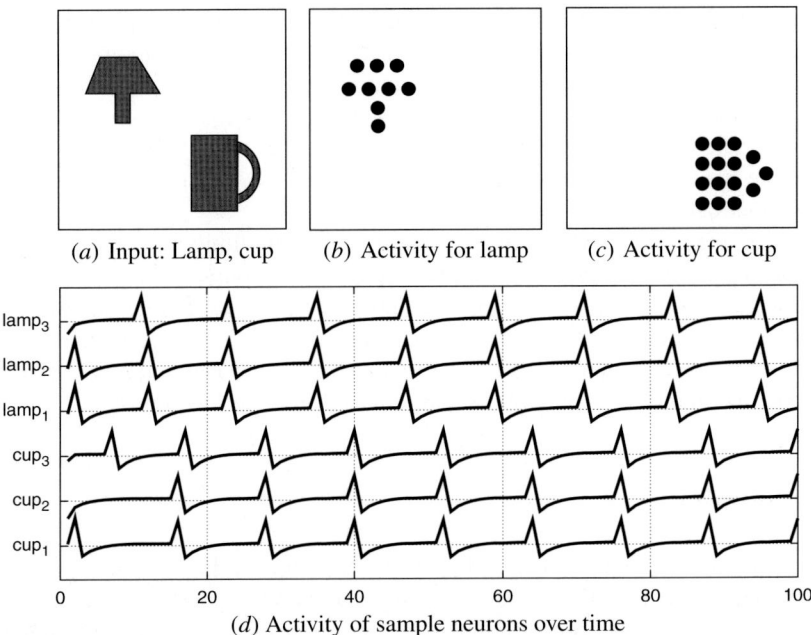

(a) Input: Lamp, cup   (b) Activity for lamp   (c) Activity for cup

(d) Activity of sample neurons over time

Fig. 3.2. Perceptual grouping through temporal coding. This schematic illustration shows how temporal coding can be used to bind and segment neural activity based on the phase of the periodic membrane potential. (a) The input image. (b) Activity of the two-dimensional map when all neurons representing the lamp are synchronized. (c) At a later time, all neurons representing the cup are synchronously turned on while the other neurons are off. (d) The membrane potential over time for three neurons representing the lamp and for three neurons representing the cup. The neurons representing the same object are synchronized and those representing different objects desynchronized. In this manner, temporal coding can be used to represent binding and segmentation. Adapted from Wang (1996).

Second, noise has been shown to help desynchronization of separate populations (Baldi and Meir 1990; Han et al. 1998; Horn and Opher 1998; Terman and Wang 1995; Wang 1995). Even if the neurons start out synchronized, noise causes some of the neurons to fire earlier or later in a random manner. The lateral interactions will then magnify these differences, eventually desynchronizing the populations.

Chapter 12 will add considerably to these results, showing that the rate of decay in the membrane potential also strongly affects synchronization, in the manner similar to conduction delay. Both binding and segmentation can be achieved at the same time, as well as simultaneous short- and long-range binding with local connections. Strong noise can indeed hurt synchronization, but it can be overcome with strong excitation and a long refractory period.

These rich temporal behaviors have been extensively used in real-world perceptual grouping tasks. For example, networks built of coupled oscillators have been applied to texture segmentation (Baldi and Meir 1990; von der Malsburg and Buhmann

1992), aerial photograph and brain-scan image segmentation (Wang 1995, 1996), and cluster analysis (Horn and Opher 1998). Networks of integrate-and-fire units have also been used in static object segmentation (Campbell et al. 1999; Eckhorn et al. 1990), moving object segmentation (Reitboeck et al. 1993), and aerial photograph and brain-scan image segmentation (Campbell et al. 1999). Such models both demonstrate how neural circuits could perform such tasks and serve as a basis for building practical applications.

However, the lateral connections in these models were limited to relatively short range, and could not learn input correlations. As a result, grouping was based only on proximity and similarity. More complicated grouping requires linking features such as oriented edges, and is not possible in these models. Part IV will demonstrate how long-range excitatory connections can be included in the model and self-organized to implement complex grouping phenomena such as contour integration.

3.3 Adaptation

Almost all computational models of the cortex are adaptive, i.e. their computations change based on the activities of the network. Adaptation is important to make such models practical; it would be difficult to design networks to perform the desired task by hand. Most importantly, cortical networks are highly adaptive. In order to understand how they develop and how they process information, it is necessary to include adaptation in the model.

Although many models of adaptation exist in artificial neural networks in general (such as backpropagation; Chauvin and Rumelhart 1995; Hecht-Nielsen 1989; Parker 1982; Rumelhart, Hinton, and Williams 1986; Werbos 1974), the models in computational neuroscience most often use some form of the synaptic adaptation mechanism proposed by Hebb (1949):

> When an axon of cell A is near enough to excite a cell B and repeatedly or persistently takes part in firing it, some growth process or metabolic change takes place in one or both cells such that A's efficiency, as one of the cells firing B, is increased.

This mechanism is called Hebbian learning, and is implemented in computational models as a weight update rule of the form

$$w'_{AB} = w_{AB}(t) + \alpha \eta_A \eta_B, \tag{3.9}$$

where w_{AB} is the old and w'_{AB} the new weight of the connection from cell A to cell B, η_A and η_B are the activations of the two cells, and α is a parameter determining the rate of learning.

Typically, Hebbian rules are combined with a mechanism to prevent the connection weights from increasing indefinitely. For instance, the weights can be decayed gradually so that the total amount of weight across all connections to or from a neuron remains approximately constant (Horn, Levy, and Ruppin 1998; Oja 1982;

Sanger 1989). More directly, the total connection strength can be normalized to have a constant sum (Rochester, Holland, Haibt, and Duda 1956; von der Malsburg 1973), thereby accurately redistributing the synaptic resources of each neuron. Such normalization is typically either subtractive or divisive (Elliott and Shadbolt 2002; Miller and MacKay 1994). In subtractive normalization, each weight is decreased by an equal amount after the weights adapt, with the amount chosen so that the total strength remains constant. All weights tend to approach either zero or some maximum strength (Miller and MacKay 1994), which does not happen in biology. In divisive normalization, each weight is instead scaled down in proportion to its original strength:

$$w'_{AB} = \frac{w_{AB} + \alpha \eta_A \eta_B}{\sum_u (w_{uB} + \alpha \eta_u \eta_B)}. \tag{3.10}$$

Such a change results in more precise weight values and thereby more precise control of the behavior of the network. Usually the normalization is done postsynaptically (i.e. the sum is taken over all input connections, as in Equation 3.10), but presynaptic normalization is also possible, with slightly different properties (Section 16.1.3).

Hebbian learning is elegant and effective, and also well supported biologically. A wealth of experimental evidence suggests that activity-dependent, correlation-based synaptic adaptation processes are involved in neural plasticity (Crair and Malenka 1995; Gustafsson and Wigström 1988; Hebb 1949; Hensch, Fagiolini, Mataga, Stryker, Baekkeskov, and Kash 1998; Hensch and Stryker 2004; Miller and MacKay 1994; see Tsien 2000 for a review). These processes can be based on long-term potentiation and depression (LTP/LTD) of synaptic connections or on growth of new connections; Hebbian learning serves as an abstraction of both mechanisms. Normalization terms were first introduced for computational reasons (Rochester et al. 1956), but recent work has uncovered a number of biological mechanisms within cells that regulate the overall synaptic strength and neural excitability during adaptation (Bourgeois, Jastreboff, and Rakic 1989; Hayes and Meyer 1988a,b; Murray, Sharma, and Edwards 1982; Pallas and Finlay 1991; Purves 1988; Purves and Lichtman 1985; see Turrigiano 1999 for a review). These "homeostatic" or "neuronal regulation" processes may implement normalization in biological systems. For instance, Turrigiano, Leslie, Desai, Rutherford, and Nelson (1998) showed that a change in a single synapse can cause the efficacy of the other synapses in the cell to change in the opposite direction. These results suggest that local change in the synaptic strength scales the strength of the other synapses in the same neuron.

Other recent experimental and theoretical work suggests ways in which the adaptation and normalization rules can be unified, based on findings of spike-timing-dependent plasticity (STDP; Fu, Djupsund, Gao, Hayden, Shen, and Dan 2002; Markram, Lübke, Frotscher, and Sakmann 1997; Panchev and Wermter 2001; Saudargiene, Porr, and Wörgötter 2004; Zhang, Tao, Holt, Harris, and Poo 1998). STDP is a variant of Hebbian learning that depends on the precise timing between presynaptic and postsynaptic spikes. Specifically, if a presynaptic neuron fires just before the postsynaptic neuron does, the strength of the connection is increased. If the presynaptic neuron fires shortly *after* the postsynaptic neuron, the weight is decreased. This rule can be implemented in a network of spiking neurons, but it has the disadvantage that it depends crucially on the individual spikes (Bohte and Mozer 2005; Song, Miller, and Abbott 2000). It is expensive to simulate and difficult to justify it in large-scale models where single units represent neural groups: Such a model would have to assume that all neurons in the group are firing in synchrony, which is unrealistic. Thus, STDP is primarily suited to modeling small networks of individual neurons.

For these reasons, standard Hebbian learning with divisive normalization will be used in the LISSOM model. It has solid biological support, and is well suited for learning correlations in visual patterns, which is what the visual cortex seems to be doing (as was discussed in Section 2.2.2). When each unit in the LISSOM cortex adapts its behavior based on the Hebbian principle, the entire adapting network forms a self-organizing map, which will be described next.

3.4 Self-Organizing Maps

As reviewed in Section 2.2, neurons in the cortex do not act in isolation; each neuron is strongly influenced by lateral inhibition and excitation in the cortical network. Even though each neuron is adapting its own connections, the activities of other neurons modulate the learning. Therefore, to understand development and plasticity, the interactions in the whole cortical network need to be taken into account.

This idea is formalized computationally in self-organizing maps. Competition and cooperation is introduced between neurons, so that only one or a few units in the network respond to each input pattern. If only these active neurons adapt, each neuron will learn to respond best to a cluster of similar patterns. Different neurons will respond to different clusters in the input space, and the network will learn a map-like representation of the inputs. Such self-organizing maps constitute the most common and most appropriate computational structure for understanding computations in the maps of the visual cortex.

The general architecture and computations in self-organizing maps will be outlined in this section, and an example of a self-organization process will be given. The properties of the architecture will be analyzed in the next two sections, focusing on how the maps represent high-dimensional input in two dimensions.

3.4.1 Variations of Map Models

Self-organizing maps are a general class of learning models, some of which are strictly directed toward understanding biological maps; others are more abstract and intended to be used in engineering applications. The general outline of this type of computational architecture is given in Figure 3.3.

The first self-organizing map model of the primary visual cortex was developed by von der Malsburg (1973) and simulated on a 1 MHz UNIVAC. He used a small two-dimensional network of neural units to model the cortex, based on the assumption that cells in a vertical column have the same response properties and can be

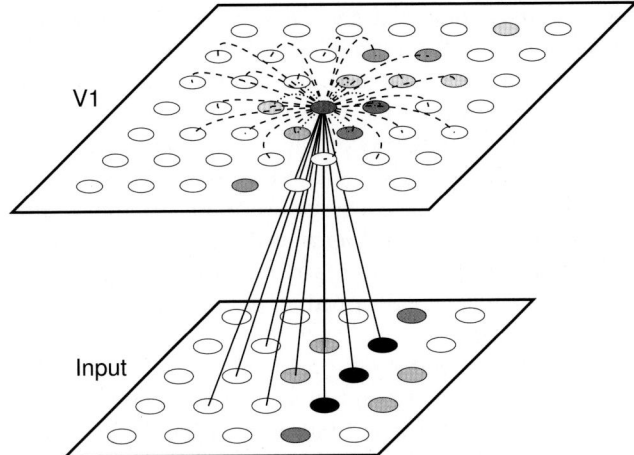

Fig. 3.3. General architecture of self-organizing map models of the primary visual cortex. The model typically consists of two sheets (also called layers, or surfaces) of neural units: input and V1. Some models also include a sheet of LGN neurons between the input and V1, or interpret the input sheet as the LGN; however, in most models the LGN is bypassed for simplicity, and the input sheet represents a receptor surface such as the retina. The input units are activated with continuous values according to the input pattern. In this example, the activations represent an elongated Gaussian, as shown in gray-scale coding from white to black (low to high). The input units are organized into a rectangular 5×5 grid; a hexagonal grid can also be used. Instead of grid input, some models provide the input features such as (x, y) position, orientation, or ocularity directly as activation values to the input units (Durbin and Mitchison 1990; Ritter et al. 1991). Others dispense with individual presentations of input stimuli altogether, abstracting them into functions that describe how they correlate with each other over time (Miller 1994). Neurons in the V1 sheet also form a two-dimensional surface organized as a rectangular or hexagonal grid, such as the 7×7 rectangular array shown. The V1 neurons have afferent (incoming) connections from neurons in their receptive field on the input sheet; sample afferent connections are shown as straight solid lines for a neuron at the center of V1. In some models the receptive field includes the entire input sheet (e.g. von der Malsburg 1973). In addition to the afferent input, the V1 neurons usually have short-range excitatory connections from their neighbors (shown as short dotted arcs) and long-range inhibitory connections (long dashed arcs). Most models save computation time and memory by assuming that the values of these lateral connections are fixed, isotropic, and the same for every neuron in V1. However, as will be shown in later chapters, specific modifiable connections are needed to understand several developmental and functional phenomena. Neurons generally compute their activation level as a scalar product of their weights and the activation of the units in their receptive fields; sample V1 activation levels are shown in gray scale. Weights that are modifiable are updated after an input is presented, using an unsupervised learning rule. In some models, only the most active unit and its neighbors are adapted; others adapt all active neurons. Over many presentations of input patterns, the afferent weights for each neuron learn to match particular features in the input, resulting in a map-like organization of input preferences over the network like those seen in the cortex.

treated as a single computational unit. Each unit had fixed excitatory lateral connections with its neighbors, and fixed inhibitory lateral connections with units farther away. Whenever an input was presented, the short-range excitation and inhibition focused activity in the best-responding areas of the network. The afferent weights of the active units were then modified according to a Hebbian rule, normalized so that the total weight of each unit was constant. When trained on simple binary patterns consisting of oriented bars, the units learned to respond preferentially to particular orientations. Furthermore, neighboring units responded to similar orientations, while next-to-neighboring units responded to nearly perpendicular orientations, so that the response profile across the network formed an orientation map qualitatively similar to that seen in the visual cortex.

Subsequently, dozens of similar self-organizing models have been proposed for different aspects of cortical self-organization, such as topography, orientation preference, and ocular dominance (Amari 1980; Ben-Yishai, Bar-Or, and Sompolinsky 1995; Berns, Dayan, and Sejnowski 1993; Bienenstock, Cooper, and Munro 1982; Bishop, Svensén, and Williams 1998; Cooper, Intrator, Blais, and Shouval 2004; Dayan 1993; Durbin and Mitchison 1990; Erwin, Obermayer, and Schulten 1995; Goodhill 1993; Goodhill and Willshaw 1990, 1994; Grossberg 1976; Hurri and Hyvarinen 2003; Kohonen 1982b; Miller 1994; Miller, Keller, and Stryker 1989; Nass and Cooper 1975; Obermayer, Blasdel, and Schulten 1992; Obermayer, Ritter, and Schulten 1990d; Olson and Grossberg 1998; Ruf and Schmitt 1998; Shouval 1995; Swindale 1980; Tanaka 1990; Willshaw and von der Malsburg 1976, 1979; Yuille, Kolodny, and Lee 1996). Among them, the self-organizing feature map (SOM; Kohonen 1982b, 2001) and the models of visual cortex based on it (Obermayer et al. 1990d, 1992) have been particularly influential, mainly because of their elegance and simplicity. In this book, "SOM" is used to refer to this particular architecture, and "self-organizing maps" to refer to the whole class of developmental models of maps.

In most of these models, the lateral interactions between neurons have been substituted with a simpler and computationally less expensive mechanism. First, each input is assumed to produce only one maximally active region in the cortex. Then, instead of using lateral interactions to find the regions of maximum activity, it is possible to simply search for the maximally active neuron, and adapt the afferent weights in a circular neighborhood around it. In this way, the models implicitly assume that the adult cortex is static, and that lateral connections are fixed or change in a simple and predetermined way. Only recently have models with specific, modifiable lateral connections started to emerge (Alexander, Bourke, Sheridan, Konstandatos, and Wright 2004; Bartsch and van Hemmen 2001; Bray and Barrow 1996; Burger and Lang 1999; Kalarickal and Marshall 2002; Weber 2001), beginning with the early versions of the LISSOM model described in this book (Miikkulainen 1991; Sirosh 1995; Sirosh and Miikkulainen 1993, 1994a). Such models can potentially account for a wider set of developmental and functional phenomena than self-organizing map models without explicit lateral connections, as will be shown in later chapters of this book.

3.4.2 Architecture and Computations

In this section, the computations that take place in the self-organizing map model of biological maps are described in detail. The discussion includes elements from several models in the literature, such as the SOM and von der Malsburg's model, and serves to define the foundation on which the LISSOM model in this book is built.

The architecture consists of a two-dimensional array of neurons representing the cortical surface, connected to an input array that represents a receptor surface such as the retina (Figure 3.3; the LGN is bypassed in most such models for simplicity). Each connection has a positive synaptic weight. Initially, these weights are random and therefore each neuron responds randomly to activity on the receptor surface. During an adaptation phase, these weights are adjusted, and the neurons gradually become more and more specific, adjusting their tuning in such a way that each neuron can only be excited by a small and spatially localized set of receptors. In the final, organized state of the network, the receptive fields of neighboring neurons are arranged such that the location of the maximal neural excitation varies smoothly with the stimulus location in the receptor surface. The neural sheet then acts as a topographic map, representing stimulus location within the receptor surface.

In the adaptation phase, input items are randomly drawn from the input distribution and presented to the network one at a time. The network responds to each vector by developing a localized activity pattern. The weight vector of the maximally responding neuron and each neuron in its neighborhood are changed toward the input vector, so that these neurons will produce an even stronger response to the same input in the future.

In other words, the map adapts in two ways at each presentation: (1) The weight vectors become better approximations of the input vectors, and (2) neighboring weight vectors become more similar. Together, these two adaptation processes eventually force the weight vectors to become an ordered map of the input space.

The process begins with very large neighborhoods, i.e. the weight vectors change in large areas. This phase results in a gross ordering of the map. The size of the neighborhood and the learning rate decrease with time, allowing the map to make finer and finer distinctions between items. Eventually, the distribution of the weight vectors becomes an approximation of the input vector distribution, and a smooth topographic map develops.

The self-organizing map can be formalized in a simple set of equations. Let us assume that there are n receptors. An input pattern consists of a set of positive activity values on the receptor array, represented as the vector $\mathbf{X} = \{\chi_1, \chi_2, \ldots, \chi_n\}$, $\chi_k \geq 0$. Each neuron (i, j) in the map array receives this same vector as its input. The neuron has n weights corresponding to the n components of $\mathbf{X}$; these weights form the vector $\mathbf{W}_{ij} = \{w_{1,ij}, w_{2,ij}, \ldots, w_{n,ij}\}$ (some of these weights may be fixed at zero to represent local receptive fields). The neuron computes its initial response as the weighted sum of the input vector and its weight vector:

$$\eta_{ij} = \sum_k \chi_k w_{k,ij} = \mathbf{X} \cdot \mathbf{W}_{ij}. \tag{3.11}$$

These initial responses of individual neurons are further modified by lateral interactions in the cortical network. For simplicity, most self-organizing map systems do
not explicitly model these lateral interactions. Instead, the interactions are assumed
to lead very rapidly to a new spatially localized distribution of activities centered
around the neuron with the maximum initial response (called the winning neuron,
e.g. neuron (r, s)). The activity of the winner is assumed to saturate at the maximum
possible activity $\eta_{\max}$. With these assumptions, the activity of the network after the
interactions can be described by a simple function, such as a Gaussian, multiplied by
the maximum activity:

$$h_{rs,ij} = \eta_{\max} \exp \left(-\frac{(r - i)^2 + (s - k)^2}{\sigma_{\mathrm{h}}^2} \right), \tag{3.12}$$

where $h_{rs,ij}$ describes the activity of neuron (i, j) when (r, s) is the winning neuron.
The function h is called the neighborhood function, because it describes how the
activity varies in the neighborhood of the winner. Its width, determined by σ_{h}, starts
out large at the beginning of the simulation, and gradually decreases to a small value
toward the end. The justification for such a decrease is that in the initial state, the
activity patterns on the network will be random and widespread, because the synaptic
weights are random. To describe such widespread activity, h should be broad as well.
As the weights adapt and form a topographic map, the rough activity pattern will
be more strongly concentrated around the winner; therefore, a narrower function is
required (Erwin, Obermayer, and Schulten 1992a,b; Kohonen 1993, 2001; Sirosh
and Miikkulainen 1997).

The input weights of the neurons are adapted according to the Hebbian rule. The
neighborhood function can be directly substituted for the network activity in this
rule; after this substitution the Hebbian Equation 3.9 becomes

$$w'_{k,ij} = w_{k,ij} + \alpha\chi_k h_{rs,ij}. \tag{3.13}$$

As the normalization mechanism, the divisive process of Equation 3.10 can be used.
It works well with high-dimensional natural inputs such as Gaussians and natural
images, which have approximately constant vector length. In such cases, the response
(Equation 3.11) depends primarily on the angle between the input and the weight
vector, which is an appropriate measure of similarity. Hebbian learning with divisive
normalization will therefore be used in the models described in this book.

However, many experiments with self-organizing maps utilize preprocessed input that tends to be low-dimensional and vary in length. To compensate, the weight
vectors are normalized to constant length, ensuring that the response is not dominated by length. The maximum activity is produced by an input vector that is identical to the weight vector and the minimum response is produced by an input that is
orthogonal to it (Kohonen 2001; Obermayer, Ritter, and Schulten 1990b; Sirosh and
Miikkulainen 1994a). After including such normalization, the weight adaptation rule
becomes

$$w'_{k,ij} = \frac{w_{k,ij} + \alpha\chi_k h_{rs,ij}}{\sqrt{\sum_u (w_{u,ij} + \alpha\chi_u h_{rs,ij})^2}}, \tag{3.14}$$

Fig. 3.4. Training a self-organizing map with Gaussian activity patterns. Each training input is a Gaussian pattern of activation on the two-dimensional array of 24×24 receptors. Four sample such patterns are shown in this figure, represented in gray-scale coding from white to black (low to high). The only dimensions of variation are the x and y positions of the Gaussian centers, and the map should learn to represent two-dimensional location, or retinotopy, as a result.

where the term within the square root represents the sum of squares of all the updated weights of neuron (i, j).

In practical applications of the self-organizing map, where propagation through afferent (or lateral) connections is not important, the above processes of response generation and weight change are often abstracted further. Instead of the weighted sum, the response is based on the Euclidean distance between the input and the weight vectors $\|\mathbf{V} - \mathbf{W}_{ij}\|$; instead of Hebbian adaptation with normalization, the weight vector is changed toward the input vector based on Euclidean difference:

$$w'_{k,ij} = w_{k,ij} + \alpha(\chi_k - w_{k,ij})h_{rs,ij}. \tag{3.15}$$

This abstracted model is, in essence, what has become known as the SOM artificial neural network architecture. Kohonen (1982a, 1989) showed that these simplifications lead to the same self-organizing behavior as the more biologically realistic version. They are, however, more straightforward to implement and more efficient to simulate, which has made many engineering applications possible (see Kaski, Kangas, and Kohonen 1998; Kohonen 2001; Oja and Kaski 1999; Oja, Kaski, and Kohonen 2003 for reviews).

The processes described above allow organizing a topographic map of the distribution of input vectors. In the next section, this process is described in detail in the case where the input varies in two dimensions; the way the map represents higher dimensional input spaces is analyzed in Section 3.5.

3.4.3 Self-Organizing Process

Self-organization of the topographic map can be best visualized when the input is inherently two-dimensional, like the map itself. To represent such input on the receptor surface, single spots of activity with a fixed width are presented at random locations (Figure 3.4). The activity of each receptor in the spot is described by the unoriented (i.e. circular) Gaussian function:

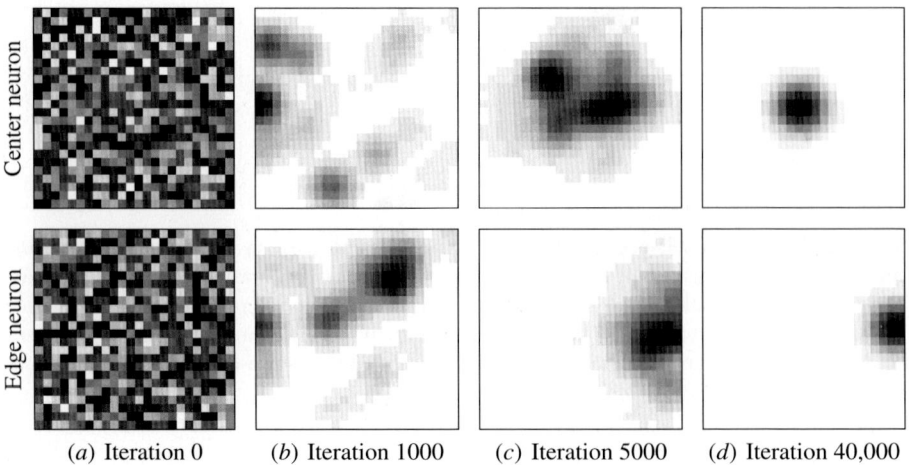

Center neuron

Edge neuron

(a) Iteration 0   (b) Iteration 1000   (c) Iteration 5000   (d) Iteration 40,000

Fig. 3.5. Self-organization of weight vectors. The weight vectors of two sample units are plotted on the receptor array at different stages of self-organization. The weight values are represented in gray-scale coding from white to black (low to high). Initially (iteration 0) the weights are uniformly randomly distributed; over several input presentations (such as those shown in Figure 3.4) the weights begin to resemble the input Gaussians in different locations of the receptor surface (iterations 1000, 5000, and 40,000). A neuron at the center of the network (top row) forms a Gaussian weight pattern at the center, while a neuron at the edge (bottom row) forms one near the edge. Such weight patterns together represent the topography of the input space, as seen in Figure 3.6.

$$\chi_k = \exp\left(-\frac{(x - x_c)^2 + (y - y_c)^2}{\sigma_u^2}\right), \tag{3.16}$$

where (x, y) specifies the location of receptor k, (x_c, y_c) the center of the activity spot, and σ_u its width. Trained with such inputs, the map should learn to represent the two-dimensional locations on the receptor surface. In other words, if the cortical sheet is interpreted as V1 and the input sheet as the retina, the model should learn a retinotopic mapping.

Such a mapping will be formed in this section with the abstract (SOM) version of the self-organizing process, where the map responds and adapts based on Euclidean distance similarity measure (Equation 3.15). The SOM is the most common version of self-organizing maps in the literature, and this simulation therefore establishes a baseline for comparison with LISSOM in the next chapter. The map consists of 40×40 units fully connected to 24×24 receptors; the weights are initially uniformly random. The input Gaussians have a width of $\sigma_u = 0.1$, and their centers are chosen from a uniform random distribution, so that they are evenly scattered over the receptor surface. The rest of the parameter values are described in Appendix E.

The weight vectors of each neuron are initially random, i.e. each value is drawn from the uniform distribution within $[0..1]$ (Figure 3.5). Over several input presentations, they gradually turn into representations of the input Gaussians at different locations. For example, the neuron at the center of the network forms a Gaussian weight pattern at the center of the receptor surface, and a neuron at the edge of the network one near the edge.

Such weight vectors form a topographic mapping of the input space. This mapping can be illustrated by first calculating the center of gravity of each neuron's weight vector (as a weighted sum of the receptor coordinates divided by the total weight; Appendix G.2). The centers can then be plotted as points on the receptor surface, as is done in Figure 3.6. The square area in each subfigure represents the receptor surface, and the centers of neighboring neurons are connected by a line to illustrate the topology of the network. Since the afferent weights are initially random, their centers of gravity are initially clustered around the middle. As inputs are presented and weights adapt, this cluster gradually unfolds, and spreads out into a smooth grid that covers the receptor surface. In the final self-organized map the topographic order of the centers matches the topographic order of neurons in the network. The network has learned to represent stimulus location accurately, and maps the input space uniformly. The plot is slightly contracted because the Gaussian weight patterns near the edges are truncated: Although the peak of a Gaussian is close to the edge, its center of gravity is always well inside the edge (Figure 3.5d, bottom row).

More generally, the distribution of the neuron weight vectors in the final map approximates the distribution of the input vectors (Ritter 1991). A dense area of the input space, i.e. an area with many input vectors, will be allocated more units in the map (Figure 3.7; simulation parameters in Appendix E). This means that such areas are magnified in the map representation, which is useful for data analysis, and also corresponds to the structure of biological maps.

In these example simulations, the only significant feature of the input is its location, and therefore the map learns to represent retinotopy. Retinotopic mappings consist of two dimensions (x and y), and because the map is also two-dimensional, such a mapping is straightforward. However, if the input patterns are elongated and oriented, or originate from two different eyes, the map will also represent those input features. Such a case is more complicated because the input has more dimensions than are available in the map. Self-organizing map models of orientation, ocular dominance, and direction selectivity all have this property, as will be discussed in Chapter 5. The next section will show how the map represents inputs with more than two dimensions of variation in a two-dimensional structure. Such an analysis allows us to understand the organization of biological maps better.

3.5 Knowledge Representation in Maps

When there are more than two dimensions of variation in the input, similar inputs may not always be represented by nearby locations on the two-dimensional map. How will the map representation approximate high-dimensional spaces? First, most often the distribution in the high-dimensional space is not uniform. The map will form a principal curve through the lower dimensional clusters of data embedded in the high-dimensional space. Second, when clusters are indeed multidimensional,

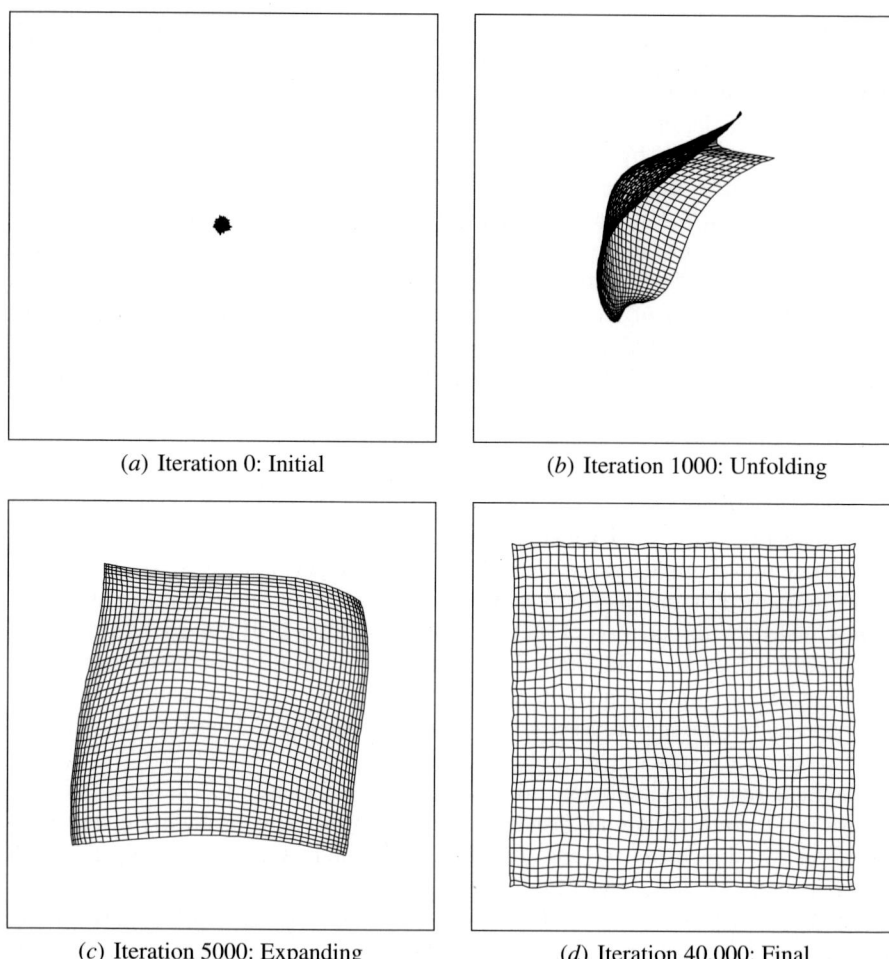

(a) Iteration 0: Initial

(b) Iteration 1000: Unfolding

(c) Iteration 5000: Expanding

(d) Iteration 40,000: Final

Fig. 3.6. Self-organization of a retinotopic map. For each neuron in the network, the center of gravity of its weight vector is plotted as a point on the receptor surface. Each point is connected to the centers of the four neighboring neurons by a line (note that these connections only illustrate neighborhood relations between neurons, not actual physical connections through which activity is propagated). Initially the weights are random, and the centers are clustered in the middle of the receptor surface. As self-organization progresses, the points spread out from the center and organize into a smooth topographic map of the input space.

the map will form hierarchical folds in the higher dimensions. The space is covered roughly uniformly, but not all similarity relations are preserved, resulting in a patchy map organization. These principles are illustrated in the two subsections below.

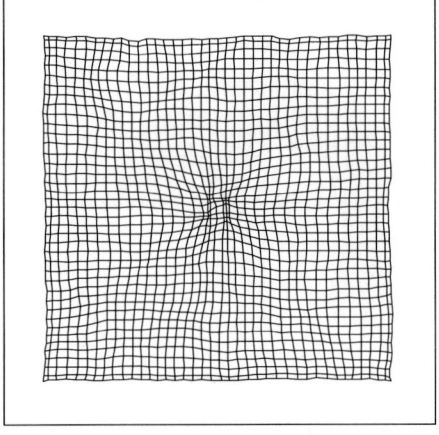
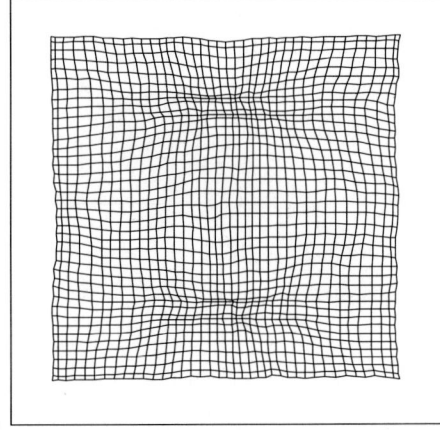

(*a*) Gaussian distribution   (*b*) Two long Gaussians

Fig. 3.7. Magnification of dense input areas. Whereas in Figure 3.6 the inputs were uniformly distributed over the receptor surface, map (*a*) was trained with inputs appearing more frequently in the middle, and map (*b*) with two such high-density areas diagonally from the middle. More units are allocated to representing the dense areas, which means that they are represented more accurately on the map. Similar magnification is observed in biological maps.

3.5.1 Principal Surfaces

Often with high-dimensional data, all dimensions do not vary independently, and only certain combinations of values are possible. For example, in a map representing (x, y) location, orientation, and direction selectivity, all locations and directions must be represented at all locations, but only directions that are roughly perpendicular to the orientation can ever be detected. Therefore, even though there are four dimensions of variation, the data are inherently three-dimensional. The first principle of dimensionality reduction in self-organizing maps is to find such low-dimensional structures in the data and map those structures instead of the entire high-dimensional space.

The standard linear method for such dimensionality reduction is principal component analysis (PCA; Jolliffe 1986; Oja 1989; Ritter, Martinetz, and Schulten 1992). PCA is a coordinate transformation where the first dimension (i.e. first principal component, or hyperplane) is aligned with maximum variance in the data; the second principal component is aligned with the direction of maximum variance among all directions orthogonal to the first one, and so on (Figure 3.8). If data need to be reduced to one dimension, the first principal component can be used to describe it. If two dimensions are allowed, the first two, and so on. In this way, as much of the variance can be represented in as few dimensions as possible.

The main problem with PCA is that if the data distribution is nonlinear, a low-dimensional hyperplane cannot provide an accurate description (Hastie and Stuetzle 1989; Kambhatla and Leen 1997; Ritter et al. 1992). This fact is illustrated in Fig-

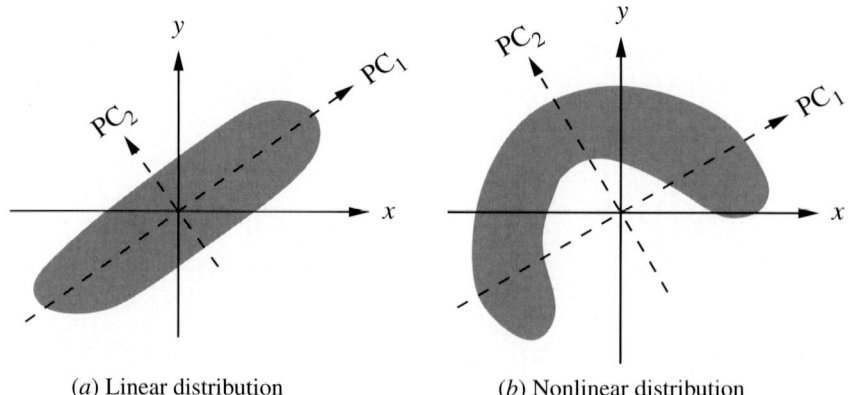

(a) Linear distribution   (b) Nonlinear distribution

Fig. 3.8. Principal components of data distributions. In principal component analysis, the data originally represented in (x, y) coordinates are transformed into the principal component coordinate system: The first principal component (PC_1) aligns with the direction of maximum variance in the data, and the second (PC_2) is orthogonal to it. The lengths of the axes reflect the variance along each coordinate dimension. (a) The two-dimensional distribution has a linear structure, and the first component alone is a good representation. However, with a nonlinear distribution (b), PCA does not result in a good lower dimensional representation, even though the distribution lies on a one-dimensional curve.

ure 3.8b: The first principal component misses the main features of the data, and provides only an inaccurate approximation.

Nonlinear distributions are best approximated by curved structures, i.e. hypersurfaces rather than hyperplanes. For such distributions, one can define principal curves and principal surfaces, in fashion analogous to principal components (Hastie and Stuetzle 1989; Ritter et al. 1992). Intuitively, the principal curve passes through the middle of the data distribution, as shown in Figure 3.9a. The center of gravity of the area enclosed by two very close normals should lie on the principal curve.

Let us consider the task of finding a principal curve of a data distribution. Let $\mathbf{X}$ be a data point and f a smooth curve in the input space, and let $d_f(\mathbf{X})$ be the distance from the data point to the closest point on the curve. The squared distance D_f of data distribution $P(\mathbf{X})$ to curve f can then be defined as

$$D_f = \int d_f^2(\mathbf{X})\, P(\mathbf{X})\, d\mathbf{X}. \tag{3.17}$$

The curve f is the principal curve of the data distribution $P(\mathbf{X})$ if D_f is minimal. Principal surfaces can be defined in the same way, by replacing curve f with a multi-dimensional surface.

It turns out that a self-organizing map is a way of computing a discretized approximation of the principal surface (Ritter et al. 1992). Assume that the principal surface is discretized into a set of vectors $\mathbf{W}_i$. For a data point $\mathbf{X}$, let $\mathbf{W}_{\mathrm{img}(\mathbf{X})}$ be the closest vector. Equation 3.17 can then be written as

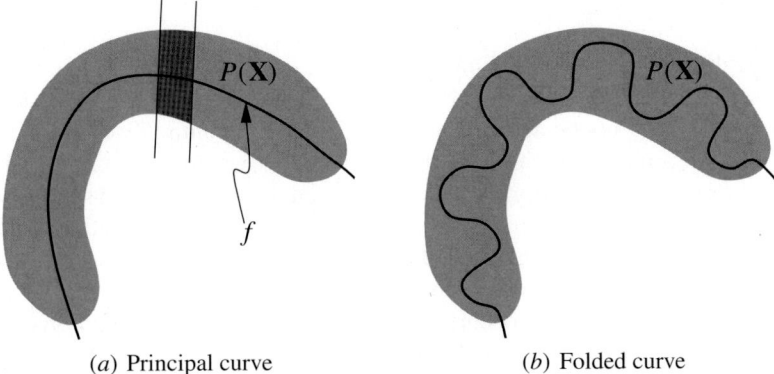

(a) Principal curve   (b) Folded curve

Fig. 3.9. Approximating nonlinear distributions with principal curves and folding. (a) The principal curve passes through the middle of the data distribution, providing a more detailed representation of nonlinear distributions than principal components. Each point on the curve is positioned at the center of gravity of the part of the distribution enclosed within two infinitesimally close normals (Ritter et al. 1992). (b) If a more detailed representation of the thickness of the distribution is desired, the curve can be folded in the higher dimension.

$$D_f = \int \|\mathbf{X} - \mathbf{W}_{\mathrm{img}(\mathbf{x})}\|^2 P(\mathbf{X}) \, d\mathbf{X}. \qquad (3.18)$$

The problem of finding the principal surface now reduces to finding a set of reference vectors $\mathbf{W}_i$ that minimizes the squared reconstruction error. It can be shown that the learning rule

$$\mathbf{W}'_{\mathrm{img}(\mathbf{X})} = \mathbf{W}_{\mathrm{img}(\mathbf{X})} + \alpha[\mathbf{X} - \mathbf{W}_{\mathrm{img}(\mathbf{X})}] \qquad (3.19)$$

minimizes this error under certain conditions (Ritter et al. 1992). This rule is identical to the abstract self-organizing map adaptation rule (Equation 3.15) when there is no neighborhood cooperation (i.e. h equals the delta function). In this case, the map places each weight vector at the center of gravity of the data points for which it is a winner. When the neighborhood function is introduced, other data points contribute also, and the center of gravity is calculated based on this larger volume.

The above analysis explains why the self-organizing map can develop efficient approximations of nonlinear high-dimensional input distributions. If there is low-dimensional nonlinear structure in the input, the map can follow the nonlinearities of the input distribution, and represent local as well as global structure. Perceptual categories as well as higher-level concepts are generally believed to organize into such low-dimensional manifolds (Kohonen, Kaski, Lagus, Salojärvi, Honkela, Paatero, and Saarela 2000; Li, Farkas, and MacWhinney 2004; Ritter et al. 1992; Roweis and Saul 2000; Seung and Lee 2000; Tenenbaum, de Silva, and Langford 2000; Tiňo and Nabney 2002). The self-organizing maps in the cortex therefore provide an efficient mechanism for representing structured sensory information in two dimensions.

3.5.2 Folding

Even though there is low-dimensional structure in the input, it may not be sufficient to represent it by a two-dimensional principal surface, i.e. the self-organizing map. For example, it may not be accurate enough to reduce the entire distribution in Figure 3.9a to the principal curve; it might also be necessary to represent how far the points are from the curve. The only way the one-dimensional curve can represent the entire area is to make tight turns across the width of the area, as well as gradually progressing along its length (Figure 3.9b). In other words, the map has to fold in the higher dimensions, in order to represent them as well as possible.

Similar folding of the map occurs when a high-dimensional distribution of points is mapped using a two-dimensional network. The results are interesting from a biological standpoint, because they help explain why patchy patterns of feature preferences, such as those for ocular dominance and orientation, form in the primary visual cortex (Kohonen 1989; Obermayer et al. 1992; Ritter et al. 1991). Let us use ocular dominance as an example. At each (x, y) location of the visual field, there are different ocular dominance values that must be represented. Let us assume that the variance in ocular dominance is smaller than the variance in the length and width dimensions. The goal is therefore to determine a dimension-reducing mapping of inputs in a flat box, where the two longer dimensions represent retinotopy, and the height dimension represents ocular dominance, onto a two-dimensional network (Figure 3.10; simulation parameters in Appendix E).

As the network self-organizes, it first stretches along the two longest dimensions of the box, and then folds in the smaller third dimension (Figure 3.10a). The folding takes place because the network tries to approximate the third dimension with the two-dimensional surface, analogous to Figure 3.9b. The weight values in the third dimension can be visualized for every neuron by coloring it with a corresponding gray-scale value, as in Figure 3.10b. The resulting pattern is very similar to the pattern of ocular dominance stripes seen in the primary visual cortex.

It is also important to understand how the mapping changes with increasing variation in the third dimension and with increasing number of dimensions. Let us first examine how the afferent patterns organize when the network is trained with input distributions of different heights. If the height is zero, all the inputs lie in a plane, and a smooth self-organized two-dimensional map will develop as in Figure 3.6. As the height is increased, fluctuations in the third dimension gradually appear, but the pattern is not stable, and keeps changing as training progresses. Beyond a threshold height z_f, however, a spontaneous phase transition occurs, and a stable folding pattern develops.

When the networks are trained with inputs with more than three dimensions of variation, the same principle can be observed in a recursive fashion. In a two-dimensional network, the map stretches along the two dimensions of maximum variance first, and folds along the dimensions of the next highest variance. A recursive folding structure then develops: The primary folds represent the dimensions of third highest variance, subfolds within the primary folds represent the dimensions with the next highest variance, and so on. Thus, the map develops a representation of the

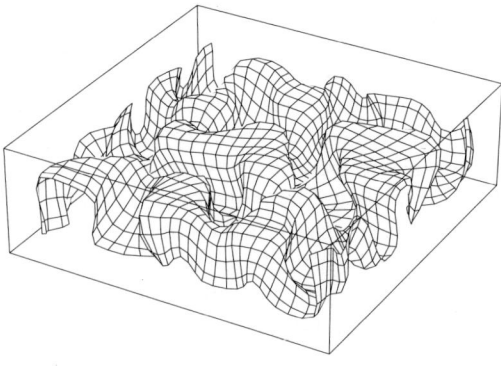

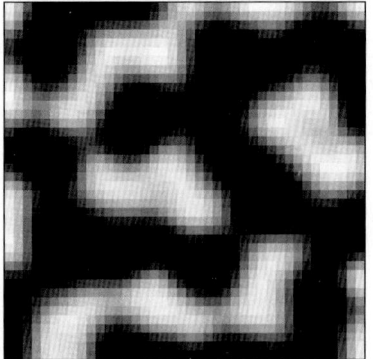

(a) Representing the third dimension by folding   (b) Visualization of ocular dominance

Fig. 3.10. Three-dimensional model of ocular dominance. The model consists of a two-dimensional map of a three-dimensional space. The first two dimensions can be interpreted as retinotopy and the third dimension as ocular dominance (Ritter et al. 1991,1992). In (a), the input space is indicated by the box outline, and the weight vectors of the map units are plotted in this space as a grid (as in Figure 3.6). The map extends along the longer retinotopic dimensions x and y, and folds in the smaller height dimension to approximate the space. (b) The weight value for the height dimension is visualized for each neuron: Gray-scale values from black to white represent continuously changing values from low to high. The resulting pattern resembles the ocular dominance stripes found in the visual cortex, suggesting that they too could be the result of a self-organized mapping of a three-dimensional parameter space.

statistically most relevant features of the input space. All the dimensions whose variance is greater than z_f are represented in the map in this recursive fashion. In this way, it is possible to capture several feature dimensions, such as retinotopy, ocular dominance, and orientation, in a single two-dimensional map. The computational model therefore offers a clear explanation for the observed overlapping map structure of the visual cortex: It is a self-organizing map representing high-dimensional input in two dimensions.

3.6 Conclusion

In understanding how the maps in the visual cortex develop and function, the cortical column has emerged as the appropriate computational unit. In most cases, the weighted-sum firing-rate model is sufficient to capture its behavior; when temporal coding is important (as in segmentation and binding), a leaky integrator model of the spiking neuron can be used, synchronizing neural activity to represent coherent bindings. These computational units adapt based on Hebbian learning, normalizing the total weight by redistributing synaptic resources.

When lateral interactions are established between such units, the self-organizing map model of the cortex is obtained. This model is a simple yet powerful learning architecture for representing the statistical structure of data distributions. If the dimensionality of the mapping surface (i.e. the network) is less than the dimensionality of the data distribution, the surface first extends nonlinearly along the most dominant dimensions of the data distribution, and then folds in the other dimensions. The folding process gives rise to local structures, such as alternating stripes resembling ocular dominance and orientation patches in the primary visual cortex. In the following chapters, such structures are shown to be not just superficially similar, but in fact good approximations of those seen in the primary visual cortex.

To make computational and mathematical analysis tractable, the self-organizing map models typically abstract away the lateral interactions in the cortex. These interactions are reintroduced in LISSOM, showing that they play a powerful and so far largely unrecognized role in cortical processing. Lateral connections between neurons can learn synergetically with the afferent connections and represent higher order statistical information in the network. This generalization results in a more accurate model of the visual cortex, accounting for development, plasticity, and many functional phenomena.

INPUT-DRIVEN SELF-ORGANIZATION

4

LISSOM: A Computational Map Model of V1

The neurobiological observations and computational principles from Part I will be brought together in this chapter to construct a laterally connected map model of the primary visual cortex. The model is based on self-organizing maps, but its connectivity, activation, and learning mechanisms are designed to capture the essential biological processes in more detail. As will be seen in later chapters of Part II, these mechanisms lead to a detailed computational account for how the visual cortex develops, adapts, and functions. In Part III, this model is further extended to include subcortical areas and higher visual areas, and in Part IV, a temporal coding mechanism and connectivity necessary for perceptual grouping. In this chapter, the motivation for the model is first reviewed, followed by a detailed description of its organization, activation, and learning processes. A simple example of self-organization is given, forming a retinotopic map of visual input.

4.1 Motivation: Cortical Maps

As was discussed in Sections 3.4 and 3.5, self-organizing maps are a powerful abstraction of learning and knowledge organization in the cortex. However, much of the behavior of the visual cortex depends on structures and processes (reviewed in Sections 2.1 and 2.2) that have been abstracted away in many self-organizing map models. In contrast, the LISSOM model (laterally interconnected synergetically self-organizing map; Miikkulainen 1991; Miikkulainen, Bednar, Choe, and Sirosh 1997; Sirosh 1995) is designed specifically to capture those processes. More specifically, LISSOM is based on five principles:

1. The central layout of the LISSOM model is a two-dimensional array of computational units, corresponding to vertical columns in the cortex. Such columns act as functional units in the cortex, responding to similar inputs, and therefore form an appropriate level of granularity for a functional model.
2. Each unit receives input from a local anatomical receptive field in the retina, mediated by the ON-center and OFF-center channels of the LGN. Such connectivity corresponds to the neural anatomy; it also allows modeling a large area of

the visual cortex and processing large realistic visual inputs, which in turn allows studying higher visual function such as visual illusions, grouping, and face detection.

3. The cortical units are connected with excitatory and inhibitory lateral connections that adapt as an integral part of the self-organizing process. Before LISSOM, the function of these lateral interactions had not been analyzed in detailed computational models. They can play a central role in establishing an efficient visual representation, and in mediating functions such as illusions and perceptual grouping.

4. The units respond by computing a weighted sum of their input, limited by a logistic (sigmoid) nonlinearity. This is a standard model of computation in the neuronal units that matches their biological characteristics well.

5. The learning is based on Hebbian adaptation with divisive normalization. Hebbian learning is well supported by neurobiological data, and recent biological experiments have also suggested how normalization could occur in animals. These mechanisms are found to be computationally both necessary and powerful in explaining functional phenomena, such as the indirect tilt aftereffect.

In other words, LISSOM takes the central idea of self-organizing maps (1), and implements it at a level of known visual cortex structures (2 and 3) and processes (4 and 5). Although each of these principles has been tested in other models (Section 3.4.1), their combination is novel and allows LISSOM to account for a wide range of phenomena in the development, plasticity, and function of the primary visual cortex. The details of the model are described next, with examples to illustrate the processing that occurs.

4.2 The LISSOM Architecture

The LISSOM model of V1 is based on a simulated network of cortical neurons with afferent connections from the external world and recurrent lateral connections between neurons. These connections adapt based on correlated activity. The result is a self-organized structure where afferent connection weights form a map of the input space, and lateral connections store long-term correlations in neuronal activity. In this section, the layout and connectivity of the LISSOM model are described in detail.

4.2.1 Overview

LISSOM is intended to model accurately the biological structures and processes that are most important for the observed developmental and functional phenomena. Other biological features are abstracted in the model, in order to reduce confounding factors and to make an efficient, systematic analysis possible.

The V1 network in LISSOM is a sheet of $N \times N$ interconnected computational units, or "neurons" (Figure 4.1). Because the focus is on the two-dimensional organization of the cortex, each neuron in V1 corresponds to a vertical column of cells

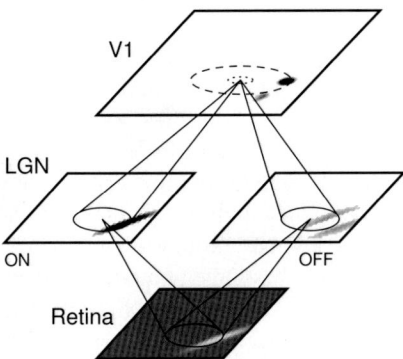

Fig. 4.1. Architecture of the basic LISSOM model. LISSOM consists of a hierarchy of two-dimensional sheets of neural units, including an array of retinal receptors, ON and OFF channels in the LGN, and a cortical network representing V1. The LGN and V1 activation is shown in gray-scale coding from white to black (low to high). The activity on the retina (a single oriented Gaussian) is presented like natural images: Light areas are strongly activated, dark areas are weakly activated, and medium gray represents background activation. This input gray scale will be used for all models that include the LGN and which can therefore process natural images. Sample connections are shown for one unit in each LGN sheet and one in V1. The LGN afferents form a local anatomical receptive field on the retina, and cause ON-center LGN units to respond to light areas surrounded by dark, and OFF-center units to dark areas surrounded by light. Neighboring LGN neurons have different but overlapping RFs. Similarly, V1 neurons have afferent receptive fields on the LGN sheets. V1 neurons also receive lateral excitatory and lateral inhibitory connections from nearby V1 neurons; these connections are shown as dotted and dashed circles around the V1 neuron, respectively. V1 activity is patchy because only those neurons respond whose feature preferences match the orientation, eye of origin, and direction of movement of the pattern currently in their receptive fields.

through the six layers of the biological cortex. This columnar organization helps make the problem of simulating such a large number of neurons tractable, and is viable because the cells in a column generally fire in response to the same inputs (Section 2.1.2). The activity of each neuron is represented by a continuous number within [0..1]; individual spiking is not modeled in basic LISSOM. Therefore, it is important to keep in mind that LISSOM neurons are *not* strictly identifiable with single cells in the biological cortex; instead, LISSOM models biological mechanisms at an aggregate level.

Each cortical neuron receives external input from two types of neurons in the LGN: ON-center and OFF-center. The LGN neurons in turn receive input from a small area of the retina, represented as an $R \times R$ array of photoreceptor cells. The afferent input connections from the retina to LGN and LGN to V1 are all excitatory. In addition to the afferent connections, each cortical neuron has reciprocal excitatory and inhibitory lateral connections with other neurons. Lateral excitatory connections have a short range, connecting only close neighbors in the map. Lateral inhibitory connections run for long distances, but may be patchy, connecting only selected neurons.

The ON and OFF neurons in the LGN represent the entire pathway from photoreceptor output to the V1 input, including the ON/OFF processing in the retinal ganglion cells and the LGN. Although the ON and OFF neurons are not always physically separated in the biological pathways, for conceptual clarity they are divided into separate channels in LISSOM. Each of these channels is further organized into an $L \times L$ array corresponding to the retinotopic organization of the LGN. For simplicity and computational efficiency, only single ON and OFF channels are used in LISSOM simulations in this book, but multiple channels could be included to represent different spatial frequencies. Also, the photoreceptors are uniformly distributed over the retina; since the inputs are relatively small in the LISSOM experiments, the fovea/periphery distinction is not crucial for the basic model.

Each neuron develops an initial response as a weighted sum (scalar product) of the activation in its afferent input connections. The lateral interactions between cortical neurons then focus the initial activation pattern into a localized response on the map. After the pattern has stabilized, the connection weights of cortical neurons are modified. As the self-organization progresses, these neurons grow more nonlinear and weak connections die off. The result is a self-organized structure in a dynamic equilibrium with the input.

The following subsections describe the specific components of the LISSOM model in more detail. They focus on the basic version of the model trained with unoriented Gaussian inputs, to highlight the basic principles as clearly as possible. In Chapter 5 this model is extended to two retinas, more complex Gaussian inputs, and natural images, in order to study how orientation preference, ocular dominance, and direction selectivity develops in the visual cortex. In Chapters 6 and 7, the model is temporarily abstracted further by bypassing the LGN, and this more efficient model is used to study plasticity and visual illusions. In Part III, the model is extended to include subcortical and higher level areas, in order to understand the synergy of genetically and environmentally driven development. Finally, Part IV will describe an extension that includes spiking neurons and long-range excitatory connections, to explain how perceptual grouping occurs.

4.2.2 Connections to the LGN

Previous models have explained how the connections from the retina to the LGN could develop from internally generated activity in the retina (Eglen 1997; Elliott and Shadbolt 1999; Haith 1998; Keesing, Stork, and Shatz 1992; Lee, Eglen, and Wong 2002a). LISSOM instead focuses on learning at the cortical level, so all connections to neurons in the ON and OFF channels are set to fixed strengths.

The strengths were chosen to approximate the receptive fields that have been measured in adult LGN cells, using a standard difference-of-Gaussians model (DoG; Cai, DeAngelis, and Freeman 1997; Rodieck 1965; Tavazoie and Reid 2000). First, the center of each LGN receptive field is mapped to the location in the retina corresponding to the location of the LGN unit (appendix Figure A.1). This mapping ensures that the LGN will have the same two-dimensional topographic organization as the retina. Using that location as the center, the weights are then calculated from

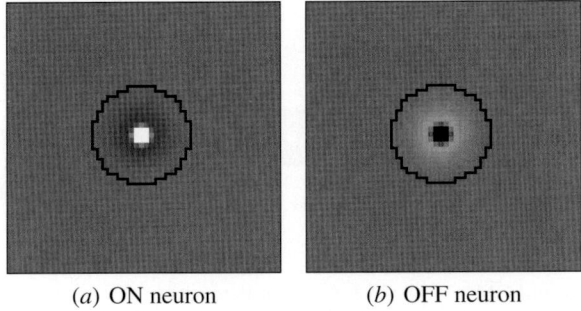

(a) ON neuron   (b) OFF neuron

Fig. 4.2. Afferent weights of ON and OFF neurons in the LGN. The receptive fields of a sample ON neuron and a sample OFF neuron, both located at the center of the corresponding 36×36 LGN sheet, are shown as gray-scale values on a 54×54 retina. The jagged black line traces the anatomical boundary of the RF, that is, connections exist only from retinal receptors located inside the boundary. Medium gray represents zero weight, dark gray stands for inhibitory connections, and light gray represents excitatory connections. Each RF shape is a difference of two Gaussians, the center and the surround. The Gaussians are normalized to have the same total strength, but the center Gaussian concentrates that strength in a much smaller region; in this example, $\sigma_c = 0.5$ and $\sigma_s = 2$. ON cells have an excitatory center and an inhibitory surround (a), and OFF cells have an inhibitory center and an excitatory surround (b), as in Figure 2.2a,b. These RFs perform edge detection at a spatial frequency determined by the width of the center; they highlight areas of the input image that have edges and lines, and do not respond to large areas of constant illumination.

the difference of two normalized Gaussians. More precisely, the weight $L_{xy,ab}$ from receptor (x, y) in the receptive field of an ON-center cell (a, b) with center (x_c, y_c) is given by

$$
L_{xy,ab} = \frac{\exp\left(-\frac{(x-x_c)^2+(y-y_c)^2}{\sigma_c^2}\right)}{\sum_{uv} \exp\left(-\frac{(u-x_c)^2+(v-y_c)^2}{\sigma_c^2}\right)} - \frac{\exp\left(-\frac{(x-x_c)^2+(y-y_c)^2}{\sigma_s^2}\right)}{\sum_{uv} \exp\left(-\frac{(u-x_c)^2+(v-y_c)^2}{\sigma_s^2}\right)}, \quad (4.1)
$$

where σ_c determines the width of the central Gaussian and σ_s the width of the surround Gaussian. The weights for an OFF-center cell are the negative of the ON-center weights, i.e. they are calculated as the surround minus the center. Figure 4.2 shows examples of such ON and OFF receptive fields.

Note that even though the OFF cells have the same weights as ON cells (differing only by the sign), their activities are not redundant. Since the firing rates in biological systems cannot be negative, each cell is thresholded to have only positive activations, as described in more detail in Section 4.3.2. As a result, the ON and OFF cells will never be active at the same cortical location. They therefore provide complementary information, both in the model and in the visual system. Separating the ON and OFF channels in this way makes it convenient to compare the model with experimental results.

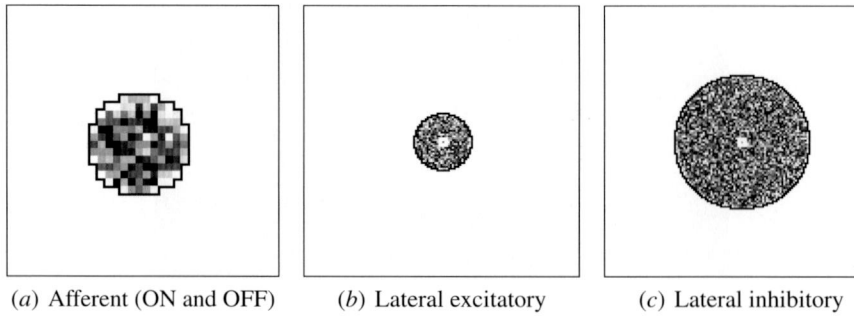

(*a*) Afferent (ON and OFF)   (*b*) Lateral excitatory   (*c*) Lateral inhibitory

Fig. 4.3. Initial V1 afferent and lateral weights. The initial incoming weights of a sample neuron at the center of V1 are plotted in gray-scale coding from white to black (low to high). (*a*) The afferent RF center for each neuron was determined by first finding the location on the LGN that corresponds to the location of the neuron in V1, then randomly scattering the center by ±1.2 retinal units (5%) from that location. The neuron was then connected to LGN units within a radius of 6.5 from the center with random normalized weights. These weights are shown on the 36×36 LGN sheet, with the jagged black line tracing the RF boundary. The weights from the ON and OFF LGN sheets were initially identical. Plots (*b*) and (*c*) similarly show the lateral weights of this neuron, plotted on the 142×142 V1 and outlined with a black line. The neuron itself is marked with a small white square in the middle. The excitatory connections were initially random within a radius of 14.2, and inhibitory within 34.5 units. Later figures will show how these connections become organized through input-driven self-organization.

4.2.3 Connections in the Cortex

In contrast to the fixed connection weights in the LGN, all connections in cortical regions in LISSOM are modifiable by neural activity. They are initialized according to the gross anatomy of the visual cortex, with weight values that provide a neutral starting point for self-organization.

Each neuron's afferent receptive field center is located randomly within a small radius of its optimal position, i.e. the point corresponding to the neuron's location in the cortical sheet. The neuron is connected to all ON and OFF neurons within radius r_A from the center (Figures 4.3*a* and A.1). For proper self-organization to occur, the radius r_A must be large compared with the scatter of the centers, and the RFs of neighboring neurons must overlap significantly, as they do in the cortex (Sirosh and Miikkulainen 1997).

Lateral excitatory connections are short range, connecting each neuron to itself and to its neighbors within a close radius. The extent of lateral excitation should be comparable to the activity correlations in the input. Lateral inhibitory connections extend in a larger radius, and also include connections from the neuron itself and from its neighbors (Figure 4.3*b,c*). The range of lateral inhibition may vary as long as it is greater than the excitatory radius. This overall center–surround pattern is crucial for self-organization, and approximates the lateral interactions that take place at high contrasts in the cortex (Section 2.2.1). Long-range excitatory connections can also be included, as will be done in Part IV when simulating perceptual grouping and completion phenomena.

Both the afferent and the lateral connection weights are initially unselective. Like prior work, the simulations in this book are generally based on random (but normalized) initial weight values. To speed up the simulation, values corresponding to a Gaussian profile are sometimes used as well. Further, to demonstrate that any differences between the ON and OFF channels (as well as the two eyes and the different lags in Chapter 5) are learned from the data, all channels of each neuron are initialized with the same set of random weights. As will be shown in Section 8.4, the specific initial values of these weights have little effect on subsequent self-organization.

4.3 Response Generation

Before each input presentation, the activities of all units in the LISSOM network are initialized to zero. The system then receives input through activation of the retinal units. The activity propagates through the ON and OFF channels of the LGN to the cortical network, where the neurons settle the initial activation through the lateral connections, as will be described in detail below.

4.3.1 Retinal Activation

An input pattern is presented to the LISSOM model by activating the photoreceptor units in the retina according to the gray-scale values in the pattern. Figure 4.4a shows a basic input pattern consisting of multiple unoriented Gaussians. To generate such input patterns, the activity χ_{xy} for photoreceptor cell (x, y) is calculated according to

$$\chi_{xy} = \max_{k} \exp\left(-\frac{(x - x_{c,k})^2 + (y - y_{c,k})^2}{\sigma_u^2}\right), \qquad (4.2)$$

where $(x_{c,k}, y_{c,k})$ specifies the center of Gaussian k and σ_u its width. At each iteration, $x_{c,k}$ and $y_{c,k}$ are chosen randomly within the retinal area; σ_u is usually constant.

As will be described in later chapters, more complex artificial patterns, with varying widths and elongations, can be generated in the same way, or input can be formed by rendering natural images directly on the photoreceptor units. However, unoriented, constant-width Gaussians will be used in this chapter to illustrate the basic properties of the model.

4.3.2 LGN Activation

The cells in the ON and OFF channels of the LGN compute their responses as a squashed weighted sum of activity in their receptive fields (Figure 4.4b). More precisely, the response ξ_{ab} of ON or OFF-center cell (a, b) is calculated as

$$\xi_{ab} = \sigma\left(\gamma_L \sum_{xy} \chi_{xy} L_{xy,ab}\right), \qquad (4.3)$$

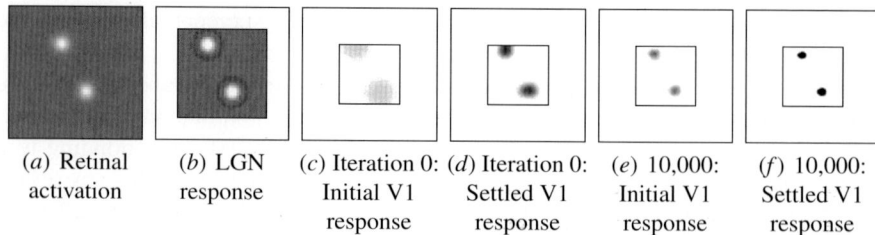

| (*a*) Retinal activation | (*b*) LGN response | (*c*) Iteration 0: Initial V1 response | (*d*) Iteration 0: Settled V1 response | (*e*) 10,000: Initial V1 response | (*f*) 10,000: Settled V1 response |

Fig. 4.4. Example input and response. At each self-organization iteration in LISSOM, the photoreceptors in the retina are activated according to the gray-scale values of the input image (as described in Figure 4.1). In this example, two unoriented Gaussians with width $\sigma_u = 3.0$ were drawn on random spatially separated locations on a 54×54 array of receptors (*a*). The 36×36 ON and OFF cell responses are plotted in the central square of (*b*) by subtracting the OFF cell responses from the ON: Dark areas represent higher activity in the OFF channel, light areas higher activity in the ON channel, and medium gray represents equal activation in both channels. The LGN responds to edges and lines in the input, with high ON cell activations where the input is brighter than the surround, and high OFF cell activations where the input is darker than the surround. Before self-organization (i.e. iteration 0), the 142×142 V1 map initially responds broadly and unspecifically to the input patterns (*c*); V1 activations are represented in the central square in gray scale from white to black (low to high). The lateral connections focus the response into discrete activity "bubbles" in (*d*), and connections are then modified. After 10,000 input presentations and learning steps, the initial and settled V1 responses are more focused, forming a sparse representation of the input (*e* and *f*). This figure also illustrates how the retina is mapped to the LGN and the LGN mapped to V1 in LISSOM. The LGN and V1 networks are drawn to the same scale as the retina (as indicated by the outside squares), so that activity at a given location in the LGN and V1 corresponds to a stimulus at the corresponding location in the retina. The retina is larger than the LGN and the LGN larger than V1 so that all LGN and V1 neurons have complete receptive fields (i.e. they are not cut off by the network boundary; Figure A.1). In subsequent activity figures, such padding is omitted and only the retinal and LGN area that matches the V1 network is shown. An animated demo of the map response can be seen at http://computationalmaps.org.

where χ_{xy} is the activation of cell (x, y) in the receptive field of (a, b), $L_{xy,ab}$ is the afferent weight from (x, y) to (a, b), and γ_L is a constant scaling factor. The squashing function $\sigma(\cdot)$ (Figure 4.5) is a piecewise linear approximation of the sigmoid activation function mentioned in Section 3.1.4:

$$\sigma(s) = \begin{cases} 0 & s \leq \theta_1, \\ (s - \theta_1)/(\theta_u - \theta_1) & \theta_1 < s < \theta_u, \\ 1 & s \geq \theta_u. \end{cases} \tag{4.4}$$

As in other models, this approximation is used because it implements the essential thresholding and saturation behavior, and can be computed more quickly than a smooth logistic function.

Changing γ_L in Equation 4.3 by a factor m is equivalent to dividing θ_1 and θ_u by m. Even so, γ_L is treated as a separate parameter to make it simpler to use the same values of θ_1 and θ_u for different networks. The specific value of γ_L is set man-

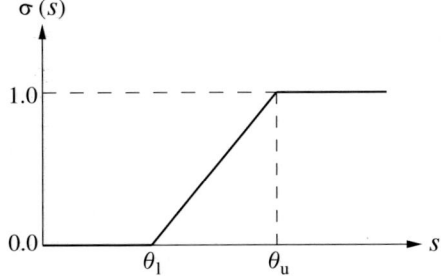

Fig. 4.5. Neuron activation function $\sigma(s)$. The neuron requires an input as large as the threshold θ_l before responding, and saturates at the ceiling θ_u. The output activation values are limited to $[0..1]$. This activation function is an efficient approximation of the logistic (sigmoid) function.

ually so that the LGN outputs approach 1.0 in the highest-contrast regions of typical input patterns. This allows each subsequent level to use similar parameter values in general, other than γ_L.

Because of its DoG-shaped receptive field, an LGN neuron will respond whenever the input pattern is a better match to the central portion of the RF than to the surrounding portion. The positive and negative portions of the RF thus have a push–pull effect (Hirsch, Alonso, Reid, and Martinez 1998a; Hubel and Wiesel 1962). That is, even if an input pattern activates the ON portion of the LGN RF, the neuron will not fire unless the OFF portion is *not* activated. This balance ensures that the neurons will remain selective for edges over a wide range of brightness levels. Section 6.2.3 will show that this push–pull effect is crucial when natural images are used as input to the model. Overall, the LGN neurons respond to image contrast, subject to the minimum and maximum activity values enforced by the activation function.

4.3.3 Cortical Activation

The cortical activation mechanism is similar to that of the LGN, but extended to support self-organization and to include lateral interactions. The total activation is computed by combining the afferent and lateral contributions. First, the afferent stimulation s_{ij} of V1 neuron (i, j) is calculated as a weighted sum of activations in its receptive fields on the LGN:

$$s_{ij} = \gamma_A \left(\sum_{ab \in \text{ON}} \xi_{ab} A_{ab,ij} + \sum_{ab \in \text{OFF}} \xi_{ab} A_{ab,ij} \right), \tag{4.5}$$

where ξ_{ab} is the activation of neuron (a, b) in the receptive field of neuron (i, j) in the ON or OFF channels, $A_{ab,ij}$ is the corresponding afferent weight, and γ_A is a constant scaling factor. The afferent stimulation is squashed using the sigmoid activation function, forming the neuron's initial response as

$$\eta_{ij}(0) = \sigma s_{ij}, \tag{4.6}$$

where $\sigma(\cdot)$ is a piecewise linear sigmoid as in Equation 4.4.

After the initial response, lateral interaction sharpens and strengthens the cortical activity over a very short time scale. At each of these subsequent discrete time steps, the neuron combines the afferent stimulation s with lateral excitation and inhibition:

$$\eta_{ij}(t) = \sigma\left(s_{ij} + \gamma_E \sum_{kl} \eta_{kl}(t-1)E_{kl,ij} - \gamma_I \sum_{kl} \eta_{kl}(t-1)I_{kl,ij}\right), \quad (4.7)$$

where $\eta_{kl}(t-1)$ is the activity of another cortical neuron (k,l) during the previous time step, $E_{kl,ij}$ is the excitatory lateral connection weight on the connection from that neuron to neuron (i,j), and $I_{kl,ij}$ is the inhibitory connection weight. All connection weights have positive values. The scaling factors γ_E and γ_I represent the relative strengths of excitatory and inhibitory lateral interactions, which determine how easily the neuron reaches full activation.

The cortical activity pattern starts out diffuse and spread over a substantial part of the map (Figure 4.4c,e). Within a few iterations of Equation 4.7, it converges into a small number of stable focused patches of activity, or activity bubbles (Figure 4.4d,f). Such settling results in a sparse final activation, which allows representing visual information efficiently (Section 14.2; Barlow 1972; Field 1994). It also ensures that nearby neurons have similar patterns of activity and therefore encode similar information, as seen in the cortex.

While the cortical response is settling, the afferent input remains constant. However, the lateral interaction is not strong enough to maintain the activity bubble when the input changes. A change in the input will cause the net input of the neurons (i.e. the sums of the afferent and lateral activations) to fall below the threshold θ_1. A new response will then form with little interference from the previous response. The LISSOM network could therefore be trained even with continuously changing inputs, without explicitly resetting the network to zero activity in between.

4.4 Learning

Self-organization of the connection weights takes place in successive input iterations, usually 5000–20,000 in total. Each iteration consists of presenting an input image, computing the corresponding settled activation patterns in each neural sheet, and modifying the weights. Weak lateral connections are periodically removed, modeling connection death in biological systems. In order to achieve smooth maps, the lateral excitation radius, sigmoid, and learning rate parameters can be gradually adjusted over the course of learning.

4.4.1 Weight Adaptation

After the activity has settled, the connection weights of each cortical neuron are modified. Both the afferent and lateral weights adapt according to the same biologically motivated mechanism: the Hebb rule (Hebb 1949) with divisive postsynaptic normalization (Section 3.3):

$$w'_{pq,ij} = \frac{w_{pq,ij} + \alpha X_{pq}\eta_{ij}}{\sum_{uv}(w_{uv,ij} + \alpha X_{uv}\eta_{ij})}, \tag{4.8}$$

where $w_{pq,ij}$ is the current afferent or lateral connection weight (either A, E or I) from (p,q) to (i,j), $w'_{pq,ij}$ is the new weight to be used until the end of the next settling process, α is the learning rate for each type of connection (α_A for afferent weights, α_E for excitatory, and α_I for inhibitory), X_{pq} is the presynaptic activity after settling (ξ for afferent, η for lateral), and η_{ij} stands for the activity of neuron (i,j) after settling, Afferent inputs (i.e. both ON and OFF channels together), lateral excitatory inputs, and lateral inhibitory inputs are normalized separately.

In line with the Hebbian principle, when the presynaptic and postsynaptic neurons are frequently simultaneously active, their connection becomes stronger. As a result, the neurons learn correlations in the input patterns. Normalization prevents the weight values from increasing without bounds; this process corresponds to redistributing the weights so that the sum of each weight type for each neuron remains constant. As was discussed in Section 3.3, such normalization can be seen as an abstraction of neuronal regulatory processes.

4.4.2 Connection Death

Modeling connection death in the cortex (Section 2.2.2), lateral connections in the LISSOM model survive only if they represent significant correlations among neuronal activity. Once the map begins to organize, most of the long-range lateral connections link neurons that are no longer simultaneously active. Their weights become small, and they can be pruned without disrupting self-organization.

The parameter t_d determines the onset of connection death. At t_d, lateral connections with strengths below a threshold w_d are eliminated. From t_d on, more weak connections are eliminated at intervals Δt_d during the self-organizing process. Eventually, the process reaches an equilibrium where the mapping is stable and all lateral weights stay above w_d. The precise rate of connection death is not crucial to self-organization, and in practice it is often sufficient to prune only once, at t_d.

Most long-range connections are eliminated this way, resulting in patchy lateral connectivity similar to that observed in the visual cortex. Since the total synaptic weight is kept constant, inhibition concentrates on the most highly correlated neurons, resulting in effective suppression of redundant activation (Section 14.2). The short-range excitatory connections link neurons that are often part of the same bubble. They have relatively large weights and are rarely pruned.

4.4.3 Parameter Adaptation

The above processes of response generation, weight adaptation, and connection death are sufficient to form ordered afferent and lateral input connections like those in the cortex. However, the process can be further enhanced with gradual adaptation of lateral excitation, sigmoid, and learning parameters, resulting in more refined final maps.

As the lateral connections adapt, the activity bubbles in the cortex will become more focused, resulting in fine-tuning the map. As in other self-organizing models (such as SOM), this process can be accelerated by gradually decreasing the excitatory radius until it covers only the nearest neighbors (Erwin et al. 1992a,b; Kohonen 1993, 2001; Sirosh and Miikkulainen 1997). Such a decrease helps the network develop more detailed organization faster.

Gradually increasing the sigmoid parameters θ_l and θ_u produces a similar effect. The cortical neurons become harder to activate, further refining the response. Also, the learning rates α_A, α_E and α_I can be gradually reduced, ensuring that a smooth global organization and well-tuned receptive fields develop.

Such parameter adaptation models the biological processes of maturation that take place independently from input-driven self-organization, leading to critical periods and loss of plasticity in later life. These biological processes are reviewed in Section 16.1.6, and their effect on self-organization is demonstrated in Sections 9.4 and 13.4.

4.5 Self-Organizing Process

In order to validate the LISSOM model, an example self-organizing process forming an ordered map of two-dimensional input is presented in this section and compared with the corresponding simulation with the SOM model of Section 3.4.3.

4.5.1 Method

The model consisted of a 54×54 retina, 36×36 ON and OFF channels, and a 142×142 V1 network. This network, like other LISSOM networks discussed in this book (except the scaled-up versions in Chapters 10 and 15) corresponds approximately to a 5 mm $\times$ 5 mm area of macaque V1. The V1 size was first chosen to match the estimated number of columns in such an area, and the other parameters were then set to simulate it realistically (Appendices A.2–A.4).

The LGN neurons were set as in Figure 4.2, and the cortical network was initialized with random weight values, as shown in Figure 4.3. The network was organized in 10,000 presentations of two unoriented Gaussian patterns. Although more than two Gaussians could be used in principle, they are too large to be distributed uniformly on the small retina without overlap (Appendix A.4). An example of these patterns is shown in Figure 4.4, together with the resulting responses in the initial and self-organized network.

4.5.2 Afferent Connections

Figure 4.6 illustrates how the afferent weights of the cortical neurons self-organize. The figure shows an example set of final weight patterns on the ON and OFF sheets, and their combined effect. The initial rough patterns shown in Figure 4.3a

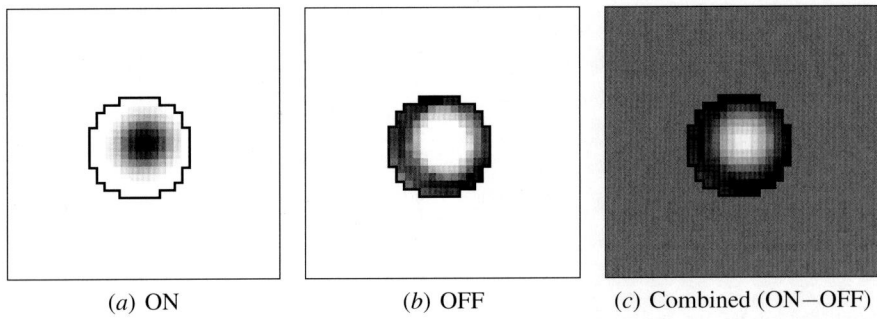

(*a*) ON   (*b*) OFF   (*c*) Combined (ON−OFF)

Fig. 4.6. Self-organized V1 afferent weights. The final afferent weights of the sample neuron at the center of the V1 network are shown on the (*a*) ON and (*b*) OFF sheets as in Figure 4.3*a*. (*c*) Their combined effect is shown by subtracting the OFF weights from the ON. From initially random weights, a smooth profile emerged on both LGN sheets, reflecting the Gaussian input patterns. Their combined effect is center–surround, resulting in a focused, edge-enhanced response in V1. Because the anatomical RF (shown with the black outline) of this neuron was positioned slightly left and below its topographically ordered position, the Gaussian weight patterns formed slightly above and to the right of the RF center. As shown in Figures 4.7*a* and 4.8*b*, these locations represent the retinotopic order well.

evolved into smooth Gaussian profiles, corresponding to the Gaussians used as input. The combined effect of the afferents is center–surround, forming a focused, edge-enhanced response in V1.

With unoriented Gaussian inputs, the only feature in the input that can be learned by the map is retinotopy. The weight profiles of different neurons indeed peaked over different parts of the LGN sheet in topographically correct locations (Figure 4.7*a*). As in Section 3.4.3, this organization can be visualized by plotting the center of gravity of each RF in the retinal space and connecting the centers of neighboring neurons by a line. The resulting Figure 4.8 shows that the centers organized from initially scattered positions into a smooth retinotopic map.

4.5.3 Lateral Connections

The lateral connections adapt together with the afferents. As the afferent receptive fields organize into a uniform map, the activity correlations within the network decrease with distance approximately like a Gaussian, with strong correlations to near neighbors and weaker correlations to more distant neurons. The lateral excitatory and inhibitory connections therefore acquire a Gaussian shape, and the combined lateral excitation and inhibition becomes an approximate difference of Gaussians (Figure 4.9). In the central area of the map, these patterns are unoriented, and become more elongated near the edges (Figure 4.7*b*).

The DoG organization allows the map to sharpen the initial response and form more focused activity patterns. As was discussed in Section 4.4.3, this process will result in gradual fine-tuning of the afferent connections and the retinotopic map. This process is especially important near the edges of the map, allowing the map

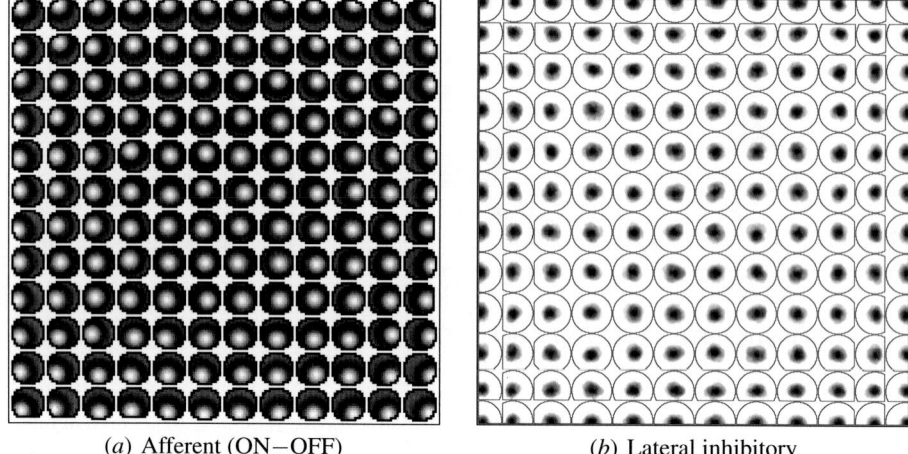

| (a) Afferent (ON−OFF) | (b) Lateral inhibitory |

Fig. 4.7. Self-organized afferent and lateral weights across V1. The self-organized combined afferent weights (as in Figure 4.6c) and the lateral inhibitory weights (as in Figure 4.9b) are shown for every 12th neuron horizontally and vertically across the V1 network array, starting at neuron (5,5) in the top left corner. (a) The Gaussian weight pattern systematically moves outward away from the center of the anatomical RF the closer the neuron is to the edge of the network, allowing the map to expand to represent the input space better, as shown in Figure 4.8. (b) Similarly, the lateral inhibitory weights of units near the edge of the network are elongated along the edge, allowing sharp responses to form at the edge.

to expand to fill the space. The next chapter will show that when the input consists of more complex features, such as orientation or ocular dominance, the connection patterns will further elongate and become patchy, representing activity correlations in the input. Such connections turn out to be crucial in developing an efficient coding of visual input (Section 14.2).

4.5.4 Differences Between LISSOM and SOM

It is important to contrast the self-organizing process above with that of the SOM model of Section 3.4.3. In SOM, each unit receives the same input, consisting of the entire visual field. The response of the network is assumed to consist of a single activity bubble. A global supervisor determines its location by finding the maximally responding unit, and its shape according to a predetermined neighborhood function. This process makes learning more regular and results in accurate maps, which is often important in practical applications where biological accuracy is not the goal.

In contrast, the units in a LISSOM map receive inputs from local areas in the retina, represented in terms of ON and OFF channels, and the response is based on purely local exchange of activation without any global supervision. Such a process more accurately models the structure of the visual cortex. Because of this structure, multiple inputs can be presented to the model simultaneously, and the map organizes

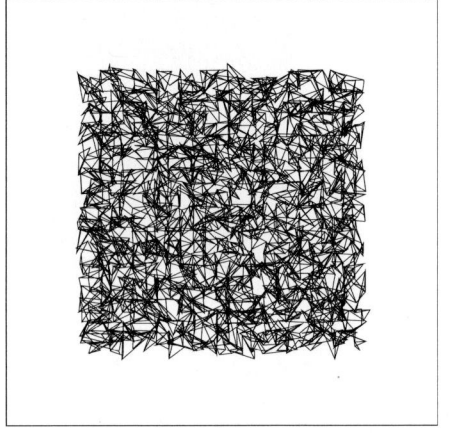

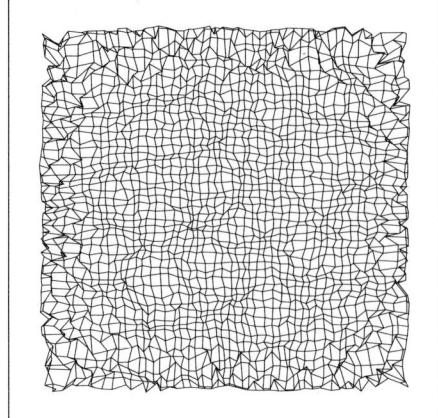

(*a*) Initial disordered map   (*b*) Final retinotopic map

Fig. 4.8. Self-organization of the retinotopic map. The center of gravity of the afferent weights of every third neuron in the 142×142 V1 is projected onto the retinal space (represented by the square outline). As in Figure 3.6, each center is connected to those of the four neighboring neurons by a line, representing the topographical organization of the map. Initially, the anatomical RF centers were slightly scattered topographically and the weight values were random (*a*). The map is contracted because the receptive fields were initially mapped to the central portion of the retina so that each neuron has full RFs (Figure A.1). As self-organization progresses, the map unfolds to form a regular retinotopic map (*b*). The map expands slightly during this process, because neurons near the edge become tuned to the peripheral regions of the input space (Figure 4.7*a*). The map does not fill the input space entirely, because the center of gravity will always be located slightly inside the space. These results show that LISSOM can learn retinotopy like SOM does, but using mechanisms more close to those in biology.

at multiple locations at the same time. This capability is crucial when modeling natural inputs and large areas of the visual field (Chapter 10).

Because the receptive fields in LISSOM are local, the map is already partially topographically ordered in the beginning. If the receptive fields instead covered the whole retina, LISSOM would self-organize an initially random map very much like the SOM model in Section 3.4.3; it would also be able to adapt to multiple inputs at once, unlike the SOM (Sirosh and Miikkulainen 1997). Therefore, the initial order is not computationally necessary in LISSOM: It is a side effect of a biologically more realistic architecture.

The LISSOM topographic map may be less regular than that formed with SOM (compare Figures 4.8 and 3.6). However, the LISSOM map stores more information about the input: In addition to the topography stored in the afferent connections, the lateral connections represent feature correlations. Such knowledge allows building an efficient representation of the input, which makes visual processing more effective (Chapter 14).

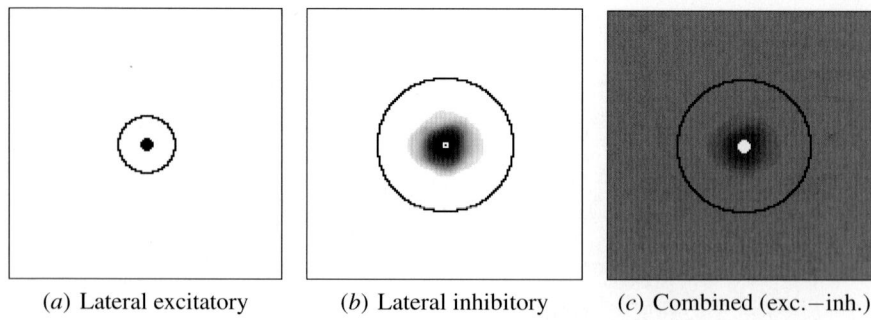

(*a*) Lateral excitatory   (*b*) Lateral inhibitory   (*c*) Combined (exc.−inh.)

Fig. 4.9. Self-organized V1 lateral weights. The lateral excitatory (*a*) and inhibitory (*b*) weights for the sample neuron at the center of V1 are plotted on V1 as in Figure 4.3*b,c*; the small white square in (*b*) marks the neuron itself and the jagged black outline traces the original connectivity before self-organization and pruning. The combined profile (*c*) shows inhibitory weights subtracted from the excitatory, and illustrates the total effect of the lateral connections: Dark indicates a net inhibitory effect, white a net excitatory effect, and medium gray no net effect. During self-organization, smooth patterns of excitatory and inhibitory weights emerged, resulting in a DoG lateral interaction profile. Near the edge of the network this profile is elongated along the edge, as shown in Figure 4.7*b*. Such profiles sharpen the response of the network and allow an accurate retinotopic organization to develop.

4.6 Conclusion

The LISSOM model is a computational implementation of the basic principles of information processing in the visual system. It is designed to approximate biological mechanisms at the columnar level, focusing on properties most important for understanding the structure, development, and function of the visual cortex. The model demonstrates how input-driven Hebbian adaptation of afferent and lateral connections can account for these phenomena.

In the remaining chapters of Part II, the network begins self-organization in a naïve state with uniform and unselective connection weights, and is trained with a range of abstract and natural inputs. This approach allows demonstrating in detail how the self-organizing process extracts and encodes visual information. Part III will focus on how the visual system can actually be constructed from realistic internal and external inputs by combining genetically and environmentally driven self-organization. In Part IV, LISSOM will be extended to model the low-level time-dependent behavior of neurons so that perceptual grouping can be studied. Chapter 15 in Part V will further demonstrate how LISSOM can be scaled to model larger areas, including the entire V1 at the columnar level.

Beyond these specific goals, LISSOM advances understanding of general cortical mechanisms in two important ways. First, because it is a computational model, its function can be described in precise mathematical terms. Such an analysis is done in Chapter 14, suggesting that the self-organized structures serve an important function: They form an efficient representation of the visual information. Second, when building a computational model, assumptions must be made about biological processes that are not well understood. Such assumptions lead to predictions that can be tested in biological experiments. The LISSOM assumptions will be reviewed in detail in Chapter 16, evaluating their plausibility and identifying the resulting predictions.

5

Development of Maps and Connections

In this chapter, input-driven self-organization in the LISSOM model will be studied systematically, showing how features of visual input can become represented in the cortex. The focus will be on orientation, ocular dominance, and direction selectivity, where enough constraints exist from biology to constrain the models. The properties of the input patterns are first varied one dimension at a time, demonstrating that LISSOM develops maps much like those found in the cortex. In a scale-up simulation, all of these dimensions are then varied simultaneously, resulting in a map with joint representations for multiple features. Later chapters in Part II will demonstrate how the same self-organizing principles can operate in the adult, resulting in neural plasticity and visual aftereffects.

5.1 Biological Background

Computational models can be highly useful for gaining insight into biological mechanisms. Before they can be trusted, such models must be validated against biological data, to make sure their structures and processes are realistic. For maps and their connections, the biological data can be categorized in several different ways: (1) qualitative descriptions based on visual plots, vs. quantitative descriptions based on Fourier transforms, gradient calculations, and histograms (Erwin et al. 1995; Swindale 1996); (2) data on normal animals, vs. animals raised in abnormal visual environments and in sensory deprivation; and (3) data on single map features, vs. data on multiple features and their interactions. Map organization, receptive fields, and lateral connections were described qualitatively in Sections 2.1 and 2.2, focusing primarily on normal animals and single map features. This section will complement that review by describing quantitative results, studies on abnormal animals, and results on feature interactions.

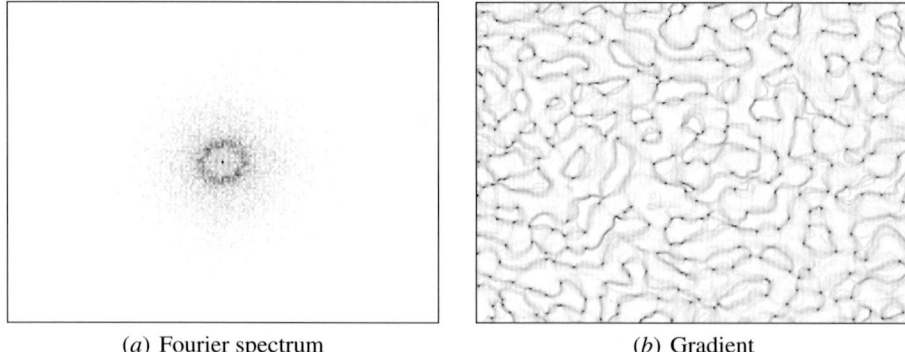

(*a*) Fourier spectrum   (*b*) Gradient

Fig. 5.1. Fourier spectrum and gradient of the macaque orientation map. Plot (*a*) shows the two-dimensional Fourier spectrum of the map in Figure 2.4, calculated using methods described by Erwin et al. (1995) on orientation map data from Blasdel (1992b). In this and subsequent Fourier spectrum figures, the center represents the DC component and the midpoint of each edge $1/2$ of the highest possible spatial frequency of the image horizontally and vertically (i.e. the Nyquist frequency; Cover and Thomas 1991); the amplitude is represented in gray scale from white to black (low to high). As typically found in animal maps, the spectrum is ring shaped, indicating that the orientations repeat in all directions with a spatial frequency that corresponds to the radius of the ring. (*b*) The orientation gradient of the same map is plotted in gray scale from white to black (low to high; calculated from Blasdel 1992b as described in Appendix G.6). The high-gradient areas (dark ridges) correspond to fractures; the pinwheel centers are usually located at the ends of fractures. The gradient map makes the global arrangement of these features easy to characterize.

5.1.1 Quantitative Descriptions of Maps and Connections

Neurons in the visual cortex respond selectively to a number of input features such as location, orientation, eye of origin, and direction of movement, and preferences for these features vary systematically across the cortex. This organization can be visualized in maps, which can then be described qualitatively, as was done in Section 2.1.2. These same visualizations can also be analyzed numerically, measuring distributions of features and how they change across the map.

For example, the two-dimensional Fourier transform of an orientation map reveals how regular or periodic the map is, e.g. how often patches for each orientation are repeated across the surface. Biological maps have ring-shaped Fourier transforms (Figure 5.1*a*), revealing that *in all directions* map features repeat regularly, with an average periodicity corresponding to the radius of the ring. Orientation preference histograms complement Fourier transforms, measuring how many neurons prefer each orientation. As will be discussed in Chapter 9, animal maps are slightly biased toward vertical and horizontal orientations, reflecting the edge statistics of the visual environment.

The gradient of a cortical map measures how much each point in the map differs from its neighbors. Regions where map properties change sharply, such as pinwheel centers and fractures, have a large gradient. As an example, Figure 5.1*b* displays the gradient of the orientation map in Figure 2.4. Fractures are seen as long ridges in this plot; pinwheel centers are often located at the ends of fractures, or appear as single dots. Although gradient plots are visually similar to selectivity plots (Figure 2.4*b*), it is important to realize that they are measuring two different properties. Low selectivity often coincides with high gradient in such plots, but this result may be an artifact of averaging the responses of several cells in a high-gradient area (Maldonado, Gödecke, Gray, and Bonhoeffer 1997). Gradient plots could thus be used to identify where selectivity measurements are unreliable.

Selectivity histograms can display useful information about gross map properties as well, but they are similarly affected by averaging. Estimates of selectivity obtained with different techniques, such as optical imaging and microelectrode recordings, differ widely (Maldonado et al. 1997). Further, as will be shown in Section 13.4, different types of training patterns result in different selectivity histograms in computational models. This measure is therefore less useful for comparing models with biological maps.

Histograms can also be used to quantify lateral connectivity patterns. As was described in Section 2.2, neurons tend to connect roughly symmetrically to their near neighbors, and to neurons farther away through patchy long-range connections. The long-range patches link primarily neurons with roughly collinear orientation preferences. One way to quantify such patterns is by measuring the angles between the orientation preferences of connected neurons. A histogram of how often various angles occur demonstrates that lateral connections indeed link neurons with similar preferences (Section 11.5.3).

5.1.2 Experimental Manipulation of Maps

As was reviewed in Sections 2.1.4 and 2.2.2, disrupting or changing the input patterns to the visual cortex during development can profoundly change the resulting maps and their connectivity. Of such manipulation studies, ocular dominance is perhaps the best known, and provides a detailed test case for computational models.

In the medical condition of strabismus (cross-eye), the eyes cannot focus on the same point in space; this condition can be induced experimentally in animals by cutting some of the eye muscles. As a result, each eye sees entirely different images, instead of the highly overlapping images in normal vision.

Under artificial strabismus, ocular dominance maps still develop, but their properties differ from normal maps (Figure 5.2). Strabismic maps have more sharply delineated ocular dominance areas, with stripes containing few connections from the opposite eye (Löwel 1994). The ocular dominance stripes are also significantly larger in the strabismic maps than in normal maps. Lateral connectivity patterns are also affected: Whereas in normal animals the lateral connections do not significantly favor one eye over the other (Bosking et al. 1997; Löwel and Singer 1992), in strabismic animals they become specific to each eye (Figure 5.2; Löwel and Singer 1992). These differences apparently result from a decrease in correlation between visual activity patterns between the two eyes.

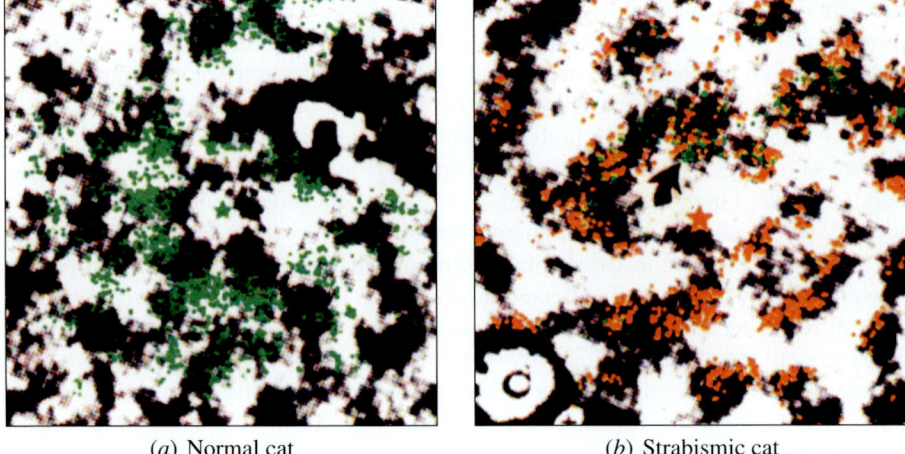

(*a*) Normal cat   (*b*) Strabismic cat

Fig. 5.2. Normal vs. strabismic cat ocular dominance maps and lateral connections. These plots show corresponding 5 mm × 5 mm portions of the V1 ocular dominance maps from a normal cat (*a*) and from a cat raised with artificial strabismus (*b*). The maps were obtained using anatomical tracers, which result in categorical eye preferences (represented by light and dark areas instead of gray scale as in Figure 2.5). Both maps contain patches specific to each eye, but the patches are larger and more sharply delineated in the strabismic case. In (*a*), the green star indicates where fluorescent tracer was injected, and the green dots show where lateral retrograde transport took them. The lateral connection patterns do not significantly depend on the ocular dominance patterns. In (*b*), the red star and the green star (pointed by the arrow) mark two separate injection sites in right-eye columns (black). The lateral connections preferentially target neurons with the same eye preference (black patches, marked with red and green dots), and avoid neurons with the opposite eye preference (white). Each injection killed the nearby cells as a side effect, and therefore the ocular dominance and connection patterns are not visible in the areas surrounding the injections. Those areas are likely to be strongly connected to the neurons at the injection site. Detail of a figure by Löwel and Singer (1992), reprinted with permission, copyright 1992 by the American Association for the Advancement of Science.

Other input features have also been manipulated experimentally, with similar results. For instance, reducing or increasing the range of orientations seen by an animal can cause corresponding changes in the orientation map (Blakemore and Cooper 1970; Sengpiel et al. 1999). Such cases will be discussed more in detail in Section 8.1.

5.1.3 Interactions Between Multiple Maps

Although selectivity to each different input feature can be mapped independently as described above, all maps in V1 are overlaid onto the same set of neurons. Each neuron thus contributes to multiple maps, and the maps of different feature dimensions interact with each other. These interactions can be visualized in combined maps of

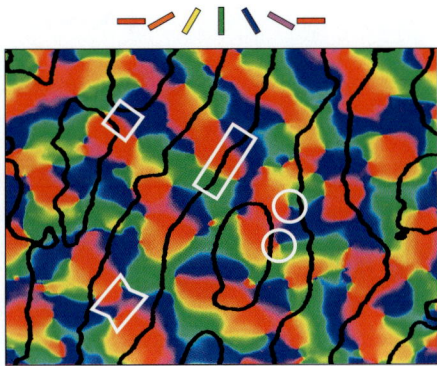

Fig. 5.3. Combined OR/OD map in the macaque. The orientation and ocular dominance maps shown separately in Figure 2.5 are overlaid in this plot. Color encodes OR preference (according to the key on top; example map features are outlined in white as in Figure 2.4), and the black outlines represent OD stripe boundaries, obtained as areas of high OD gradient. The features in the two maps are systematically organized. The OD boundaries intersect the OR boundaries of linear zones (long rectangle) at right angles, and rarely follow an OR boundary. Pinwheel centers (circles) are usually found well inside the OD stripes, and rarely near their boundaries. Note that, unlike the small patches seen in the cat OD maps (Figure 5.2), the OD patches in the monkey typically consist of long stripes (as also shown in Figure 2.5b). Reprinted with permission from Blasdel (1992b), copyright 1992 by the Society for Neuroscience; annotations added, OD contours replotted from data by Blasdel (1992a).

orientation, ocularity, and direction; to facilitate discussion of such maps, abbreviations OR, OD, and DR will be used for these feature dimensions in this and later chapters.

For instance, Figure 5.3 shows that the boundaries between ocular dominance stripes tend to intersect the boundaries between orientation patches at right angles (Blasdel 1992b). Orientation pinwheel centers are also typically found near the centers of ocular dominance stripes, and rarely intersect their boundaries.

Orientation and direction preferences also interact. Neurons have spatiotemporal receptive fields selective for both orientation and motion direction (Figure 5.4a; DeAngelis, Ohzawa, and Freeman 1993, 1995; Shmuel and Grinvald 1996). These RFs are formed by specific excitatory (ON) and inhibitory (OFF) subregions that vary over time, providing selectivity for both orientation and motion direction.

These preferences are arranged into maps of orientation and direction selectivity in the cortex. As an example, Figure 5.4 shows such maps in the ferret V1. Direction maps have a similar structure to orientation maps: Nearby neurons prefer similar directions, and the map has linear zones, pairs of pinwheels, saddle points, and fractures (Figure 5.4b; Weliky et al. 1996). Moreover, the direction map tends to be aligned with the orientation map, with orientation and direction preferences generally meeting at right angles (Figure 5.4c). Iso-orientation patches are also often subdivided into patches for each direction of motion.

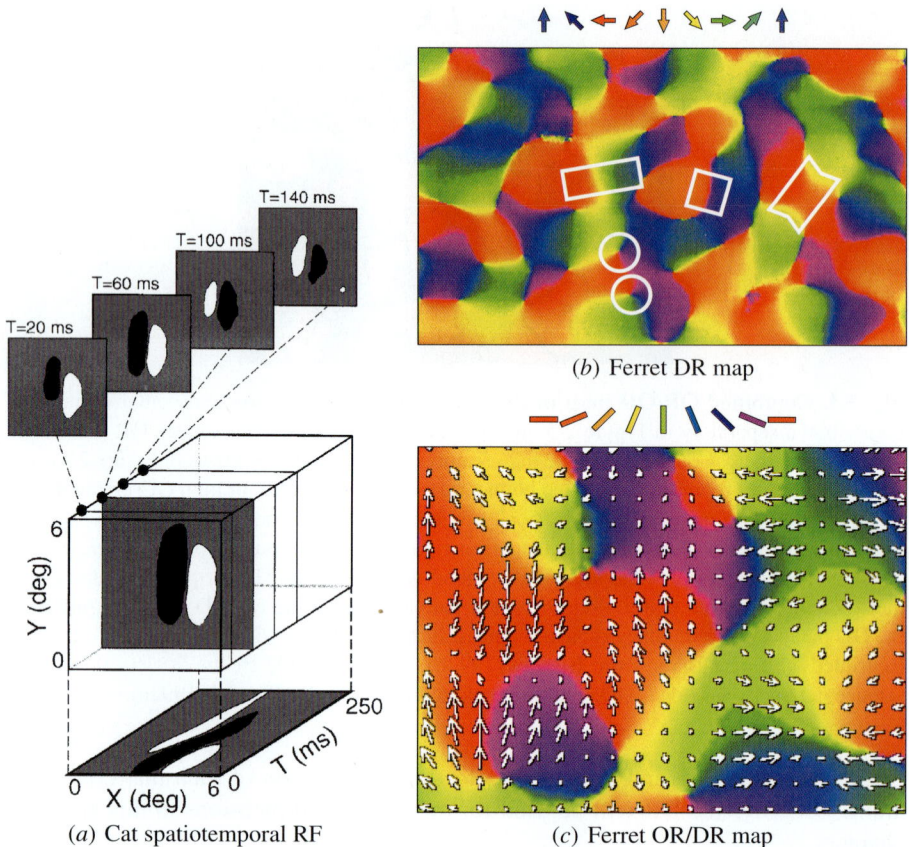

(a) Cat spatiotemporal RF   (b) Ferret DR map

(c) Ferret OR/DR map

Fig. 5.4. Spatiotemporal receptive fields, direction maps, and combined OR/DR maps in animals. In addition to orientation and eye of origin, neurons in V1 are selective for direction of motion. These spatial and motion preferences can be described as spatiotemporal RFs, representing the sequence of patterns that would most excite the neuron. (a) A sample such RF for a V1 cell from the cat, measured through microelectrode recording (DeAngelis, Ghose, Ohzawa, and Freeman 1999; reprinted with permission, copyright 1999 by the Society for Neuroscience; gray scale added). Sample RFs in the two-dimensional visual space at times 20, 60, 100, and 120 ms are shown on top, and a continuous integration of the RFs along the vertical (which is the preferred orientation of the neuron) is drawn in the bottom plane. The neuron's spatial preferences change systematically over time, giving it a spatiotemporal preference for a black vertical line moving horizontally to the right. (b) Spatial arrangement of such preferences in a 3.2 mm × 1.6 mm area of ferret V1: Nearby neurons prefer similar directions in a manner similar to orientation maps (measured through optical imaging and displayed using the color arrow key on top; Weliky et al. 1996, reprinted with permission, copyright 1996 by Nature Publishing Group; annotations added and DR arrows removed by interpolation). Example map features are outlined in white as in Figure 2.4. (c) Interaction of direction preferences with the orientation map (Weliky et al. 1996; reprinted with permission, copyright 1996 by Nature Publishing Group; arrows changed from black to white). The 1.4 mm × 1.1 mm subarea of V1 around the right edge of the square in (b) is colored according to orientation preference (using the color bar key above the plot). Each arrow points in the preferred direction, and its length indicates how selective the neuron is for that preference. Direction and orientation preferences tend to be perpendicular, and orientation patches are often subdivided for opposite directions of motion.

The lateral connections in normal animals follow mostly orientation preferences (Section 2.2.1). However, as described above, in strabismic animals eye preference instead becomes the dominant feature, suggesting that the connection patterns develop to reflect the most dominant features of the map. How less prominent input features, such as direction of motion, interact with the dominant patterns is not yet known.

The biological observations outlined in this section are constraints that computational models of the visual cortex must satisfy and explain. Considerable progress has indeed been made in understanding this data computationally, as will be reviewed next.

5.2 Computational Models

Computational models have been crucial for understanding how orientation preference and ocular dominance maps develop, and they have recently been used to gain insight into motion preference maps as well. Most such models are based on self-organizing maps, reviewed in Section 3.4 in the context of computational models above and below the map level. This section focuses specifically on map-level models of the visual cortex, outlining how their architectures and predictions differ from LISSOM's.

Computational map models range from more realistic to more abstract along several dimensions, including: (1) models that learn incrementally through individual image presentations, vs. models that represent long-term development as an abstract process; (2) models that include specific, patchy lateral connections, vs. models based on abstract lateral interaction functions; and (3) models that can process photographic images of natural stimuli, vs. models that work only with artificial image stimuli. Models built so far have represented some of these dimensions realistically and abstracted others; LISSOM is the first incremental model that develops specific, realistic patchy lateral connectivity from natural images. The review below is organized along the first two dimensions, separately identifying models that have been tested with natural images (see Erwin et al. 1995; Swindale 1996 for further comparisons).

5.2.1 Non-Incremental Models

Models that are based on abstract representations of developmental processes can be conceptually elegant and computationally efficient. On the other hand, it is not possible to account for the same level of detail as with models that learn from individual images. The two main non-incremental approaches are spectral models and correlation-based learning, both of which suggest that large-scale structures, such as orientation and ocular dominance patches, may arise as artifacts of neural processing rather than a principled way of representing visual input.

Spectral Models

Spectral models of maps do not include neurons or connections. The goal is simply to reproduce the patterns in biological maps, as opposed to explaining how maps contribute to visual function (Grossberg and Olson 1994; Niebur and Wörgötter 1993; Rojer and Schwartz 1990). It turns out that patterns similar to ocular dominance and orientation maps can be obtained by simply filtering two-dimensional random noise using a well-chosen function. For instance, ocular-dominance-like stripes can be produced by starting with an array of random numbers, convolving the array using a band-pass filter, and thresholding the result (Rojer and Schwartz 1990). Thus, spectral models suggest that orientation and ocular dominance patterns may not be functionally significant, arising only as an an artifact of biological processes unrelated to information processing.

However, the maps derived in this way are only superficially similar to biological maps, and differ in several crucial respects (Swindale 1996). For instance, in the orientation maps produced of Rojer and Schwartz (1990) each orientation vector is perpendicular to the gradient of the map at that location, which is not seen in animal maps. It is not yet clear whether spectral models could be modified to overcome these difficulties. It is also not clear what specific biological processes could implement such highly abstracted computations, making the spectral models difficult to verify or refute experimentally.

Correlation-Based Learning Models

More closely tied to biological processes are the correlation-based learning (CBL) models (Erwin and Miller 1998; Linsker 1986a; Miller 1994; Miller et al. 1989; Tanaka 1990; Yuille, Kammen, and Cohen 1989). Models in this category rely on the assumption that the visual system is essentially linear. Under this assumption, the developmental process can be represented as a simple set of functions representing long-term correlations in the response to input patterns. This approximation speeds up the calculations considerably compared with incremental learning, and makes it possible to analyze the model mathematically.

Most CBL models focus on individual neurons or small groups of neurons, with the exception of those of Miller (1994), Miller et al. (1989), and Erwin and Miller (1998). The overall architecture of the CBL map models is similar to that of LISSOM, with a retina, LGN with ON-center and OFF-center neurons, and V1. Initially the afferent connections have random strength within a circular receptive field, and lateral interactions have a DoG profile. Unlike in LISSOM, however, lateral interactions are fixed and non-recurrent, and weight normalization is subtractive. Also, because the activation functions are linear, individual image presentations can be replaced with long-term averages of activity correlations. Instead of implementing Hebbian learning of individual input patterns, CBL models compute what Hebbian learning in a linear system would produce over many presentations.

Miller (1994) showed that such correlation-based learning results in maps of ocular dominance and of orientation. Monocular cells develop because subtractive normalization eventually leads to inputs from one eye becoming completely dominant, with zero-strength connections from the other eye (Miller and MacKay 1994). These cells are grouped into ocular dominance stripes because lateral interactions cause nearby neurons to have similar responses. Orientation selectivity develops due to similar competition between ON and OFF-center inputs, but within a single cell's receptive field. With appropriate parameters and correlation functions, cells develop both ON and OFF subregions, making them selective for orientation. Lateral interactions then organize these preferences to orientation patches.

However, a number of key predictions of CBL models are inconsistent with recent biological data (Erwin et al. 1995): (1) The width of biological ocular dominance columns does not depend on the width of lateral interactions, but on the input correlations (Löwel 1994); (2) the RF shapes in CBL are typically only weakly selective for orientation, unlike in animal maps; (3) the Fourier spectrum of the CBL orientation map is concentrated around the origin, i.e. at low frequencies, instead of being shaped like a ring (Swindale 1996); (4) the subtractive normalization mechanism in the model forces synapses either to their maximum weight or zero, which is biologically unrealistic; and (5) the model cannot account for adult plasticity and dynamic reorganization (as reported by e.g. Gilbert 1998; Pettet and Gilbert 1992); once the synapses reach the extreme values, further adaptation is very difficult.

In general, CBL models suggest that large-scale features, such as orientation and ocular dominance maps, develop primarily as artifacts of neural connectivity patterns. In contrast, in the incremental learning approach they emerge as a principled way of representing visual input. As a result, incremental models are more complex and computationally intensive, but can produce more realistic results and explain a broader range of phenomena.

5.2.2 Incremental Models with Fixed Lateral Connections

A large number of incremental models of the visual cortex have been proposed in the last 30 years; almost all of them are based on fixed, isotropic lateral connectivity. As was described in Section 3.4.1, the earliest model of this type was built by von der Malsburg (1973); similar models have since then been developed using the SOM algorithm, the elastic net algorithm, and similar architectures. Von der Malsburg's model developed oriented receptive fields and pinwheels (before such patterns had been found experimentally), demonstrating the basic computational processes underlying development in the visual cortex. Later models have shown how orientation, ocular dominance, and direction maps can form, and also how receptive fields selective for each of these dimensions develop (Barrow and Bray 1992; Dong and Hopfield 1992; Durbin and Mitchison 1990; Elliott, Howarth, and Shadbolt 1996; Elliott and Shadbolt 1999; Farkas and Miikkulainen 1999; Goodhill 1993; Grossberg 1976; Grossberg and Olson 1994; Grossberg and Seitz 2003; Hyvärinen and Hoyer 2001; Miyashita, Kim, and Tanaka 1997; Miyashita and Tanaka 1992; Obermayer et al. 1990d; Obermayer, Sejnowski, and Blasdel 1995; Olson and Grossberg 1998; Osan and Ermentrout 2002; Piepenbrock and Obermayer 2002; Shouno and Kurata 2001; Shouval, Intrator, and Cooper 1997; Stetter, Müller, and Lang 1994; Swindale 1992; Wimbauer, Wenisch, van Hemmen, and Miller 1997b). So far, the SOM and related models have produced the best description of such maps, measured by analytical comparisons with experimentally observed maps (Swindale 1996).

Nearly all of these models have focused only on a single feature dimension (e.g. orientation or ocular dominance). A few models have included multiple dimensions in the same simulation, such as OR and OD (Erwin and Miller 1998; Goodhill and Cimponeriu 2000; Grossberg and Seitz 2003; Obermayer et al. 1995; Olson and Grossberg 1998; Osan and Ermentrout 2002) or OR and DR (Blais, Cooper, and Shouval 2000; Farkas and Miikkulainen 1999; Miyashita et al. 1997; Shouno and Kurata 2001; Wimbauer et al. 1997b). Simulating multiple dimensions at once ensures that parameters are not tuned for only one feature, and interactions between dimensions allow validating the model more extensively. To our knowledge, no published model has developed joint OR, OD, and DR maps; the first study of such maps will be presented in Section 5.6.

Most of these models have been tested only with artificial inputs such as oriented Gaussian patterns or pure random noise, which can be strictly controlled to obtain strong results. However, many phenomena studied in later chapters of this book depend crucially on properties of natural images. Such images have wide ranges of contrast and large areas of activation, and it is therefore necessary to include the processing steps in the ON and OFF channels of the retina and the LGN into the model. Although a few such models have been proposed (Barrow and Bray 1992; Hyvärinen and Hoyer 2001; Shouval et al. 1997; Weber 2001), most models do not include these requisite processing stages.

Crucially, it is not yet clear whether the output from the existing natural image models actually preserves the essential information in the images. So far, the models have been used only to investigate how the map forms, not how it performs visual processing. This issue is important because several of the approximations common in models with fixed lateral connectivity, such as choosing a single winning location for adaptation across the entire cortex, do not scale up to natural images with a realistic size. Development is often driven by only a few strong features in each image, but visual performance requires multiple features of different contrasts to be represented simultaneously. One important goal of the experiments in Chapter 10 is to verify that LISSOM indeed preserves the information necessary for higher levels of processing, such as face perception areas. It is not known whether other models can be used for such studies.

5.2.3 Incremental Models with Modifiable Lateral Connections

The early map models were developed before specific, patchy lateral connections were discovered in the visual cortex, and assumed that lateral interactions would have a fixed, uniform shape. Nearly all subsequent models have relied on similar assumptions, primarily because it is computationally very expensive to store and simulate individual connections. However, it is crucial to include such connections to account for a number of key experimental results (as reviewed in Section 2.2), and such connections are also important for information processing (as will be demonstrated in Chapter 14). Explaining how the lateral connections self-organize into the characteristic patchy patterns is a crucial part of understanding the development and function of topographic maps.

After LISSOM was first introduced a decade ago, several models with modifiable lateral connections have been developed (Alexander et al. 2004; Bartsch and van Hemmen 2001; Bray and Barrow 1996; Burger and Lang 1999; Weber 2001). Of these, the models by Alexander et al. (2004) and Burger and Lang (1999) form patchy long-range lateral connections, but the rest do not. The Alexander et al. (2004) model relies on abstract binary input patterns like those of von der Malsburg (1973), and it is not clear how to extend it to support gray-scale and natural-image inputs, or to develop multi-lobed receptive fields. The Burger and Lang (1999) model is very similar to LISSOM, including ON and OFF channels, modifiable lateral connections, and support for natural images. Mathematically, their most significant difference from LISSOM is the activation function: The Burger and Lang model includes a linear function (sum of all inputs), whereas LISSOM's activation function is nonlinear, and also iteratively incorporates lateral influences. The nonlinearities make the LISSOM map more selective for input features and less dependent on contrast, which is crucial for simulating responses to realistic inputs. The Burger and Lang model is also based on a different method for normalizing the afferent weights, although both methods should lead to similar results. So far, only a small range of cortical phenomena have been investigated with the Burger and Lang model, but with certain modifications it should be possible to obtain results similar to LISSOM's. Other models would need to be extended significantly to study the same range of phenomena, e.g. by including specific lateral connections and mechanisms for processing natural images.

Although considerable progress has been made in understanding the biological data reviewed in Section 5.1, most of the models have focused on a limited set of phenomena. LISSOM is the first where they are all brought together under a single common principle of adaptation and organization. In this chapter, LISSOM is first evaluated as a model of each separate abstract input feature dimension, and then as a joint model that self-organizes to represent multiple features in natural images together.

5.3 Orientation Maps

In this section, a LISSOM model of how orientation maps and lateral connections develop based on input-driven self-organization is presented. The model described in Chapter 4 is extended to oriented input and shown to develop oriented receptive fields and a global orientation map. The resulting maps and lateral connections are then analyzed numerically. In later sections, ocular dominance and direction maps are studied separately, and combined maps of all three input feature dimensions are developed both from abstract and natural image inputs.

5.3.1 Method

The LISSOM model of orientation maps is otherwise identical to that described in Chapter 4 (Figure 4.1) except different training inputs are used. In the main sim-

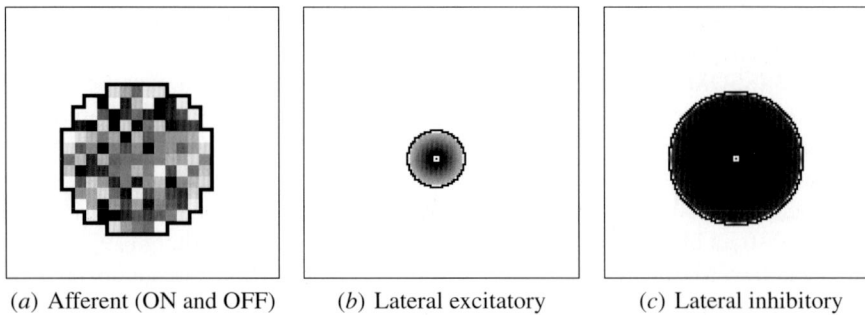

(*a*) Afferent (ON and OFF)   (*b*) Lateral excitatory   (*c*) Lateral inhibitory

Fig. 5.5. Initial V1 afferent and lateral weights. These plots show the initial weights of the V1 neuron at location (79,68) in the 142 × 142 V1 map. As in Figure 4.3, each set of weights is outlined in black and plotted in gray scale on the neural region from which they originate; however, the padding in the LGN is omitted so that the area of (*a*) corresponds to that of Figure 5.6*b*. The neuron itself is marked with a small white square in (*b*) and (*c*). The afferent RFs were initially random, as shown in (*a*); the ON and OFF channel weights were identical for each neuron. Plots (*b*) and (*c*) display the lateral weights of this neuron; initially they had a Gaussian profile. Later figures will show how these connections become selective and patchy through self-organization.

ulation, the inputs consist of images of elongated Gaussian spots, instead of the unoriented Gaussians of Equation 4.2. The activity of each retinal receptor χ_{xy} is calculated according to

$$\chi_{xy} = \max_{k} \exp\left(-\frac{[(x-x_{c,k})\cos(\phi) - (y-y_{c,k})\sin(\phi)]^2}{\sigma_a^2}\right.$$
$$\left. -\frac{[(x-x_{c,k})\sin(\phi) + (y-y_{c,k})\cos(\phi)]^2}{\sigma_b^2}\right), \qquad (5.1)$$

where σ_a and σ_b determine the width along the major and minor axes of the Gaussian, and ϕ its orientation, chosen randomly from the uniform distribution in the range $0 \leq \phi < \pi$. Such inputs are abstractions of elongated features in images and in spontaneous neural activity and allow demonstrating the model properties clearly. In Section 5.3.5, the model is trained with a range of other patterns, and the resulting differences in organization analyzed.

The self-organization proceeds as described in Chapter 4. The way the connections are initialized does not have a large effect, as long as they are roughly isotropic. Since self-organization of scattered RF centers and random initial lateral weights was already demonstrated in Section 4.5, to make the results easier to interpret the afferent connections are initially random around topographically ordered centers, and the lateral connections have initially a Gaussian profile (Figure 5.5). At each input presentation, multiple Gaussian spots are presented on the retina at random orientations and random, spatially separated locations (Figure 5.6*a*). The activation propagates through the LGN and the afferent connections of V1 and produces an initial response in the V1 network (Figure 5.6*c*). The initial response is typically diffuse and widespread, but by recurrent lateral excitation and inhibition, it rapidly evolves

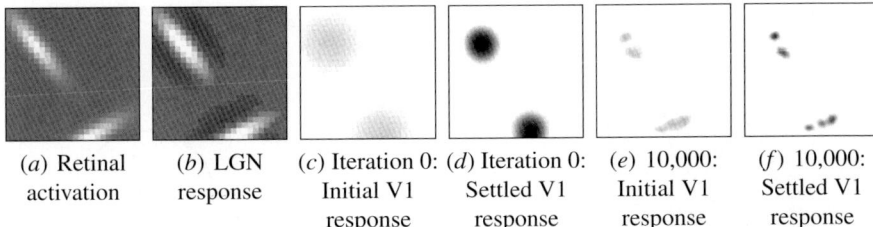

| (a) Retinal activation | (b) LGN response | (c) Iteration 0: Initial V1 response | (d) Iteration 0: Settled V1 response | (e) 10,000: Initial V1 response | (f) 10,000: Settled V1 response |

Fig. 5.6. Example input and response. A sample input on the retina, the LGN response, and the initial and settled cortical response before and after training are shown as in Figure 4.4, except the padding in retina and LGN is omitted so that all plots represent the same retinal area. To train the orientation map, two oriented Gaussians were drawn with random orientations and random, spatially separated locations on the retina (a). As discussed in Appendix A.4, while more than two spots could be used, they are too large to be distributed uniformly on the small retina. The LGN responses are plotted in (b) by subtracting the OFF cell responses from the ON. The LGN responds strongly to the edges of the oriented Gaussians. Initially, the responses of the V1 map are similar for all orientations (c and d). After 10,000 input presentations, the V1 response extends along the orientation of the stimulus, and is patchy because neurons that prefer similar positions but different orientations do not respond (e and f). An animated demo of the map response can be seen at http://computationalmaps.org.

into stable, focused patches of activity (Figure 5.6d). After the activity has settled, the strength of each synaptic connection is updated. A new set of oriented Gaussians is then generated at random positions and orientations, and the process repeats for 10,000 iterations. Appendix A.5 lists the details of the simulation parameters. Small variations of the parameters result in roughly similar maps; a representative example is analyzed in detail in the sections that follow.

5.3.2 Receptive Fields and Orientation Maps

Initially, the activation patterns are very similar even for different orientations, allowing the map to develop a global retinotopic order. Gradually, oriented receptive fields start to form, and the lateral connections start to follow receptive field properties. As Figure 5.7 shows, the final RFs and lateral connections are very similar to those found in biology (Section 2.1.2; Bosking et al. 1997; Hubel and Wiesel 1965, 1968; Sincich and Blasdel 2001). Afferent RFs are Gabor-shaped with separate ON and OFF lobes, making them strongly selective for orientation. Lateral connections are patchy, and the long-range connections originate from neurons with similar orientation preferences. Across the map (Figure 5.8), the RFs have a variety of shapes; most are highly selective for inputs of a particular orientation, and others are unselective. Lateral connections tend to follow the RF shape, linking neurons with similar RFs.

The global organization of the RFs can be visualized similarly to biological orientation maps, by labeling each neuron by the preferred angle and degree of selectivity for inputs at that angle. To determine these labels, responses to sine gratings of various orientations were measured and recorded (as described in Appendices G.1.3

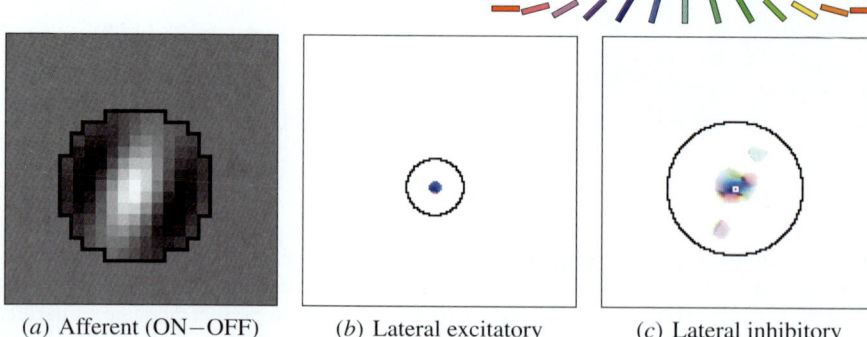

(*a*) Afferent (ON−OFF)   (*b*) Lateral excitatory   (*c*) Lateral inhibitory

Fig. 5.7. Self-organized V1 afferent and lateral weights. The weights of the neuron in Figure 5.5 are shown after self-organization. In (*a*), the OFF weights were subtracted from the ON weights, as in Figure 4.6*c*. This neuron prefers a line oriented at 60°, i.e. diagonal from bottom left to top right, and responds most strongly to a white line overlapping the light portion of its RF, surrounded by black areas overlapping the dark portions. Other neurons developed similar RFs with different preferred orientations (Figure 5.8). This type of RF structure is commonly seen in biological V1 neurons (Figure 2.2*d*; Hubel and Wiesel 1962, 1968). In the lateral weight figures (*b*, *c*, and other later such figures), the following convention is used: The hue (i.e. color) represents the orientation preference of the source neuron, according to the key along the top. The saturation of the color (i.e. its fullness, or intensity) represents how selective the source neuron is for this orientation; unselective neurons are shown in gray. The value of the color (i.e. its brightness) indicates the strength of the connection, with nonexistent or zero-weight connections shown as white. The jagged black outline traces the original lateral connections, and a small white square (in *c*) identifies the neuron itself. Using such a scale, plot (*b*) displays the lateral excitatory weights of this neuron. All connected neurons are strongly colored blue or purple, i.e. orientations similar to the orientation preference of this neuron. The lateral inhibitory weights are plotted in (*c*). After self-organization and connection pruning, only connections from neurons with similar orientations remain, and they are extended along the preferred orientation of this neuron. The connection pattern is patchy, because connections from neurons with opposite preferences are weak or have been pruned away entirely. Such patchy, orientation-specific connection patterns are also seen in biological V1 neurons (Figure 2.7; Bosking et al. 1997; Sincich and Blasdel 2001).

and G.3. These values were then used to plot the orientation map shown in Figure 5.9. Initially, all the afferent weights are random. As a result, the orientation preferences of the RFs appear random and unselective (top row). As self-organization progresses and afferent weights develop oriented receptive fields, a complex orientation map develops (bottom row). Even with such abstract inputs, the map is a good match to those measured in animals, and contains structures such as linear zones, pinwheels that often occur in matched pairs, saddle points and fractures (compare Figure 5.9 with Figure 2.4). The maps can be further analyzed with numerical techniques, showing quantitatively the same structures as the primary visual cortex.

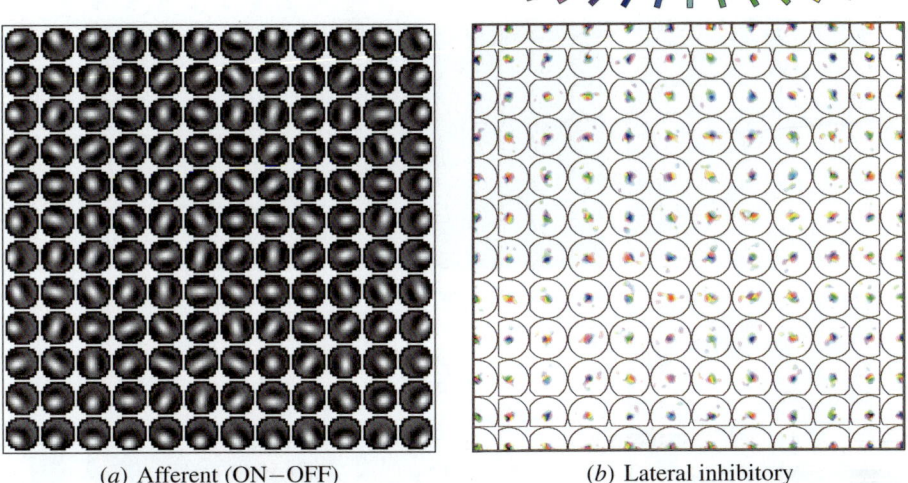

| (a) Afferent (ON−OFF) | (b) Lateral inhibitory |

Fig. 5.8. Self-organized afferent and lateral weights across V1. This plot shows the range of afferent and lateral weights developed by the neurons in the orientation map, by plotting them for every 12th neuron horizontally and vertically, using the conventions introduced in Figure 5.7. (a) A number of two- and three-lobed receptive fields exist with strong orientation preferences. Some neurons, however, have ring-shaped RFs and respond to all directions equally. (b) These neurons receive lateral inhibitory connections from all nearby neurons, but from distant neurons only if they have similar OR preferences and are located along the preferred orientation. The lateral excitatory connections of each neuron (not shown) come from all nearby neighbors, and thus are all nearly circular.

5.3.3 Analysis of the Orientation Maps

The LISSOM orientation map was analyzed using the numerical techniques described for biological maps in Section 5.1. The histogram of orientation preferences is flat (Figure 5.9d), showing that the architecture does not have biases for any particular orientations (as it would, for instance, if it had square receptive fields). It therefore reflects the uniform distribution of orientations during training. If LISSOM was instead trained with natural images, where certain angles are over-represented, the resulting histogram would be more similar to those found in animals (Sections 5.3.5 and 9.3).

The Fourier spectrum of the orientation map has the typical ring-shaped structure of biological maps, indicating that the orientation patches repeat at regular intervals in all directions (Figure 5.10a). Because the network has fewer units than the animal optical imaging data have pixels, the plot is fuzzier, but the overall shape is similar. The map gradient is also similar to that of cortical maps (Figure 5.10b). Discontinuities are represented by high gradient: Pinwheel centers are seen as high points and fractures as more linear ridges connecting pinwheel centers. In LISSOM, high gradient generally coincides with low selectivity, because neurons self-organize to

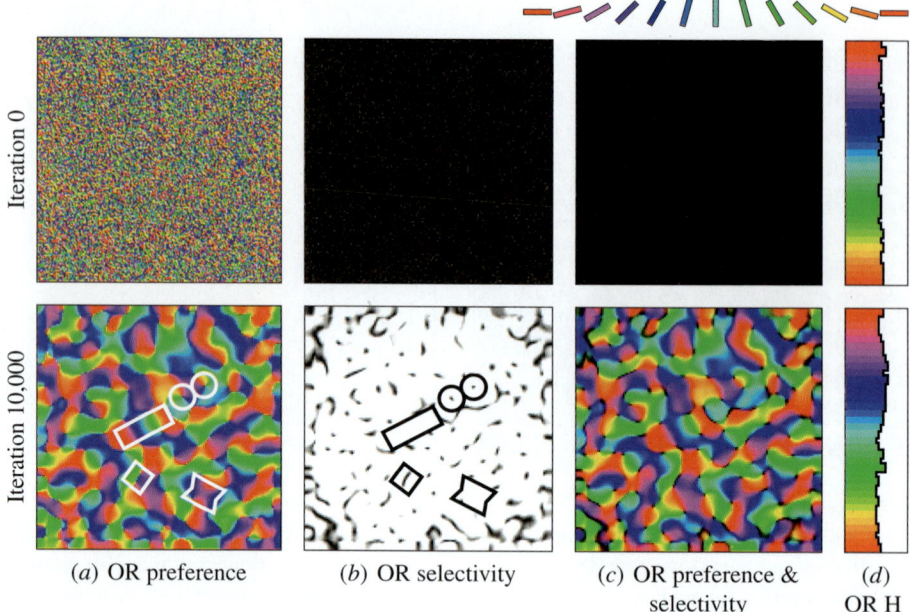

(*a*) OR preference	(*b*) OR selectivity	(*c*) OR preference & selectivity	(*d*) OR H

Fig. 5.9. Self-organization of the orientation map. The orientation preference and selectivity of each neuron was computed before (top row) and after self-organization (bottom row). The preferences are color coded and selectivity represented in gray scale as in Figure 2.4. (*a*) The orientation preferences were initially random, but over self-organization, the network developed a smoothly varying orientation map. The map contains all the features found in animal maps, such as linear zones, pairs of pinwheels, saddle points, and fractures (outlined as in Figure 2.4). (*b*) Before self-organization, the neurons are unselective (i.e. dark), but nearly all of the self-organized neurons are highly selective (light). (*c*) Overlaying the orientation and selectivity plots (by representing selectivity with color saturation as in Figure 5.7) shows that regions of low selectivity in the self-organized map tend to occur near pinwheel centers and along fractures. (*d*) Histograms of the number of neurons preferring each orientation (OR H) are essentially flat because the initial weight patterns were random, the training inputs included all orientations equally, and LISSOM does not have artifacts that would bias its preferences. These plots show that LISSOM can develop biologically realistic orientation maps through self-organization based on abstract input patterns. An animated demo of the self-organizing process can be seen at http://computationalmaps.org.

respond together with their neighbors. Whether this is true of animal maps as well is still controversial (as was discussed in Section 5.1.1).

The orientation discontinuities also affect the retinotopic mapping from the retina to V1 (Figure 5.11). The large-scale organization corresponds to the retina: for example, neurons in the upper left of the cortex respond to activity in the upper left of the retina. On small scales, however, this mapping is distorted because the orientation map represents both position and orientation smoothly across the same surface. Such distortions also occur in cat orientation maps, where orientation gradient is found to

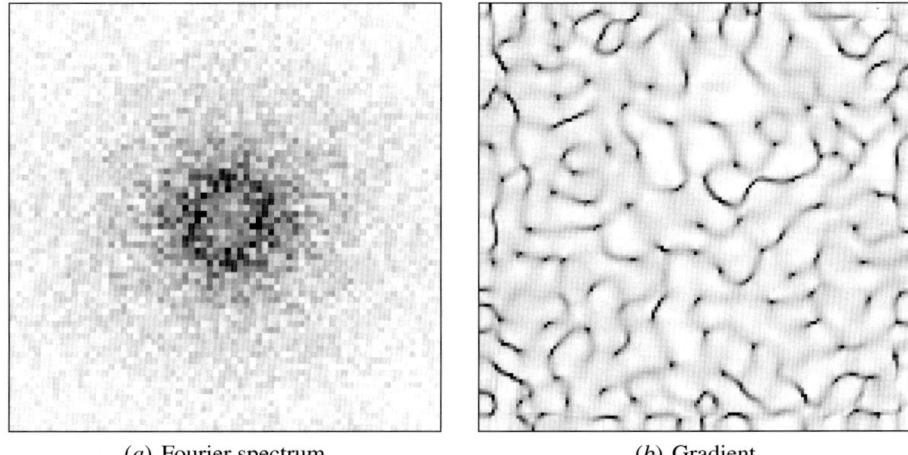

(*a*) Fourier spectrum   (*b*) Gradient

Fig. 5.10. Fourier spectrum and gradient of the orientation map. (*a*) The Fourier spectrum
is ring shaped as it is for biological maps (cf. Figure 5.1*a*), indicating that all directions are
represented at regular intervals. (*b*) The global arrangement of high- and low-gradient areas
is similar to that in biological maps: Regions with high gradient coincide with discontinuities
such as pinwheel centers and fractures, and the fractures tend to connect the pinwheel centers
(cf. Figure 5.1*b*).

correlate with distance between RF centers (Das and Gilbert 1997). Such a correlation exists in LISSOM maps as well, although it is weaker and the relationship is
more complex.

The above analytical comparisons demonstrate that the afferent structures in LIS-
SOM essentially replicate the afferent organization in the cortex. The next section
will show that the patterns of lateral connections also compare well to those observed
in biology.

5.3.4 Lateral Connections

The lateral connection weights self-organize at the same time as the orientation map.
Initially, the connections are spread over long distances and cover a substantial part
of the network (Figure 5.5). As the lateral weights self-organize, the connections
between uncorrelated regions grow weaker, and after pruning only the strongest
connections remain (Figure 5.7). The surviving connections of highly orientation-
selective cells, such as the one illustrated in Figure 5.12*a*, link areas of similar orientation preference, and avoid neurons with different orientation preferences. Furthermore, the connection patterns are elongated along the direction in the map that
corresponds to the neuron's preferred stimulus orientation. This organization reflects
the activity correlations caused by the elongated Gaussian input pattern: Such a stimulus activates primarily those neurons that are tuned to the same orientation as the
stimulus, and located along its length. At locations such as fractures, where a cell is

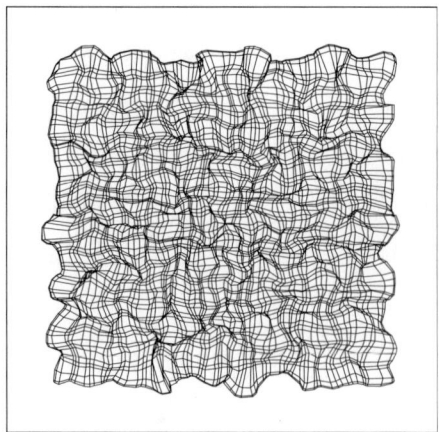

Fig. 5.11. Retinotopic organization of the orientation map. The center of gravity of the afferent weights of every second neuron was calculated and plotted in the retinal space, and those of neighboring neurons connected with lines (as in Figure 4.8). The overall organization of the map is an evenly spaced grid with local distortions. These distortions result from mapping both orientation and retinal position smoothly into the same two-dimensional surface; such distortions have been found experimentally on animal maps as well (Das and Gilbert 1997).

sandwiched between two orientation patches of very different orientation preference, the lateral connections are elongated along the two directions preferred by the two adjacent patches (Figure 5.12d). The lateral connections of unselective cells, such as those at pinwheel centers, come from all orientations around the cell (Figure 5.12b). Connections at saddle points are similar to those at fractures, in that they include the two orientations of the saddle, but they also include intermediate orientations that typically match the orientation preference of the saddle neuron itself (Figure 5.12c). Thus, the pattern of lateral connections of each neuron closely follows the global organization of receptive fields, and represents the long-term activity correlations over large areas of the network.

These results were originally discovered independently in the LISSOM model (Sirosh et al. 1996a), and some of them have already been confirmed in recent neurobiological experiments. In the iso-orientation patches of the tree-shrew cortex, horizontal connections were found to be distributed anisotropically, extending farther and giving rise to more terminals along the preferred orientation axis of the neuron in visual space (Bosking et al. 1997; Fitzpatrick et al. 1994; Sincich and Blasdel 2001). Most of these terminals also connected to cells with the same orientation preference. The connection patterns at pinwheel centers, saddle points, and fractures have not been studied experimentally so far; the LISSOM model predicts that they will have unselective, broad unimodal, and biaxial distributions, respectively.

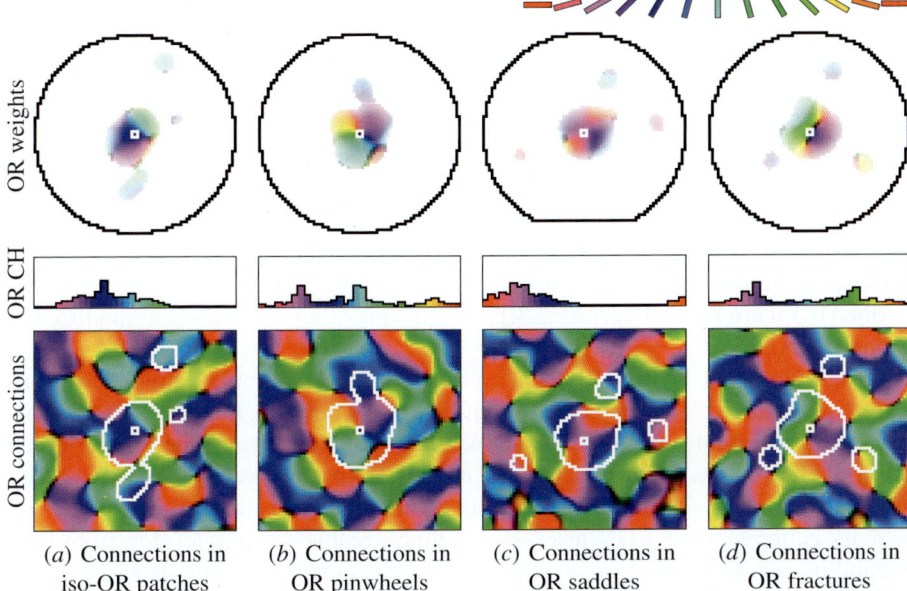

(a) Connections in iso-OR patches   (b) Connections in OR pinwheels   (c) Connections in OR saddles   (d) Connections in OR fractures

Fig. 5.12. Long-range lateral connections in the orientation map. The lateral inhibitory connection weights of four sample neurons from the marked regions in Figure 5.9 are shown in the top row, situated in the orientation map as shown in the bottom row. The small white square in both figures identifies the neuron; the black outline on top indicates the extent of these connections before self-organization, and the white outline on the map plot shows their extent after self-organization and pruning. On top, the color coding represents the connected neuron's orientation, selectivity, and connection strength, as in Figure 5.7; the map encodes orientation and selectivity as in Figure 5.9c. The connection histogram (CH) in the middle shows how many connections come from neurons of each orientation. For every neuron, the strongest connections originate from the neuron's nearby neighbors, as indicated by the large, bright central area in each weight plot. The long-range connection patterns differ depending on where the neuron is located in the orientation map. (a) Neurons in the middle of an OR patch receive connections from neurons with similar preferences, aligned along the orientation preference of the neuron (for this neuron, about 65°, i.e. blue). (b) At pinwheel centers, the connections come from all directions and orientations and are nearly isotropic. The histogram is nearly flat, with small peaks near orientations that happen to be overrepresented in the pinwheel. (c) Connections at saddle points extend along the two orientations of the saddle, in this case red (0°) and blue (65°). The neuron also makes connections with intermediate orientations and directions; these connections match its own OR preference (30°, purple), and result in one broad peak in the histogram. (The connections of this neuron are cut off along the bottom because it is located near the bottom of the map.) (d) Connections of neurons at fractures are also elongated along the two directions of the neighboring orientation patches. The neuron plotted in (d) is on a fracture between yellow–green (130°) and blue–purple (40°), and makes connections with both of these orientations. In contrast to saddle points, it does not connect with intermediate orientations and directions, resulting in two distinct peaks in the orientation histogram. While the connection patterns in iso-orientation patches have already been confirmed in biology, the patterns at the other map features are predictions for future experiments.

5.3.5 Effect of Input Types

The preceding analysis focused on a single LISSOM model trained with oriented Gaussian patterns, which made the results clear and unambiguous. In a series of simulations, the model was trained with other patterns that have been hypothesized to contribute to orientation selectivity, including spontaneous neural activity and natural visual images. Spontaneous activity was modeled after retinal waves and represented as noisy disk patterns (as will be described in detail in Section 9.2.1). Natural images included retina-size closeups of natural objects and landscapes closely matched with natural input (from a dataset by Shouval, Intrator, Law, and Cooper 1996, described in Figure 8.4 and Section 9.3.1). The details of these simulations are listed in Appendix A.5.

The orientation maps that result from each of these training patterns are compared in Figure 5.13. In each case, orientation maps develop, but the maps and RFs are well ordered only for patterns that contain spatial structure. The Fourier transform of these maps is ring-shaped, as it is of animal orientation maps. Oriented Gaussians are not required, as long as the inputs have edges that produce elongated activity patterns at the LGN level. For instance, realistic maps develop from both natural images and retinal waves. Importantly, the simulations with disks and noisy disks demonstrate for the first time that full-fledged orientation maps can develop from large, unoriented activity patterns, which has been believed difficult to achieve (Miller 1994).

The receptive field types that develop depend strongly on the input patterns. For instance, if all patterns are brighter than their surround (the row labeled "Gaussians"), most of the resulting RFs have an ON-center with two flanking OFF lobes. If also patterns that are darker than the surround are included, both ON and OFF-center RFs develop ("Plus/Minus"). Note that in either case, the background illumination is not important, because the LGN responds only to brightness differences, not absolute levels. With disks, most RFs have two lobes, because the input contained edges but no thin lines. A variety of RF types develop in simulations with noisy disks and natural images, reflecting the wide variety of input patterns seen during training.

Orientation maps develop even with random noise inputs, because even random patterns have local clusters that are brighter or darker than their surround. These clusters lead to patches of activity in the ON channel adjacent to patches in the OFF channel, and orientation-selective neurons develop. However, the resulting map is not well ordered: Many neurons are only weakly selective and the RFs do not have well-developed profiles. The Fourier transform is disk-shaped, indicating that the map consists of orientation patches of all sizes, instead of a largely uniform size seen in animal maps and in maps formed with other input types. The conclusion is that spatial structure is necessary for realistic orientation maps to form.

Overall, these results suggest that realistic orientation maps can develop based on a wide variety of spatially coherent stimuli, and that the choice of these stimuli more strongly affects the RFs than the maps. Thus, LISSOM predicts that the types of RFs observed in different species are at least partially due to the patterns the animals see during development.

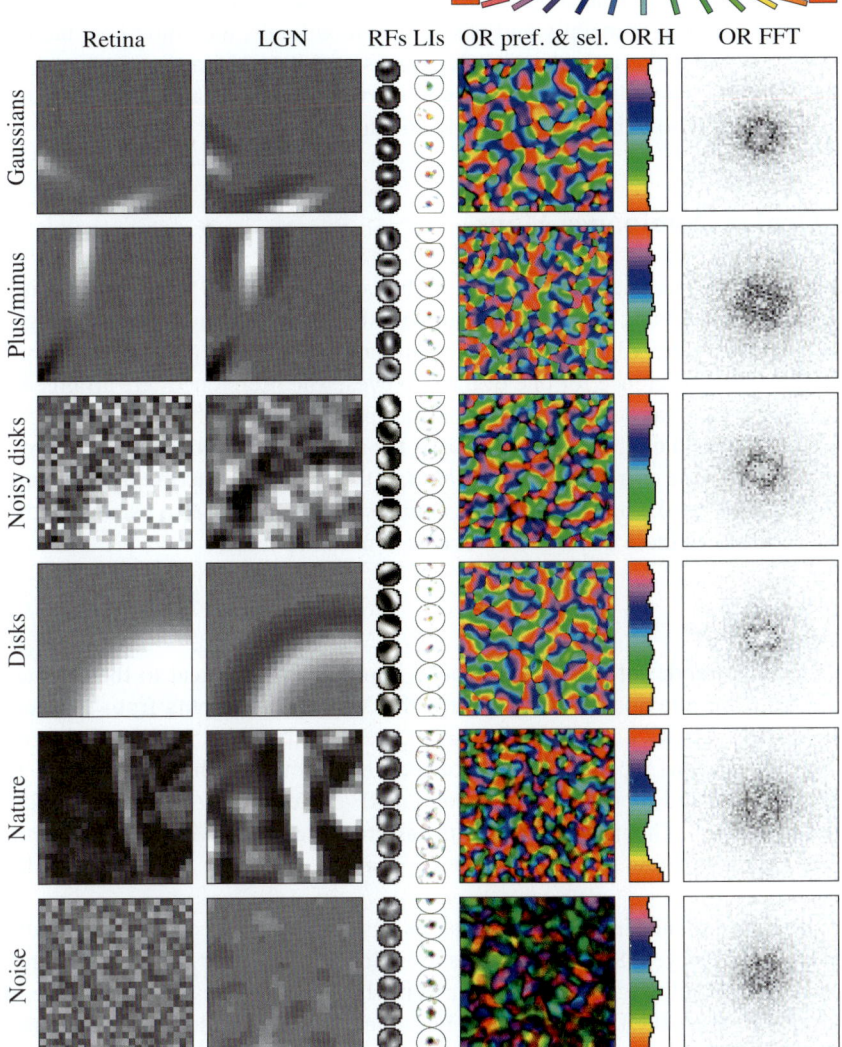

Fig. 5.13. Effect of training patterns on orientation maps. In this and later similar figures, the rows represent different self-organization experiments. Each row typically shows a sample retinal activation, the LGN response to that activation, final receptive fields (ON−OFF) of sample neurons, their lateral inhibitory connections (LIs), the orientation preference and selectivity map, the orientation preference histogram, and the fast Fourier transform (FFT) of the orientation preferences. The RFs and LIs are drawn to a smaller scale than LGN and V1. For clarity, most OR models are based on abstract input patterns like the oriented Gaussians in the top two rows. However, OR maps develop robustly with a wide variety of input patterns, including large circular patterns (middle rows) and natural images (second row from the bottom; image from a dataset by Shouval et al. 1996, 1997). Maps develop even with random noise (bottom row), although such maps are relatively unselective and the RFs do not have realistic profiles. Spatial structure is therefore necessary in LISSOM for biologically realistic maps to form.

In Chapter 9, these results are put together into a detailed model of how the orientation maps develop prenatally and postnatally in animals, including how internally generated activity like retinal and PGO waves and natural images may each contribute. The more detailed model also predicts what RF types are likely to be found in newborns, and how those will differ from adult RFs.

5.4 Ocular Dominance Maps

In this section, a second retina will be included in the LISSOM model, and it will be self-organized using unoriented Gaussian-shaped inputs on both retinas. Under these conditions, the V1 neurons develop either binocular receptive fields, or receptive fields with preference for one of the two eyes. The preferences are arranged into global maps of ocular dominance, with stripes preferring the left eye alternating with stripes preferring the right. Simultaneously, the lateral connections organize into patterns that reflect the correlations in the input. If the inputs are uncorrelated, the model replicates biological data on strabismic animals. The model is also used to study the effect of disparity more generally on ocular dominance maps.

5.4.1 Method

The LISSOM model of ocular dominance is otherwise identical to the orientation model from the previous section, except the V1 receives inputs from two retinas (Figure 5.14). The ON and OFF channels are set up the same way for both retinas, and the V1 neurons receive afferent connections from both channels through local receptive fields and topographically ordered RF centers (see Appendix A.1 for the model equations). The initial values of the afferent weights are random and identical for both eyes, to show that self-organization is not driven by initial weight differences.

Ocular dominance in LISSOM develops based on differences in activity patterns between the two eyes. Such differences are a likely source for ocular dominance in animals as well, although it is not yet known what types of activity patterns are most important for normal OD development. Possible candidates include spontaneous retinal waves that occur independently in each eye, correlated LGN activity originating in the brainstem (e.g. during sleep), position or brightness differences of corresponding visual image features in each eye, or combinations of all these factors. As an abstraction of a variety of such factors, brightness differences between matching patterns in each eye will be used in the main LISSOM experiment. In Section 5.4.4, more indirect sources such as small position differences will be evaluated as well.

The training inputs consisted of unoriented (circular) Gaussian spots of Equation 4.2, multiplied by a brightness factor s_b. Two spots were drawn in each eye in randomly chosen spatially separate locations. In the normal case these locations were constrained to be the same in the two eyes, and in the strabismic case they were independent in each eye. The brightness s_b of each spot in the left eye was chosen randomly at each iteration from the range $[0..1]$, and the brightness of the

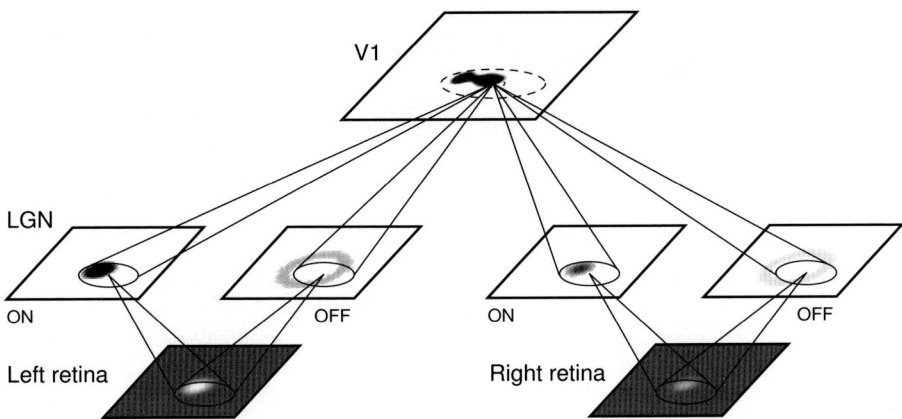

Fig. 5.14. LISSOM model of ocular dominance. The architecture is otherwise identical to that of the LISSOM OR network, except now there are two retinas, leading to two ON and OFF channels. Each V1 neuron receives afferents from corresponding positions on the ON and OFF LGN sheets for each eye.

corresponding spot in the right eye was computed as $1.0 - s_b$ so that the total input activation remained constant. With such input, the LGN responds to both the left and the right spot, but more strongly to the brighter one. V1 responds as it would to a bright spot in one eye, but will learn stronger connections from the eye with the brighter input. The network was trained for 10,000 iterations like the orientation map of Section 5.3. The rest of the simulation parameters are described in Appendix A.6.

The OD maps that emerged in normal and strabismic self-organization are described in the next two subsections, and the effect of input disparity, i.e. the degree to which the inputs are independent in the two eyes, is analyzed in the last subsection.

5.4.2 Normal Ocular Dominance Maps

Through self-organization, the network developed eye-specific receptive fields and responses (Figure 5.15). The RFs are not significantly selective for orientation, because the training inputs were small and unoriented. Different neurons prefer one eye over the other to different degrees, but as in animals, nearly all neurons are binocular to some degree.

The global arrangement of the eye preferences was visualized by recording the response of each neuron to patterns presented in each eye individually, as described in Appendix G.4. Figure 5.16 shows the resulting ocular dominance map, consisting of alternating stripes in irregular patterns across the network. Selectivity in that map measures how strongly the neurons favor inputs from one eye. The most strongly binocular neurons are found near the OD stripe boundaries and the most strongly monocular ones are at the center of such stripes. Overall, most neurons are binocular to some degree. Similar graded functional patterns of OD are seen e.g. in the macaque monkey using optical imaging techniques (Figure 2.5).

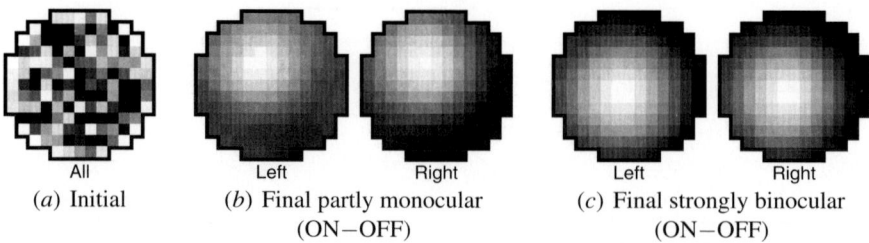

| (a) Initial | (b) Final partly monocular
(ON−OFF) | (c) Final strongly binocular
(ON−OFF) |

Fig. 5.15. Self-organization of afferent weights into OD receptive fields. (a) The afferent weights of a sample neuron, located as shown in Figure 5.17a, are plotted before self-organization (as in Figure 5.5). Initially these weights are random and identical for both eyes and both channels in each eye. (b) The final receptive fields of the same neuron are visualized for each eye by subtracting the OFF weights from the ON weights (as in Figure 5.7). Over the course of self-organization, most neurons develop a preference for one eye or the other, although they retain significant connections from both eyes. Many of this neuron's connections from the left eye are weak (indicated by medium gray), so it responds more strongly to input in the right eye. (c) On the other hand, neurons near the OD stripe boundaries, like the one in Figure 5.17b, become strongly binocular, with smooth, isotropic RFs that are nearly identical in each eye. The ocular dominance stripes shown in Figure 5.16 are based on such subtle eye preferences, as they are in animal OD maps.

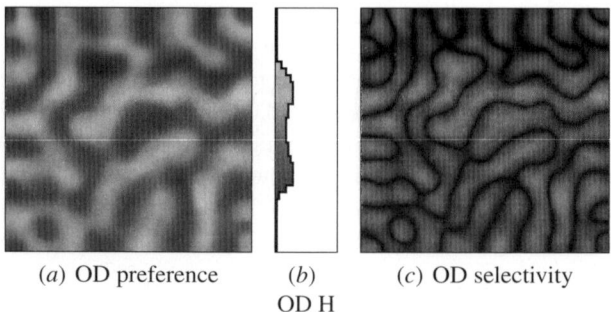

| (a) OD preference | (b)
OD H | (c) OD selectivity |

Fig. 5.16. Self-organized ocular dominance map. Light areas in (a) indicate neurons that prefer the left eye, dark areas those that prefer the right eye, and medium gray indicates no net preference. The histogram (b) shows how these preferences are distributed, with left monocular neurons at the top, binocular neurons in the middle, and right monocular neurons at the bottom. Most neurons are binocular, slightly preferring one eye or another, as they do in animals (Figure 2.5). Plot (c) illustrates how selective the neurons are for ocularity, with light areas indicating monocular neurons and dark areas those that are binocular. Less selective regions fall between ocular dominance stripes, as in animal maps.

The stripes form based on the push–pull effect of the lateral connections: Local excitation ensures that nearby neurons will respond to similar stimuli and thus have correlated activity (causing eye-specific regions to develop), and long-range inhibition causes activity to be anti-correlated over larger distances (causing the eye-specific regions to alternate).

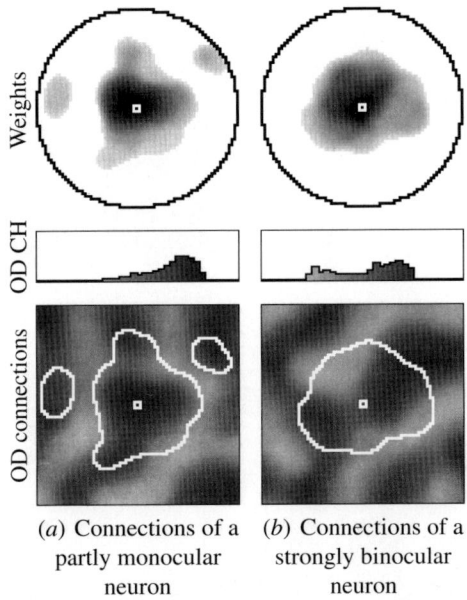

(*a*) Connections of a   (*b*) Connections of a
partly monocular   strongly binocular
neuron   neuron

Fig. 5.17. Long-range lateral connections in the ocular dominance map. The inhibitory lateral connection strengths for the two neurons in Figure 5.15 are plotted on top in gray scale, and their local neighborhood is shown in the map below (as in Figure 5.12). In both cases the strongest connections come from the neuron's near neighbors. The connections of the partly monocular neuron (*a*) follow the ocular dominance map structure, with strongest connections from neurons with the same eye preference (dark). As a result, the connection histogram (middle) is biased toward the right eye (dark). In contrast, the connections of strongly binocular neurons (*b*) are not influenced by the OD map, and their connection histograms mirror the histogram of the OD map (Figure 5.16*b*). Similar patterns have been found experimentally in cats (Löwel 1994; Löwel and Singer 1992).

In the normal OD case, the lateral connections are not particularly patchy, and most neurons receive connections from neighbors of both eye preferences (Figure 5.17). Such connectivity arises because most neurons are only partly monocular: Their activation is correlated even with neighbors that prefer the opposite eye, and such connections remain active. Only the connections of the most strongly monocular neurons follow the ocular dominance stripes. Such neurons are activated predominantly by inputs from one eye, and their activity patterns are more strongly correlated with other similar neurons.

5.4.3 Strabismic Ocular Dominance Maps

As discussed in Section 5.1.2, the maps and lateral connections in strabismic animals differ significantly from those of normal animals. Similarly, a LISSOM OD map organized with uncorrelated inputs is very different from the normal case (Figure 5.18). By the end of strabismic development, nearly all neurons have become

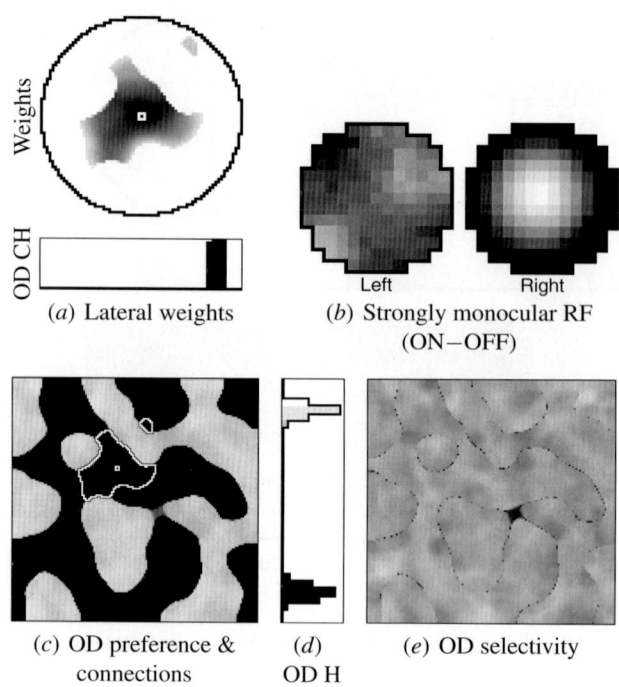

(a) Lateral weights

(b) Strongly monocular RF
(ON−OFF)

(c) OD preference &
connections

(d)
OD H

(e) OD selectivity

Fig. 5.18. Ocular dominance and long-range lateral connections in the strabismic ocular dominance map. The strabismic simulation was otherwise identical to the normal case of Figures 5.15–5.17, except the inputs were presented at random positions in each eye. Compared with the normal case, the OD stripes are wider (c), and nearly all neurons are highly selective (e) and highly monocular (indicated by the wide separation of the peaks in the histogram d). As an example, the neuron shown with the small white square in (c) has the receptive fields shown in (b); the connections from the left eye are poorly organized and weak (indicated by medium gray). In (c), the white outline delineates the strongest lateral inhibitory connections to this neuron. Unlike in the normal case, these connections include only monocular neurons responding to the same eye (visible in the weight histogram in (a)), and strictly follow the ocular dominance stripes. The connections are strongest in the immediate vicinity of the neuron, but not much weaker even near the stripe boundaries (a). Overall, strabismus changes the map organization, RFs, and lateral connections much like it does experimentally in animals (Löwel 1994; Löwel and Singer 1992; Figure 5.2).

strongly monocular, and the boundaries between the stripes for each eye are sharply defined. The lateral connections also correlate strongly with the ocular dominance stripes, i.e. nearly all of the lateral connections come from neurons that prefer the same eye. Similar patterns have been found in cats using anatomical tracing techniques (Löwel 1994; Löwel and Singer 1992).

These results are due to the uncorrelated inputs in the two eyes. Corresponding receptive fields on the two eyes are rarely active at the same time, so all active neurons in a small patch of cortex are likely to be receiving input from the same eye. Through Hebbian learning, the afferent connections of all these neurons to that eye are strengthened, as are the lateral connections between them. Over the course of self-organization, these weight changes result in sharply defined boundaries between stripes and between lateral connection patches, each selective for one eye or the other.

The OD stripes are also significantly wider in the strabismic LISSOM model than in the normal model, matching experimental findings in cats (Löwel 1994) and in previous computational models (Goodhill 1993). In LISSOM, nearby neurons learn similar preferences if they frequently respond to the same inputs. When the inputs are uncorrelated, two neurons respond together if they prefer the same eye and their RFs on that eye overlap. Since the RFs of nearby neurons overlap significantly, the OD stripes tend to be wide, alternating between large groups that prefer one eye or the other. In the normal case, however, the inputs in the two eyes are correlated. Even neurons that prefer different eyes often respond together because they represent the same retinotopic location. As correlation increases, retinotopy determines more of the response, and the OD stripes become less selective and narrower. In the limit, the inputs are fully correlated, and the OD stripes disappear altogether.

The strabismic patterns of ocular dominance and lateral connections that LISSOM develops closely match observations in biology. The LISSOM model shows how such connections and global organization can develop, based on a network that extracts structure from correlations in the visual input.

5.4.4 Effect of Input Disparity

Abstracting the possible sources of ocular dominance, the LISSOM OD model above was based on inputs that differ in brightness in the two eyes (like the models of Bauer, Brockmann, and Geisel 1997; Riesenhuber, Bauer, Brockmann, and Geisel 1998). Some previous models have shown that OD maps might instead result from retinal disparity, i.e. differences in feature position in each eye (Burger and Lang 1999; Sirosh and Miikkulainen 1994a). Disparity itself is a separate feature dimension from ocular dominance. Neurons can be fully binocular yet prefer different positions in each eye, which is important for stereo vision. Even so, disparity does reduce the activity correlations between the eyes, and could thus contribute to ocular dominance as well.

Disparity was modeled in LISSOM by randomly placing a spot at (x, y) in the left retina, and then placing a corresponding spot within a radius of $s_s R$ of (x, y) in the right retina, where R is the width of the retina. The scatter parameter $s_s \in [0..1]$ specifies the spatial correlations between spots in the two retinas. When $s_s = 0$, the inputs are in perfectly matched positions, and when $s_s = 1$ they are scattered independently (as in strabismus). By adjusting s_s, it is possible to simulate different degrees of disparity between the images in the two eyes (see Appendix A.6 for the rest of the simulation parameters).

In Figure 5.19, networks trained with brightness differences alone ($s_s = 0$), with mild disparity ($s_s = 0.2$), with moderate disparity ($s_s = 0.4$), and with perfect strabismus ($s_s = 1.0$) are compared. All these networks developed OD maps, but those with disparity are less realistic. With only mild disparity, the eye preferences

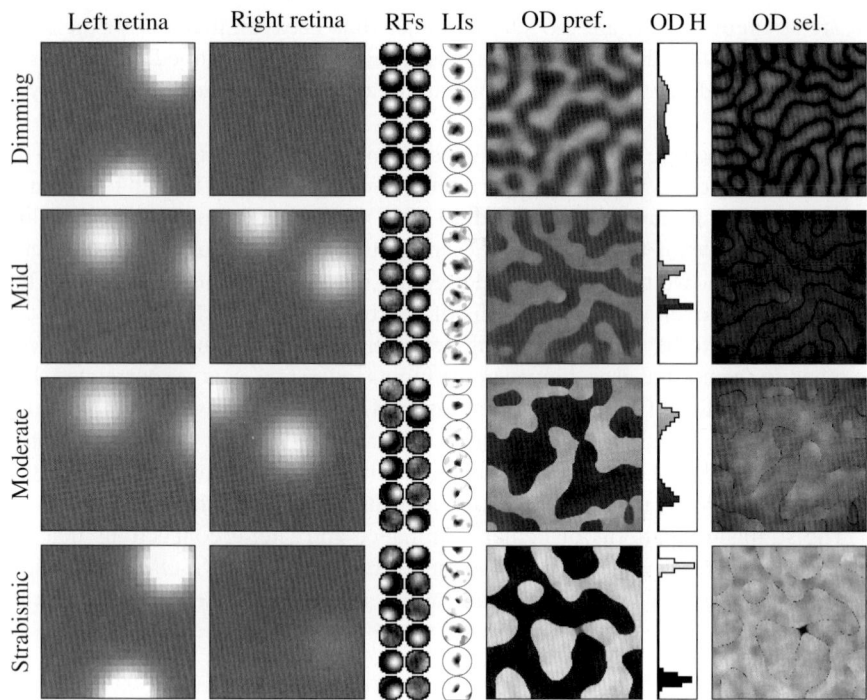

Fig. 5.19. Effect of disparity on ocular dominance maps. Each row presents a different simulation using the same network but a different degree of disparity between the inputs in the two eyes. From left to right, each row shows a retinal activation (left and right eyes), final RFs of a set of sample neurons, their lateral connections, the OD preference map, its histogram, and the OD selectivity map. For comparison, the results from brightness differences are reproduced in the top row (labeled "Dimming"), and the results from strabismic maps in the bottom row. The main result is that OD maps can be obtained from disparity differences (with no brightness differences), but the results do not match animal maps well. Small amounts of disparity (scatter $s_s = 0.2$) result in unrealistically clear boundaries between stripes even with relatively weak OD preferences (row "Mild"), as is evident in the histogram. Moderate disparity ($s_s = 0.4$; row "Moderate") approaches the strabismic results, with strongly monocular RFs, sharp stripe boundaries, and connections only to neurons that prefer the same eye, unlike in normal animals. These results suggest that ocular dominance patterns can result from differences in either position or brightness, but brightness differences lead to maps that more closely match those found in animals.

are weak, yet there is already a clear boundary between OD stripes, instead of the smooth boundaries seen in animal OD maps. Lateral connections are also patchy and specific to one eye, unlike in animals. As the disparity is increased further, the maps form sharply defined monocular stripes, more closely matching strabismic animal OD maps than normal maps (compare the moderate case with the strabismic case).

The reason for such sharp stripes is that the neurons have sharply defined ON and OFF subregions, and are therefore highly sensitive to position. Even a small difference in position between the eyes activates entirely different sets of neurons, and thus responses to each eye are effectively uncorrelated (or even anti-correlated). These results suggest that input disparities must be small relative to the RF size, or else strabismic-like OD maps will develop.

Interestingly, realistic maps can be obtained even with disparity-based inputs if the ON and OFF channels are bypassed (not shown; Sirosh and Miikkulainen 1994a). In such a case, any input that a V1 neuron receives is excitatory, nearby inputs always excite nearby V1 neurons, and smooth OD preferences are obtained. Such self-organization may be possible in early prenatal development, based on internally generated inputs like those postulated for orientation in Chapter 9. As discussed in more detail in Section 17.2.3, such computational constraints predict what kind of receptive fields and input pattern correlations are likely to be found in early development.

In this section, LISSOM OD maps were trained based on unoriented Gaussians only. Most other input types, like those tested in Figure 5.13, would result in an overlaid OR map as well. The effect of other input types such as natural images on OD will therefore be studied in Section 5.6 as part of combined OR/OD/DR maps.

5.5 Direction Selectivity Maps

In this section the LISSOM model is extended to moving inputs. Spatiotemporal receptive fields develop as a result, and they are organized into a map according to preferred direction of motion. These results show that activity-dependent self-organization extends to time-varying input as well, which is a crucial step toward making the model biologically realistic. The last step, organizing multiple stimulus features at once, will be discussed in the next section.

5.5.1 Method

Direction selectivity in LISSOM arises from LGN output that arrives at the cortex with a variety of delays (as in some previous models, e.g. Wimbauer, Wenisch, Miller, and van Hemmen 1997a; Wimbauer et al. 1997b). In the LGN, most cells fire soon after the retinal stimulus. However, other neurons, called lagged cells, respond only after a fixed delay (Saul and Humphrey 1992). The delay times vary between cells over a continuous range up to hundreds of milliseconds (Mastronarde, Humphrey, and Saul 1991; Wolfe and Palmer 1998). V1 neurons can use these timing differences to develop spatiotemporal receptive fields (Humphrey, Saul, and Feidler 1998).

The LISSOM architecture for direction selectivity is shown in Figure 5.20. The model is similar to those presented in previous sections, consisting of a hierarchy of two-dimensional sheets of neural units. Retinal receptors feed input to several paired sheets of ON-center and OFF-center LGN units (with a different lag for each pair), which in turn activate cortical neurons in V1 (see Appendix A.1 for the model equations). Because direction preferences are generally defined in terms of oriented

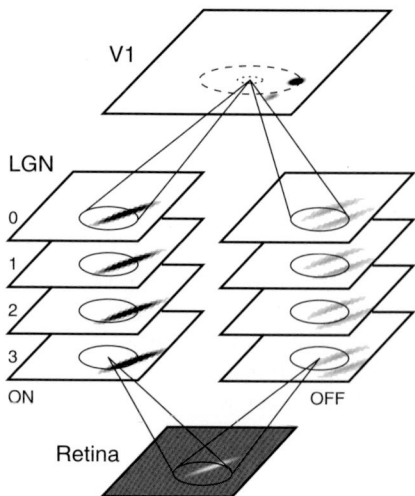

Fig. 5.20. LISSOM model of orientation and direction selectivity. The architecture is similar to the LISSOM OR network, except the ON and OFF channels consist of sheets of neurons with different lag times (from 0 to 3 in this case). Moving input patterns are drawn on the retina in discrete timesteps, like frames of a movie. At the first timestep, the ON and OFF LGN cells with time lag 3 compute their activity. At each subsequent timestep, the input pattern is moved slightly and LGN cells with lags 2, 1, and 0 each compute their activity in turn. Once all LGN cells have been activated, the initial V1 response is computed based on the responses on the eight LGN sheets. The activity then spreads laterally within V1 as usual in LISSOM.

lines, the model is self-organized to represent both direction and orientation at once, using moving oriented patterns as input.

The input consists of short movies of oriented Gaussians moving across the retina at random locations and directions. Each movie is presented as a sequence of frames. At each timestep t, the frame t is drawn on the retina, and the activity levels of all LGN cells with lag t are calculated. After all the LGN neurons have been activated, each V1 neuron computes its initial response based on the activation on all LGN sheets. After the initial response, the V1 activity settles through short-range excitatory and long-range inhibitory lateral interaction, and afferent and lateral weights are modified as described in previous sections. The model is then ready for the next input movie presentation.

In the experiments in this section, there were four 36×36 ON-center and four 36×36 OFF-center sheets, and they received input from a single 54×54 retina with a single moving oriented Gaussian pattern. Single inputs were used to avoid overlap and bias in the input distributions. The network was organized in 20,000 presentations of moving input patterns, so that the total number of inputs was the same as in OR and OD simulations. The rest of the parameters are described in Appendix A.7. The orientation and direction maps that resulted are analyzed next, followed by a comparison of maps obtained at different input speeds.

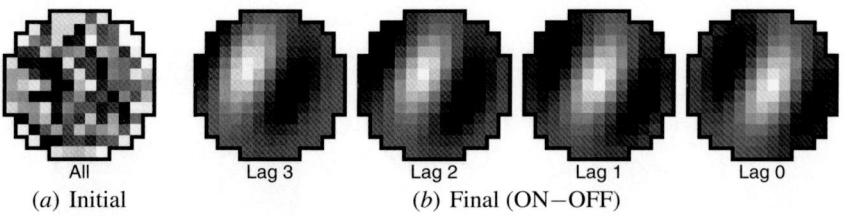

All	Lag 3	Lag 2	Lag 1	Lag 0
(a) Initial		(b) Final (ON−OFF)		

Fig. 5.21. Self-organization of afferent weights into spatiotemporal RFs. (a) The lag-0 weights for a sample neuron, located as shown in Figure 5.24a, are plotted before self-organization (as in Figure 5.5). Initially, all four lags in both channels have the same random weights; these weights are different for each neuron. (b) The final afferent weights for the same neuron are visualized by subtracting the OFF weights from the ON weights (as in Figure 5.7). Together, these plots show that the most effective stimulus for this neuron is a diagonal light bar moving diagonally down and to the right. More specifically, this neuron will be highly active at time t if there was a light bar aligned with the ON subregion in the "Lag 3" RF at time $t - 3$, a bright bar aligned with the ON subregion of the "Lag 2" RF at time $t - 2$, and so on. Visual cortex neurons in animals have similar spatiotemporal properties (Figure 5.4a; DeAngelis et al. 1995).

5.5.2 Direction Maps

Figure 5.21 shows the self-organized afferent weights for a representative neuron in V1. Nearly all neurons developed spatiotemporal receptive fields strongly selective for both direction and orientation. Each neuron responds best to a line with a particular orientation moving in a direction perpendicular to that orientation. Such receptive fields are similar to those found experimentally in the cortex (Section 5.1.3; DeAngelis et al. 1993, 1995).

The orientation preferences form a smoothly varying orientation map (Figure 5.22). As in the LISSOM OR-only model, the map contains realistic features such as iso-orientation patches, linear zones, pairs of pinwheels, saddle points, and fractures. The techniques used to analyze the orientation map in Section 5.3 (selectivity measures, Fourier transform, gradient, retinotopic mapping, and OR preference histogram), lead to essentially the same results. This outcome verifies that OR-only simulations are a valid approach to understanding orientation maps.

The same neurons are also selective for motion direction, forming a direction map (calculated as described in Appendix G.5). Direction preferences are also smoothly organized across the cortex, and contain similar features. Regions of lower direction selectivity occur near pinwheel centers and along fractures, as in the OR map. The orientation and direction preference histograms are essentially the same (i.e. flat), both reflecting the distribution of edges in the training input.

The interaction between the orientation and direction maps is illustrated in Figure 5.23, and is similar to what has been observed in animals (Section 5.1.3; Shmuel and Grinvald 1996; Weliky et al. 1996). For instance, a patch of neurons highly selective for one orientation and direction of motion will usually have an adjacent or contiguous patch selective for the same orientation but opposite direction. In LIS-

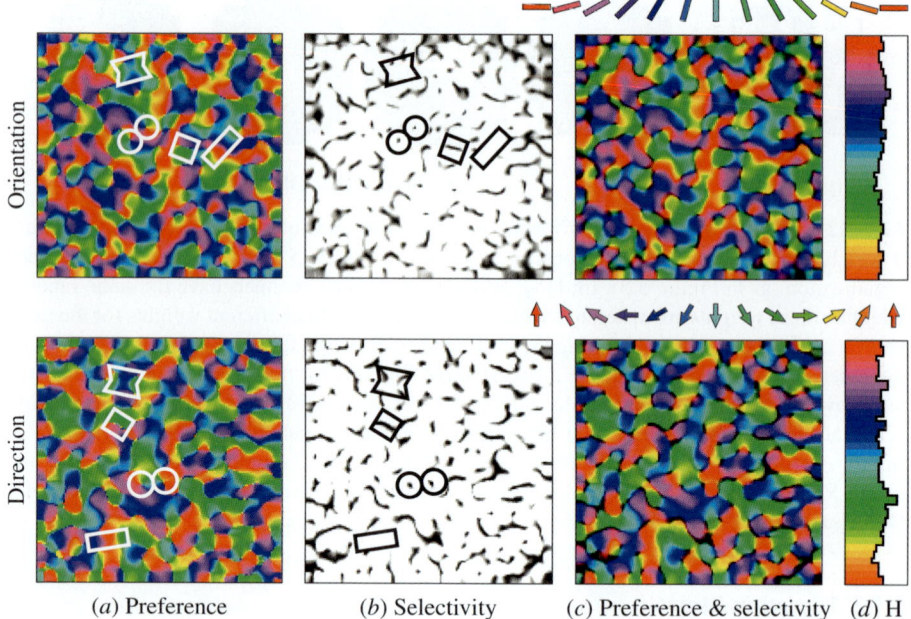

<table>
<tr><td>(a) Preference</td><td>(b) Selectivity</td><td>(c) Preference & selectivity</td><td>(d) H</td></tr>
</table>

Fig. 5.22. Self-organized OR/DR map. The orientation (top row) and direction (bottom row) maps in the LISSOM OR/DR model were computed separately after self-organization. The orientation preferences are coded using the color bar key on top, and the direction preferences using the color arrow key in the middle. Selectivity is shown in gray scale in both cases, with black indicating low selectivity (as in Figure 5.9). (*a*) The network represents both orientation and direction in smoothly varying maps that contain all the features found in animal maps, such as linear zones, pairs of pinwheels, saddle points, and fractures (outlined as in Figure 2.4). (*b*) Most neurons become selective for specific orientation and direction of motion, and are therefore nearly white in the selectivity plots. (*c*) Overlaying the preference and selectivity plots shows that regions of low selectivity occur near pinwheel centers and along fractures in both maps. (*d*) The histograms are essentially flat because the training inputs were unbiased. These plots show that LISSOM can develop biologically realistic orientation and direction maps through self-organization based on abstract input patterns.

SOM, these patterns develop because neurons that prefer similar orientations but opposite directions have more similar RFs (and thus responses) than neurons that prefer different orientations. As a result, opposite direction preferences are often grouped together on the map.

Lateral connections within the map follow its global organization, primarily linking neurons with similar orientation and direction preferences (Figure 5.24). Such connection patterns reflect the activity correlations during self-organization. Neurons with similar orientation and direction preferences are often active together, and become more strongly connected. Connections of neurons that are highly selective for orientation and direction extend along the orientation preference, not direction, and avoid orthogonal orientations and opposite directions. Neurons in DR fractures

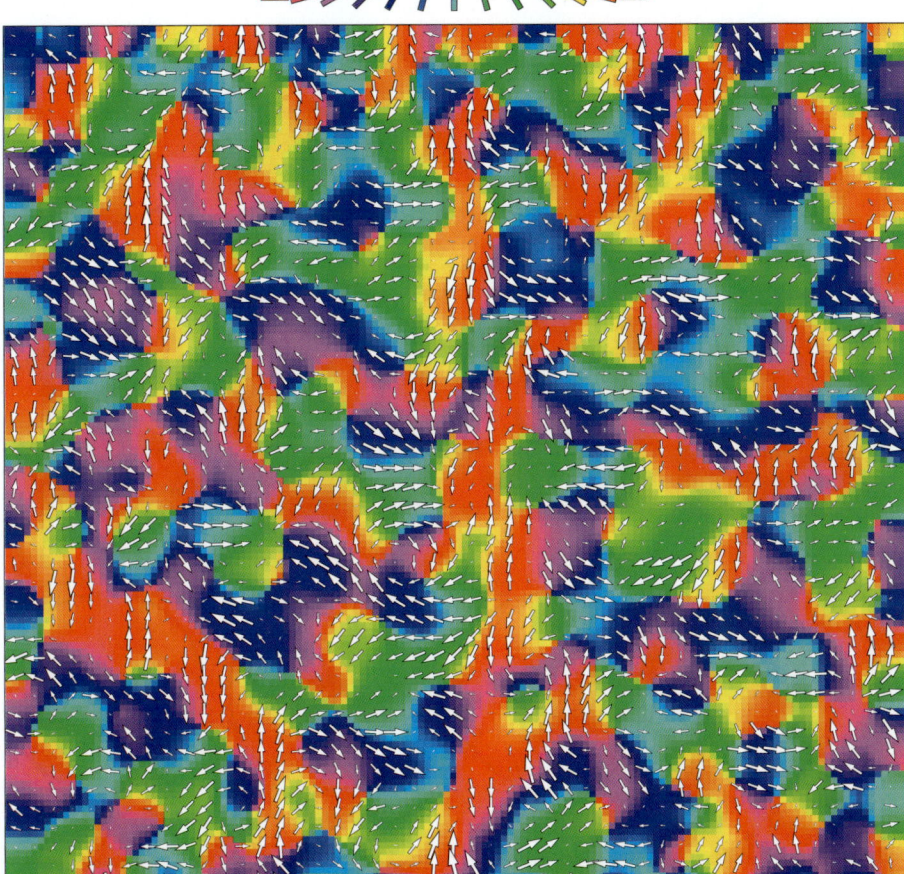

Fig. 5.23. Combined OR/DR map. Using the plotting conventions from Figure 5.4, each of the 142×142 neurons in the LISSOM OR/DR map is colored with its preferred orientation, and the direction preferences for every third neuron are plotted as arrows overlaid on the orientation map. The direction preferences are generally perpendicular to the preferred orientation, and large iso-orientation patches are often divided into two areas with opposite direction preferences. Such an organization matches experimental data well (Figure 5.4; Weliky et al. 1996). An animated version of this plot can be seen at http://computationalmaps.org.

connect with neurons of both directions; they all have similar orientation preferences, and the connections extend along that orientation. At DR pinwheels the connections come from all direction preferences, and at DR saddles they correspond to the direction preferences in the saddle. Both DR pinwheels and DR saddles can occur at a variety of OR map features.

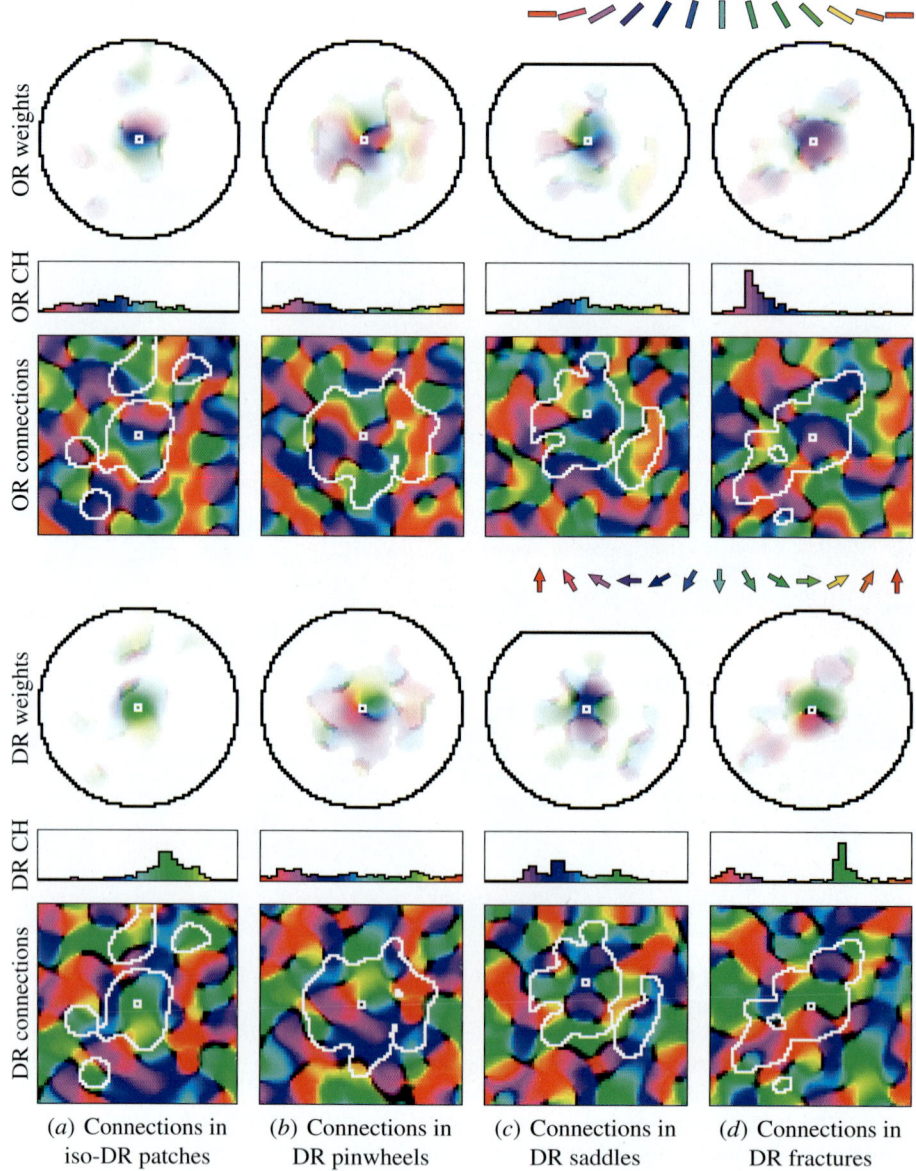

(a) Connections in iso-DR patches   (b) Connections in DR pinwheels   (c) Connections in DR saddles   (d) Connections in DR fractures

Fig. 5.24. Long-range lateral connections in the combined OR/DR map. The inhibitory lateral connections of four sample neurons from the marked regions in the DR map of Figure 5.22 are shown both on the OR (top) and DR (bottom) preference and selectivity maps, as in Figure 5.12. (*a*) The neuron from Figure 5.21 receives connections from neurons with similar OR and DR preferences, and its connections extend along the OR axis (60°), not the DR axis (330°). (*b*) Neurons in pinwheels receive connections from neurons of all DR preferences; their OR histograms vary depending on their location in the OR map. (*c*) Saddle neurons receive connections corresponding to the DR preferences of the saddle, and their OR histograms vary. (*d*) At fractures, connections originate primarily from neurons of the two fracture preferences, and extend along their common OR preference. These OR- and DR-specific lateral connections are predictions that can be tested in future biological experiments.

While lateral connections are known generally to follow orientation preferences, how such connectivity is further affected by direction selectivity has not been analyzed. The LISSOM results constitute predictions for future such biological experiments.

5.5.3 Effect of Input Speed

Compared with the orientation-only simulations, the OR/DR map introduces one new parameter: the speed of the input patterns, i.e. how many retinal units the pattern moves between lags. Figure 5.25 shows how this parameter affects the formation of direction maps.

For speed zero, the inputs are stationary and no direction map develops. The RFs to each lag become identical, and are essentially the same as in the OR-only map (Section 5.3). As the speed is increased, the neurons become more selective for direction of motion. Interestingly, the spacing between DR patches becomes larger as the neurons become more selective; this trend is also visible in the ring diameter in the Fourier plots (Figure 5.25). At the same time, the ring diameter in the OR map increases slightly, and by speed 2 the features in the OR map become smaller than those in the DR map.

Thus, with fast enough inputs, the DR map becomes the largest-scale organization, with smaller patches for orientation. Although the results from speed 1 are most similar to the existing animal results (from the ferret), the results from higher speeds are predictions for maps in other species with greater motion sensitivities. LISSOM also predicts that animals raised in environments with more motion during the animals' critical period will develop direction maps whose spatial scale is larger than that of their orientation maps.

Overall, the results in this section show that LISSOM can account for spatiotemporal preferences in addition to spatial ones. Although Gaussian images were used in this section for clarity, similar results can be obtained with natural images (as will be demonstrated in the next section).

So far in this chapter we have seen how orientation selectivity, direction selectivity, and lateral connectivity each develop synergetically in the model, and the results match data from animal experiments. The next step is to show how they interact in a single unified model of feature preferences in the cortex.

5.6 Combined Maps of Multiple Features

The LISSOM OR, OD, and DR models introduced in the previous sections are all based on the same basic architecture, differing only in the number of eyes and LGN sheets simulated and the type patterns used to train the model. Such uniformity makes it possible to combine them into a comprehensive model that self-organizes all three feature preferences at once. The model is first validated against biological data on how orientation and ocular dominance interact. It is then trained with moving images, and predictions are made on how all three features self-organize together. In the final

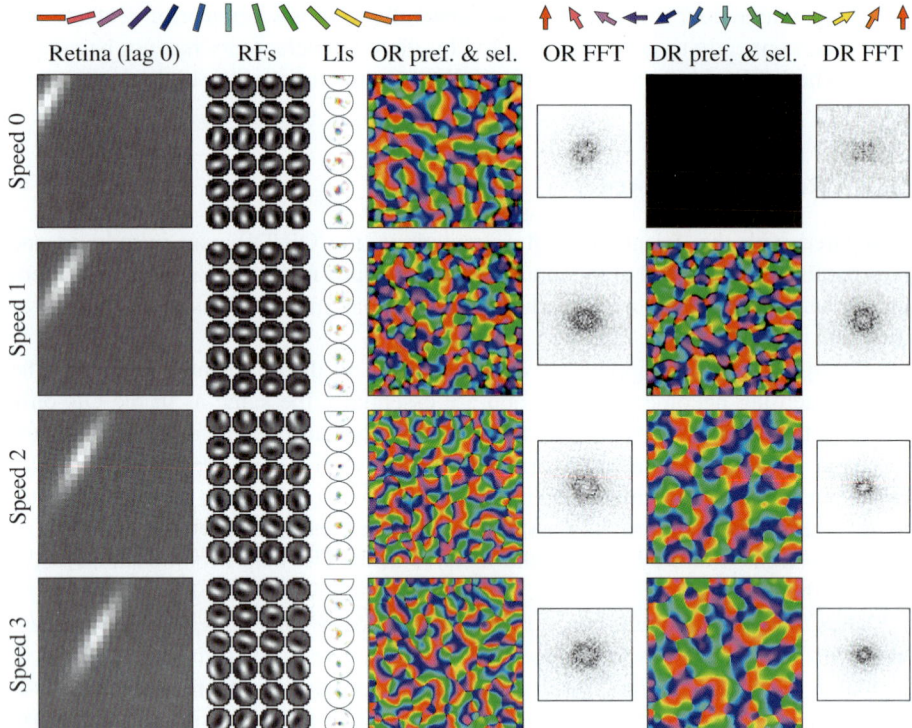

Fig. 5.25. Effect of input speed on direction maps. From left to right, each row shows a sample retinal activation at lag 0, final receptive fields to LGN regions with lags 3, 2, 1, and 0 (left to right) of six sample neurons, the inhibitory lateral connections of those six neurons, the orientation preference and selectivity map, the Fourier transform of the OR preferences, the direction preference and selectivity map, and the Fourier transform of the DR preferences. Orientation and direction histograms are not shown because they are all nearly flat. Each row shows the result from using training inputs moving at a different speed, ranging from zero (stationary) to moving three retinal units between each group of lagged LGN cells. For the example input shown, the lag 3 input was always the one shown in the top row (labeled "Speed 0"), and by lag 0 it had moved to the position shown in each row. When the inputs were stationary (i.e. all lags had the same input patterns), no direction map or direction-selective units developed, and the "DR pref. & sel." map is entirely dark. As the speed increases, more units become direction selective, and direction becomes the largest-scale organization in the map. This increase in feature size is visible in the DR Fourier transform plots, where a smaller spatial frequency (larger feature spacing) leads to smaller rings as speed is increased. These results are predictions for maps in animals with different retinal motion sensitivities or those raised in environments with different speeds of visual motion.

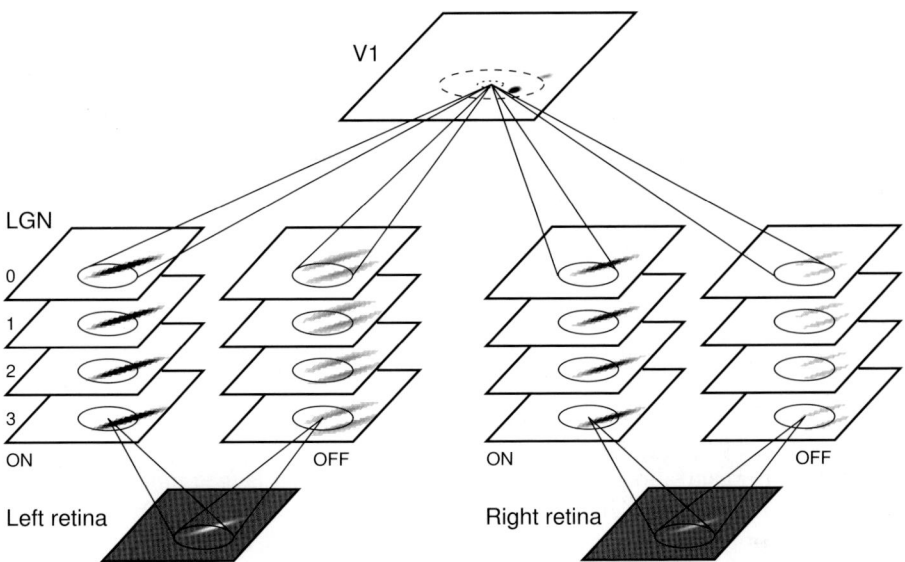

Fig. 5.26. LISSOM model of orientation, ocular dominance, and direction selectivity. The architecture is a combination of the LISSOM OD network of Figure 5.14 and the DR network of Figure 5.20. For each input movie, one retina is chosen randomly to be dimmer than the other. The frames are then drawn on the retina one timestep at a time, and propagated to the LGN sheets with the appropriate lag. The V1 neurons receive input from all 16 LGN sheets simultaneously, and settle the activation through lateral connections as usual in LISSOM.

section, the effects of organizing it with natural images instead of oriented Gaussians are considered. This stepwise approach makes it clear which aspects of the input patterns and architecture are crucial for each type of feature preference.

5.6.1 Method

The LISSOM model of simultaneous OR/OD/DR is a combination of the ocular dominance network of Section 5.4 and the direction selectivity model of Section 5.5. It contains 16 LGN sheets, consisting of four ON and four OFF sheets for each of the two eyes (Figure 5.26; see Appendix A for the model equations).

The model is trained with moving oriented Gaussian patterns at randomly chosen locations and directions. As in the DR network, one Gaussian is used per iteration, but multiple Gaussians would be used for larger retinas. As in the OD network, the brightness s_b of the Gaussian in the left eye is chosen randomly in the range $[0..1]$ for each movie, and the brightness of the corresponding spot in the right eye is computed as $1.0 - s_b$. Also as in the DR network, each of the four sheets in the four LGN channels receives a frame of the input movie with a different lag. Once all 16 LGN sheets have been activated, V1 neurons calculate their initial activation based on the total LGN activation, settle it through lateral connections, and adapt their weights as in previous models.

While there are significant biological data on how OR and OD interact in biological maps, little is known at present on how DR confounds these interactions. Moreover, although lagged LGN cells have been found in cats (Hubener, Shoham, Grinvald, and Bonhoeffer 1997; Löwel, Bischof, Leutenecker, and Singer 1988), they may not exist in monkey LGN (Saul and Humphrey 1992). Therefore, in the first experiment in this section, only the LGN sheets with lag 0 will be used, and the resulting OR/OD map will be compared with OR/OD results in monkeys. The parameters of this experiment are set as in the OD-only simulation (Appendix A.6). In the second experiment, the full OR/OD/DR model is trained for 20,000 iterations as described above, and predictions are made on how these features interact in animals like cats. The parameters for the OR/OD/DR simulation are described in Appendix A.9.

5.6.2 Combined Orientation / Ocular Dominance Maps

Through self-organization, the combined network without lagged inputs developed realistic maps for both orientation and ocular dominance (Figure 5.27). Both maps are similar to the single-feature maps of Sections 5.3 and 5.4, and contain the features typical of animal maps. The OR map contains linear zones, pairs of pinwheels, saddle points, and fractures, and its selectivity, Fourier transform, gradient, retinotopic mapping, and OR preference histogram are similar to those in previous LISSOM OR maps. The OD preference histogram and the OD selectivity map are similar to the previous LISSOM OD map. These results demonstrate that LISSOM is capable of forming realistic maps of independently varying feature dimensions simultaneously.

The maps interact in patterns similar to those seen in monkeys (compare Figure 5.27 with 5.3). Pinwheel centers and fractures rarely overlap with the OD stripe boundaries; ocular dominance boundaries tend to cross orientation patches in linear zones at right angles, and OD boundaries rarely follow an orientation boundary. Neurons that are least selective for orientation are thus most selective for the eye of origin, and vice versa. Such patterns emerge in LISSOM because the self-organizing process favors an organization where the responses are spread evenly across the cortex (Section 3.4.3). As a result, the different regions of the map become selective for different features. Overall, these patterns are consistent with biological data, suggesting that LISSOM is a valid model of how such features develop in animals like monkeys.

Interestingly, when disparity is used to create differences between the eyes (instead of dimming), the joint OR/OD map that develops is not as realistic (not shown). In this case, each eye develops independent orientation maps, and the boundaries between orientation patches often follow the OD boundaries (unlike in animals). In effect, the neurons responding to each eye develop independently, because a given neuron rarely receives activation from both eyes at once. This result provides further computational evidence that processes like dimming, not disparity, are the source of OD maps.

The receptive fields for each eye are similar to those in the LISSOM OR network. They are all binocular to some degree, with slightly stronger connections from one eye or the other. The lateral connections are strong locally, and in the longer range

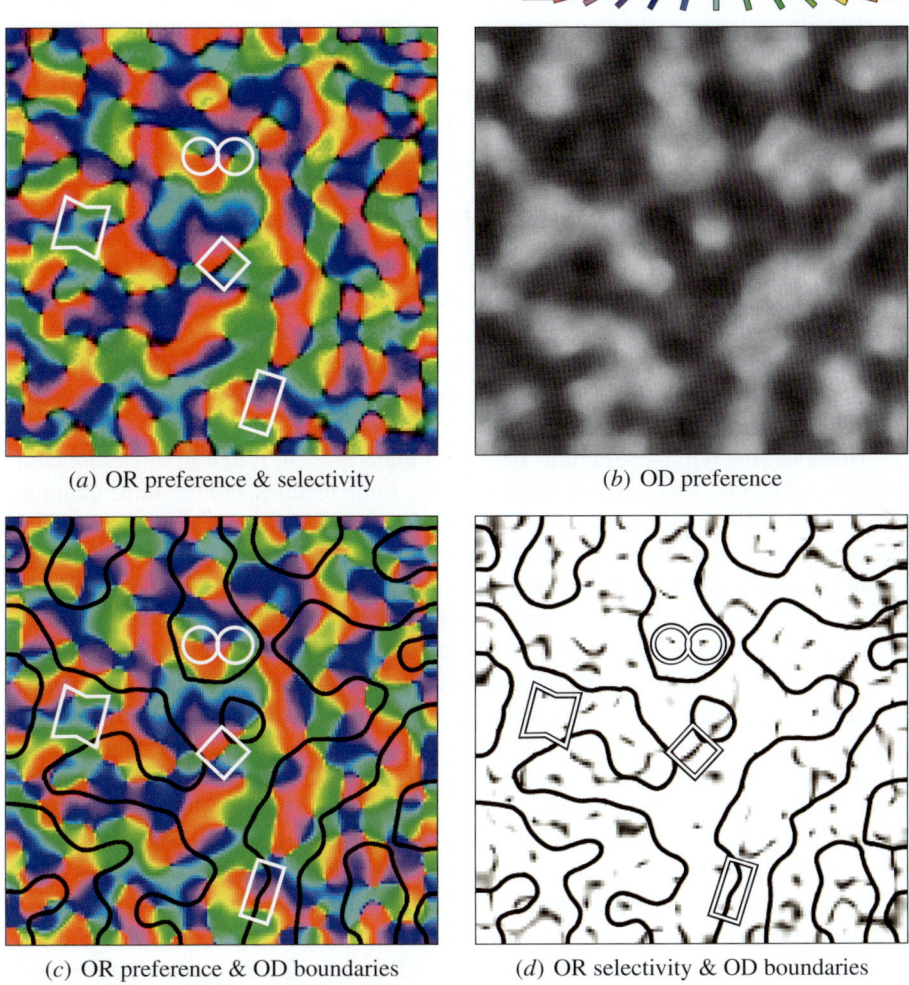

(a) OR preference & selectivity   (b) OD preference

(c) OR preference & OD boundaries   (d) OR selectivity & OD boundaries

Fig. 5.27. Self-organized OR/OD map. Based on oriented Gaussian patterns with different brightnesses in each eye, LISSOM develops realistic orientation (a) and ocular dominance (b) maps in the same area of cortex. The orientation map features are outlined in (a), (c) and (d) as in Figure 2.4. In (c), the orientation preferences are overlaid with the ocular dominance gradient: High gradient (black) marks the boundary between OD stripes. These boundaries rarely overlap pinwheel centers or fractures, they intersect OR boundaries in linear zones at right angles, and they rarely follow OR boundaries. These relationships are further highlighted in (d), where regions of low orientation selectivity (pinwheel centers and fractures) are plotted in dark gray, overlaid with the ocular dominance boundaries. Again, these features rarely intersect, suggesting that the map organization results from distributing selectivity for different features evenly across the cortex. Similar interaction between orientation and ocular dominance is seen in biological maps (Figure 5.3; Blasdel 1992b)

follow the orientation preferences, as they do in animals (Figure 5.28). The neurons most selective for one eye also prefer connections from that eye, but the preference is not absolute like it is in strabismic animals (Section 5.4.3). These results show that the lateral connection patterns seen in the simpler separate maps also exist in the combined maps, with preferences for multiple features overlaid onto the same set of neurons.

5.6.3 Combined Orientation / Ocular Dominance / Direction Maps

To model all three feature types in the same network, the entire model of Figure 5.26 was trained with movies of moving oriented Gaussians, randomly dimmed in one eye. The network developed realistic maps for orientation, ocular dominance, and direction selectivity (Figure 5.29). The combined map is more complex than those developed by the single-feature networks, because the network is representing three overlaid, interacting feature maps.

However, the typical structures of animal maps are still found in this more complex map. The OR, OD, and DR maps have the same features and the same quantitative measures as the individual LISSOM OR, OD, and DR maps. Orientation patches are often divided into patches selective for opposite directions of motion, and orientation pinwheels and fractures avoid boundaries of the ocular dominance stripes. The receptive fields are analogous to those in single-feature LISSOM networks, i.e. oriented and binocular, with a slight preference for one eye or the other (Figure 5.32). The lateral connections are strong in a local neighborhood, but at longer distances mostly follow the orientation and direction preferences, as was found in the individual LISSOM OR, OD, and DR networks (Figure 5.32). These results demonstrate that LISSOM can form realistic maps and lateral connections based on multiple features; on the other hand, they also show that the approach of studying each feature separately is valid and leads to insights that carry over to more complex maps.

Interestingly, some relationships that were clear in the OR/OD network are not as uniform in the OR/OD/DR network. For instance, the ocular dominance boundaries in linear zones do not intersect orientation patches at right angles as often, and pinwheel centers are not always at the center of OD stripes. Such variation corresponds to differences between maps in different animal species. The patterns observed in the OR/OD network have been characterized primarily in monkeys, and it is not known if monkeys have lagged cells at the LGN level (Saul and Humphrey 1992). Conversely, the lagged cells in the OR/OD/DR network are similar to those found in cats, where the OR/OD intersection patterns are less clear (Hubener et al. 1997; Löwel et al. 1988), similar to the OR/OD/DR LISSOM model. The two versions of the combined LISSOM model therefore suggest that the differences seen in the intersection patterns may be due to the differences in how these species represent time-varying input.

5.6.4 Effect of Input Types

The oriented Gaussians used in the above experiments allow obtaining maps where the feature interactions are most clear. The last step is to extend these results to

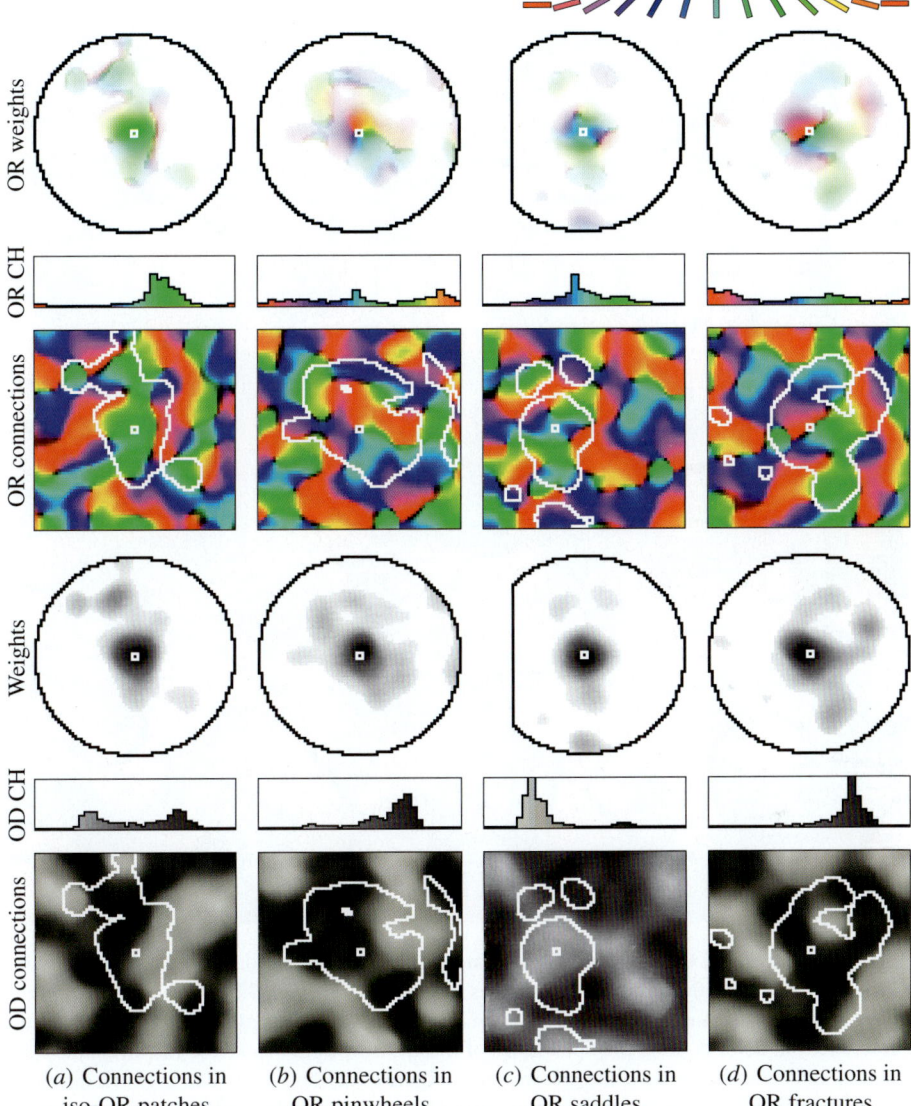

(*a*) Connections in
iso-OR patches

(*b*) Connections in
OR pinwheels

(*c*) Connections in
OR saddles

(*d*) Connections in
OR fractures

Fig. 5.28. Long-range lateral connections in the combined OR/OD map. The inhibitory lateral connections for four sample neurons from the regions marked in Figure 5.27 are shown situated on the OR preference and selectivity map (top) and the OD map (bottom) as in Figures 5.12 and 5.17. The connection patterns on the OR map are similar to those in the orientation-only map. The most selective iso-OR neurons (*a*) are located near the OD stripe boundaries, and receive connections equally from neurons of both eye preferences. Neurons in iso-OR regions away from the OD boundaries connect more strongly to one eye (not shown). OR fractures (*b*) and pinwheels (*d*) tend to occur near the centers of OD stripes and receive connections primarily from the same eye preference. Saddle points (*c*) can occur either in the middle or near the boundaries of OD stripes, and thus can have either monocular or binocular connection patterns; the example neuron is in a monocular OD region and connects primarily to the left eye. These connection patterns further extend the predictions from Figure 5.12, showing how the connection patterns are shared between the OD and OR maps.

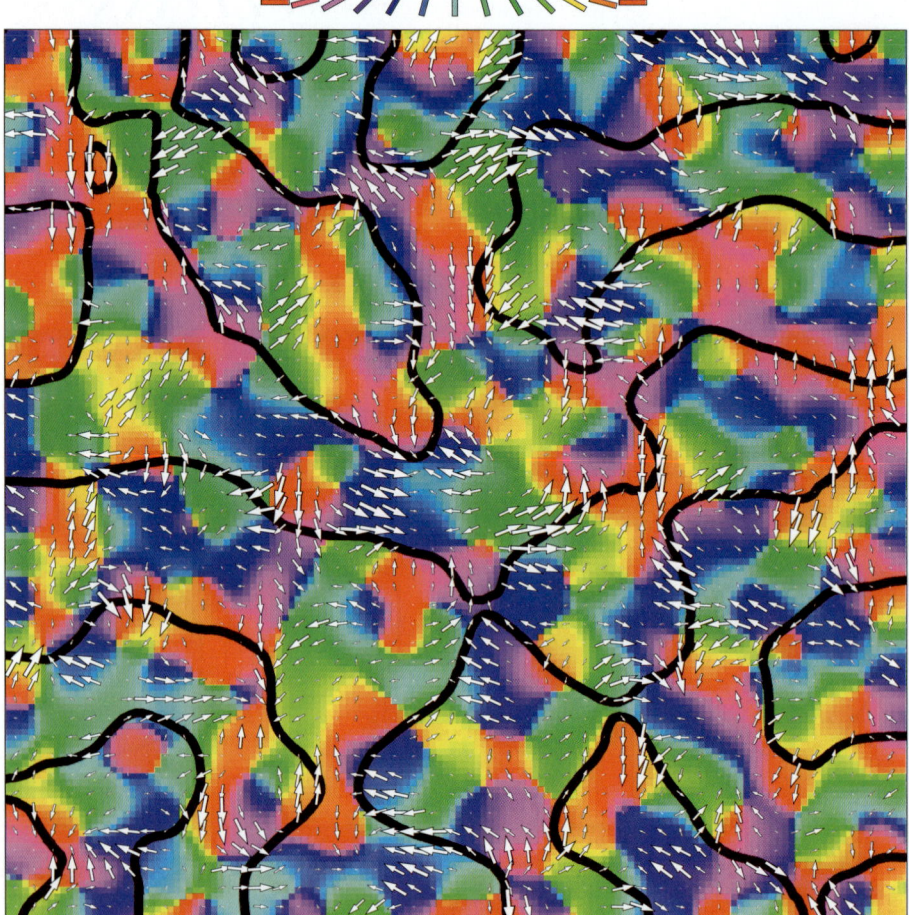

Fig. 5.29. Combined OR/OD/DR map trained with Gaussians. Based on oriented, moving Gaussian patterns with different brightnesses in each eye, LISSOM develops overlaid orientation, ocular dominance, and direction maps simultaneously. This plot shows the orientation preferences in color coding, the boundaries of the OD stripes in black, and the direction preferences and selectivities as white arrows, as in Figures 5.23 and 5.27. The network develops a realistic orientation and direction map, with OR patches subdivided into areas preferring the opposite directions of motion. Ocular dominance boundaries tend to cross linear zones at right angles, rather than following the orientation map. These results are similar to the ones with individual input dimensions, complicated by the fact that multiple dimensions are being mapped at once. Similar results have been observed experimentally with the cat visual cortex (Hubener et al. 1997; Löwel et al. 1988).

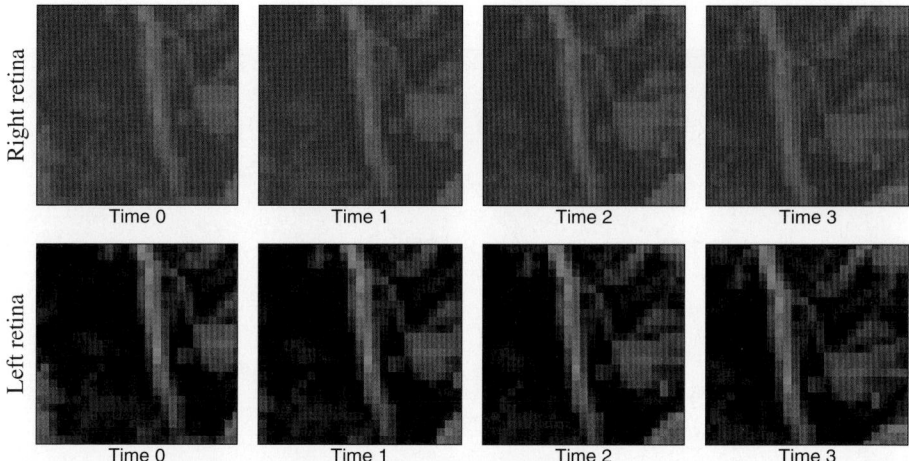

Fig. 5.30. Example natural image input for training the OR/OD/DR map. The top row shows a sequence of images presented to the right eye and the bottom row the corresponding images presented to the left eye. The two sequences are identical except for brightness; in this example, the right eye was randomly chosen to be darker than the left. The sequence represents a short movie where the image is moving to the left and slightly downward. The original image is from a dataset by Shouval et al. (1996, 1997).

other types of inputs, including noisy disks and natural images. These input types will be used in Part III as internal and external training patterns, in order to obtain maps most comparable to biology. Identifying how the complexity of natural internal and external input changes the observed map structures is an important step toward understanding how biological maps develop.

The simulations were run otherwise with the same parameters as the Gaussian OR/OD/DR experiment except for the training inputs, which consisted of the noisy disk and natural image patterns of Section 5.3.5 (see Appendix A.9 for details). A single noisy disk per iteration was used in the first experiment, located randomly on the retina and moving in a random direction. In the second, an input pattern was selected randomly from the dataset of images (by Shouval et al. 1996, 1997) and swept across each eye in a random direction as shown in Figure 5.30. In both simulations, the left and right eyes had identical inputs, differing only in brightness.

The maps developed with natural image inputs (Figure 5.31) were similar to the Gaussian input case (Figure 5.29), although more variable and less smoothly organized. Again, OD boundaries tend to cross OR linear zones at right angles, and a patch of neurons highly selective for a direction often has a nearby patch selective for the same orientation but the opposite direction. Neurons highly selective in one dimension receive connections primarily from neurons with similar preferences for that dimension, while neurons with low selectivity (e.g. binocular neurons) receive connections from neurons with a wide range of preferences.

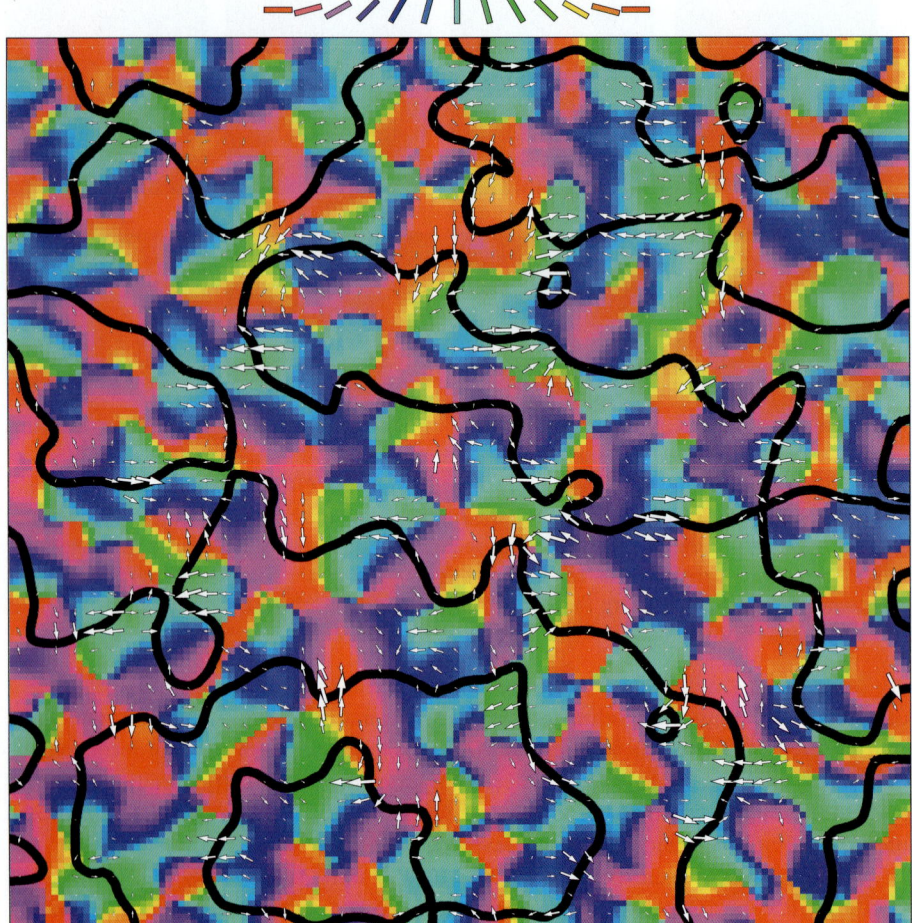

Fig. 5.31. Combined OR/OD/DR map trained with natural images. The combined LIS-SOM map from natural images is similar to the one in Figure 5.29. More of the units are selective for orientations near horizontal and vertical, so the map has more red and cyan than with artificial training inputs. Not as many neurons are highly selective for direction, but many of the selective patches are located next to other patches selective for the same orientation but opposite direction. Overall, these results are similar to those from artificial stimuli, with greater variability reflecting the more complicated feature correlations in natural images.

Natural image input affects the RFs and lateral connections more strongly than the maps (Figure 5.32). As in the single-feature simulations, the afferent RFs develop a variety of shapes, including both two-lobe and three-lobe RFs, in contrast to the uniformly three-lobed RFs of the Gaussian-trained map. Because natural images have correlations with longer range, the lateral connections are also wider and patchier, as they are in the single-feature natural image maps.

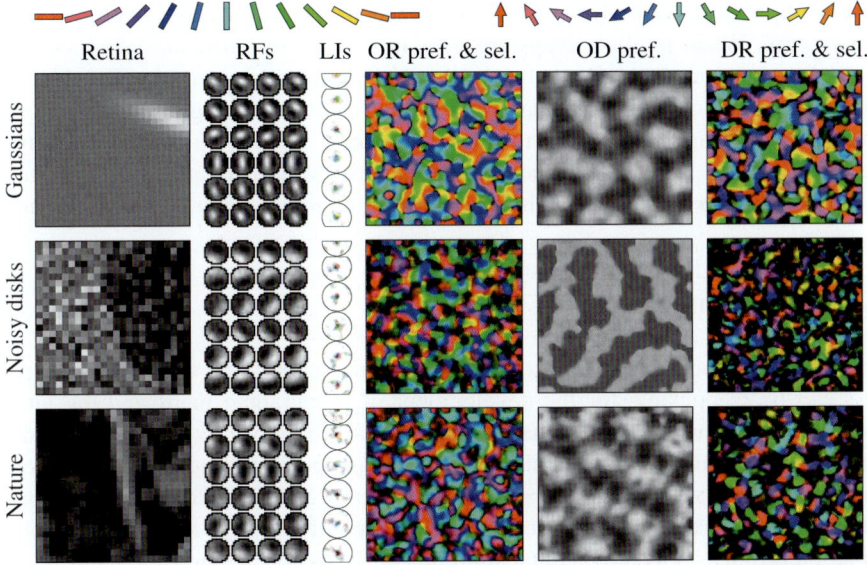

Fig. 5.32. Effect of training patterns on OR/OD/DR maps. From left to right, each row shows a sample retinal activation, final spatiotemporal receptive fields of sample neurons with lags 3, 2, 1, and 0 from left to right, their inhibitory lateral connections, the orientation preference and selectivity map, the ocular dominance map, and the direction preference and selectivity map. The top row shows a network trained with oriented Gaussians, the middle with noisy disks, and the bottom with natural images. All networks develop realistic orientation maps. Maps with Gaussian inputs and noisy disks develop only smooth, two- or three-lobed RFs, whereas networks with natural images develop a wide variety of RF types, corresponding to the wide range of patterns seen in natural images. In each case, the lateral connection patterns follow the features in the map. These results show that joint OR/OD/DR maps can develop from a variety of abstract and realistic input stimuli.

The simulations with noisy disk patterns show that joint maps of ocular dominance, orientation, and direction can also develop based on spontaneous activity patterns (Figure 5.32). Again, the RFs and lateral connections differed significantly from the Gaussians case, with primarily two-lobed RFs and more long-range lateral connections. The map was less selective for direction than with the other input types, primarily because the neurons became selective for moving curved edges whereas selectivity was measured using straight sine gratings. The ocular dominance map is more sharply delineated (as it is in strabismus) than with other inputs, primarily because the background noise was often stronger than the dim input in the other eye, reducing the overall correlation. Overall, these results suggest that if spontaneous activity patterns have enough motion and differ enough between the eyes, maps of all three preference types can develop.

Together, the results presented in this section show computationally for the first time how the OR, OD, and DR input features interact during development, and suggest a simple explanation for how the combined structures observed in animal maps can emerge. When multiple maps are overlaid, the global structure becomes more difficult to interpret, which is why the single-feature maps constitute a useful abstraction. The combined results are realistic and allow matching of anatomical differences with map organization in different species (as will be discussed in more detail in Section 17.2.2). The model also suggests which complexities are due to mapping multiple features and which are a result of complexities in natural inputs.

5.7 Discussion

The LISSOM model shows how a single local and unsupervised self-organizing process can be responsible for the development of topographic maps and the lateral connection structures in the primary visual cortex. The model is the first to account for the self-organization of lateral connections together with orientation maps, ocular dominance maps, and direction selectivity maps. Starting from random-strength connections, neurons develop oriented spatiotemporal receptive fields and lateral interaction profiles cooperatively and simultaneously. When the input varies in several dimensions at once, the cells and lateral connections become selective for multiple features, using the same architecture and learning rules that apply to individual feature dimensions. Such self-organization stores long-range activity correlations between feature-selective cells in the lateral connections. As will be demonstrated in more detail in Section 14.2, these correlations can be used to eliminate redundant information during visual processing, and to make cortical cells more selective.

LISSOM makes several testable predictions about specific lateral connection patterns in the cortex. The LISSOM orientation map predicts that the long-range connections at pinwheel centers, saddle points, and fractures have unselective, broad unimodal, and biaxial distributions, respectively. The direction map predicts that long-range lateral connections primarily link neurons with similar orientation and direction preference, extend along the orientation preference, and avoid orthogonal orientations and opposite directions. At DR fractures they connect with both directions and extend along their common orientation, in DR saddles they connect with the directions of the saddle, and in DR pinwheels they connect with all directions. The OD preferences are overlaid with the OR and DR preferences: The neurons most selective for one eye prefer connections from that eye, but such a preference is absolute only in strabismic animals. These predictions can be tested experimentally by combining optical imaging and injected tracers (like Bosking et al. 1997; Löwel 1994; Löwel and Singer 1992; Malach et al. 1993; Sincich and Blasdel 2001), as will be discussed in Section 16.4.2.

For clarity, the simulations shown in this chapter were each based on a single type of input pattern, such as Gaussians or natural images. The biological visual cortex, however, may be exposed to multiple sources of activity during development, including spontaneous, internally generated patterns and visual, externally evoked inputs. Biological development is thus likely to depend on a complex combination of such patterns. As will be shown in detail in Chapter 9, modeling both prenatal internal activity and postnatal visual images allows the model to account for both the primitive orientation maps seen at birth in animals and the more refined adult maps. Such more complex simulations are a natural extension of the single-input experiments presented in this chapter, providing a solid basis for understanding how V1 develops in animals.

LISSOM simulations serve to identify the types of input patterns that lead to normal and abnormal development of maps and connections, which can in turn help focus future biological experiments. For instance, realistic LISSOM ocular dominance maps develop from strength or brightness differences between patterns in the two eyes, but not from large position differences. This result leads to interesting predictions about the roles of internal and external input in shaping the OD maps. In newborn mammals, a rudimentary OD map exists at birth, presumably organized based on prenatal internally generated input such as retinal waves (Crair, Horton, Antonini, and Stryker 2001; Horton and Hocking 1996). If retinal waves are the only source, the LISSOM model predicts that the map would be strabismic, because the retinal waves are uncorrelated between the two eyes. This prediction is difficult to verify directly, but because normal adult maps are not strabismic, two interesting possibilities follow: (1) If the newborn maps are not strabismic, the LISSOM results suggest that the internally generated input that constructs the newborn map must be correlated between the two eyes. Since the retinal waves are not, there must be some additional source of internally generated patterns, like the PGO waves, that plays this role. (2) If newborn maps are indeed strabismic, visual experience must play a crucial role in shaping a normal adult map. In effect, the OD map is then constructed in two phases, prenatal and postnatal, like the orientation map. Possibilities for identifying such patterns and processes experimentally, and for modeling them computationally, are described in more detail in Section 17.2.3.

The LISSOM direction selectivity model is based on a simple but effective model of moving inputs, i.e. translation in a random direction within the image plane at each presentation. In the future it can be made more realistic by interleaving input frames with propagation and settling. Such a process is more complicated but should lead to similar results. More importantly, it is not yet clear what types of moving input drive the development of direction-selective neurons and direction maps in animals. Possible candidates include drifting retinal waves before birth, moving objects in the environment (relative to a stationary background), and optic flow due to eye or head movements. Each of these types of natural motion have different statistical properties, which is important to take into account in a more detailed model. As suggested above for OD and OR, multiple types of internal and external time-varying input are likely to contribute to the developmental process. Future models will depend on measuring and characterizing sources of these inputs in detail, as will be discussed in Section 17.2.3.

In addition to the lagged LGN cells modeled in LISSOM, different species may utilize other mechanisms for establishing direction selectivity and motion preferences (see Clifford and Ibbotson 2002 for a review). For instance, some connection pathways within V1 (e.g. between cortical layers in a column) may have delays long enough to serve as a memory of previous activity levels, much like the lagged afferent inputs do. Once they are characterized in sufficient detail, such mechanisms can be implemented in LISSOM, and they can potentially help account for a wider variety of motion preference results within the same basic framework (as will be discussed in Section 17.1.5).

In a similar manner, the LISSOM model can be instrumental in understanding differences between species more generally. The simulations in this chapter were necessarily based on pooled data from several different species because comparable experimental data do not yet exist across species. As such data are obtained, the model can be parameterized to account for a specific species in detail. As will be discussed in Section 17.2.2, such computational experiments should allow us to understand which differences are significant and what their origins are.

The LISSOM results in this chapter focused on orientation, ocular dominance, and direction maps, which are experimentally the best understood feature dimensions. Future models may also be trained with input that vary in spatial frequency, color, and disparity. Such simulations should result in preferences and maps for these additional features without requiring significant changes to the LISSOM model itself. Because little is known about these dimensions in biology, the simulations could be use as a guideline for further experimental studies. Such simulations will be discussed in more detail in Section 17.2.1.

5.8 Conclusion

The results in this chapter demonstrate that a single local and unsupervised self-organizing process can develop both the afferent and lateral connection structures in the primary visual cortex. This model is the most complete computational simulation of V1 to date. The model suggests that the afferent connections represent visual features with the highest variance, such as topography, orientation, ocularity, and direction. The lateral connections represent correlations between such features, and implement an efficient coding of visual information.

The same self-organizing process may continue to operate in the adult, serving a different role: It may be responsible for plasticity and adaptation after damage, as will be described in the next chapter.

6

Understanding Plasticity

So far we have seen how LISSOM organizes visual information in the afferent receptive fields and lateral connections and develops a representation of visual features and their correlations. The resulting structures are similar to those seen in the visual cortex, and the model can be used to investigate phenomena in the adult cortex that are difficult to understand experimentally. One of the most intriguing such phenomena is that the maps in the adult sensory cortex are highly plastic, i.e. adaptable. How can the cortex stay plastic in the adult, yet maintain the structures necessary for visual processing? This chapter will demonstrate how the same self-organizing process that builds the cortical organization during early development can also maintain it in a state of continuously adapting dynamic equilibrium in the adult.

Two types of cortical plasticity have been observed in biology: reorganization following lesions in the receptor surface, and reorganization after lesions in the cortex. These experiments will be modeled with LISSOM, showing how the network compensates for lesions in the retina, why the reorganization behavior after retinal lesions produces dynamic changes in receptive fields, how the network reorganizes after cortical lesions, and how the reorganization mechanisms suggest techniques to accelerate recovery following cortical surgery or stroke. The results suggest that the same self-organizing mechanism that allows the cortex to learn and represent input information also makes it robust against damage.

6.1 Biological and Computational Background

Reorganization after retinal and cortical lesions is well documented in biological experiments, as will be reviewed in this section. A number of modeling studies also suggest how these processes might take place computationally.

6.1.1 Reorganization After Retinal Lesions

Early studies of cortical development had suggested that most of the cortical structures develop only during a critical period just after birth and are hardwired afterward

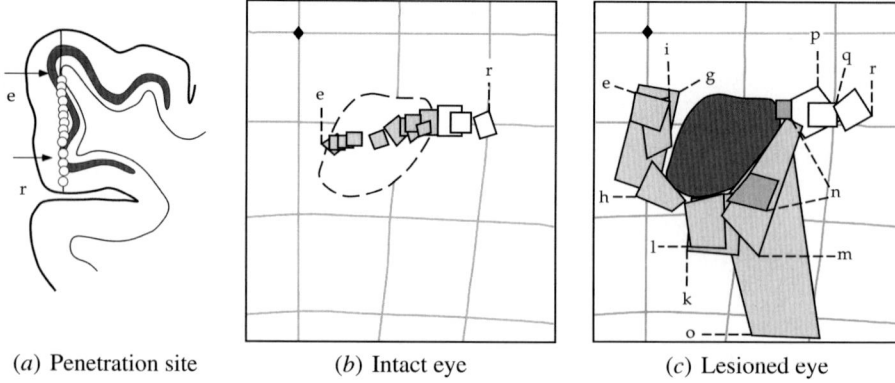

(a) Penetration site   (b) Intact eye   (c) Lesioned eye

Fig. 6.1. Reorganization of receptive fields after a retinal lesion. The receptive field distributions in the intact and the lesioned eye of a single adult cat are plotted in visual space 4–8 hours after a retinal lesion in one eye. (a) The electrode penetration site (vertical line) is shown along with the locations (open circles) of neurons whose receptive fields are plotted in (b) and (c). The arrows on the left mark the boundary of the cortical region that represents the retinal lesion. The neurons are labeled "e" to "r" from top to bottom and span a distance of about 5 mm. (b) The receptive fields in the visual space of the intact eye are shown as rectangles labeled "e" to "r" according to the neuron's position along the penetration site. The grid lines are spaced 10° apart and the black diamond marks the area centralis. The dashed contour in the middle marks the area that corresponds to the scotoma in the lesioned eye; the receptive fields within this region are colored gray. The receptive fields show an orderly progression from left to right. In another penetration before the lesion (not shown), a distribution similar to that in (b) was observed in both eyes; therefore, the distribution in the intact eye suggests how the receptive fields in the lesioned eye were located before the lesion. (c) In the lesioned eye, the scotoma is shown as a dark gray area. The receptive fields that used to respond to the area inside the scotoma (the gray rectangles labeled "e" to "o") have moved outward, and now represent the perilesion area. Several of them (e.g. "g", "h", and "m") have also aligned with the scotoma boundary. As a result, the unresponsive area in the cortex (a perceptual blind spot) has disappeared, even though damage persists in the retina. Reprinted with permission from Calford et al. (1999), copyright 1999 by the Royal Society of London.

(Albus and Wolf 1984; Bonds 1979; Braastad and Heggelund 1985; Hubel et al. 1977; Wiesel 1982). The unspoken assumption was that all the information-carrying pathways are firmly and immutably formed during this early period and the plasticity necessary for learning and compensating for damage exists only in higher cortical areas. However, subsequent results showed that the primary cortical areas are capable of substantial change even in the adult brain (see Buonomano and Merzenich 1998; Chklovskii, Mel, and Svoboda 2004; Kaas 1991 for reviews). For example, after a small artificial blind spot or a physical lesion (i.e. a scotoma) is induced in the retina, an unresponsive area appears in the visual cortex. Over time, this area gradually recovers function and starts to respond to stimuli outside the scotoma (Figure 6.1; Calford, Wang, Taglianetti, Waleszczyk, Burke, and Dreher 2000; Chino, Kaas, Smith, Langston, and Cheng 1992; Chino, Smith, Kaas, and Cheng 1995; Darian-Smith and

Gilbert 1995; Waleszczyk, Wang, Young, Burke, Calford, and Dreher 2003). Consequently, the blind spot becomes invisible, even though the damage persists in the retina. The receptive fields on each eye organize largely independently, i.e. monocular and binocular lesions lead to similar results (Section 6.5).

The cortical changes following a retinal lesion can be quite rapid, as experiments by Pettet and Gilbert (1992) showed. They simulated a retinal scotoma by artificially masking a portion of the visual field and stimulating only the surround of the mask. The mask prevented a part of the cortex from receiving visual stimulation, producing a corresponding cortical scotoma. After the surround had been stimulated for a period of several minutes, the retinal scotoma was removed, and the receptive fields of the unstimulated neurons were found to have enlarged. Psychophysical experiments further showed that after removing the scotoma, a stimulus at the edge of the scotoma appeared to be shifted toward the center (Kapadia et al. 1994). Similar results were later observed with physical lesions to the retina, including detaching it partially (Schmid, Rosa, and Calford 1995) and removing photoreceptors with a focal laser (Calford et al. 1999). In both cases, the receptive fields were observed to expand rapidly.

Prima facie, such a dynamic expansion of receptive fields of the unstimulated neurons and the perceptual shift accompanying it appear incompatible with Hebbian self-organization, which suggests that only active neurons should change substantially. It has therefore been difficult to explain these observations. However, recent experimental results indicate that corticocortical lateral interactions may play a critical role in such postlesion reorganization. For example, if the activities of neurons outside the cortical scotoma are artificially suppressed, the reorganized neurons become silent as well (Calford, Wright, Metha, and Taglianetti 2003), suggesting that lateral activation may be necessary for reorganization. In this chapter, the LISSOM model will be used to provide a concrete computational explanation for this phenomenon, based on lateral interactions and the reorganization of afferent receptive fields.

6.1.2 Reorganization After Cortical Lesions

Lesions to the visual cortex may result, for example, from a stroke or head injury: Neurons in a small area of the cortex become unresponsive to input. Similar to retinal lesions, the cortex gradually reorganizes and recovers some of the function that was initially lost.

Recovery following cortical lesions involves three phases (Merzenich, Recanzone, Jenkins, and Grajski 1990):

1. Immediately after the lesion, there is an immediate partial compensation for the cortical damage, as the receptive fields of neurons surrounding the lesion become larger (Wurtz, Yamasaki, Duffy, and Roy 1990) and expand inward (Sober, Stark, Yamasaki, and Lytton 1997). As a consequence, less function is lost than expected based on the prelesion map (Figure 6.2).
2. Over the next several days, the reorganization includes mixed effects: Some receptive fields shift outward (Merzenich et al. 1990), in addition to those that

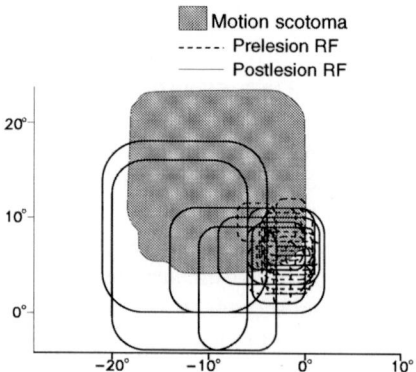

Fig. 6.2. Reorganization of receptive fields after a cortical lesion. The receptive fields measured through microelectrode recording before and after lesion in the middle temporal (MT) area of a macaque monkey are shown in visual space; the axes indicate angles in the visual field. The retinal area represented by the lesioned cortex is shown in gray (labeled "Motion scotoma"). Before the lesion, receptive fields of neurons immediately outside the cortical lesion boundary (dashed rounded rectangles) responded to the lower right corner of the gray area. However, 3 hours to 3 days after the lesion, the receptive fields (solid rounded rectangles) became larger and covered a larger portion of the scotoma, suggesting that neurons outside the lesion boundary were compensating for the loss of function. Reprinted with permission from Sober et al. (1997), copyright 1997 by the American Physiological Society.

expand inward. Also, in some cases the receptive fields became smaller (Eysel and Schweigart 1999; Sober et al. 1997) in addition to those that enlarged. Some of the immediate compensation is lost, because some perilesion neurons do not respond as strongly as before, and in some cases more function is lost than expected based on the prelesion map (Merzenich et al. 1990). A similar pattern of transient performance improvement, followed by a decline, is also seen in motor cortex (Nudo, Wise, Fuentes, and Milliken 1996), suggesting that it is a general trend in cortical recovery.

3. Over subsequent weeks, the receptive fields expand inward again, as the perilesion neurons begin to respond to part of the receptor surface that used to be represented within the lesion (Eysel and Schweigart 1999; Wurtz et al. 1990; Zepeda, Sengpiel, Guagnelli, Vaca, and Arias 2004). Gradually, the cortex compensates for the lesion and regains much of the lost function.

The mechanisms underlying such reorganization, especially during the regressive phase in the days following the lesion, are not well understood (Sober et al. 1997; Zepeda et al. 2004). Apparently, reorganization following cortical lesions is based on different mechanisms than that following retinal lesions. With retinal lesions, the distribution of input to the cortex changes, and the recovery consists of reorganizing the afferents. With cortical lesions, the input distribution remains unchanged but the local network structure is altered, and this process is likely to depend crucially on lateral connections. Using the LISSOM model, this hypothesis will be tested computationally in Section 6.4.

6.1.3 Computational Models

During the last few years, there have been several efforts to develop computational models of cortical map reorganization (Goodall, Reggia, Chen, Ruppin, and Whitney 1997; Grajski and Merzenich 1990; Kohonen 1989; Obermayer, Ritter, and Schulten 1990a,c; Pearson, Finkel, and Edelman 1987; Sober et al. 1997; Sutton, Reggia, Armentrout, and D'Autrechy 1994). Most of these were designed either to model isolated plasticity phenomena or to demonstrate general principles of plasticity, and did not aim at a computational theory of how the observed cortical structures organize and reorganize. For example, Sober et al. (1997) showed how changing the parameters in an otherwise non-adaptive model could induce a variety of changes with retinal lesions, including both receptive field expansion and contraction.

A more general theory is provided by models based on self-organizing maps (Andrade, Muro, and Morán 2001; Goodall et al. 1997; Obermayer et al. 1990a,c; Sirosh and Miikkulainen 1994b, 1996a). For example, Obermayer et al. (1990a,c) used a SOM-based model to study how the somatosensory cortex reorganizes following the amputation of a finger. In animal experiments, such a lesion is analogous to reorganization following retinal lesions: Soon after the amputation, the unresponsive cortical region gradually regains function and responds to the skin of the remaining digits (Merzenich, Nelson, Stryker, Cynader, Schoppmann, and Zook 1984). The skin surface was modeled with a set of receptors scattered in the shape of a hand, and the somatosensory cortex with a SOM network. Circular Gaussian spots were presented on the receptor surface as input stimuli, and the SOM self-organized into a topographic map of the hand. Subsequently, the receptors of one digit were deactivated, and the self-organization continued. Because the input distribution had changed, the map reorganized based on the stimuli on the remaining fingers. The receptive fields of the neurons in the inactive region gradually moved outward from the lesioned finger, and began to respond best to other stimuli. In the end, all neurons that originally had responded to the lesioned finger represented the surviving fingers.

The SOM-based model illustrates how a topographic map can reorganize, but it does not account for phenomena such as the dynamic expansion of receptive fields seen after peripheral lesions. Moreover, if a cortical lesion were to be simulated in a SOM model, it would reorganize monotonically and recover function completely, unlike the cortex. In addition, Goodall et al. (1997) demonstrated that for proper recovery to take place, the area surrounding the cortical lesion needs to activate more easily than before the lesion. The reason for each of these observations is that the models are driven by afferent adaptation only; as shown by Andrade et al. (2001) and Sirosh and Miikkulainen (1994b, 1996a), specifically adapting lateral connections can overcome these problems in principle, although their models did not include fully detailed map structures.

Building on these foundations, the remaining sections will demonstrate how a large-scale orientation map model with explicit, adapting lateral interactions can pro-

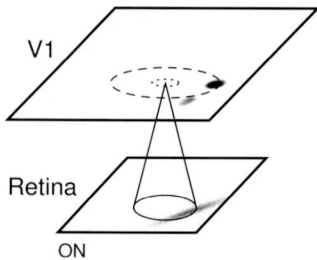

Fig. 6.3. Architecture of the reduced LISSOM model. As long as Gaussian patterns are used as input, it is not necessary to include separate ON and OFF channels in the model. Instead, the input can be directly presented as ON channel activations. For simplicity and compatibility with other similar models, this channel is called the Retina, and its neurons are referred to as photoreceptors. The afferent connections form a local anatomical receptive field directly on the redefined retina; the lateral connections are similar to those in other LISSOM models. The reduced model is more efficient to simulate computationally than the full LISSOM, with equivalent results (as shown in Figures 6.4 and 6.5).

vide an accurate and general account for the observed reorganizing behavior with both retinal and cortical lesions.

6.2 The Reduced LISSOM Model

Any of the trained LISSOM networks discussed in Chapters 4 and 5 can be used to study plasticity. However, the fine details of those maps are not necessary, and a simplified, computationally more efficient model can be used as well. Since the focus is on reorganization of the map only, the model can be trained with artificial inputs, and the ON/OFF channels can be bypassed (Figures 6.3 and 6.4). Such a reduced model will first be described in detail below, and demonstrated to develop an OR map equivalent to the LISSOM model of Section 5.3. The role of ON/OFF channels is then analyzed experimentally, and shown to be important for natural images, but unnecessary for the artificial inputs used in this chapter.

6.2.1 Method of Self-Organization

As was mentioned in Section 5.3.5, the ON and OFF channels in LISSOM allow forming similar responses despite differences in background illumination in the input. This property is important with natural images, but not necessary when artificial patterns such as Gaussians are used as input. In such cases, the model can be made computationally more efficient by including only the ON channel. Further, this channel can be combined with the retina into a single sheet of neurons, activated by the input image like the ON sheet in the LGN. For simplicity and compatibility with other similar models (Section 3.4.1), this sheet will be called the retina, and its

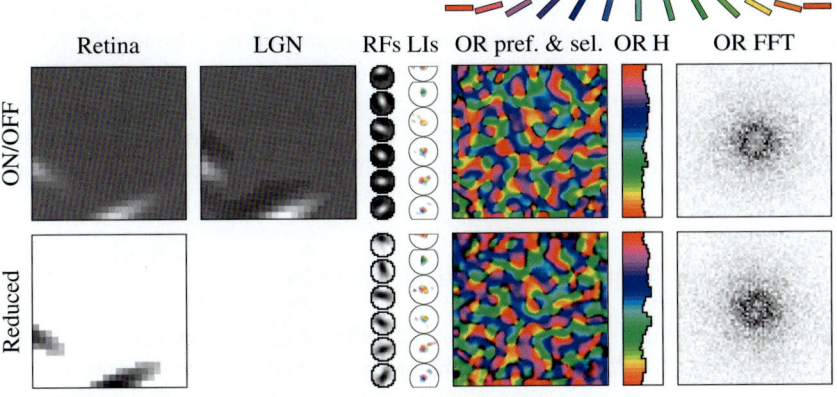

Fig. 6.4. Effect of ON/OFF channels on orientation maps. The two rows show the results for two LISSOM networks trained with the same stream of Gaussian inputs. The top network is the LISSOM OR map from Section 5.3, and the bottom network is the reduced LISSOM model of Figure 6.3. As in Figure 5.13, each row includes a sample retinal activation, the LGN response (for the ON/OFF LISSOM network), the final receptive fields of sample neurons, their inhibitory lateral connections, the orientation preference and selectivity map, and the histogram and the Fourier transform of the orientation preferences. For the ON/OFF model, the inputs consisted of photograph-like images of Gaussians such as those used in Chapters 4 and 5, shown in gray scale from black to white (low to high), with medium gray representing background activation. In contrast, the reduced LISSOM inputs were similar to the activations in the ON channel, i.e. gray scale from white to black (low to high), with white background. The reduced LISSOM RFs are shown in gray scale like ON weights from white to black (low to high), whereas the ON/OFF LISSOM RFs are combined by subtracting the OFF weights from the ON, as e.g. in Figure 4.6. The RF orientations, lateral connections, and map organization are almost identical in the two models. The RFs on the ON/OFF channels have multiple ON and OFF lobes, and become slightly more oriented. As a result, the ON/OFF map is somewhat more selective. The histogram of each orientation map is nearly flat for both networks, because the inputs were uniformly distributed. These results show that as long as the maps are trained with the same stream of Gaussian inputs, functionally similar maps develop with or without the LGN. However, Figure 6.5 will show that the ON/OFF channels of the LGN are necessary for processing natural images.

neurons will be referred to as photoreceptors. The resulting reduced LISSOM architecture is otherwise similar to that of Section 5.3, except the cortical neurons receive input directly from such a redefined retina (Figure 6.3).

In the reduced LISSOM simulations, the cortical network consisted of an array of 142×142 neurons, and a retina of 36×36 receptors. The neurons in the cortical sheet received afferent connections from broad overlapping circular patches on the retina. The center of the anatomical receptive field of each cortical neuron was placed at the location in the central 24×24 portion of the retina corresponding to the location of the neuron in the cortex, so that every neuron had a complete set of afferent connections (Figure A.1). The connection strengths were initially random in a circular area within six units from the RF center. The lateral weights were initially set to a smooth Gaussian profile. The rest of the parameters are described in Appendix B.

The network was organized in 10,000 input presentations of two randomly located and oriented Gaussians. The input sequence was exactly the same as for the LISSOM OR map of Section 5.3. As will be demonstrated in more detail in Section 8.4, such identical training makes it possible to compare the two architectures in detail.

6.2.2 Orientation Maps with and without the LGN

In the self-organizing process, a well-formed orientation map emerged, with the typical oriented receptive fields, topographic order, and patchy lateral connections (Figure 6.4). In fact, this map is almost identical to that in Section 5.3, with the same iso-orientation patches and other map features roughly in the same locations. The similarities extend to the individual neuron level as well: The RF orientations and lateral connection patterns are usually very similar between corresponding neurons in the two maps.

When the inputs consist of oriented Gaussians, the V1 activity patterns are the same with or without ON and OFF channels (Figure 6.5). Since the self-organizing process is driven by these activity patterns, the same map results in both networks. In other words, with such inputs, the reduced model is functionally equivalent to LISSOM with ON/OFF channels.

This equivalency explains why models with and without ON and OFF cells have both been able to develop realistic orientation maps. It also suggests that if the computational experiment focuses on the organization of the map and its lateral connections, and is based on artificial inputs, the simulations can be made more efficient by bypassing the LGN. This simplification will be utilized in the remainder of Part II, as well as in Parts IV and V of the book.

6.2.3 The Role of ON/OFF Channels

It is also important to point out how the LISSOM models with and without the ON/OFF channels differ. Most obviously, although they have the same orientation, the RF shapes are drastically different. These shapes are determined by both the V1 and LGN activities, and the LGN activities in the OFF channel differ greatly from those in the ON channel.

The difference is not important as long as the inputs consist of oriented Gaussians: The maps still respond to the same input with a similar activation pattern. However, the LGN plays a crucial role in suppressing spurious activation with other types of inputs, such as natural images and retinal wave patterns, which cover substantial parts of the retina. In such large patterns, there are often large active areas, large gradual changes in brightness, and nonzero mean levels of illumination. The ON and OFF cells suppress spurious activation in such cases, and allow the map to respond based on orientation (Figure 6.5).

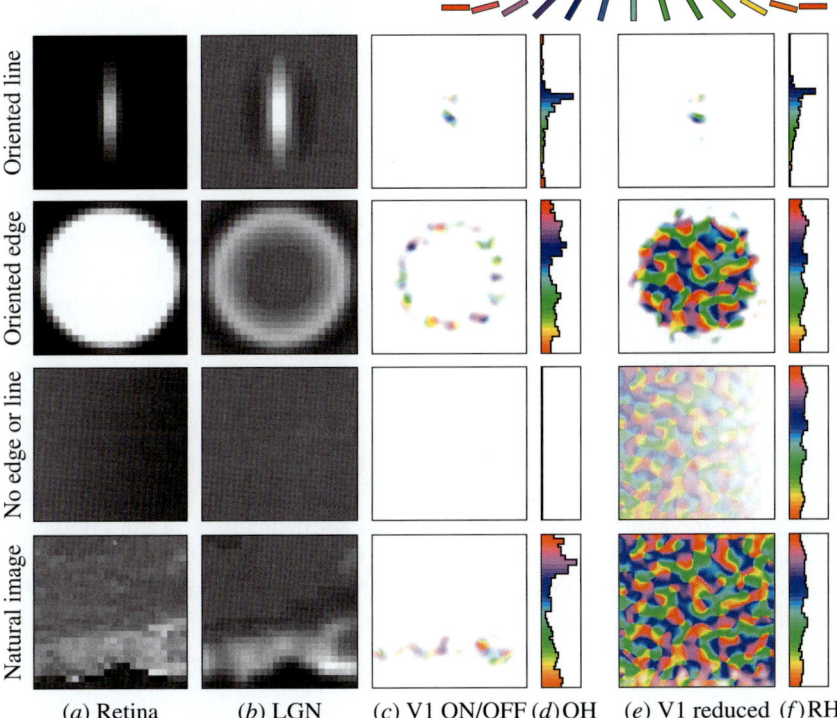

(a) Retina   (b) LGN   (c) V1 ON/OFF (d) OH   (e) V1 reduced (f) RH

Fig. 6.5. Role of ON/OFF channels in processing various kinds of inputs. Each row shows a sample retinal activation, the LGN response, the V1 response and its histogram (OH) for the ON/OFF LISSOM network, and the V1 response and its histogram (RH) for the reduced LISSOM network. The sample inputs are plotted in gray scale from black to white (low to high) and the LGN activations by subtracting the OFF cell responses from the ON. In the V1 plots (c,e), orientation preferences of those neurons that respond are color coded according to the key on top, and color saturation represents the activation level (selectivity is not shown to match the perceived orientation measure; Section 7.2.1). The two networks respond similarly to an oriented Gaussian input on a blank background (top row), which is why very similar orientation maps developed in Figure 6.4. As seen in the histograms, only neurons with orientation preferences matching the input line respond. However, the networks behave very differently for other types of input. The ON and OFF channels filter out nonzero background levels and smooth, gradual changes in brightness, which ensures that V1 ON/OFF responds only to oriented patterns and sharp edges (second row; the response is strongest on the bright side of the edge because only bright Gaussians were used in training, as shown in the top row of Figure 5.13). In contrast, overall background illumination with no edges is ignored by the ON/OFF network (third row), whereas it activates nearly *all* of the V1 neurons in the reduced model. Without the LGN, the response to most patterns is determined by the total amount of brightness in the input, rather than by the orientation preference of the V1 neurons. Nonzero background levels, gradual changes in illumination, and large, bright objects are all common in natural images (bottom row), and thus the ON and OFF channels are crucial for preserving orientation selectivity when processing such images. On the other hand, the ON and OFF channels can be omitted for networks that process only schematic patterns on a blank background. The natural image is a retina-size detail (as shown in Figure 8.4(e)) from National Park Service (1995).

Natural scenes and retinal waves contain many such features, and thus including ON and OFF cells in LISSOM is crucial for experiments that utilize such inputs, like those in Chapter 5 and in Part III.

6.2.4 Methods for Modeling Plasticity

The plasticity simulations were performed using the reduced LISSOM orientation map network described above. After self-organization has reached the settled state shown in Figure 6.4, both the lateral and afferent connections are in a dynamic equilibrium with the input distribution. They adapt each time an input is presented, but the overall organization does not change significantly, as long as the architecture stays intact and the input distribution does not change. This model forms the foundation for studying plasticity.

To study the effect of retinal lesions, the dynamic equilibrium is disrupted by introducing an artificial scotoma: The activity in a square region of the photoreceptor array is set to zero for all subsequent inputs (Figure 6.6). A cortical lesion can be introduced in the same way: The outputs of a set of neurons in the middle of the cortical network are set permanently to zero (Figure 6.9). In both cases, the lesion disrupts the dynamic equilibrium and forces the network to adapt, as described in detail in the next two sections.

6.3 Retinal Lesions

As a result of the retinal scotoma, neurons in the center of the corresponding cortical area no longer receive input (Figure 6.6). Their response is lost, the dynamic equilibrium is disturbed, and the self-organizing process adapts the map and the receptive fields accordingly.

6.3.1 Reorganization of the Map

As shown in Figure 6.7, the reorganization in LISSOM proceeds in the same manner as observed in the biological cortex (Section 6.1.1; Chino et al. 1992). Initially, the afferent RFs are laid out across the retina relatively uniformly, with local distortions due to OR patches (as described in Section 5.3.3). After the lesion, the afferent RFs of the central, unstimulated neurons remain in the same location as before, but those of the surrounding neurons move outward. These neurons receive input from the receptors surrounding the scotoma, but no stimulation from inside it. Through Hebbian adaptation, the connections from the outside become stronger and those from the inside weaker, resulting in the observed shift outward.

Gradually, almost all neurons that receive afferent input shift their afferent weights outside the scotoma. Most of the initially unresponsive neurons of the network now respond to the periphery of the scotoma. How complete this process is depends on the size of the scotoma. If the lesion is large enough, a set of central

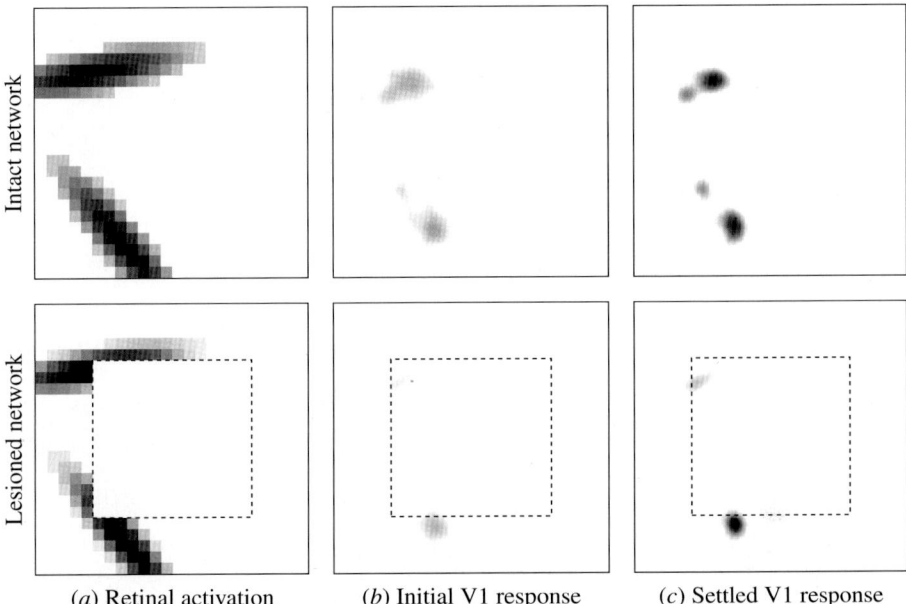

Intact network

Lesioned network

(*a*) Retinal activation   (*b*) Initial V1 response   (*c*) Settled V1 response

Fig. 6.6. Retinal activation and V1 response before and after a retinal scotoma. The initial and settled responses of the intact network (top row) and the lesioned network (bottom row) to input in (*a*) are shown in (*b*) and (*c*). The activations are displayed in gray scale from white to black (low to high; the orientation preferences of active V1 neurons are not shown). The retinal lesion is simulated by setting the activity of a set of receptors to zero, as shown in the bottom row of (*a*). The dotted line in (*a*) marks the lesioned area on the retina (the retinal scotoma), and in (*b*) and (*c*) marks the corresponding portion of V1 (the cortical scotoma). The cortical scotoma is approximately as wide as the lateral connections, matching artificial scotomas in biological experiments. Many of the neurons that responded to the intact input do not receive sufficient activation in the lesioned network and remain silent (because the topography of the retinal preferences is not uniform around the edges, some neurons inside the cortical scotoma still respond). Such changes in activity disrupt the dynamic equilibrium, forcing the network to reorganize.

neurons are never stimulated again; they retain their old receptive fields (as shown in Figure 6.7) and appear as a dark area in the visual field. With smaller lesions, the combined effect of the afferent input and lateral excitation is enough to cause them to reorganize as well, eventually making the blind spot in the retina invisible. The LISSOM model therefore suggests a mechanism for the reorganization in response to the retinal scotoma, and allows predicting its extent.

Corresponding changes can be seen in the orientation map (Figure 6.7). At the center of the scotoma the map remains unchanged, but near the edge, where the neurons' receptive fields have shifted, significant reorganization can be observed. Many neurons near the boundary of the scotoma become selective for the orientation of the boundary. In response, the neurons farther away from the boundary adapt so

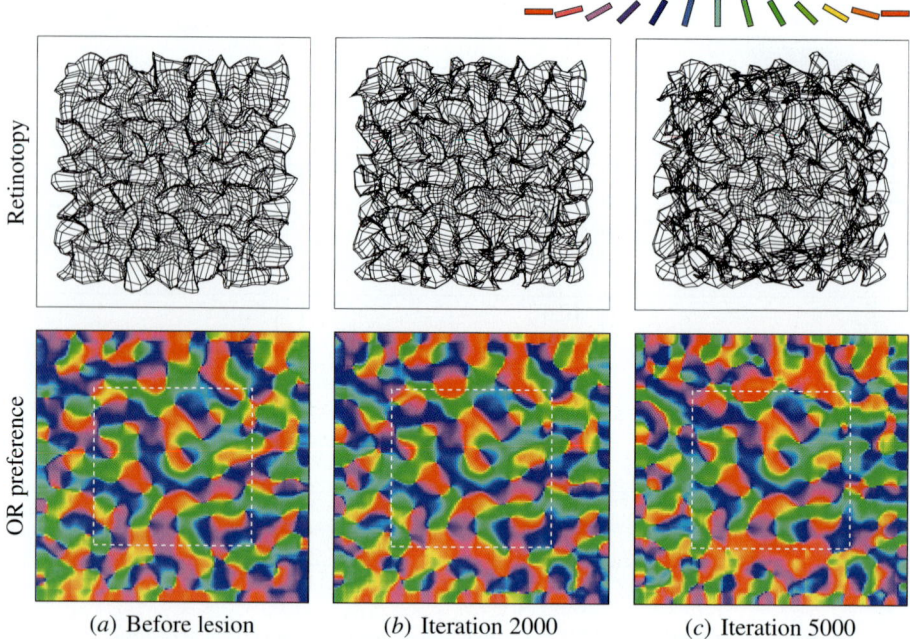

(*a*) Before lesion   (*b*) Iteration 2000   (*c*) Iteration 5000

Fig. 6.7. Reorganization of the orientation map after a retinal scotoma. In the top row, the RF centers of every third neuron in the network are plotted as a grid in the retinal space; the bottom row displays the corresponding map of orientation preferences (selectivity is not shown). The RF centers in the grids are calculated from the settled response (instead of the afferent weights as e.g. in Figure 5.11; Appendices G.2 and G.3), because the lesioned map is not in equilibrium with the input. The dotted white line shows the cortical scotoma, i.e., the region of V1 corresponding to the lesioned area of the retina. (*a*) Before the scotoma, the RF centers are organized into a retinotopic map with orientation-based distortions, as in Figure 5.11. (*b*) Shortly after the scotoma, neurons whose RFs were entirely covered by the scotoma retain their old RFs, but the surrounding neurons start to reorganize their afferent weights into the periphery of the scotoma. (*c*) Five thousand iterations after the scotoma, most of the receptive fields have moved out into the periphery of the lesion (with corresponding inward changes in perception as demonstrated in Figure 6.8); how many remain in the center depends on how large the scotoma is compared with the RFs and the lateral connections. The orientation map is unchanged within the central region of the scotoma, but along the cortical scotoma boundary (in white) many neurons have become selective for the orientation of the boundary, and the rest of the map has adapted to these changes. The reorganization of the retinotopic map provides a detailed computational account for the outward shift in the RF center found by Chino et al. (1992; Section 6.1.1), while the changes in the orientation map constitute predictions for future experiments. An animated demo of the reorganization process can be seen at http://computationalmaps.org.

that the orientation map remains smooth. Such reorganization of the orientation map has not been studied in biology, although the receptive fields have been observed to align with the lesion boundary (Figure 6.1). The results from the LISSOM model therefore constitute predictions for future biological experiments.

6.3.2 Dynamic Receptive Fields

As observed in biology (Section 6.1.1; Pettet and Gilbert 1992), the reorganization in the LISSOM map also causes rapid changes in the receptive field size of the central, *unstimulated* neurons. As the neurons surrounding the cortical scotoma reorganize their RFs to the periphery, they become insensitive to the center of the retinal scotoma. If the scotoma is now removed, and an input is presented in the scotoma region, only the previously unstimulated neurons (which did not reorganize) respond vigorously to the new input; the surrounding ones do not (Figure 6.8c). Therefore, there is considerably less lateral inhibition from the surrounding neurons to the central neurons. Whereas the central neurons previously responded only weakly to stimuli at the periphery of the scotoma, such responses are now unmasked. Consequently, their RFs have become larger. The expansion is greatest along the preferred orientation because the strongest afferent weights lie in this direction (Figure 5.12a), and any decrease of inhibition unmasks responses mainly in that direction.

This explanation for dynamic receptive fields could be verified in a simple biological experiment. If inhibition to the unresponsive region of the cortex were to be suppressed (by selectively blocking inhibitory neurotransmission using a chemical such as bicuculline), the influence of the surround would be removed from its response. The receptive fields would then have the same size before and after the scotoma. On the contrary, if lateral inhibition is not responsible for the expansion, the dynamic changes of receptive field size would still occur.

The reorganization in LISSOM can account for the psychophysical result of inward shift (Section 6.1.1; Kapadia et al. 1994) as well. The neurons whose receptive fields have moved outward now respond to inputs farther from the center than before. Therefore, an input at the edge of the retinal scotoma stimulates many neurons inside the cortical scotoma that previously would not have responded, and the response pattern is shifted inward (Figure 6.8c). After the scotoma is removed and the normal stimulation reestablished, the reorganized RFs gradually return to the normal state, and the shift disappears.

The LISSOM model thus shows how the same self-organizing processes and lateral interactions that shape the receptive fields during early development can maintain them in a continuously adapting, dynamic equilibrium with the visual environment in the adult. Damage to such a system then results in map reorganization and dynamic receptive fields, giving a detailed computational account for these biological phenomena.

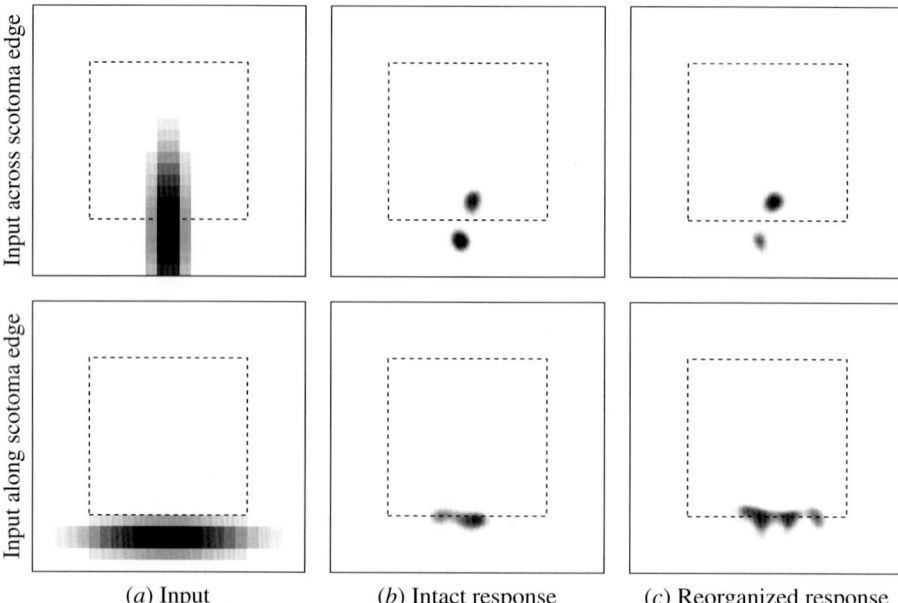

(*a*) Input   (*b*) Intact response   (*c*) Reorganized response

Fig. 6.8. Dynamic RF expansion and perceptual shift after a retinal scotoma. In the top row, the response of the network to a single vertical input across the bottom edge of the retinal scotoma (*a*) is shown before the lesion (*b*) and after the cortex reorganized and the scotoma was removed (*c*). The lower activity patch, due to neurons just outside the cortical scotoma, has almost disappeared in the reorganized response, because these neurons now prefer horizontal inputs (as seen in the OR map of Figure 6.7*c*). As a result, these neurons do not inhibit the neurons inside the scotoma as strongly as before, and the inside neurons now have larger effective RFs, as indicated by the slightly larger and more intense top activity patch. The inward perceptual shift is most clearly seen when the input is just outside the retinal scotoma and parallel to its boundary, like the horizontal input below the scotoma in the bottom row. The reorganized response is much larger than the initial response because most neurons near the bottom boundary now prefer horizontal inputs. In addition, the RFs of these neurons have shifted outward (as seen in the retinotopy plot of Figure 6.7*c*), which results in a corresponding small shift of the response pattern inward. These results replicate the dynamic RF size expansion and the corresponding inward shift in the perceived location found in biological experiments (Section 6.1.1; Kapadia et al. 1994; Pettet and Gilbert 1992); the magnification of boundary orientations is a prediction of the model.

6.4 Cortical Lesions

The reduced LISSOM network of Section 6.2 was used as the starting point for the cortical plasticity experiments as well. A cortical lesion was induced in the final self-organized network (Figure 6.9), and the self-organizing process adapted accordingly.

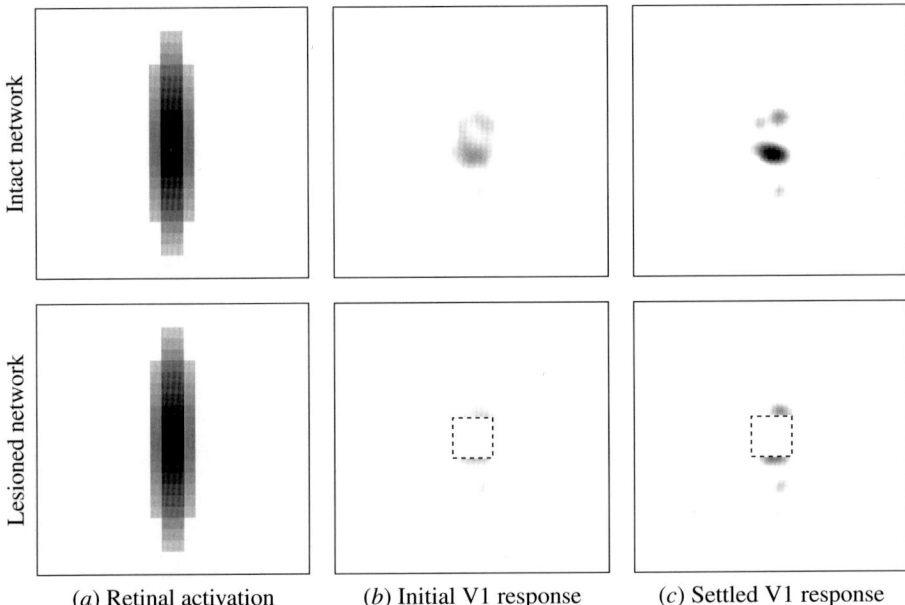

Intact network

Lesioned network

(*a*) Retinal activation   (*b*) Initial V1 response   (*c*) Settled V1 response

Fig. 6.9. Retinal activation and V1 response before and after a cortical lesion. The initial and settled responses of the intact network (top row) and the lesioned network (bottom row) to the retinal activation in (*a*) are shown in (*b*) and (*c*), as in Figure 6.6. The cortical lesion is simulated by keeping the input intact but setting the activity of cortical neurons to zero in a central region of the map (indicated by the dotted line in *b* and *c*). As with a retinal scotoma, the changes in activity disrupt the dynamic equilibrium and force the network to reorganize.

6.4.1 Reorganization of the Map

Three phases of reorganization were observed, as in animal experiments (Section 6.1.2; Merzenich et al. 1990). Immediately after the lesion, the receptive fields of neurons in the perilesion zone became larger. The lesioned neurons no longer inhibit the surrounding neurons, and activation that was previously suppressed is now unmasked. This result can be seen by comparing the response to a typical input before and after the lesion (Figure 6.10*a,b*). The postlesion activity extends farther outside the lesioned area than before. In effect, the perilesion neurons now respond to a new part of the input space: They took over part of the job of the lesioned area, and the apparent loss of representation is smaller than expected based on the prelesion map.

As soon as the equilibrium was disrupted, both lateral and afferent connections of the active neurons started to adapt. There is a lopsided distribution of activity close to the lesion boundary, with no activity inside and normal activity outside. Therefore, neurons close to the boundary encounter an imbalance of lateral interaction. By Hebbian adaptation, the lateral weights of these neurons strengthen to the active regions outside the lesion, and eventually become concentrated in the perilesion zone (Figure 6.11).

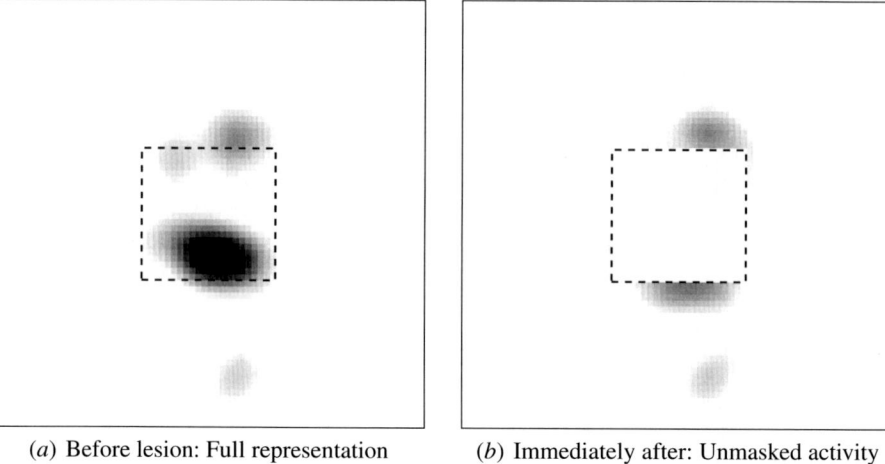

(*a*) Before lesion: Full representation	(*b*) Immediately after: Unmasked activity

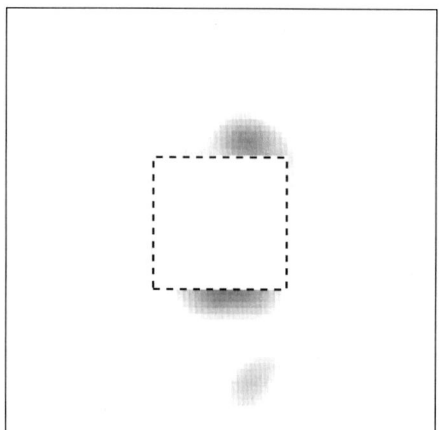 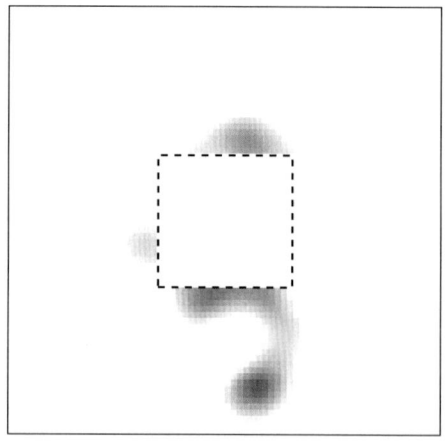

(*c*) Iteration 750: Increased inhibition	(*d*) Iteration 5000: Partial recovery

Fig. 6.10. Cortical response after a cortical lesion. The settled activity of neurons for the central 70×70 region of V1 is shown for the input in Figure 6.9*a* before the lesion (*a*), immediately after (*b*), several hundred adaptation iterations later (*c*), and after complete reorganization, i.e. in the new dynamic equilibrium (*d*). The lesioned area is marked as a dotted line in each plot. Immediately after the lesion, the activity spreads out to neurons that were previously strongly inhibited by the lesioned neurons. For instance, most of the activity just below the lesioned area in (*b*) did not exist in (*a*). These neurons partially compensate for the loss of function, which is less severe than expected. As lateral connections reorganize (Figure 6.11), this unmasked activity decreases slightly because lateral inhibition increases: For example, the active area just below the lesion becomes narrower and lighter (*c*). In the long term, after the afferent weights reorganize (Figure 6.12), the activity outside the lesioned area strengthens again (*d*). Though lateral inhibition is still stronger in the perilesion area, the afferent input overcomes the inhibition, and neurons at the boundary of the lesion become strongly responsive to inputs previously stimulating lesioned neurons. Similar stages are seen in biological lesion experiments (Section 6.1.2; Merzenich et al. 1990).

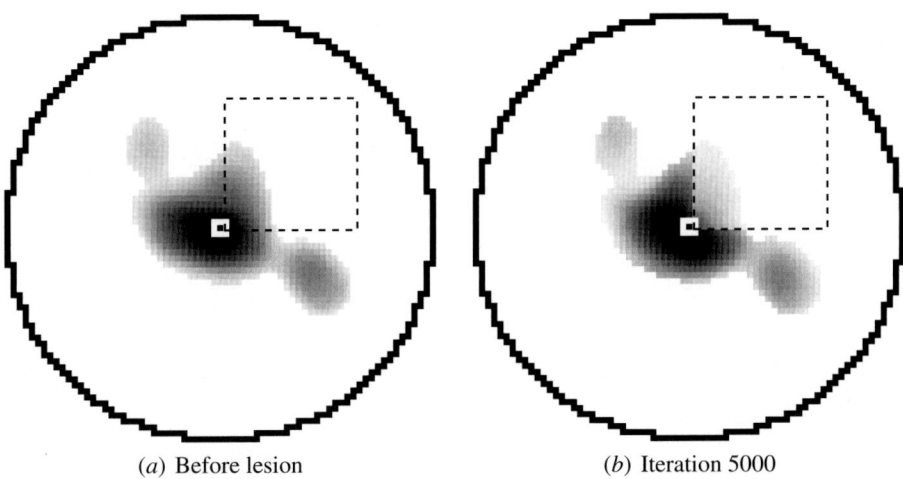

(*a*) Before lesion   (*b*) Iteration 5000

Fig. 6.11. Reorganization of lateral inhibitory weights after a cortical lesion. The lateral inhibitory weights of a neuron at the bottom-left corner of the lesion are plotted in gray scale from white to black (low to high; orientation preferences or selectivity are not shown). The small white square marks the neuron and the jagged black outline indicates the connectivity before self-organization and pruning, as in Figure 5.12. (*a*) The connections before the lesion follow the neuron's orientation preference as usual. (*b*) Through Hebbian adaptation after the lesion, the connections from neurons in the lesioned area approach zero, because those neurons are no longer active. Because the total inhibitory weight is kept constant by weight normalization, the inhibition concentrates in the connections outside the lesioned zone. This inhibition decreases the responses of the perilesion neurons, giving a computational account for the regressive phase in biology (Section 6.1.2; Merzenich et al. 1990).

Neurons just outside the lesion are most strongly affected by such reorganization: Because they have the largest number of connections from the lesion zone, their lateral inhibition concentrates in the few remaining connections. As a result, during the second phase of reorganization, the lateral inhibition becomes strong outside the lesion, and the previously unmasked activity is suppressed again (Figure 6.10c), resulting in an increased loss of function.

Even after the lateral connections reorganize, the perilesion neurons respond to inputs previously stimulating the lesioned zone. Therefore, through Hebbian adaptation, the afferent weights of these neurons increase in areas that were previously represented within the lesion. Such a reorganization is clearly seen in the topographic map of the afferent receptive fields (Figure 6.12). Gradually, the receptive fields move inward, and the representation of the receptor surface within the lesion zone is partially taken over by the neurons around it. Thus, the network partly compensates for the cortical lesion, and some of the lost function is regained, as in biology (Figure 6.10d). The compensation is not uniform but depends on the orientation preferences. The neurons that prefer orientations perpendicular to the lesion boundary change the most, because they had a large number of lateral connections from the lesioned area. On the other hand, those with preferences parallel to the boundary

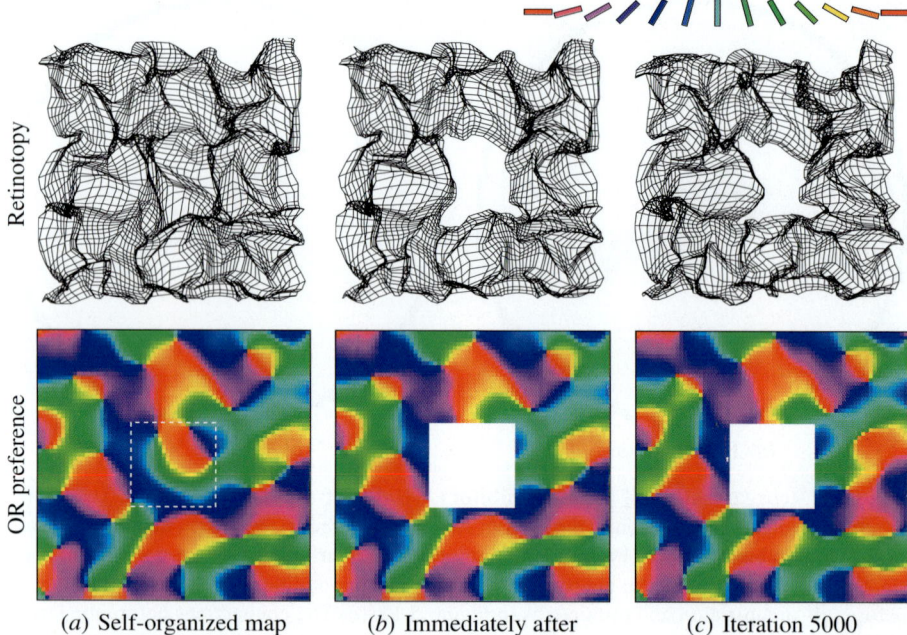

(a) Self-organized map   (b) Immediately after   (c) Iteration 5000

Fig. 6.12. Reorganization of the orientation map after a cortical lesion. These plots show
the retinotopic organization (calculated from settled responses as in Figure 6.7) and the orientation preferences of the central 70×70 region of the 142×142 cortex. The dotted white line
in (a) shows the area that will be lesioned. Immediately after the lesion, the map spreads out
slightly into the lesioned area, because the neurons near the lesion boundary respond to inputs
previously represented by the lesioned neurons. This expansion can be observed by comparing corresponding areas around the lesion in (a) and (b), such as the lesion's top boundary.
Over time (c), the map expands farther into the lesioned area, regaining some of the lost function. Neurons whose preferred orientations are perpendicular to the lesion boundary change
the most because they have the most connections cropped by the boundary. For example,
along the top of the lesion, the neurons colored cyan and green on the right side (with vertical
preferences) shift their RFs significantly inward, whereas the red, orange, and purple neurons
on the left side (with nearly horizontal preferences) do not change much. Thus, the model
gives a possible computational explanation for the observed reorganization processes in biology (Section 6.1.2; Merzenich et al. 1990). It further predicts that the specific patterns of
expansion depend on the orientation preferences of the neurons around the lesion, and that
the extent of recovery depends on how large the lesion is compared with lateral excitation and
the afferent receptive fields. An animated demo of the reorganization process can be seen at
http://computationalmaps.org.

are less affected by the lesion. Such effects of orientation have not been studied in
biology, and are predictions of the model.

6.4.2 Limits of Reorganization

The extent to which the topographic map can reorganize depends on how large the lesion is compared with the extent of the lateral excitatory connections and the anatomical receptive fields. If the receptive fields of neurons on each side of the lesion overlap and the lateral excitation spans the lesion, the receptive fields can shift all the way to the center and cover the region completely. Full functionality would be regained in such a case.

In general, connections passing through the lesioned area do not survive in lesions caused by stroke or surgery, and a complete recovery is not possible in these cases. However, with isolated cell death, such as that in old age, a complete recovery should be expected. This observation could explain why humans can lose a large fraction of their cortical neurons with age without losing any specific representations or functions.

The effects of cortical lesions are reversible in LISSOM, and the original topographic representation will be relearned if the lesioned neurons are restored to normal. Therefore, just as in the case of retinal lesions, the topographic map is dynamic and is maintained in a dynamic equilibrium with external inputs.

6.4.3 Medical Implications

Efforts to understand recovery following acute focal brain damage have traditionally been based on either clinical studies or animal models. LISSOM provides a well-specified computational model to complement these approaches. The computational model can be used to elucidate the precise mechanisms of recovery and reorganization, which otherwise would not be clear from observing the clinical symptoms. Once these mechanisms are known, it is possible to propose effective treatments and therapeutic strategies.

The LISSOM model suggests two techniques to accelerate recovery following surgery or stroke in the sensory cortices. Normally, the recovery time after surgery would include some immediate recovery, a phase of regression due to reorganized inhibition, and a gradual and slow compensation for the functional loss afterward. The regression phase could be ameliorated by reducing the suppression due to lateral inhibition. Such a technique would selectively deactivate inhibitory interactions around the surgical area using a transient blocker of inhibitory neurotransmitters. Neurons around the area would then fire intensively because of reduced inhibition, and the afferent connections would adapt rapidly to compensate for the lesion. By the time the blocker is absorbed, a substantial number of afferent receptive fields would have shifted and compensated for the lesion. Although the inhibition would still strengthen after the blockade, the recovery would be faster.

Second, the receptive fields of perilesion neurons could be forced to shift and the topographic map reorganized (as in Figure 6.12c) even before surgery. Such an effect could be achieved by intensive and repetitive stimulation of the area expected to lose sensation and by the sensory deprivation of its surroundings. Driven by the excessive stimulation, neurons outside the surgical zone would shift their receptive fields inward, and the area expected to lose sensation would be represented in a much larger area of the cortex. Then, after surgery, the receptive fields would have to move much less to reach their final state and the recovery would again be faster.

In the future, therapy that anticipates functional loss using computational models and compensates for it in advance could form a prelude to neurological surgery, minimizing damage and making recovery as fast and effective as possible.

6.5 Discussion

The LISSOM model shows how receptive fields can be maintained dynamically by excitatory and inhibitory lateral interactions in the cortex. The combined effect of afferent input, lateral excitation, and lateral inhibition determines the neural responses. When the balance of excitation and inhibition is disrupted, neural response patterns change dynamically and receptive fields expand or contract and change their shape. If the perturbations are transient, the weight changes are minimal, and the receptive field properties do not change significantly. If the perturbations persist, the weight changes accumulate and the receptive fields reorganize substantially. Such an ability to reorganize makes the cortex fault tolerant. Apart from injuries, natural cell death occurs in the retina and the brain due to age and disease. If the cortical synapses remain plastic, the remaining neural resources automatically compensate for such losses.

The LISSOM retinal lesion simulations were based on a single eye only. Strictly speaking, such results correspond to experiments where lesion is induced in both eyes (Chino et al. 1995; Gilbert and Wiesel 1992), or in one eye with simultaneous enucleation of the other eye (Chino et al. 1992; Kaas, Krubitzer, Chino, Langston, Polley, and Blair 1990). However, in several recent studies, the other eye was left intact (Calford et al. 1999, 2000, 2003; Waleszczyk et al. 2003), adding an interesting dimension to the experiment and the interpretation of the results. Importantly, Calford et al. (1999, 2000) showed that with monocular retinal lesions, binocular neurons in the corresponding cortical scotoma area reorganize their receptive fields on the lesioned eye as they do with binocular lesions, while their receptive fields on the intact eye remain largely unchanged. This observation is useful because it is difficult to measure both prelesion and postlesion receptive fields of the same neurons; comparing the receptive fields in the two eyes allows obtaining an approximate measure of how they changed. It also suggests that the LISSOM simulations are a valid approximation of monocular lesions as well. However, the LISSOM plasticity studies can also be extended to two eyes, using the model of ocular dominance introduced in Section 5.4. Such a model could be instrumental in understanding in detail how the afferent and lateral contributions from the two eyes interact in the recovery from monocular retinal lesions.

Consistent with recent biological observations (Buonomano and Merzenich 1998; Gilbert 1998; Karni and Bertini 1997), the simulations with the LISSOM model suggest that the cortex is continuously learning and adapting to new visual environments.

Receptive fields and lateral connections constantly adjust in order to match the statistical properties of the visual world. How fast and how locally such adaptation takes place depends on the survival requirements for the animal. If the statistics of the visual world are highly nonstationary, it is important for an animal to adapt constantly to new circumstances, and a high degree of plasticity is required. If the environment is relatively constant, the cortex is probably less plastic or becomes fixed early on. Computational simulations help making such theories concrete, leading to specific predictions that can be tested in further biological experiments.

6.6 Conclusion

The LISSOM model demonstrates that simultaneous adaptation of afferent and lateral connections can explain not only how cortical structures develop, but also how they remain plastic in the adult. The model shows how phenomena such as map reorganization and dynamic receptive fields are produced by lateral interactions and Hebbian learning. The simulated reorganizations are reversible, and demonstrate how a topographic map can be maintained in a dynamic equilibrium with extrinsic input. The model allows predicting the extent and time course of the reorganization, and suggests how recovery after cortical surgery could be hastened by blocking lateral inhibition locally in the cortex and by forcing the topographic map to reorganize prior to surgery.

These computational demonstrations are significant because they show that plasticity does not have to be a special mechanism added for the sole purpose of making the cortex more robust, but can be part of a unified mechanism that the cortex uses to learn and adapt. The next chapter will show how such adaptation and learning can also play a role in normal functional phenomena such as illusions and aftereffects.

7

Understanding Visual Performance: The Tilt Aftereffect

The results in the previous chapter suggest that the same self-organization processes that determine how the visual cortex develops keep it plastic in the adult. This hypothesis is extended further in this chapter: Such adaptation may be part of the normal function of the visual cortex, resulting in short-term functional phenomena such as the tilt aftereffect (TAE). In this chapter, the psychophysical data and current theories of the TAE are first reviewed, and the simulation method for testing the TAE in LISSOM is described. Adapting lateral connections in the model are shown to result in a TAE that matches human data well, and suggest how the indirect TAE can arise from synaptic normalization. The conclusion is that the TAE is not a flaw in an otherwise well-designed system, but an unavoidable result of a self-organizing process that aims at producing an efficient, sparse encoding of the input through decorrelation. This result establishes a crucial link between developmental processes and adult visual function.

7.1 Psychophysical and Computational Background

In general, humans and animals can accurately estimate the orientation of visual contours such as lines and edges. However, contours presented close together or one after the other in the same location can interact, causing distortions in their apparent orientations. When the lines are presented simultaneously, this effect is known as the tilt illusion, and when they are presented successively, it is known as the tilt aftereffect (Gibson and Radner 1937). This chapter will focus on the tilt aftereffect; possibilities for understanding tilt illusions computationally are briefly discussed in Section 7.5.

7.1.1 Psychophysical Data

The tilt aftereffect is similar to an afterimage from staring at a bright light, but it is due to changes in orientation perception rather than in perceived color or brightness. The effect can be produced in a simple experiment (Figure 7.1). After staring at

Fig. 7.1. Demonstration of the tilt aftereffect. Fixate your gaze on the circle inside the central diagram for at least 30 seconds, moving your eye slightly inside the circle to avoid developing strong afterimages. Now fixate on the diagram at the left. The vertical lines should appear slightly tilted clockwise; this phenomenon is called the direct tilt aftereffect. If you instead fixate upon the horizontal lines at the right, they should appear barely tilted counterclockwise, due to the indirect tilt aftereffect. Adapted from Campbell and Maffei 1971; reprinted from Bednar and Miikkulainen 2000b.

a pattern of tilted lines (the adaptation stimulus), subsequently viewed lines (the test stimulus) appear to have a slight tilt. The direction of the tilt depends on the orientation of the test stimulus: Those with orientations similar to the adaptation lines appear tilted away from them, while those nearly orthogonal to the adaptation stimulus appear to be tilted toward them.

The TAE has been studied extensively in the laboratory (Campbell and Maffei 1971; Gibson and Radner 1937; Mitchell and Muir 1976; Muir and Over 1970). The experimental methods vary somewhat, but subjects are usually tested in a darkened room using oriented stimuli such as lines, bars, or gratings (i.e. multiple parallel bars, as in Figure 7.1). In a typical experiment, subjects are first asked to estimate the orientation of a test stimulus by adjusting another stimulus elsewhere in the visual field until it appears to have the same orientation. They are then asked to fixate on a specific point, and shown an adaptation stimulus for 30–45 seconds. Third, they are shown the test stimulus again, and asked to estimate its orientation. The magnitude of the TAE at that angle between the adaptation and the test stimulus is the difference between the estimated orientation of the test stimulus before and after adaptation.

The results of such experiments are consistent overall, although the detailed shape of the TAE curve varies somewhat between different subjects and different measurement paradigms (Figure 7.2). The maximum angle expansion, called the direct tilt aftereffect, is usually found between 5° and 20° of difference between adaptation and test stimulus orientations (Howard and Templeton 1966). Larger differences begin to show less of an effect, and eventually reach zero somewhere between 25° and 50° (Campbell and Maffei 1971; Mitchell and Muir 1976; Muir and Over 1970). Even larger differences (up to 90°) result in a less pronounced angle contraction effect called the indirect tilt aftereffect, which is largest between 60° and 85° (Campbell and Maffei 1971; Mitchell and Muir 1976; Muir and Over 1970).

The TAE is found at all orientations, although the variance for oblique test angles is higher than for horizontal and vertical orientations (Mitchell and Muir 1976). The

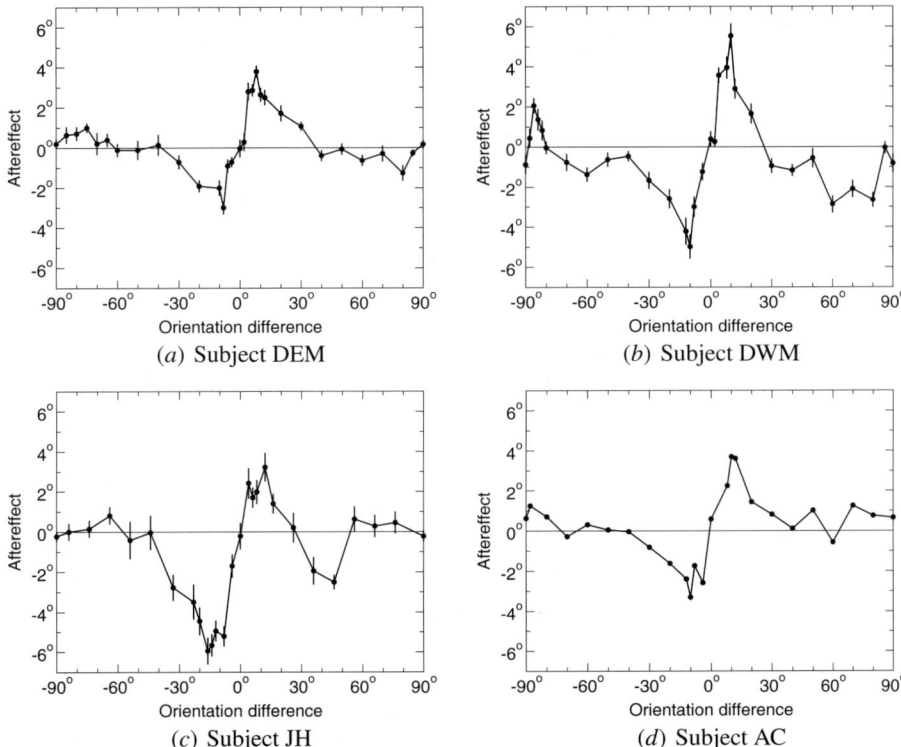

Fig. 7.2. Tilt aftereffect in human subjects. These plots show one TAE curve for each of the four subjects in Mitchell and Muir (1976): (*a*) DEM, (*b*) DWM, (*c*) JH, and (*d*) AC. The data were computed by averaging 10 trials before and 10 trials after adaptation. Error bars represent ±1 standard error of the mean (SEM); none were published for subject AC. Each trial consisted of a 3-minute adaptation to a sinusoidal grating, followed by a brief exposure to a test grating. The perceived orientation of the test grating was measured by having the subject adjust the orientation of a test line (presented in an unadapted portion of the visual field) until it appeared parallel to the test grating. For a given orientation difference counterclockwise between test and adaptation gratings, the TAE magnitude was then computed as the difference between the perceived orientations of the test grating before and after adaptation. In each case, a 0° orientation difference represents the orientation of the adaptation grating. For DEM the adaptation grating was horizontal, for AC it was vertical, and for DWM and JH it was oblique (135°). Similar direct effects were observed in all subjects, i.e. they perceived small orientation differences larger than they actually were; indirect effects varied more, but all subjects reported some contraction of large orientation differences.

magnitude of the effect increases logarithmically with increasing adaptation time and decreases logarithmically with time elapsed since the adaptation period (Gibson and Radner 1937). The direct effect saturates at approximately 4° (Campbell and Maffei 1971; Greenlee and Magnussen 1987; Magnussen and Johnsen 1986; Mitchell and Muir 1976). The largest documented indirect effect in central (foveal) vision is approximately 2.5°, and reaches up to about 60% of the magnitude of the direct effect for a given subject and paradigm (Campbell and Maffei 1971; Mitchell and Muir 1976; Muir and Over 1970). The TAE is spatially localized: Adaptation for an input in one retinal location has no measurable effect on test inputs in other locations sufficiently distant (Gibson and Radner 1937).

The TAE has been studied extensively because it is reliable and easy to measure. It is also important in that it may provide a window into how V1 adapts to visual input, as will be reviewed next.

7.1.2 Theoretical Explanations

Over the years, a wide variety of explanations for the TAE have been put forth (Barlow 1990; Coltheart 1971; Dong 1995; Dragoi, Sharma, Miller, and Sur 2002; Dragoi, Sharma, and Sur 2000; Field 1994; Földiák 1990; Gibson and Radner 1937; Köhler and Wallach 1944; Levine and Grossberg 1976; Tolhurst and Thompson 1975; Vaitkevicius, Karalius, Meskauskas, Sinius, and Sokolov 1983). Modern TAE theories are based on changes in how orientation-selective cells respond in V1 and other visual cortex areas. Hubel and Wiesel (1959, 1962, 1965, 1968) first observed that if the same oriented visual pattern is presented repeatedly, the neurons that at first responded strongly become more difficult to excite. This desensitization, called pattern adaptation, persists even after the pattern is removed. Conversely, neurons with orientation preferences orthogonal to the adaptation input actually respond *below* their resting level during this process, and may therefore become facilitated instead of desensitized (Hubel and Wiesel 1967). Based on these results, TAE theories have been developed from two perspectives: (1) neural fatigue, i.e. individual neurons become less responsive, and (2) lateral inhibition, i.e. interactions between multiple neurons change.

Neural Fatigue

The neural fatigue theory of the TAE is based directly on the results of Hubel and Wiesel (1959, 1962, 1965, 1968): If neurons with orientation preferences close to the adaptation input become fatigued, neurons that prefer more distant orientations will respond most strongly for similar input during testing. Assuming that the perceived orientation depends on the orientation preferences of the most activate neurons, a direct TAE results (Sutherland 1961). Similarly, the indirect TAE could be due to the (weaker) facilitation of neurons orthogonal to the adaptation input (Muir and Over 1970), or to fatigue of neurons with cross-shaped receptive fields (Coltheart 1971).

Although the fatigue theory was plausible originally, cortical neurons do not actually fatigue in this manner. They can be activated repeatedly in vitro with direct electrical simulation, without decreasing their output or reducing their sensitivity to input (Thomson and Deuchars 1994). Moreover, firing is not crucial for pattern adaptation: Cells continue to adapt even when their firing is prevented pharmacologically, and forcing the cells to fire does not cause them to adapt (Vidyasagar 1990). Thus, pattern adaptation, and therefore the TAE, is unlikely to be caused by fatigue from repeated firing.

Lateral Inhibition

In contrast, according to the lateral inhibition theory, pattern adaptation is based on changes in the lateral interactions between multiple neurons, instead of changes within single neurons (Barlow 1990; Blakemore, Carpenter, and Georgeson 1970; Carpenter and Blakemore 1973; Dong 1995; Field 1994; Földiák 1990; Tolhurst and Thompson 1975). Increasing lateral inhibition would make them appear to fatigue, yet they would still respond to direct electrical stimulation.

The inhibition theory also accounts for the phenomenon of disinhibition (Magnussen and Kurtenbach 1980). If two adaptation inputs each alone cause a direct TAE for the same test input, the effect would be expected to be even stronger if they are presented together during adaptation (Blakemore and Carpenter 1971). More detectors would be activated, and thus more would be fatigued. However, such adaptation actually *reduces* the magnitude of the TAE (Magnussen and Kurtenbach 1980). The inhibition theory predicts that the two inputs inhibit each other, and thus the total activation near the test line is lower. As a result, less adaptation occurs than before, and the TAE will be weaker.

The inhibition theory does not specify exactly how the increased inhibition results in the direct TAE. One possible explanation is that inhibitory transmitters accumulate in the target cell (Gelbtuch, Calvert, Harris, and Phillipson 1986; Masini, Antonietti, and Moja 1990; Tolhurst and Thompson 1975). However, Vidyasagar (1990) was able to excite or inhibit cells in the visual cortex of cats by locally applying excitatory or inhibitory transmitters, but the cells did not show adaptation effects when tested with a visual pattern. Yet, when the cells adapted to actual visual patterns, the neighboring cells delivered those same transmitters to the target cell. The adaptation must therefore occur somewhere else, outside the target cell. As suggested by Barlow (1990), changing connection strengths between neurons could result in such adaptation. The effects would only be seen when multiple nearby neurons are activated simultaneously, as they would be for the patterns typically used in TAE experiments (Vidyasagar 1990; Wilson and Humanski 1993).

Also, the theory does not directly suggest how indirect effects could arise from increased inhibition in V1; such effects have been proposed to occur at higher cortical levels instead (van der Zwan and Wenderoth 1995; Wenderoth and Johnstone 1988). One of the main contributions of this chapter is to show how changes in lateral inhibition, paired with a local synaptic resource conservation mechanism, can lead to realistic direct and indirect tilt aftereffects within V1, without involving other visual areas. In this manner, by filling in missing components of the theory, computational modeling can play a crucial role in developing a thorough understanding of the TAE.

7.1.3 Computational Models

Because it is difficult to measure the many local changes that contribute directly to the TAE, computational models have been used to make sense of this process. However, no model has yet demonstrated how both direct and indirect tilt aftereffects could arise from local neural processing in V1.

The most complete previous TAE model was built by Dong (1995, 1996). This model is formally based on the information processing principle of decorrelation, i.e. systematically reducing redundancy in images and other sensory data. All natural images are redundant to some extent, containing components that are highly correlated. In principle, both components could be represented by just one of them, making the other redundant. A network or algorithm that compacts the image by removing such redundancies is said to decorrelate. After full decorrelation, an input image would consist of purely white noise, where every bit of information is independent of the others. Dong (1995) showed analytically that decorrelation results in direct tilt aftereffects similar to those found in humans. This mathematical result complements the more detailed LISSOM simulations, which will show how decorrelation can be implemented in a biologically plausible network.

A more detailed model based on linear differential equations was implemented by Wilson and Humanski (1993). This model is a proposed circuit for achieving contrast independence via divisive inhibitory gain control (similar to that of HLISSOM, Section 8.2.3). The model had been shown to predict contrast sensitivity before and after adaptation, and with the same parameters it was found to exhibit direct tilt effects that have an angular function somewhat similar to the human direct effects. However, Wilson and Humanski (1993) did not discuss indirect effects, and it is difficult to see how those would occur in their model.

Thus, although the previous models demonstrated the mathematical principles of the direct TAE, they did not account for the indirect effect, and did not suggest a detailed neural processing mechanism that could be responsible for the TAE in V1. The LISSOM results will show that both effects can arise out of the same processes responsible for self-organization of orientation-selective neurons in V1. Thus, the model is not just a mathematical fit to the human data: It establishes a crucial link between developmental processes and adult function.

7.2 Method

The reduced LISSOM orientation model from Section 6.2 was used for the TAE experiments as well. As was discussed in Section 6.2, as long as the inputs consist of elongated Gaussian patterns (as they do in this chapter), the reduced model is functionally equivalent to the full LISSOM model, but simpler and more efficient to simulate because it does not include the ON and OFF channels of the LGN. In order to represent orientation in the cortex more accurately, the V1 network was increased to 192×192 units; other minor parameter differences are listed in Appendix B.2. Otherwise the same simulation method was used as in studying plasticity. In addition, a method needed to be developed for measuring perceived orientation and the tilt aftereffect in LISSOM, as will be described in this section. For historical reasons, and also to make positive and negative angles more intuitive, in this chapter the $0°$ angle represents the vertical direction (12 o'clock), as opposed to the horizontal direction (3 o'clock) used elsewhere in the book.

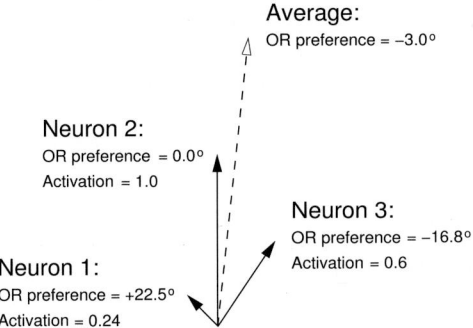

Fig. 7.3. Measuring perceived orientation as vector sum. The activations and orientations of three neurons are shown as vectors (solid lines). The angle of each vector is twice the orientation preference of that neuron, in order to make the vector orientation for 0° and 180° identical. The magnitude of each vector represents the activation level, to ensure that the encoding primarily reflects those neurons that are most active. The sum of these vectors is shown as a dashed line. The perceived orientation is half of the angle of the dashed vector.

7.2.1 Quantifying the Perceived Orientation

To compare the TAE in LISSOM with human psychophysical experiments, it is necessary to determine what orientation the model "perceives" for any given input. That is, given a set of activation values in V1, a numerical estimate of the input's orientation needs to be obtained. How this process occurs in humans is not yet known (Parker and Newsome 1998). However, V1 neurons in monkeys have the right properties to estimate the orientation nearly statistically optimally (Geisler and Albrecht 1997). For networks like LISSOM, a statistically optimal method is the vector sum, proposed by Coltheart (1971) and proven optimal by Snippe (1996). The vector sum will therefore be used with LISSOM, but it is not necessary to assume that this process is implemented directly in biological systems; any other nearly optimal method would lead to similar results.

The vector sum is a weighted average of orientation preferences that takes into account that orientations repeat every 180°. For example, two nearly horizontal preferences (e.g. −85° and +85°) should average to represent a horizontal orientation (±90°). However, the arithmetic average of −85° and +85° is 0°, which is clearly not perceptually correct. Therefore, each neuron is represented by a vector with its orientation scaled by a factor of two. Adjacent orientations are thus always represented as adjacent vector angles. Further, since each neuron should contribute only to the extent that it is active, vector magnitude is used to represent activation level. The perceived orientation is then computed as the vector sum of all active neurons, with its angle divided by two, as shown in Figure 7.3.

Note that the selectivity of the neuron to the orientation is not taken into account in this calculation. Selectivity could be used to scale the magnitude of the vector further, but it is unclear how such scaling should compare to scaling due to activity. In any case, because most neurons are highly selective in the LISSOM OR map, selectivity should not have a large effect in perceived orientation; it is therefore ignored in the vector sum method.

To verify that such measurements lead to accurate results, the perceived orientation of the LISSOM OR map was measured for input lines of every orientation spaced 5° apart; examples are shown in Figure 7.4. The estimates were always correct to within ±5°, indicating that the vector sum method is accurate on this LISSOM model. Further, to make sure that any random biases for particular patterns do not affect the results, all perceived orientation measurements in this chapter are stated in terms of differences in perceived angles, rather than in terms of the actual orientation on the retina.

Based on this method for quantifying the perceived orientation, a method for measuring the TAE in the model can be developed, as will be described next.

7.2.2 Measuring the Tilt Aftereffect

The TAE in the model can be computed by measuring the responses to different test lines after briefly training it with an adaptation line. The final, organized state of the reduced LISSOM model of Section 6.2.1 was taken as a starting point for each independent experiment in this section, so the results roughly correspond to testing a single human subject under different conditions.

To simulate adaptation, a single input line was presented to the model and its self-organizing process was run for a number of iterations. To permit detailed analysis of behavior at short time scales, the learning rates were reduced from those used during self-organization to $\alpha_A = \alpha_E = \alpha_I = 0.000005$. In separate experiments, designed to illustrate the separate contributions of adapting the afferent, lateral excitatory, and lateral inhibitory weights, learning was turned off for all but one type of connection in turn (Section 7.4). All other parameters remained as in Section 7.2, including the size and shape of the oriented Gaussian inputs.

Before adaptation and as adaptation progressed, test lines at various orientations were presented without modifying any weights. For each test line, the perceived orientation was measured as described in Section 7.2.1. The magnitude of the TAE was defined as the perceived orientation of the test line after adaptation minus the perceived orientation before adaptation.

This procedure is similar to that used for human subjects, described in Section 7.1.1. In human experiments, learning cannot be turned off, and only a single test input can be used for each adaptation. Any further test inputs would be affected by adaptation to the first test input. In contrast, learning can be disabled with LISSOM, and multiple test orientations can be presented for each adaptation episode. As a result, much more comprehensive data can be collected for the model than for humans, as shown in Section 7.3. Also, in human experiments usually only a single test stimulus is used, and the adaptation stimulus is varied. Because the angular function of the TAE has been demonstrated to be similar for all orientations (Mitchell and Muir 1976), the two procedures should lead to equivalent results.

Because the TAE curves differ substantially between individuals, particularly in the zero crossing between direct and indirect effects, it can be misleading to average

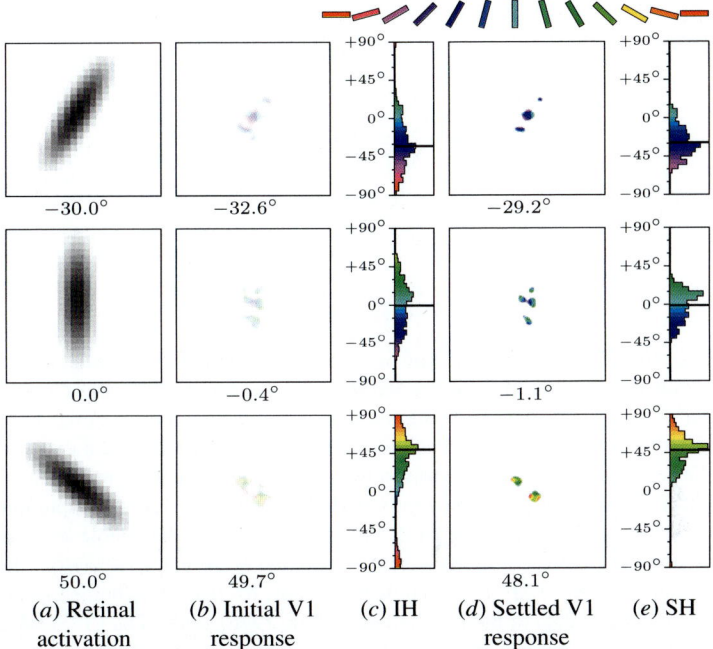

Fig. 7.4. Cortical response and perceived orientation. The V1 responses of the LISSOM orientation map are shown for three oriented Gaussians, at $-30°$, $0°$, and $+50°$ from vertical. As in the activity plots of Figure 6.5, the colors in the response plots (b,d) represent the orientation preference of each activated neuron, and the saturation (i.e. brightness) represents its activation level. The numbers below each of these plots indicate the perceived orientation, computed from orientation and activation values using the vector sum method. These values are also shown using black horizontal lines in histograms (c,e). The initial V1 response (b) is based on the afferent weights only, before the lateral interactions. The initial response histograms (IH; c) show that a broad range of orientations are activated, and that this distribution is approximately centered around the input's orientation. The location of the response in the activity plots corresponds to the location of the pattern on the retina. After settling, the activity is more focused both spatially and in orientation, as seen in the map activity plots (d) and their histograms (SH; e). The perceived orientation remains an accurate estimate of the actual orientation of the input pattern. An animated demo of this process can be seen at http://computationalmaps.org.

results from different subjects or testing paradigms. For instance, if the zero crossings vary over some range, a null area will show up in the graph around that region, even though no individual exhibited a null area. However, because the data are generally too erratic to interpret from a single run, multiple runs from testing the same individual in a single session are usually averaged (e.g. Mitchell and Muir 1976). To obtain similar measurements for the LISSOM model, a single orientation map was tested separately at nine different positions. These positions formed a 3×3 grid that covered a retinal area 6×6 units wide around the center of the retina, and included

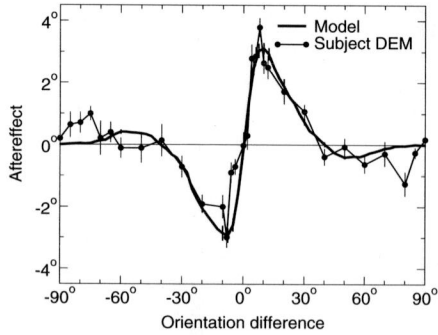

Fig. 7.5. Tilt aftereffect in humans and in LISSOM. The LISSOM network adapted to a vertical (0°) line at the center of the retina for 90 iterations, and the TAE was measured for test lines oriented at each angle. The thick line shows the average magnitude of the TAE over nine trials, as described in Section 7.2.2. Positive values of aftereffect denote a counterclockwise change in the perceived orientation of the test line. The graph is roughly anti-symmetric around 0°, i.e. the TAE is essentially the same in both directions relative to the adaptation line. The error bars indicate ±1 SEM; in most cases they are too small to be visible because the TAE was highly consistent between different runs (Appendix B.2). For comparison, the thin line with circles represents the TAE, averaged over 10 trials, for the single human subject (DEM) with the most complete data in the Mitchell and Muir (1976) study. The LISSOM TAE curve closely resembles the human TAE curve, showing both direct and indirect tilt aftereffects. Reprinted from Bednar and Miikkulainen (2000b).

many different orientation preferences. Averaging over these positions reduces the random fluctuations in TAE magnitude, but it does not change the basic shape of the curves presented in the following sections. This method will therefore be used to measure the TAE in the experiments that follow.

7.3 Results

The magnitude of the TAE in LISSOM for the various differences between the adaptation and the test line orientation is similar to those in humans. The TAE magnitude also increases over prolonged adaptation as in humans, but does not saturate in the model.

7.3.1 Magnitude Versus Orientation Difference

In the main LISSOM TAE experiment, the number of adaptation iterations was first adjusted (to 90) so that the direct TAE magnitude in the model peaked at 3°, as it usually does in human subjects. The TAE testing procedure described above was then used to determine the TAE magnitude for the whole range of differences between adaptation and testing orientations in the model.

The LISSOM results are strikingly similar to the psychophysical results (Figure 7.5). For the range 5° to 40°, human subjects exhibit angle expansion effects

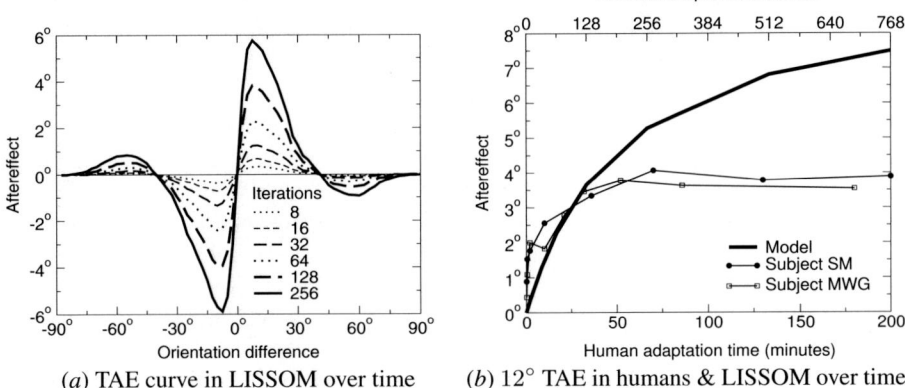

(a) TAE curve in LISSOM over time   (b) $12°$ TAE in humans & LISSOM over time

Fig. 7.6. Tilt aftereffect over time in humans and in LISSOM. Each curve in (a) shows the average TAE of the LISSOM model with a different amount of adaptation. The TAE has the same S-shape throughout, and its magnitude increases monotonically. Similar comprehensive data are not available on humans, but a vertical slice corresponding to the peak in this graph has been measured, as is shown in (b): The thin lines depict the direct TAE for $12°$ orientation difference vs. adaptation time for two human subjects SM and MWG, averaged over five trials (Greenlee and Magnussen 1987). For comparison, the thick line shows the corresponding TAE for the LISSOM model, averaged over nine trials. The adaptation time in LISSOM is measured in iterations, scaled to match the human plots as well as possible. The direct TAE increases approximately logarithmically in both LISSOM and humans; however, it does not saturate in LISSOM like it does in humans, suggesting that human adaptation faces additional limitations for long adaptation times. Reprinted from Bednar and Miikkulainen (2000b).

nearly identical to those found in the LISSOM model; the subject shown is the one whose TAE measurements are the most complete. The magnitude of this direct TAE increases very rapidly to a maximum angle expansion at $8–10°$, falling off somewhat more gradually to zero as the differences in orientation increase.

The simulations with larger orientation differences (from $40°$ to $85°$) show a smaller angle contraction, i.e. the indirect effect. Although the magnitude of this effect varies in human subjects, the LISSOM results are well within the range of human data. The indirect effect for the subject shown was typical for the Mitchell and Muir (1976) study, although some subjects showed effects up to $2.5°$ (Figure 7.2).

7.3.2 Magnitude over Time

In addition to the angular changes in the TAE, the magnitude of the TAE in humans increases regularly with adaptation time (Gibson and Radner 1937). The equivalent of time in the LISSOM model is an iteration, i.e. a single cycle of input presentation, activity propagation, settling, and weight modification. Figure 7.6a shows how the TAE varies for each orientation difference as the number of adaptation iterations is increased. The TAE curve has the same S-shape throughout, but its magnitude increases monotonically with adaptation.

Because obtaining human data for even a single curve is extremely time consuming, equivalently comprehensive data for human subjects are not available. Human time course data so far consist of only single vertical slices through the direct TAE region (Gibson and Radner 1937; Greenlee and Magnussen 1987; Magnussen and Johnsen 1986). Such data are obtained by presenting an oriented adaptation input for an extended period, interrupted at intervals by presentations of test and comparison lines. The test presentations are very short to minimize their effect on the TAE.

Such measurements show that the direct TAE in humans increases approximately logarithmically in time, and eventually reaches saturation (Figure 7.6b). The corresponding curve in LISSOM is similar to the human data, but does not saturate: The TAE magnitude keeps increasing until eventually the growing inhibition prevents the network from responding to the adaptation input altogether (the TAE magnitude is about $20°$ at that point). This difference suggests that biological systems may include additional constraints that limit the amount of learning that can be achieved over the time scale of the TAE. This issue is explored further in Section 7.5, and possible ways to model such saturation in LISSOM are proposed.

7.4 Analysis

The LISSOM TAE must result from changes in the connection strengths between neurons, because there is no other component of the model that changes as adaptation progresses. In particular, the neurons do not become more difficult to activate, and their activation levels do not decrease. Thus, there is nothing in the model that could correspond to single-cell neural fatigue. In this section, lateral inhibitory connections are identified as the source of the TAE in the LISSOM model, and how their adaptation results in direct and indirect TAE is analyzed.

7.4.1 Afferent and Lateral Contributions

Three sets of weights adapt synergetically in LISSOM: afferent, lateral excitatory, and lateral inhibitory weights. According to the lateral inhibition theory, the inhibitory weights are primarily responsible for the TAE. To test this theory in LISSOM, the contribution of each of the weight types was evaluated in three separate adaptation experiments (Figure 7.7). In each experiment, one of the weight types was adapting with an elevated rate (0.00005) and the others were fixed. As predicted, lateral inhibition determines the shape of the TAE curve. Lateral excitation and afferent contributions have only a minor effect, and it is almost precisely opposite to the combined TAE curve.

Interestingly, the inhibitory effects dominate even though all of the learning rates were identical. There are many more inhibitory connections than excitatory connections, and the combined strength of all the small inhibitory changes outweighs the excitatory changes. If excitatory learning is turned off altogether, the magnitude of the TAE increases slightly, but the shape of the curve does not change significantly (Figure 7.7).

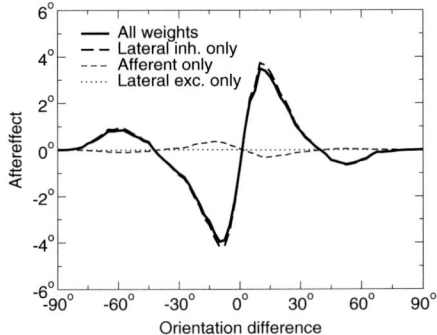

Fig. 7.7. Components of the tilt aftereffect due to each weight type. The solid line represents the magnitude of the TAE for a single trial from Figure 7.5. This trial was at the center of the retina, and is typical of the effect seen at the other 15 locations. The other curves illustrate the contribution from each different weight type separately. Other than the learning rates for these weights, the parameters were identical in each case. The line with short dashes represents the contribution from the afferent weights ($\alpha_A = 0.000005$; $\alpha_E = \alpha_I = 0$). This contribution is minor and in the direction opposite to the overall TAE curve. The dotted line represents the contribution from the lateral excitatory weights ($\alpha_E = 0.000005$; $\alpha_A = \alpha_I = 0$). It is in the same direction as that of the afferent weights, but so small it can hardly be seen (the x axis is not shown because it would have covered up this line). The line with long dashes represents the inhibitory contribution ($\alpha_I = 0.000005$; $\alpha_A = \alpha_E = 0$). These weights clearly determine the shape of the overall curve, although it is slightly reduced in magnitude by the afferent contribution.

7.4.2 Mechanisms

How do the adapting lateral connections cause the TAE? To make the analysis clear and unambiguous, the simplest case that shows a realistic TAE was studied in detail. The analysis focuses on a single trial with a Gaussian adaptation pattern at the exact center of the retina and only the lateral inhibitory weights adapting (i.e. $\alpha_I = 0.00005$ and $\alpha_A = \alpha_E = 0$). A longer adaptation period (256 iterations) was also used, in order to exaggerate the effect and make its causes more clearly visible. The analysis will show that Hebbian adaptation of the lateral inhibitory connection weights, combined with normalization of their total strength, systematically alters the response and causes the tilt aftereffects.

Each of the TAE measurements results from changes in many connections in many different neurons. However, the changes are systematic, and it is possible to understand the process by following the changes in a single neuron. Figure 7.8 shows the differences between the inhibitory connections of a typical neuron in the central region of the cortical sheet before and after adaptation. Adaptation changes the weight profile in two ways. First, connections from neurons with orientation preferences similar to the adaptation line become stronger (the blue areas in Figure 7.8d). Second, because the total strength for all inhibitory weights is normalized (Equation 4.8), the connections from other, more distant orientations (the yellow and red areas) become weaker (Figure 7.8e).

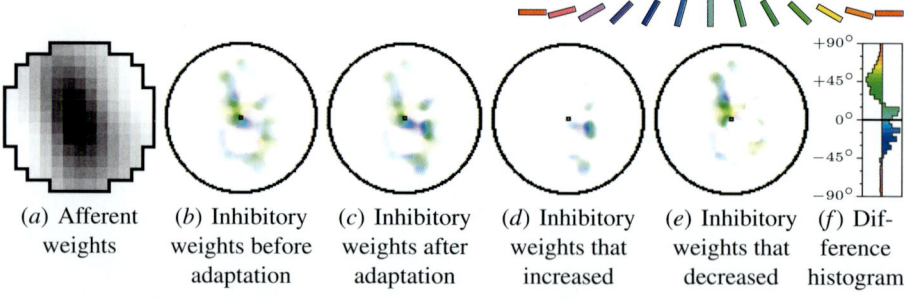

| (a) Afferent weights | (b) Inhibitory weights before adaptation | (c) Inhibitory weights after adaptation | (d) Inhibitory weights that increased | (e) Inhibitory weights that decreased | (f) Difference histogram |

Fig. 7.8. Changes in lateral inhibitory weights due to adaptation. The strengths of the inhibitory weights to the neuron marked with a small black square are shown in color coding as in Figure 5.12. This neuron prefers an orientation of $+15°$, as seen from its RF in (a), plotted in gray scale as in Figure 6.4. Before adaptation, the inhibitory connections mostly come from neurons with similar preferences along this same orientation (b). As a result of adapting to the vertical input in the center, the blue areas corresponding to the response to that input become stronger (c). This effect is summarized in (d), computed by subtracting the weights in (b) from those in (c), scaling the positive values up to a visible level, then labeling each connected neuron with the color corresponding to its orientation preference. Connections increased only from neurons with vertical preferences, i.e. those that were active in the settled response (Figure 7.9c,e, top row). As a result of normalization, the rest of the connections decreased, as seen by plotting the negative differences in (e). These connections include all orientations other than vertical. The orientation-specific changes are summarized in the difference histogram (f), which shows that the net connection strength to neurons with preferences around vertical increased, while connections to other orientations decreased. Together these changes give rise to the direct and indirect TAE, as shown in Figure 7.9.

All neurons that were activated by the adaptation input undergo similar changes. As a result, they become more strongly connected, and thus their response is lower than before adaptation. At the same time, other neurons respond more strongly than before adaptation, because they receive less inhibition. The net result is that the response to the adaptation line becomes broader, covering more area and orientations, but the overall response is still an accurate encoding of the input orientation (first row of Figure 7.9).

For other input orientations, the perception is no longer accurate. For orientations close to the adaptation line, the changes result in a shift away from the adaptation orientation. Neurons that responded during adaptation now respond less, but neurons with orientations further away respond more strongly: The result is the direct TAE (second row of Figure 7.9).

The indirect TAE occurs for lines so different in orientation that their response activates only a few of the neurons activated by the adaptation input (third row of Figure 7.9). These neurons represent orientations between the test and adaptation orientations. They respond more strongly than before adaptation, because they now receive less inhibition from other active neurons than before. As a result, the per-

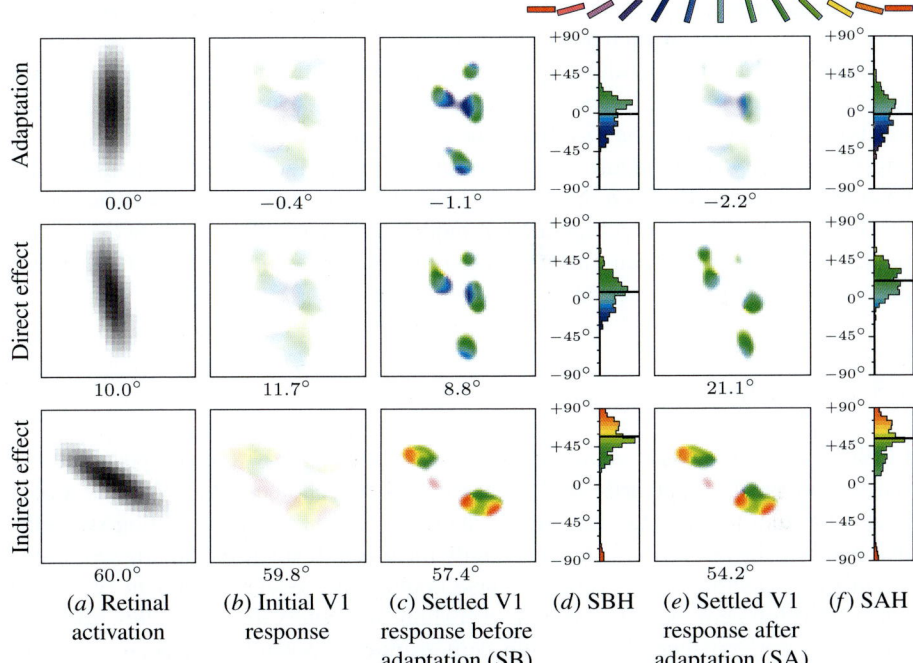

Fig. 7.9. Cortical response during adaptation and during direct and indirect tilt after-effect. Using the same plotting conventions as in Figure 7.4, each row shows (*a*) an example input, (*b*) the initial response of the central 64×64 region of V1 to that input (before the lateral connections), (*c*) the response of that region settled through the lateral connections but before adaptation to a vertical input line (SB), (*d*) the histogram of SB (SBH), (*e*) the correspond-ing settled response after a very long period of adaptation (SA), and (*f*) the histogram of SA (SAH). The top row (labeled "Adaptation") shows these responses to the same input as used for adaptation. After adaptation, the settled response is weaker, broader, and includes a wider range of orientations, but the perceived orientation stays approximately the same (compare the black lines in each histogram). For an input with a slightly different orientation (row "Direct effect"), more units encode orientations greater than $10°$ (green areas), and fewer encode those less than $10°$ (blue areas) in the settled response after adaptation than before. The net effect is a direct TAE, with the perceived orientation shifting away from the adaptation orientation, from $8.8°$ to $21.1°$ (compare the black lines in each histogram). For an input with an orien-tation very different from the adaptation pattern, the changes are more subtle (row "Indirect effect"). Only the neurons around $0°$ were activated during adaptation. Their inhibition from other vertical-preferring neurons increased, but decreased from those not active during adapta-tion. As a result, the green-colored neurons nearest $0°$ are now less inhibited by the rest of the neurons responding than before adaptation, and so they respond more strongly. The net effect is an indirect TAE, with the perceived orientation shifting toward the adaptation orientation, from $57.4°$ to $54.2°$ (compare the black lines in each histogram). Animated demos of these examples can be seen at http://computationalmaps.org.

ceived angle appears contracted. This effect is smaller than the direct effect because only a small number of such neurons are activated.

Thus, the LISSOM model shows computationally how both the direct and indirect effects can be caused by the same activity-dependent adaptation process. Further, this process is the same one that drives the development of the map in the first place, and its plasticity in the adult. The explanation for the indirect effect is novel and emerges automatically from the model, which was not tuned in any way to model the TAE. In this manner, computational modeling in general, and LISSOM in particular, can be used to obtain insights into functional phenomena that are otherwise difficult to explain at the neural level.

7.5 Discussion

LISSOM is the first computational model to account for how the complete angular function of the TAE can arise from interactions between neurons in V1. The model is also the first to simulate the spatial patterns of activity and connection strengths that underlie the effect and make it spatially localized. It suggests that the same self-organizing principles that result in sparse coding and reduce redundant activation during development may also operate over short time intervals in the adult, affecting perceptual performance. This finding demonstrates a potentially crucial computational link between development, structure, and function.

The TAE experiments were based on the reduced version of the LISSOM model, which allowed demonstrating the main effects clearly and efficiently. Further properties of the TAE could be replicated using the more complex versions of the model. For instance, maps trained on natural images would exhibit TAEs with a higher variance at oblique orientations (Mitchell and Muir 1976), because such maps have fewer neurons selective for oblique orientations (Section 9.3.2). Maps trained with moving patterns would result in a TAE that is specific to the direction of motion, and a version with frequency-selective neurons (to be discussed in Section 17.2.1) should limit the TAE to the spatial frequency of the adaptation input, both well-known features of the TAE in humans (Carney 1982; Ware and Mitchell 1974). Orientation maps trained with multiple retinas would allow transferring the TAE from one eye to the other (Campbell and Maffei 1971; Gibson and Radner 1937): Because the neurons that are most selective for orientation are binocular (Section 5.6), adapting one eye causes equal effects upon test lines in either eye. In each case, the properties of the neurons in the map should directly result in the observed properties of the TAE, without additional modifications to the model.

The LISSOM explanation of the indirect effect as a result of weight normalization is novel. Although normalization was originally included for computational reasons, recent experimental work suggests that such processes actually occur in biological neurons as they adapt (Section 3.3). Such regulation changes the efficacy of each synapse while keeping their relative weights constant, as in the LISSOM model. The process depends on adapting connections like the direct effect, but may operate at a different time scale (as suggested by psychophysical results; Wenderoth and Johnstone 1988).

The LISSOM results may also help explain why the shape and magnitude of the indirect effect varies so much between subjects, compared with the direct effect (Mitchell and Muir 1976). The TAE in LISSOM depends on how strongly those neurons activated by the adaptation input are connected to those activated by the test input. There are many connections between neurons with similar orientation preferences (Section 5.3), which makes the direct effect reliable. In contrast, there are fewer such connections between distant orientations, resulting in a less reliable and more variable indirect effect.

The LISSOM model of the TAE leads to several suggestions for future biological experiments. For example, LISSOM predicts that if inhibitory neurotransmitters are blocked, the sign of the TAE will be reversed and the effect will be significantly weaker (Figure 7.7). The activity patterns during the TAE could also be observed using optical imaging, directly verifying the model's predictions. As will be discussed in more detail in Section 16.4.3, it may be possible to test these predictions experimentally in the future, establishing a thorough understanding of the mechanisms underlying the TAE.

As was discussed in Section 7.3.2, the TAE in humans saturates near 4–5°, whereas in LISSOM it continues to increase as long as the network responds to the adaptation input. This difference suggests that adult humans have only limited ability to adapt to visual patterns. If these limits can be identified and measured electrophysiologically, it should be possible to extend the model to include them so that it would saturate at similar levels.

The current model also does not recover normal perception in darkness like humans do (Magnussen and Johnsen 1986). In the model, weights only change when neurons are active, and thus no recovery can occur without input. In humans, the TAE adaptation may be temporary, perhaps based on small additive or multiplicative changes on top of more permanent connection strengths. Such a mechanism could be simply a fast, limited, and temporary version of the self-organizing process that captures the long-term correlations (see von der Malsburg 1987; Zucker 1989) for possible mechanisms and uses for temporary plasticity). Part IV will explore how such temporary plasticity in the lateral connections could also contribute to perceptual grouping and segmentation.

In addition to the tilt aftereffect, the LISSOM model could be used to give a computational account to tilt illusions between simultaneous inputs, and aftereffects of other types such as those of motion, spatial frequency, size, position, curvature, and color, and even aftereffects in other modalities. Opportunities for future work in this area are outlined in more detail in Sections 17.2.5 and 17.2.6.

7.6 Conclusion

The computational experiments reported in this chapter strongly support the theory that tilt aftereffects result from Hebbian adaptation of the lateral connections between neurons, and provide a novel explanation of the indirect effect based on synaptic normalization. Importantly, these effects emerge from the same decorrelating process that determines how the cortex initially develops, and keeps it plastic in the adult. This process tends to deemphasize constant features of the input, resulting in short-term perceptual anomalies such as aftereffects.

A crucial component of understanding the TAE in the LISSOM model consisted of direct visualizations of cortical activity and weight changes as they were occurring in the simulated cortex. This method made it clear exactly which processes contributed to the effect. In general, computational models can demonstrate many visual phenomena in detail that are difficult to measure experimentally, thus presenting a view of the cortex that is otherwise not available. Such an analysis can complement both high-level theories and experimental work with humans and animals, significantly contributing to our understanding of the cortex.

CONSTRUCTING VISUAL FUNCTION

HLISSOM: A Hierarchical Model

Part II of the book demonstrated how input-driven self-organization can be responsible for the characteristic structure of the visual cortex, including feature-selective columns and patchy lateral connections, how this same self-organizing process can adapt to damage to the retina and the cortex, and how it results in functional phenomena such as the tilt aftereffect. In each simulation, the cortex was initially randomly organized and the complexity of the input patterns was carefully controlled, so that the properties of the self-organization process could be clearly demonstrated. In this third part, the time course of cortical development and the source and properties of the input patterns to the visual system will be analyzed in more detail. Although the self-organization process is powerful enough to build the necessary structures from natural input and random starting point, animals are born with significant cortical structure already in place. The hypothesis, tested computationally in Part III, is that such structure is due to genetically directed self-organization, which is then refined by environmentally driven self-organization. Such a synergy of nature and nurture makes the construction of visual function robust and efficient, and results in species-specific biases such as the human newborn preference for facelike visual input.

In this chapter, the synergetic approach is motivated and the LISSOM model from Part II is expanded outward into a multi-level model that allows testing the approach computationally. This model, HLISSOM (hierarchical LISSOM; Bednar 2002; Bednar and Miikkulainen 1998, 2005), can process both genetically determined internal input (from the developing retina and the brainstem) as well as external input, and it includes two areas of cortical processing (V1 and a face-selective area) where their synergy can be observed. By bringing all these components together, HLISSOM demonstrates computationally how perceptual abilities may be constructed, and allows detailed comparisons with experimental data.

8.1 Motivation: Synergy of Nature and Nurture

Recent experimental and computational results, such as those reviewed in Section 2.1.4 and presented in Part II, suggest that much of the structure and function of the visual system is constructed by a general-purpose learning process, driven by inputs from the environment. On the other hand, there is also considerable evidence that many aspects of the visual system are hardwired, i.e. constructed from a specific blueprint encoded in the genome. The conflict between these two positions is generally known as the nature–nurture debate, which has been raging for centuries in various forms (Diamond 1974).

The idea of a specific blueprint does seem to apply to the largest scale organization of the visual system, at the level of areas and their interconnections. These patterns are largely similar across individuals of the same species, and their development does not generally depend on neural activity, visually evoked or otherwise (Miyashita-Lin, Hevner, Wassarman, Martinez, and Rubenstein 1999; Rakic 1988; Shatz 1996). But at smaller scales, such as maps and lateral connections within them, there is considerable evidence for both internally and environmentally controlled development. Thus, debates center on how this seemingly conflicting evidence can be reconciled. Orientation processing is the clearest example, and will be the topic of Chapter 9; in this section, evidence for genetic and environmental influences on OR is summarized, suggesting how computational modeling can be used to account for both. Similar evidence for other perceptual features will be discussed in Section 17.2.3 and for higher levels such as face detection in Section 10.1.

Experiments since the 1960s have shown that the environment can have a large effect on the structure and function of the early visual areas (see Movshon and van Sluyters 1981 for a review). For instance, Blakemore and Cooper (1970) found that if kittens are raised in environments consisting of only vertical contours during a critical period, most of their V1 neurons become responsive to vertical orientations. Similarly, orientation maps from kittens with such rearing devote a larger area to the orientation that was overrepresented during development (Sengpiel et al. 1999). Even in normal adult animals, the distribution of orientation preferences is slightly biased toward horizontal and vertical contours (Chapman and Bonhoeffer 1998; Coppola, White, Fitzpatrick, and Purves 1998). Such a bias would be expected if the neurons learned orientation selectivity from typical environments, which have a similar orientation bias (Switkes, Mayer, and Sloan 1978). Conversely, kittens who were raised without patterned visual experience at all, e.g. by suturing their eyelids shut, have few orientation-selective neurons in V1 as an adult (Blakemore and van Sluyters 1975; Crair et al. 1998). Thus, visual experience can clearly influence how orientation selectivity and orientation maps develop.

The lateral connectivity patterns within the map are also affected by visual experience. For instance, kittens raised without patterned visual experience in one eye (by monocular lid suture) develop nonspecific lateral interactions for that eye (Kasamatsu, Kitano, Sutter, and Norcia 1998). Conversely, lateral connections become patchier when inputs from each eye are decorrelated during development (by artificially inducing strabismus, i.e. squint; Gilbert et al. 1990; Löwel and Singer 1992; Section 5.1.2).

In ferrets it is possible to reroute the connections from the eye that normally go to V1 via the LGN, so that instead they reach auditory cortex (see Sur et al. 1999; Sur and Leamey 2001 for reviews). As a result, the auditory cortex develops orientation-selective neurons, orientation maps, and patchy lateral connections, although these structures are not as pronounced as in normal maps. Furthermore, the ferret can use the rewired neurons to make visual distinctions, such as discriminating between two grating stimuli (von Melchner, Pallas, and Sur 2000). Thus, afferent input to the cortex can profoundly affect its structure and function.

Therefore, the evidence from altered environments and rewiring experiments suggests that the structure and function of V1 could simply be learned from experience with the visual environment. This experience is presumably mediated by neural activity, and experiments have shown that blocking neural activity, or blocking activity-dependent plasticity, prevents the development of orientation selectivity (Chapman and Stryker 1993; Ramoa, Mower, Liao, and Jafri 2001). Together, these findings suggest that the visually driven neural activity patterns in the LGN might be sufficient to direct the development of V1 and other cortical areas.

However, there is also significant evidence on the contrary, suggesting that visual cortex structure is genetically determined. For example, it has been known for a long time that individual orientation-selective cells exist in newborn kittens and ferrets even before they open their eyes (Blakemore and van Sluyters 1975; Chapman and Stryker 1993). Psychological studies further suggest that human newborns can already discriminate between patterns based on orientation (Slater and Johnson 1998; Slater et al. 1988). Recent advances in experimental imaging technologies have even made it possible to measure the full map of orientation preferences in young animals. Such experiments show that large-scale orientation maps exist prior to visual experience, and that these maps have many of the same features found in adults (Figure 9.2a; Chapman, Stryker, and Bonhoeffer 1996; Crair et al. 1998; Gödecke et al. 1997). The lateral connections within the orientation map are also already patchy before eye opening (Gödecke et al. 1997; Luhmann et al. 1986; Ruthazer and Stryker 1996).

Furthermore, the global pattern of orientation preferences in the maps changes very little with normal visual experience, even as the individual neurons gradually become more selective for orientation, and lateral connections become more patchy (Chapman and Stryker 1993; Crair et al. 1998; Gödecke et al. 1997). Thus, despite the clear influence of environmental input on visual cortex structure, normal visual experience primarily preserves and fine-tunes the existing structures, rather than drives their development.

How can the same circuitry be both genetically hardwired, yet also capable of significant learning and adaptation based on the environment? As was discussed in Section 2.3, new experiments are finally starting to shed light on this question: Many of the structures present at birth could result from learning of spontaneous, internally generated neural activity, such as retinal waves and PGO waves. The same activity-dependent learning mechanisms that can explain postnatal learning may simply be functioning before birth, driven by activity from internal instead of external sources. In this way, "hardwiring" may actually be learned. This explanation provides a possible way to reconcile the evidence in the nature vs. nurture debate in orientation processing. The answer is not simply that both components have an effect; there is only one developmental process, and it consists of a synergy of nature and nurture.

The goal of Part III is to test this hypothesis in detailed computational experiments. Two developmental phenomena are studied: (1) How orientation maps develop, as described above, and (2) how human infants prefer facelike visual input already at birth and how these preferences change in early life. Although the experimental data about both of these phenomena are rather confusing and even contradictory, they make perfect sense under the synergy hypothesis. The computational experiments are based on the LISSOM model, expanded outward to include lower and higher level areas of visual processing.

8.2 The Hierarchical Architecture

HLISSOM consists of the key areas of the visual system that are necessary to model how orientation and face processing develops based on internal and external inputs. The architecture is illustrated in Figure 8.1. HLISSOM extends LISSOM in three ways: (1) It includes input patterns arising from the brainstem (the PGO generator) in addition to retinal input; (2) it includes a higher level cortical face-selective area (FSA) in addition to V1; and (3) it includes divisive normalization on the afferent input. As described in Section 6.2.3, the ON and OFF channels of the LGN are necessary to process natural images.

8.2.1 Brainstem Input Area

As was reviewed in Section 2.3, the photoreceptors are not the only source of neural activity in the visual system. Spontaneous waves of activity in the retina and the PGO waves generated in the brainstem have both been shown to affect the development of the LGN, and may also influence how V1 develops. To investigate the role of such patterns, HLISSOM includes the PGO generator as an additional input area representing spontaneous activity from the brainstem.

The PGO pathway has not yet been mapped in detail in animals, but the activity that results from the PGO waves appears to be similar to that from visual input (Marks et al. 1995). Thus, for simplicity, the PGO pathway is modeled with an area like the retina, connecting to the LGN in the same way (Figure 8.1).

8.2.2 Face-Selective Area

It is easy to see how internally generated patterns could influence how orientation maps develop. However, behavioral tests with human infants (discussed in detail in Section 10.1) also suggest that internal activity may be important for the development of high-level processing, such as face perception. As was reviewed in Section 2.1.3, in adults there are face-selective cortical regions that receive input from V1. HLISSOM includes such a face-selective region called the FSA so that the development of face perception can be studied.

The FSA represents the first region in the ventral processing pathway above V1 that has receptive fields spanning approximately 45° of visual arc, i.e. large enough

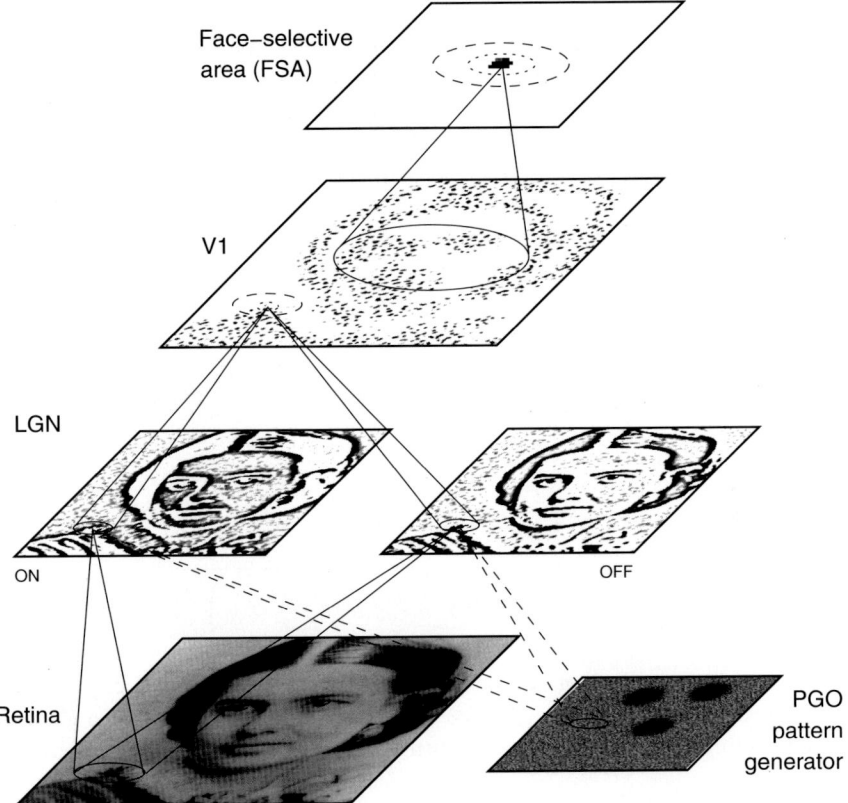

Fig. 8.1. Architecture of the HLISSOM model. Each sheet of units in the model visual pathway is shown with a sample activation pattern and the connections to one example unit. The activities are shown in gray scale as in Figure 4.1. Visual inputs are presented on the retina, and the resulting activity propagates through afferent connections to each of the higher levels. Internally generated PGO input propagates similarly to visual input. Activity in the model at any one time originates either in the PGO sheet or the retina, but not both at once. As in the LISSOM model, the activity in the cortical levels (V1 and FSA) is focused by lateral connections, which are initially excitatory between nearby neurons (dotted circles) and inhibitory between more distant neurons (dashed circles). The final patterns of lateral and afferent connections in the cortical areas develop through an unsupervised self-organizing process, as in LISSOM. After self-organization is complete, each stage in the hierarchy represents a different level of abstraction. The LGN responds best to edges and lines, suppressing areas with no information. The V1 response is further selective for the orientation of each contour; the response is patchy because neurons preferring other orientations do not respond. The FSA represents the highest level of abstraction — a neuron in the FSA responds when there appears to be a face in its receptive field on the retina.

to span a human face at close range. Although the infant connectivity patterns are not known, areas V4v (ventral V4) and LO (lateral occipital area) match this description based on adult patterns of connectivity (Haxby et al. 1994; Kanwisher et al. 1997; Rodman 1994; Rolls 1990). The generic term "face-selective area" is used rather than V4v or LO to emphasize that the results do not depend on the region's precise location or architecture, only on the fact that the region has receptive fields large enough to allow face-selective responses. Through self-organization, neurons in the FSA become selective for patterns similar to faces, and do not respond to most other objects and scenes.

8.2.3 Afferent Normalization

Compared with LISSOM's afferent stimulation function (Equation 4.5, Appendix A), HLISSOM adds an additional parameter γ_n to allow divisive (shunting) normalization:

$$s_{ij} = \frac{\gamma_A \left(\sum_{ab \in \text{ON}} \xi_{ab} A_{ab,ij} + \sum_{ab \in \text{OFF}} \xi_{ab} A_{ab,ij} \right)}{1 + \gamma_n \left(\sum_{ab \in \text{ON}} \xi_{ab} + \sum_{ab \in \text{OFF}} \xi_{ab} \right)}, \tag{8.1}$$

where ξ_{ab} is the activation of neuron (a, b) in the receptive field of neuron (i, j) in the ON or OFF channels, $A_{ab,ij}$ is the corresponding afferent weight, and γ_A is a constant scaling factor. An analogous normalization is done on the inputs from the V1 to FSA.

Equation 8.1 divides the afferent stimulation of the neuron by the total activation in its receptive fields, i.e. it normalizes the response according to total input. If the unit has a strong afferent connection to an input location and that location is active, the normalization increases the neuron's overall activation; if it has a weak connection to that location, it decreases the activation. This push–pull effect is an abstraction of contrast invariant responses in biology (Sections 16.1.4 and 17.1.2). As seen in Figures 8.2 and 8.3, afferent normalization helps ensure that the cortex responds uniformly even to large natural images, which have a wide variety of contrasts at different locations. As a result, all afferent weights can be excitatory, and adapt based on Hebbian learning as in LISSOM.

Artificial input patterns and inputs that cover only a small area of the visual field have relatively uniform contrasts. In simulations that use such input, afferent normalization can be omitted and γ_n left at zero. This was the case for the LISSOM simulations in Part II, and also for most other self-organizing models of the visual cortex. In the face perception simulations, however, large natural images are used, and afferent normalization is necessary.

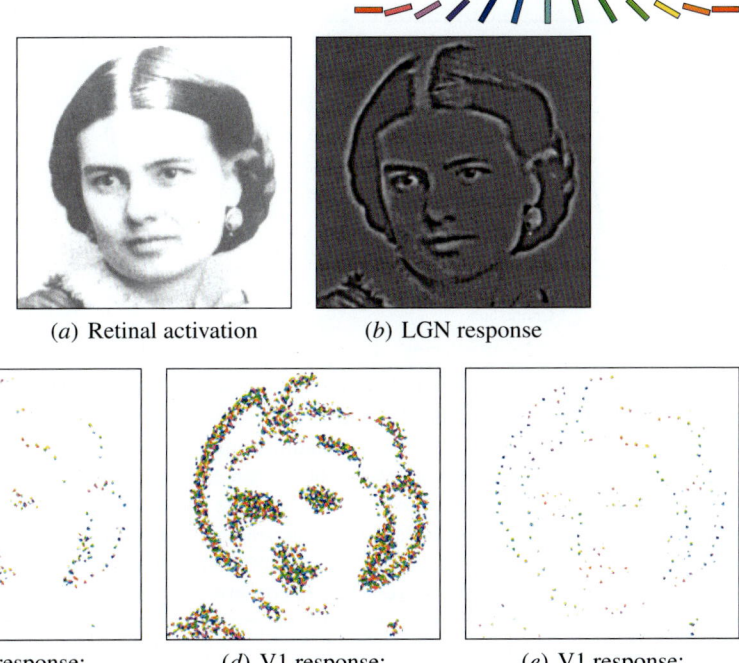

(a) Retinal activation   (b) LGN response

(c) V1 response: (d) V1 response: (e) V1 response:
$\gamma_n = 0$, $\gamma_A = 3.25$ $\gamma_n = 0$, $\gamma_A = 7.5$ $\gamma_n = 80$, $\gamma_A = 30$

Fig. 8.2. Effect of afferent normalization on V1 responses. The LGN response (b) to the activation in (a) is visualized by subtracting the OFF channel activation from the ON, and the V1 responses (c–e) by color coding each neuron according to how active it is and what orientation it prefers (as in Figure 6.5, except this network, from Section 10.2, is much larger). (c) Without afferent normalization ($\gamma_n = 0$), the network can respond only to the strongest contrasts in the image (as in Figure 6.5): The low-contrast oriented lines, such as those along the bottom of the chin, are lost. (d) When the afferent scale (γ_A) is increased, the network begins to respond to these lines as well, but its activation resulting from the high-contrast contours becomes widespread and unselective. (e) With normalization ($\gamma_n = 80$, $\gamma_A = 30$), the responses are largely invariant to input contrast, and instead are determined by how closely the input pattern matches the receptive field pattern of each neuron. The activations preserve the important features of the input, and the V1 activation pattern can be used as input to a higher level map for tasks such as face processing. Afferent normalization is therefore crucial for producing meaningful responses to natural inputs, which vary widely in contrast. Figure 8.3 shows how afferent normalization affects the responses of single neurons, which underlie these differences in the V1 response.

8.3 Inputs, Activation and Learning

Let us review the inputs and the activation and learning processes in HLISSOM, focusing on how they differ from LISSOM. As in LISSOM, learning is driven by input patterns drawn on the input sheets, which in HLISSOM consist of either the retina or the PGO generator, but not both at once (Figure 8.1). Since in Part II the goal was

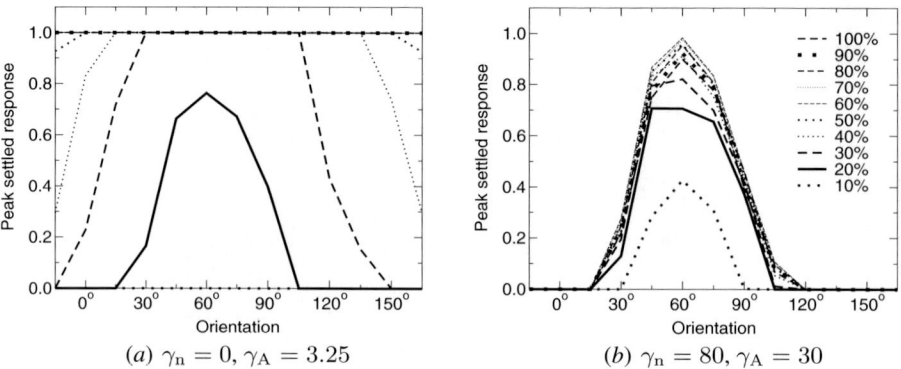

$(a)\ \gamma_n = 0,\ \gamma_A = 3.25$ $(b)\ \gamma_n = 80,\ \gamma_A = 30$

Fig. 8.3. Effect of afferent normalization on V1 neuron tuning. The differences in V1 population activities shown in Figure 8.2 are due to changes in how individual neurons respond at different contrasts. These plots show orientation tuning curves of the neuron at the center of the cortex, which prefers stimuli oriented at $60°$. Each curve shows the peak settled responses of this neuron to sine gratings whose orientations are indicated in the x-axis and contrast specified in the legend at right. In each case, the sine grating phase was used that resulted in the largest response. (*a*) Without afferent normalization, the neuron becomes less selective for orientation as contrast increases. Given enough contrast (above 50%), the neuron responds at full strength to inputs of all orientations, and thus no longer provides information about the input orientation. (*b*) With normalization, the tuning curve is the same over a wide range of contrasts, allowing the neuron to respond only to inputs that match its orientation preference. The curves are similar at 20% contrast (solid line), but the neuron now responds selectively to other contrasts as well. Afferent normalization is therefore crucial for preserving orientation selectivity over a wide range of contrasts.

to demonstrate which essential features of the input are responsible for orientation, ocular dominance, and direction selectivity, each simulation used only a single type of input. In contrast, with HLISSOM the goal is to understand the how internally generated inputs and environmental inputs together influence self-organization, and thus HLISSOM simulations include the input activity patterns thought to occur at each developmental age.

Internally generated activity has been observed in several locations in the developing visual system (Section 2.3). Given the current evidence, retinal waves are the most likely source for prenatal self-organization of V1 orientation maps (as will be discussed in Section 16.2.1). One such pattern is reproduced in Figure 8.4*a*, showing activity in retinal cells of the ferret before photoreceptors have developed. Although their precise origin still needs to be determined, retinal waves activate the ON and OFF neurons differently (Section 16.2.1; Myhr, Lukasiewicz, and Wong 2001). Such patterns can be modeled as "noisy disks", i.e. large active areas (modeling ON channel activation) and large inactive areas (modeling OFF channel activation) with oriented edges in a noisy background (Figure 8.4*b*). These patterns will be used to

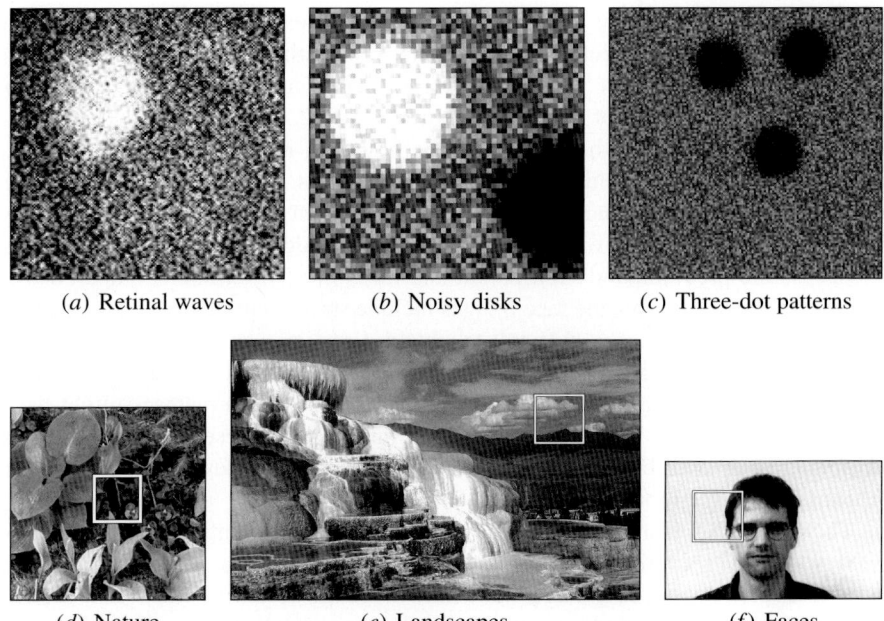

(a) Retinal waves   (b) Noisy disks   (c) Three-dot patterns

(d) Nature   (e) Landscapes   (f) Faces

Fig. 8.4. Internally generated and environmental input patterns. The three top images depict prenatal input patterns on the retina and the PGO pathway in gray scale from black to white (low to high). (a) A sample retinal wave pattern from the ferret (see also Figure 1.2) is used to motivate the actual patterns in HLISSOM experiments. (b) The "noisy disk" representation of retinal waves is used to organize the orientation map prenatally. A light disk models activity in the ON channel and a dark disk that in the OFF channel. (c) A PGO activity configuration of three dark noisy disks, corresponding to the two eyes and the nose/mouth area, is proposed to underlie prenatal development of face preferences. The three bottom images are samples of visual inputs, including those of (d) nature, (e) landscapes, and (f) faces. Randomly located retina-size segments (such as those shown by white squares) are used to train and test V1, and full face images to train and test the FSA, measuring how the variation in postnatal training affects the orientation map and how the face preferences develop postnatally. Sources: (a) Feller et al. (1996), (d) Shouval et al. (1996, 1997), (e) National Park Service (1995), (f) Achermann (1995), copyright 1995 by University of Bern.

organize the HLISSOM orientation map prenatally, i.e. before the onset of visual experience. [1]

Other sources of internally generated activity share features with retinal waves; however, their structure has not been characterized in enough detail to date to model directly. The hypothesis tested in the face preference experiments is that PGO patterns consisting of triplets of such waves (Figure 8.4c), corresponding roughly to the

---

[1] For convenience, the terms "prenatal" and "postnatal" are used to refer to the phases before and after the onset of visual experience. In humans this onset indeed coincides with birth, but in animals such as ferrets and cats it roughly matches eye opening, which happens several days or weeks after birth.

dark outlines of the two eyes and the nose and mouth area, could explain why human newborns are drawn to facelike visual inputs. Like retinal waves, PGO waves are not the only possible source for such patterns, but they are the most likely cause for prenatal self-organization of higher levels (Section 16.2.2).

Postnatal training, on the other hand, is based on natural visual inputs in the retina, including photographic images of natural objects, landscapes, and faces (Figure 8.4d–f). Each of these datasets has a slightly different distribution of orientations: Whereas objects have more horizontal and vertical edges than other orientations, landscapes are predominantly horizontal and faces mostly vertical. Compared with the prenatal PGO patterns, the face images include strong outlines and more detailed internal features. While it is difficult to obtain a realistic set of such patterns that would match the experience of an infant (as will be discussed in Section 17.3.3), it is possible to demonstrate what effects the variations of such patterns might have. In several experiments in Chapters 9 and 10, variations of these patterns as well as the prenatal ones will be tested to evaluate how strongly the developmental process depends on specific input features.

The ON and OFF sheets of the LGN in HLISSOM are identical to those in LISSOM, with DoG receptive fields (Equation 4.1) that filter out large, uniformly bright or dark areas, leaving only edges and lines (through Equation 4.3). The cortical sheets, i.e. V1 and the FSA, are similar to V1 in LISSOM. Each consists of initially unselective, laterally connected units that become selective through learning. V1 receives input from the ON and OFF cells of the LGN, while the FSA receives input from the laterally settled response of V1. The mapping from V1 to the FSA is constructed just like the mapping from the LGN to V1 in Figure A.1(a), so that no FSA neuron has a receptive field that is cropped by the border of V1.

HLISSOM simulations with large natural images generally start with an initial normalization strength γ_n of zero, because neurons are initially unselective. As neurons become selective over the course of training, γ_n is gradually increased. To prevent the net responses from decreasing, the scaling factor γ_A is set manually to compensate for each change to γ_n. The goal is to ensure that the afferent response ζ will continue to have values in the full range [0..1] for typical input patterns, regardless of the γ_n value. At the same time, γ_n ensures that the cortex responds to all areas of the input, not just the areas with the highest contrast.

After the afferent normalization, the initial cortical response is calculated from the afferent response using a sigmoid activation function (Equations 4.4–4.6). Activation then settles due to the lateral connections (Equation 4.7), and each weight is updated as in LISSOM. Through this process, HLISSOM develops realistic ordered maps, receptive fields, and lateral connections.

8.4 Effect of Input Sequence and Initial Organization

Before analyzing how the different training sequences affect self-organization in HLISSOM, it is necessary to verify that the resulting organization is indeed primarily determined by the inputs, and not by the initial random state of the network. This hypothesis is experimentally verified in this section using HLISSOM, but it applies to all LISSOM models and also to other similar self-organizing models. There are two types of variability between runs with different random numbers: the random order, location, and position of the individual input patterns at each iteration, and the random initial values of the connection weights. Each one can be independently varied while the other is kept constant, and the resulting differences can be observed.

A series of orientation map simulations similar to those in Section 5.3 were run in this way (Figure 8.5; Appendix C.1). The results demonstrate that the map shape does not depend on the random initial values of the weights, as long as the initial weights are drawn from the same random distribution. This observation is consistent with those on the SOM model (Cottrell, de Bodt, and Verleysen 2001), confirming that self-organization is less sensitive to initial conditions than e.g. backpropagation learning (Kolen and Pollack 1990). Instead, the self-organized orientation map pattern in HLISSOM depends crucially on the stream of inputs. Two different streams lead to different orientation maps, even if the streams are drawn from the same distribution. The overall properties of the maps (such as the distance between orientation patches, the number of pinwheels, etc.) are very similar, but different input streams lead to different arrangements of patches and pinwheels.

HLISSOM is insensitive to initial weights for three reasons, all of which are common properties of most incremental Hebbian models. First, because input patterns vary smoothly, the receptive fields are relatively large, and early in self-organization the afferent weights are uniformly random, the initial scalar product responses between the input and weight vectors (Equation 4.5) are similar regardless of the specific weight values. Second, because these responses settle through lateral excitation (Equation 4.7), the final activity levels are even more similar. Third, with a high enough learning rate, the initial weight values are soon overwritten by the Hebbian learning based on the final responses (Equation 4.8). The net result is that as long as the initial weights are generated from the same distribution, their precise values do not significantly affect map organization. Similar invariance to the initial weights should be found in other Hebbian models that compute the scalar product of the input and a weight vector, particularly if they include lateral excitation and use a high learning rate in the beginning of self-organization.

In animals, maps are also similar between members of the same species, but they differ in the specific arrangements of orientation patches and pinwheels (Blasdel 1992b). The HLISSOM model predicts that the specific orientation map pattern in the adult animal depends primarily on the order and type of activity seen by the cortex in early development, and not on the details of the initial connectivity. This result also means that it is very important to study how different input streams affect the self-organization process, as will be done in the next two chapters.

8.5 Conclusion

The HLISSOM model includes the retina and a brainstem pattern generator, the LGN (both ON and OFF channels), V1, and a higher level face-selective region. It can be

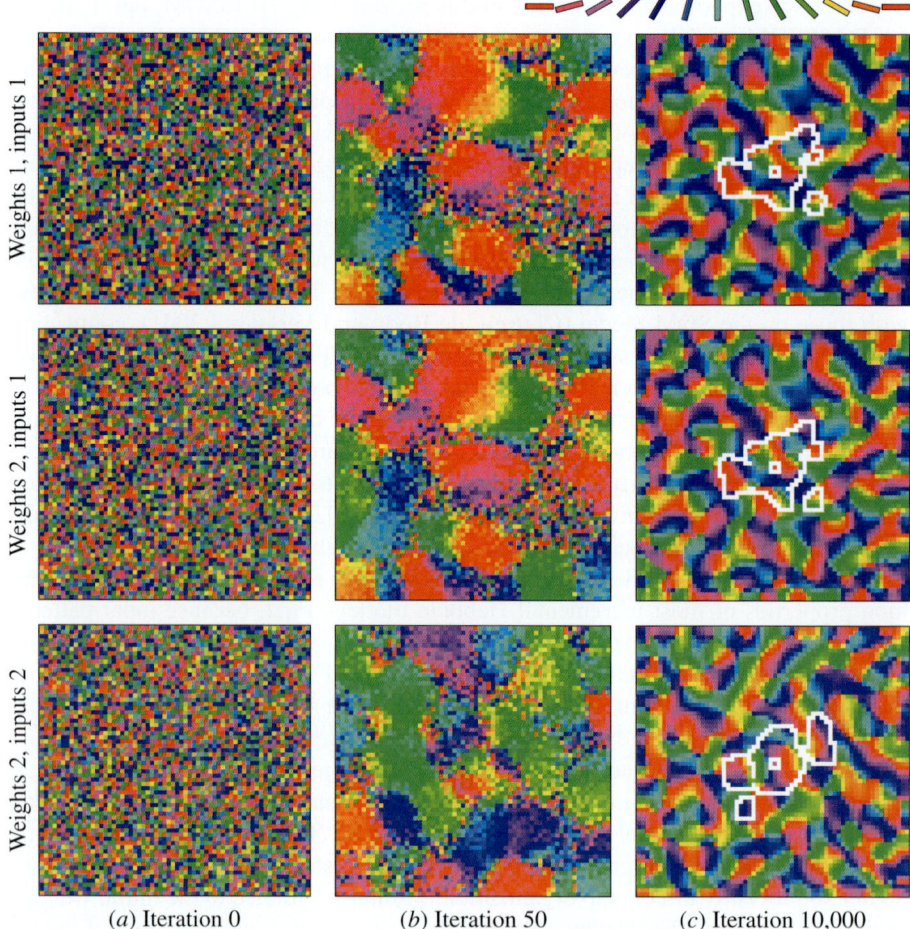

Fig. 8.5. Effect of different input streams and initial organizations on the self-organizing process. Using a different stream of random numbers for the weights (top two rows) results in different initial maps of orientation preference (*a*), but has almost no effect on the final self-organized maps (*c*), nor the lateral connections in them. (The lateral connections are shown in white outline for one sample neuron, marked with a small white square; orientation selectivity is not plotted in this Figure to make the preferences visible in the initial map.) The final result is the same because lateral excitation smooths out differences in the initial weight values, and leads to similar large-scale patterns of activation at each iteration. This process can be seen in the early map (*b*): The same large-scale features are emerging in both maps despite locally different patterns of noise caused by the different initial weights. In contrast, changing the input stream (bottom two rows) produces very different early and final map patterns and lateral connections, even when the initial weights are identical. Thus, the input patterns are the crucial source of variation, not the initial weights. An animated demo of these examples can be seen at http://computationalmaps.org.

trained with both internally generated patterns and natural images, and the resulting organization depends on the input sequence, not the initial unordered state. These components and properties allow HLISSOM to simulate developmental processes crucial for orientation and face processing in young animals and infants, as will be shown in Chapters 9 and 10. The results can be compared with experimental data and often lead to specific predictions for future experiments.

9

Understanding Low-Level Development: Orientation Maps

Using the HLISSOM model introduced in the previous chapter, this chapter will demonstrate how genetic and environmental influences can interact in developing biologically realistic orientation maps, V1 receptive fields, and lateral connections. The focus will be specifically on orientation maps because of the wealth of experimental data now available about their development. Patterns resembling retinal waves are first shown to have the right properties for developing rudimentary maps like those seen in newborns. These maps are then refined in postnatal learning with natural images, allowing the map to adapt to the statistical properties of the environment. The simulations show how HLISSOM can account for much of the complex process of orientation map development in V1, and also serve as a well-grounded test case for the methods used in the face perception experiments in the next chapter.

9.1 Biological Motivation

The LISSOM simulations in Section 5.3.5 showed that orientation maps can form based on a variety of input patterns, as long as the patterns have sufficient spatial structure. The properties of the map are slightly different in each case, reflecting the features of the input. The first goal of this chapter is to analyze what kind of features internally generated input should have to explain the rudimentary orientation map structure seen in newborns.

In particular, Miller (1994) suggested that retinal wave patterns (discussed in Section 2.3.3) might be too large and too weakly oriented to drive the development of orientation preferences in V1. However, these patterns contain spots of activity that have oriented edges. As long as these spots are large relative to V1 receptive fields, their shape and size should not matter, nor should the background noise; the oriented edges should be enough for V1 neurons to learn to represent orientation. This hypothesis is indeed verified in Section 9.2 in computational experiments with HLISSOM: Retinal waves do have sufficient structure to allow orientation maps and selectivity to develop; further, training on such patterns results in maps that match newborn maps better than those trained with idealized inputs or with random noise.

This result allows asking the next question: How are such internal inputs combined with external ones during development? None of the map models discussed in Section 5.2 has yet demonstrated how the orientation map can smoothly integrate information from these two sources. A number of the models have simulated spontaneously generated activity (e.g. Burger and Lang 1999; Linsker 1986a,b,c; Mayer, Herrmann, and Geisel 2001; Miller 1994; Piepenbrock, Ritter, and Obermayer 1996), and a few models have shown self-organization based on natural images (as reviewed in Section 5.2). Yet, to our knowledge, the only orientation map model to be tested on a prenatal phase with spontaneous patterns followed by a postnatal phase is the Burger and Lang (1999) model. They found that if a map organized based on uniformly random noise was subsequently trained on natural images (actually, patches from a single natural image), the initial structure was soon overwritten. As was discussed in Section 8.1, this is a curious result because animal maps instead maintain the same overall structure during postnatal development.

Section 9.3 will demonstrate how the HLISSOM prenatal map is smoothly refined in postnatal training with natural images. The prenatal map organization is not very different from that of a naïve network, i.e. an initially random map trained only with natural images. Therefore, postnatal training with natural images will only locally adjust the map, not replace it with something else. In this way HLISSOM will show how internal inputs and natural images can interact to construct realistic orientation maps, an important finding that has not been explained by previous models.

After demonstrating that prenatal and postnatal training together can account for the experimental data, the question is: Why are there two phases? There is indirect evidence that the visual cortex keeps getting trained with internally generated inputs at least for several weeks after birth (Crair et al. 1998). Would it be possible to obtain the refined adult orientation map in such continued training with internal inputs? Conversely, is the prenatal phase necessary, or would an adult map form just as well through postnatal training with natural images only?

Further simulations in Section 9.4 demonstrate that accurate adult maps can be obtained with internally generated patterns alone, and with natural images alone. However, there are good reasons why both phases exist: Prenatal training is an advantage because it allows the animal to have a functional visual system already at birth, and its further development will be more robust. Postnatal adaptation, on the other hand, allows it to form an accurate representation of the environment that it actually encounters during its life. Therefore, both phases serve a distinctly different role in constructing the visual system. How these two processes could continue interacting throughout the animal's life, balancing the need to adapt to the environment and the need to maintain stable visual abilities, is in important further research question, discussed in Section 17.2.4.

9.2 Prenatal Development

In this section, HLISSOM is trained with internally generated patterns to develop a rudimentary orientation map similar to e.g. those of newborn kittens. Maps trained with the noisy disks model of retinal waves match biological data better than maps trained with idealized versions of internal inputs and inputs consisting of only noise.

9.2.1 Method

The prenatal HLISSOM network consisted of a 96×96 cortex, 36×36 LGN, 54×54 PGO sheet, and 108×108 retina. In three separate experiments, this same network was trained with three different kinds of inputs to match the newborn orientation maps as well as possible. The goal was to understand whether the large, noisy spots seen in retinal waves are sufficient for forming newborn orientation maps, and what role spatial correlation and noise might each play in their self-organization.

In the main experiment, retinal waves were modeled with light and dark noisy disks (Figure 8.4a; Section 8.3). To generate such input, a spot-like structure was first rendered based on a circular disk with smooth Gaussian fall-off in brightness around the edges. Although the retinal wave patterns are often elongated, making them circular in this experiment shows that elongation is not necessary for orientation selectivity to develop. Uniformly distributed random noise was added to represent neural activities realistically. Although some of the noise in the observed retinal wave patterns is most likely due to measurement error, it is reasonable to assume that at least some of it is due to genuine neural activity, and should be included in the model.

More specifically, each input pattern contained one noisy disk, and was specified by the brightness of the disk related to the background (either light or dark), the location of the disk center (x_c, y_c), the radius r_d of the full-intensity central portion of the disk, and the width σ_d for Gaussian smoothing of its edge. To calculate the activity for each retinal location (x, y), the Euclidean distance d of that location from the disk center is first measured as

$$d = \sqrt{(x - x_c)^2 + (y - y_c)^2},$$
(9.1)

and the activity χ for receptor (x, y) is then calculated as

$$\chi_{xy} = \begin{cases} 1.0 & \text{if } d < r_d, \\ \exp\left(-\frac{(d-r_d)^2}{\sigma_d^2}\right) & \text{otherwise.} \end{cases}$$
(9.2)

The centers (x_c, y_c) were chosen randomly and the brightness of each pattern was either positive or negative relative to the mean brightness, chosen randomly. Noise was then included in this disk pattern by adding a uniformly distributed value in the range ± 0.5 to each pixel.

To evaluate the contributions of spatial correlations and noise in the results, in two separate experiments a similar network was trained with the disk-like patterns without the added noise, and with patterns that consisted of noise only (with each input pixel a random number within $[0..1]$).

In each experiment, the network was trained for 1000 iterations, since this amount of training was found experimentally to represent prenatal self-organization well. If the prenatal phase was concluded earlier and training continued with natural images, the postnatal training would override the prenatal organization; if instead the prenatal phase lasted longer than 1000 iterations, the postnatal phase would have little effect in refining the maps. The rest of the simulation parameters are detailed in Appendix C.1.

In the next two subsections, the map organization, receptive fields, and lateral connections resulting from noiseless and noisy disks and from noise alone will be compared.

9.2.2 Map Organization

The main result from the prenatal experiments is that the HLISSOM model trained with patterns modeling retinal waves develops an orientation map very similar to that found in newborn ferrets and binocularly deprived kittens (Figures 9.1 and 9.2; Chapman et al. 1996; Crair et al. 1998). This result shows that even simple internally generated inputs can be responsible for the observed prenatal self-organization. This result also explains how identical orientation maps can form for both eyes even without shared visual experience (using reverse lid suture; Gödecke and Bonhoeffer 1996). The development is driven by the orientations of small patches around the edge of the circular spots. These oriented edges are visible in the LGN response to the disk pattern in Figure 9.1.

The map develops lateral connection patterns that are oriented and patchy, although to a lesser extent than in the adult. They are a good match with animal data (such as those of Ruthazer and Stryker 1996). Oriented receptive fields with ON and OFF subregions also develop. Both two-lobed and three-lobed receptive fields are common for simple cells in adult V1 (Hubel and Wiesel 1968), but the RF types in newborns are not known. In the HLISSOM simulation with noisy disks, most neurons develop two-lobed receptive fields because the input patterns consisted of edges only (and no lines or bars). These results suggest that if orientation map development in animals is driven by large, spatially coherent spots of activity, newborns will primarily have two-lobed V1 receptive fields.

9.2.3 Effect of Training-Pattern Variations

Comparing the above results with noiseless and noise-only versions of the retinal wave patterns leads to several insights. The noiseless patterns result in a more regular map and smoother RFs, making the neurons highly selective (as seen in the middle row of Figure 9.1). They are actually more selective than newborn maps. Adding spatially uncorrelated noise, as was done in the "Noisy disks" simulation (top row), makes it harder for the neurons to become highly selective, resulting in maps that faithfully replicate newborn maps.

Interestingly, orientation maps develop even from uniformly random noise (bottom row; this result and the longer simulation in Section 5.3.5 replicate that of Linsker (1986a,b,c) in a biologically more detailed model). However, the resulting V1 map is significantly less organized than typical animal maps, even at birth. Most neurons are also only weakly selective for orientation, as can be seen in the sample RFs, most of which would be a good match to many different oriented lines. These

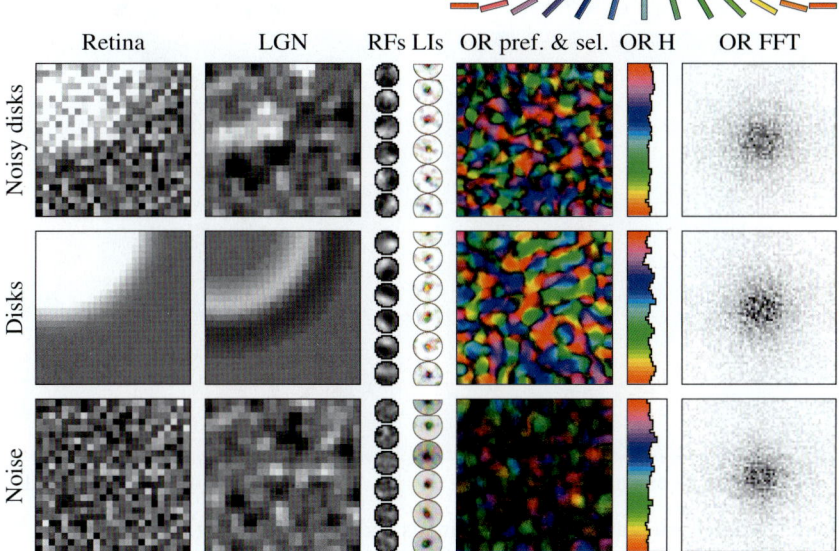

Fig. 9.1. Effect of internally generated prenatal training patterns on orientation maps.
Three different networks were trained for 1000 iterations to match newborn orientation maps
as well as possible. The networks and training parameters were otherwise identical except
different training inputs were used. As in Figure 5.13, the columns show a sample retinal activation, the LGN response to that activation, self-organized receptive fields for sample neurons,
lateral inhibitory weights of these same neurons, the organization of the orientation map with
selectivity superimposed in gray scale, and the histogram and the Fourier transform of the
OR preferences. Overall, the features seen in the corresponding fully organized maps of Figure 5.13 have already started to emerge in each of these maps, although they are less distinct
at this stage. They contain linear zones, pairs of pinwheels, saddle points, and fractures, and
their retinotopic organization and gradient (not shown) are roughly similar to adult maps. The
ring-like shape of the Fourier transform is also starting to emerge with disk and noisy disk
inputs. The map obtained with noisy disks is the best match with animal maps (Figure 9.2).
Note that nearly all of the resulting receptive fields have two lobes (i.e. they are edge-selective)
rather than three (line-selective), predicting that a similar distribution would also be found in
newborns. With noiseless patterns (middle row), the RFs are very smooth, and the neurons
become highly selective for orientation, unlike neurons seen in newborn maps. On the other
hand, with uncorrelated random noise (bottom row), the neurons become significantly less selective and the RFs do not have regular shapes like they do in animals. The "Noisy disks" map
therefore constitutes the most realistic model of prenatal self-organization, and will be used as
a starting point for postnatal training.

results suggest that the inputs need to be spatially coherent for realistic receptive
fields and maps to develop; noise alone is not sufficient.

In summary, the noisy disks model of internal training patterns leads to orientation maps that are a good match with those seen in newborns. These patterns have
enough oriented edges to drive self-organization, and enough noise to prevent the

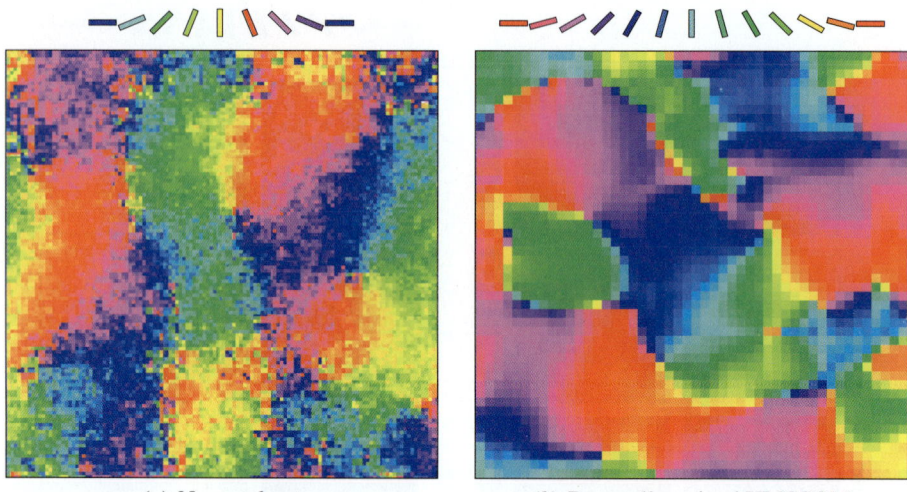

(a) Neonatal cat   (b) Prenatally trained HLISSOM

Fig. 9.2. Prenatal orientation maps in animals and in HLISSOM. (a) A 1.9 mm × 1.9 mm section measured through optical imaging in a 2-week-old binocularly deprived kitten, i.e. a kitten without prior visual experience. The map is not as smooth as in the adult, and many of the neurons are not as selective (not shown), but the map already has iso-orientation patches, linear zones, pairs of pinwheels, saddle points, and fractures (detail of a figure by Crair et al. 1998, reprinted with permission, copyright 1998 by the American Association for the Advancement of Science). (b) The central 30 × 30 region of the "Noisy disks" orientation map from Figure 9.1. The overall organization is very similar in the two maps, suggesting that prenatal training with internally generated patterns may be responsible for the observed maps at birth.

map from becoming too selective. Such maps form a good starting point for further refinement with natural images, as will be demonstrated next.

9.3 Postnatal Development

This section shows how a prenatally trained HLISSOM can continue learning with natural images. Instead of overwriting the prenatal order, the map gradually gets more refined, and eventually represents the statistical distribution of features in the training images.

9.3.1 Method

The prenatal HLISSOM model trained for 1000 iterations with noisy disks was used as a starting point for the postnatal simulations. In 9000 further iterations, 108 × 108 segments of natural images were presented to the retina, and the network was allowed to self-organize with the learning parameters listed in Appendix C.1.

In the main experiment, the network was trained postnatally with a dataset most closely matched with natural input (dataset "Nature"). This set consists of 25 256×256-pixel images of naturally occurring objects taken by Shouval et al. (1996, 1997). Nearly all of these images are short-range closeups, although there are a few wide-angle landscapes showing the horizon. All orientations are represented, but overall this dataset includes slightly more horizontal and vertical contours than other orientations. The main research question is then answered by observing how the map becomes gradually more refined during self-organization with these inputs.

The second goal was to understand what role postnatal training might play in helping the animal cope with its environment. As was discussed in Section 8.1, when animals are raised in artificial environments with only vertical lines, the numbers of orientation-selective cells in V1 will reflect this bias (Blakemore and Cooper 1970). The orientation maps of such animals also have enlarged domains for the overrepresented orientations (Sengpiel et al. 1999). Even when raised in normal environments, the maps become smoother and more selective through postnatal experience (Crair et al. 1998). To understand these phenomena computationally, the HLISSOM model was trained on two other natural image datasets as well.

The second postnatal training set, "Landscapes", consisted of 58 stock photographs from the National Park Service (1995). Nearly half of the images in this set are wide-angle photographs showing the horizon or other strong horizontal contours; a few also include man-made objects, such as fences. Therefore, this dataset has significantly more horizontally oriented contours than other contours. The third postnatal set, "Faces", consists of 30 frontal photographs of upright human faces (Achermann 1995), which contain more vertical orientations than do the other two sets. Example images from these three postnatal datasets are shown in Figure 9.3. Each of these three sets of natural images has different distributions of oriented edges, and the resulting self-organized maps should differ accordingly.

9.3.2 Map Organization

Starting from the rough prenatal orientation map, postnatal training with natural images gradually refines the map (Figure 9.6, top row). The neurons become more selective and the organization of the map changes slightly. Note, however, that the overall shape of the postnatal map remains similar to the prenatal map, as has been found to be the case in animals, but not in previous models of prenatal and postnatal development of orientation maps (Burger and Lang 1999).

The final adult map matches animal data very well (Figure 9.4). The overall organization of features on this map is similar to measurements from e.g. monkeys, cats, and ferrets. Whereas the prenatal map has a roughly uniform distribution of orientation preferences, the final map is biased for horizontal and vertical orientations. This is important because a similar bias has been found in adult animals (Figure 9.5; Chapman and Bonhoeffer 1998; Coppola et al. 1998).

Most of the RFs in the final map are orientation selective (Figure 9.3, top row), as found in V1 of animals. However, they are still less selective than those in maps

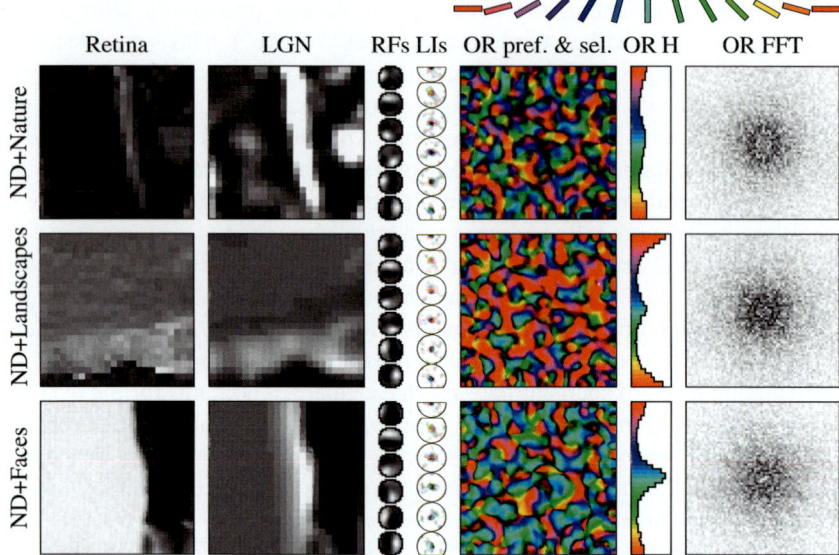

Fig. 9.3. Effect of environmental postnatal training patterns on orientation maps. Each simulation started with the same initial map, trained prenatally for 1000 iterations on noisy disks (ND) as shown in the top row of Figure 9.1. Postnatally, this map was trained for 9000 iterations under the same parameters but with retina-size segments of three different kinds of natural image inputs (the full images for these examples are shown in Figure 8.4*d–f*). In each case, maps with realistic features, RFs, lateral connections, and Fourier transforms developed. The final maps are less selective than those trained with artificial stimuli (Section 5.3), matching biological maps well. They also differ significantly on how the preferences are distributed. The network in the top row was trained on images of natural objects and primarily close-range natural scenes from Shouval et al. (1996, 1997). Like biological maps, this map is slightly biased toward horizontal and vertical orientations (as seen in the histogram), reflecting the edge statistics of the natural environment. The network in the second row was trained with stock photographs from the National Park Service (1995), consisting primarily of landscapes with abundant horizontal contours. The resulting map is dominated by neurons with horizontal orientation preferences (red), with a lesser peak for vertical orientations (cyan), which is visible in both the map plot and the histogram. The network in the bottom row was trained with upright human faces, by Achermann (1995). It has an opposite pattern of preferences, with a strong peak at vertical and a lesser peak at horizontal (bottom row). Thus, postnatal self-organization in HLISSOM depends on the statistics of the input images used, explaining why horizontal and vertical orientations are more prominent in animal maps, and how this distribution can be disturbed in abnormal visual environments. It also suggests that postnatal learning plays an important role in how visual function develops: It allows the animal to discover what the most important visual features are and allocate more resources for representing them.

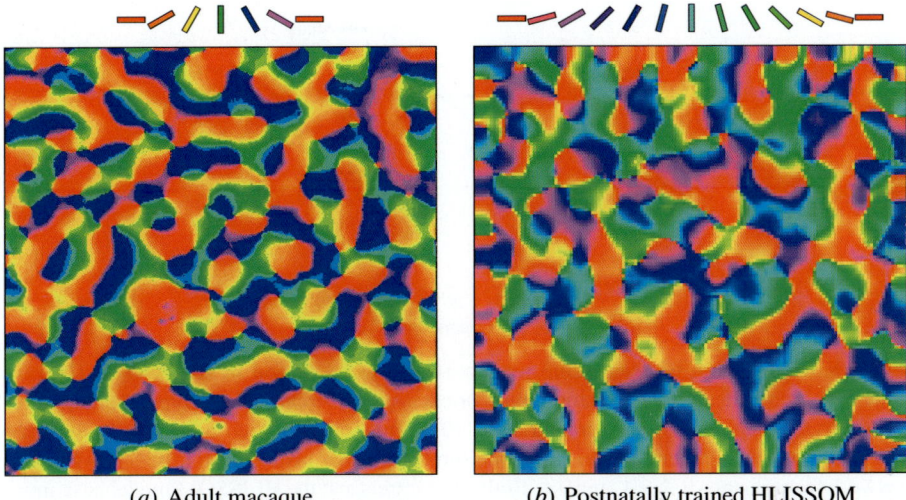

(a) Adult macaque   (b) Postnatally trained HLISSOM

Fig. 9.4. Postnatal orientation maps in animals and in HLISSOM. (*a*) A 5 mm × 5 mm area
of the orientation preference map in adult macaque (detail of Figure 2.4*a*, reprinted with permission from Blasdel 1992b, copyright 1992 by the Society for Neuroscience). After postnatal
training on natural images, the HLISSOM map (*b*) replicates its structure very well. Thus, the
HLISSOM model shows how both the prenatal and adult orientation maps can develop based
on internally generated and environmental stimuli.

trained with artificial stimuli (i.e. Section 5.3), which is realistic and expected because the natural images contain many patterns other than pure edges. The receptive
fields have a realistic multi-lobe structure similar to those observed in simple cells
of monkeys and cats (Hubel and Wiesel 1962, 1968). Lateral connection patterns are
patchy and oriented, as they are in the adult animal (Bosking et al. 1997; Sincich and
Blasdel 2001).

Thus, postnatal training with natural images can explain how orientation maps
develop during early life. The model can also help us understand why such postnatal
learning is useful, as will be discussed next.

9.3.3 Effect of Visual Environment

The HLISSOM model suggests a computational explanation for the horizontal and
vertical biases in the orientation preferences. As was discussed in Section 9.3.1, the
"Nature" image set has slightly more horizontal and vertical edges than edges in
other orientations. Because self-organizing maps allocate resources according to the
input distribution (as was discussed in Section 3.4.3), these orientations become more
prominent in the map. Since vertical and horizontal contours are overrepresented in
the natural environment as well (Switkes et al. 1978), HLISSOM suggests a possible
mechanism for how the observed biases could result.

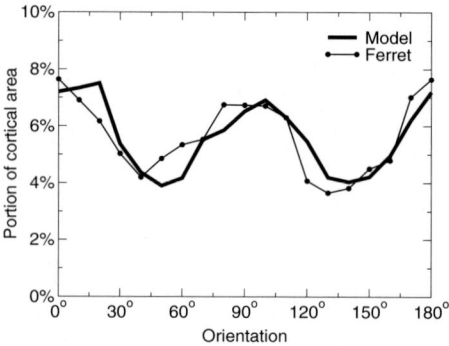

Fig. 9.5. Distribution of orientation preferences in animals and in HLISSOM. The thin line with circles delineates a histogram of orientation preferences for a typical adult ferret visual cortex (replotted from Coppola et al. 1998; measured through optical imaging in an oval 8.4 mm × 3.3 mm area). The thick line shows a similar histogram for the "ND+Nature" network from Figures 9.3, 9.4, and 9.6. Both adult ferrets and the HLISSOM model have more neurons representing horizontal or vertical than oblique contours, reflecting the statistics of the natural environment. HLISSOM maps trained on internally generated patterns alone instead have an approximately flat distribution, as seen in the histograms of Figure 9.1.

Moreover, if the statistical properties of the inputs are altered, the resulting maps should reflect this change. To demonstrate this idea computationally, the HLISSOM networks trained with the three different postnatal natural image datasets (as described in Section 9.3.1) can be compared.

The postnatal development of the HLISSOM map during the first few hundred iterations turned out very similar regardless of the training patterns used. This result is in line with Crair et al.'s (1998) finding (discussed in more detail in the next section) that visual experience typically has only a small effect on early map development in kittens.

Yet, in continued HLISSOM training the maps start to diverge (Figure 9.3). Compared with the slight horizontal and vertical biases of the "ND+Nature" map, the final "ND+Landscapes" map is strongly biased toward horizontal contours, with a much weaker bias for vertical. These biases are visible in the orientation plot, which is dominated by red (horizontal) and, to a lesser extent, cyan (vertical). The opposite pattern of biases is found for the "ND+Faces" map, which is dominated by cyan (vertical) and, to a lesser extent, red (horizontal). These results are analogous to those with animals raised in artificial environments over the long term (Section 8.1; Blakemore and Cooper 1970; Sengpiel et al. 1999). They suggest that the visual system learns to encode the edge statistics of the visual environment in the orientation map, a result that to our knowledge has not been demonstrated computationally before.

The HLISSOM model therefore shows how postnatal learning can contribute to building an effective visual system. The most common contours in the environment are the best represented in visual cortex, which will result in more effective processing of typical visual input.

9.4 Prenatal and Postnatal Contributions

Although prenatal and postnatal training together can account for the experimental data, are both phases necessary in order to construct a realistic orientation map? Somewhat surprisingly it turns out that training with internally generated patterns and training with natural images alone are both sufficient in principle. However, the animal would either not be able to perform visually at birth, or would not be able to adapt its performance to fit the environment better.

9.4.1 Method

To understand whether prenatal and postnatal phases are both necessary, three further experiments were run. First, the newborn network was trained for another 9000 iterations (with the same parameters as the postnatal network) with noisy disk patterns to determine whether training with internally generated patterns only could result in an organization similar to the adult map.

Second, another network with the same architecture and learning parameters, called the "Nature" network, was trained for 10,000 iterations with the "Nature" set of inputs, but starting from an unordered, random initial organization at iteration 0. That is, both the prenatal and postnatal phases used the same dataset of natural images. This network is used to test whether HLISSOM can self-organize from natural images alone, and whether its final organization will be different from a network trained with a prenatal map as a starting point.

Third, another randomly initialized network called "Blank+Nature" was trained with the same set of natural image inputs starting from iteration 1000, i.e. after some of the maturation processes had already taken place (Sections 4.4.3 and 16.1.6). That is, the simulation parameters had been changed according to the schedule for the first 1000 iterations, even though all inputs were blank and no training had actually occurred. The initial network therefore had a shorter excitation radius, steeper sigmoid, and slower learning rate than it would have had at iteration 0 (Appendix C.1). The purpose was to see whether there was a critical period after which training from natural images only would fail to generate a realistic map. Together with the main experiment combining prenatal and postnatal learning, these three experiments allow identifying distinct roles for the prenatal and postnatal phases of orientation map development.

9.4.2 Effect of Training-Regime Variations

When trained with only internally generated inputs, an orientation map develops that is qualitatively and quantitatively very similar to HLISSOM maps that were trained fully or partly with natural images (Figure 9.6). This result is important because it suggests that a continual generation of internal inputs during early life could be responsible for the development, instead of actual visual experience.

There is indeed evidence that internally driven self-organization continues after birth (or eye opening). Crair et al. (1998) found that similar maps develop in kittens

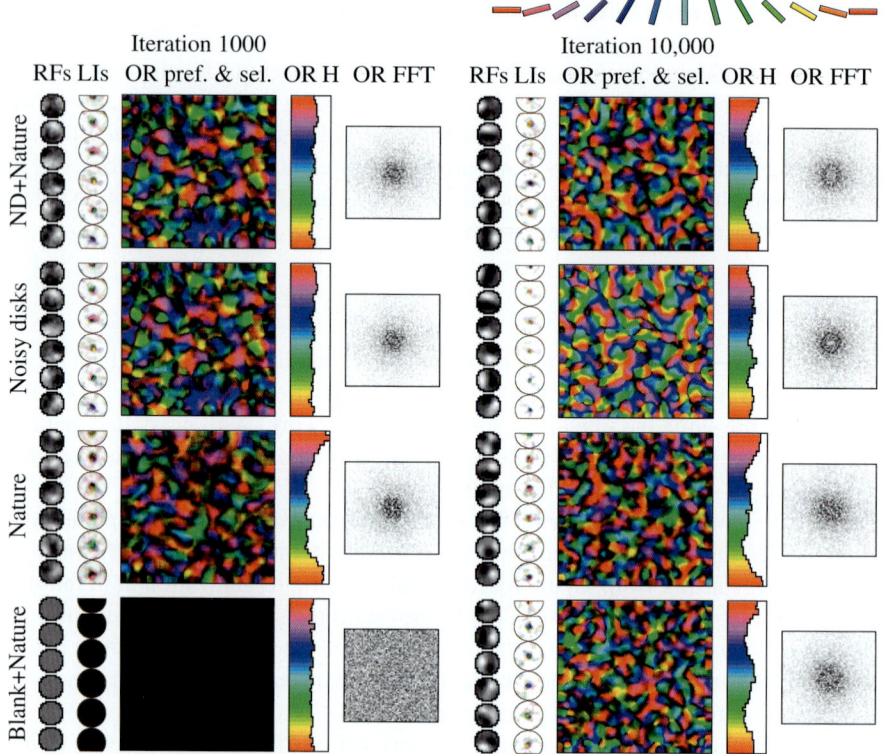

Fig. 9.6. Effect of prenatal and postnatal training on orientation maps. The different rows illustrate how the prenatal training phase affects the final self-organized maps. The state of each network at iteration 1000 is shown on the left half, and the final state at iteration 10,000 on the right half. In the "ND+Nature" simulation (the same as in Figures 9.1 and 9.3), postnatal training makes more neurons sensitive to horizontal and vertical contours and more selective in general. However, the overall map shape remains similar, as found experimentally in animals (Chapman et al. 1996; compare individual orientation patches between pairs of maps on the top row). However, even without any prenatal training (bottom row), or when the network is trained with natural images also prenatally (third row), HLISSOM develops a qualitatively similar final map. In these cases, its organization depends only on the properties of the natural images, not on the internally generated patterns under genetic control. Conversely, even when natural images are replaced by internally generated ones in postnatal training (second row), orientation maps still develop. However, they are not a good match to the visual environment: For example, the orientation histogram is essentially flat. These results suggests that prenatal training is useful mostly because it allows animals to have a functional visual system already at birth, forming a robust starting point for further development. Postnatal training, on the other hand, allows the animal to adapt to the actual visual environment.

for several weeks whether their eyes are open (i.e. when self-organization is driven by visual inputs) or whether they are sutured shut (i.e. when it is presumably still driven by internal patterns). At least for a while, therefore, postnatal self-organization may be influenced by internal inputs. How long this process continues and what its long-term effects are is not known; however, eventually visual inputs will have a significant effect, resulting in representations that match the input statistics, as was discussed in the previous section. It is possible that the role of internally generated inputs gradually changes from self-organization to maintenance, i.e. counteracting excessive adaptation to noisy inputs from environmental input that might otherwise take place. This possibility is discussed in more detail in Section 17.2.4.

When trained with natural images only, the map develops a very similar final organization as the prenatally trained HLISSOM maps (Figure 9.6). This result is interesting because it suggests that prenatal training is not necessary to obtain functional adult maps. However, such training can still be very useful for the animal. The animal will have a functioning orientation detection system immediately at birth, giving it a survival advantage. Prenatal learning may also make postnatal learning more robust against variations in parameter values and random fluctuations in the inputs. The prenatal training patterns are simpler and well separated from each other (Feller et al. 1996), suggesting that well-organized maps will develop under a greater range of conditions than they could for natural images. Prenatal training may also be important for the development of higher areas connected to V1 (e.g. V2 and V4), because it ensures that the map organization in V1 is approximately constant after birth (Figure 9.6). As a result, the higher areas can begin learning appropriate connection patterns with V1 even before eye opening.

Still, it is interesting that orientation maps can develop without any initial order under such widely varying input conditions: oriented patterns, large unoriented patterns, wide variety of natural images, and to some extent, even just noise. It does not seem likely that internally generated patterns would exist only in order to organize low-level maps, if they can be obtained so robustly. Instead, it is possible that the primary effect of prenatal training is to bias the system so that high-level functions are easier to develop. This is the hypothesis studied in detail in the next chapter.

9.5 Discussion

The results in this chapter show that prenatal training on internally generated activity followed by postnatal training on natural images can account for how orientation maps, orientation selectivity, receptive fields, and lateral connections in V1 develop. The same activity-dependent learning rules can explain development based on both internally and externally generated activity. The two types of activity serve important but different roles in this developmental process, and both are crucial for replicating the experimental data.

Comparing orientation maps and RFs trained on random noise vs. those trained on images, disks, or Gaussians suggests that oriented features are needed for realistic receptive fields. Even though rough maps develop without such features, the receptive fields do not match those typically measured in animals. A similar result was recently found independently by Mayer et al. (2001) using single-RF simulations. However, they conclude that natural images are required for realistic RFs, because they did not consider patterns like noisy disks. The results in this chapter suggest that any pattern with large, coherent spots of activity will suffice, and thus that natural images are not strictly required for RF development.

In animals, the map that exists at eye opening has more noise and fewer selective neurons than the prenatally trained maps in Figures 9.1 and 9.2 (Chapman et al. 1996; Crair et al. 1998). As a result, in animals the postnatal improvement in selectivity is larger than that shown here for HLISSOM. The difference may result partly from measurement noise, but also partly from the immature receptive fields in the developing LGN (Tavazoie and Reid 2000). Using a more realistic model of the LGN would allow the map to improve more postnatally, but it would make the model significantly more complex to analyze. Neurons may also appear less selective at birth because the cortical responses vary more in infants. Such behavior could be modeled by adding internal noise to the prenatal neurons, which again would make the model more complex to analyze but would not fundamentally change the self-organizing process.

A recent study has also reported that the distribution of orientation-selective cells matches the environment even in very young ferrets, i.e. that horizontal and vertical orientations are over-represented in orientation maps at eye opening (Chapman and Bonhoeffer 1998). One possible explanation for this result is that the retinal ganglion cells along the horizontal and vertical meridians are distributed nonuniformly (Coppola et al. 1998), which could bias the statistics of internally generated patterns. Even if such a prenatal bias exists, HLISSOM shows how biased visual experience is sufficient for the map to develop preferences that match the visual environment.

9.6 Conclusion

The HLISSOM results show that internally generated activity and postnatal learning can together explain much of the development of orientation preferences. Either type of activity alone can lead to orientation maps, but only with realistic prenatal activity and postnatal learning with real images can the model account for the full range of experimental results. The model also suggests a distinct role for both kinds of inputs: Prenatal learning allows the animal to have a functional visual system at birth, forming a robust starting point for further development, and postnatal learning allows refining it to represent the environment better.

In this chapter, the HLISSOM model was tested in a domain that has abundant experimental data for validation. The next chapter will utilize this map as the first cortical processing stage, and will use the prenatal and postnatal simulation techniques to model how the cortical circuitry develops in the much less well-studied domain of face processing.

Understanding High-Level Development: Face Detection

The previous chapter showed that internally generated patterns and visual experience can explain how orientation preferences develop prenatally and postnatally in V1, a process that is well documented and allows validating the model with neurobiological data. In this chapter, the same ideas will be applied to face detection, which has been extensively studied psychophysically, but where little neurobiological data exist. The simulations will demonstrate that internally generated patterns result in face preferences similar to those observed in newborns. When the system is trained further with real images, it learns faster and more robustly. The time course of learning matches that of human infants, showing a weaker response to schematic patterns and a stronger response to familiar faces. These results complement those for orientation processing, showing how prenatal and postnatal learning could also combine genetic and environmental influences in constructing higher visual function. The psychophysical data and existing theories on infant face detection are first reviewed below, followed by the description of HLISSOM prenatal and postnatal learning experiments.

10.1 Psychophysical and Computational Background

Although the neurobiological foundations of infant face detection are still unknown, it has been studied extensively using psychophysical methods. The experiments have inspired several computational models and theories, which will be reviewed and evaluated in this section.

10.1.1 Psychophysical Data

Although much of the biological data on the visual system comes from cats and ferrets, face-selective neurons or regions have not yet been documented in these animals, either adult or newborn. Even in primates, the data are sparse: The youngest primates that have been tested and found to have face-selective neurons are 6-week-old monkeys (Rodman 1994; Rodman, Skelly, and Gross 1991). Six weeks is a significant amount of visual experience, and it has not yet been possible to measure neurons or regions in younger monkeys. Thus, it is unknown whether the cortical regions that are face selective in adult primates are also face selective in newborns, or whether they are even fully functional at birth (Bronson 1974; Rodman 1994). As a result, how these regions develop remains highly controversial (see de Haan 2001; Gauthier and Logothetis 2000; Gauthier and Nelson 2001; Nachson 1995; Slater and Kirby 1998; Tovée 1998 for reviews).

While measurements at the neuron and region levels are not available, behavioral tests with human infants suggest that face detection develops like orientation maps. In particular, internal, genetically determined factors are also important for face detection. The main evidence comes from a series of studies showing that human newborns turn their eyes or head toward facelike stimuli in the visual periphery longer or more often than they do so for other stimuli (Goren et al. 1975; Johnson, Dziurawiec, Ellis, and Morton 1991; Johnson and Mareschal 2001; Johnson and Morton 1991; Mondloch, Lewis, Budreau, Maurer, Dannemiller, Stephens, and Kleiner-Gathercoal 1999; Simion, Valenza, Umiltà, and Dalla Barba 1998b; Valenza, Simion, Cassia, and Umiltà 1996). These effects have been found within minutes or hours after birth. Figure 10.1 shows how several of these studies have measured the face preferences, and Figure 10.2 shows a typical set of results. Whether these preferences represent genuine preference for faces is controversial, in part because measuring pattern preferences in newborns is difficult (Cohen and Cashon 2003; Easterbrook, Kisilevsky, Hains, and Muir 1999; Hershenson, Kessen, and Munsinger 1967; Kleiner 1987, 1993; Maurer and Barrera 1981; Simion, Cassia, Turati, and Valenza 2001; Slater 1993; Thomas 1965). Newborn preferences for additional patterns will be reviewed in Section 10.2, which also shows that HLISSOM exhibits similar face preferences when trained on internally generated patterns.

Early postnatal visual experience also affects face preferences, as it does how orientation maps develop. For instance, an infant only a few days old will prefer to look at its mother's face, relative to the face of a female stranger with "similar hair coloring and length" (Bushnell 2001) or "broadly similar in terms of complexion, hair color, and general hair style" (Pascalis, de Schonen, Morton, Deruelle, and Fabre-Grenet 1995). A significant mother preference is found even when non-visual cues such as smell and touch are controlled (Bushnell 2001; Bushnell, Sai, and Mullin 1989; Field, Cohen, Garcia, and Greenberg 1984; Pascalis et al. 1995). The infant presumably prefers the mother because he or she has learned the mother's appearance. Indeed, Bushnell (2001) found that newborns look at their mother's face about 1/4 of their time awake over the first few days, which provides ample time for learning.

Pascalis et al. (1995) found that the mother preference disappears when the external outline of the face is masked, and argued that newborns are learning only face outlines, not faces. They concluded that newborn mother learning might differ qualitatively from adult face learning. However, HLISSOM simulation results in Section 10.3 will show that learning of the whole face (internal features and outlines)

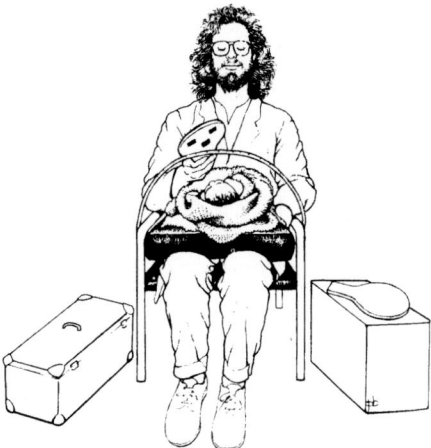

Fig. 10.1. Measuring newborn face preferences. A few minutes or hours after birth, human infants are presented schematic stimuli, measuring how far to the side their eyes or head track each stimulus. The experimenter does not see the specific pattern shown, and neither does the observer who measures the baby's responses. Face preferences have been found even when the experimenter's face and all other faces seen by the baby were covered by surgical masks. Reprinted with permission from Johnson and Morton (1991), copyright 1991 by Blackwell.

can also result in mother preferences. Importantly, masking the outline in HLISSOM also erases these preferences, even though outlines were not the only parts of the face that were learned. Thus, HLISSOM predicts that newborns instead learn faces holistically, as has been suggested for adults (Farah, Wilson, Drain, and Tanaka 1998).

Experiments with infants over the first few months reveal a surprisingly complex pattern of face preferences. Newborns up to 1 month of age continue to track facelike schematic patterns in the periphery, but older infants do not (Figure 10.3; Johnson et al. 1991). Curiously, in central vision, schematic face preferences are not measurable until about 2 months of age (Maurer and Barrera 1981), and they decline by 5 months of age (Johnson and Morton 1991). Section 10.3 will show that in each case such a decline can result from learning real faces, coupled with the different rate of maturation of fovea and periphery in the retina.

In summary, much of the neural basis of face processing is still unclear, in the adult and especially in newborns. However, behavioral experiments suggest that human newborns can detect faces already at birth, and their performance develops postnatally as they experience real faces. These experiments suggest that face-selective neurons develop based on both prenatal and postnatal factors, like neurons in the orientation map. Both cases can be explained based on internally generated neural activity: The system develops through input-driven self-organization both before and after birth.

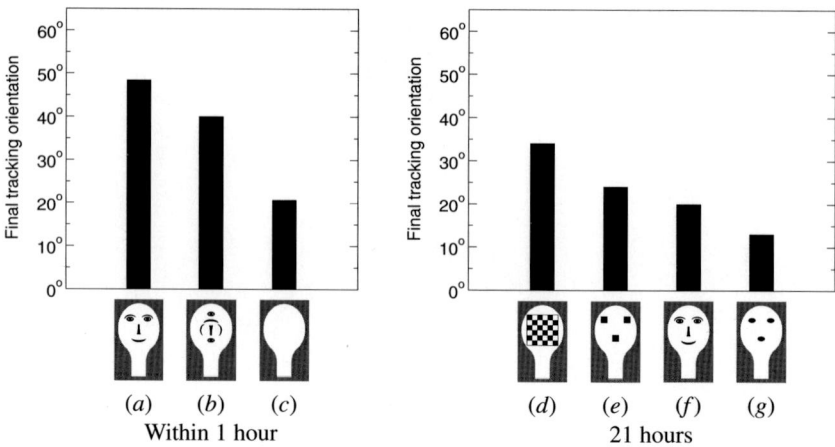

Fig. 10.2. Face preferences in newborns. Using the procedure from Figure 10.1, Johnson et al. (1991) measured responses of human newborns to a set of head-sized schematic patterns. The graph at left gives the result of a study conducted within 1 hour after birth; the one at right gives results from a separate study with newborns an average of 21 hours old. Each bar indicates how far the newborns tracked the image pictured below with their eyes on average. Because the procedures and conditions differed between the two studies, only the relative magnitudes should be compared. Overall, the study at left shows that newborns respond to facelike stimuli (*a,b*) more strongly than to simple control conditions (*c*); all comparisons were statistically significant. This result suggests that face processing is in some way genetically coded. In the study at right, the checkerboard pattern (*d*) was tracked significantly farther than the other stimuli, and pattern (*g*) was tracked significantly less far; no significant difference was found between the responses to (*e*) and (*f*). The ovals are not as visible to the newborn as the square dots, and the checkerboard stimulates newborn's low-level visual system extremely well. These results suggest that simple three-dot patterns can invoke face preferences much like facelike patterns do, but low-level visual stimulation can also have a significant effect. Replotted from Johnson et al. (1991).

10.1.2 Computational Models of Face Processing

The models discussed in this book so far have simulated visual processing only up to V1 and did not include any of the higher cortical areas that are thought to underlie face-processing abilities. Most computational systems that include face processing were not intended as biological models, but instead focus on specific engineering applications such as face detection or face recognition (e.g. Bartlett, Movellan, and Sejnowski 2002; Burton, Bruce, and Hancock 1999; Graham and Allinson 1998; Ko and Byun 2003; Lawrence, Giles, Tsoi, and Back 1997; O'Toole, Millward, and Anderson 1988; Rao and Ballard 1995; Rowley, Baluja, and Kanade 1998; Viola and Jones 2004; Wiskott and von der Malsburg 1996; Yilmaz and Shah 2002; see Phillips, Wechsler, Huang, and Rauss 1998; Yang, Kriegman, and Ahuja 2002 for reviews). A few biologically motivated face processing models exist, but like the engineering systems they either bypass the circuitry in V1 and below, or treat it as a

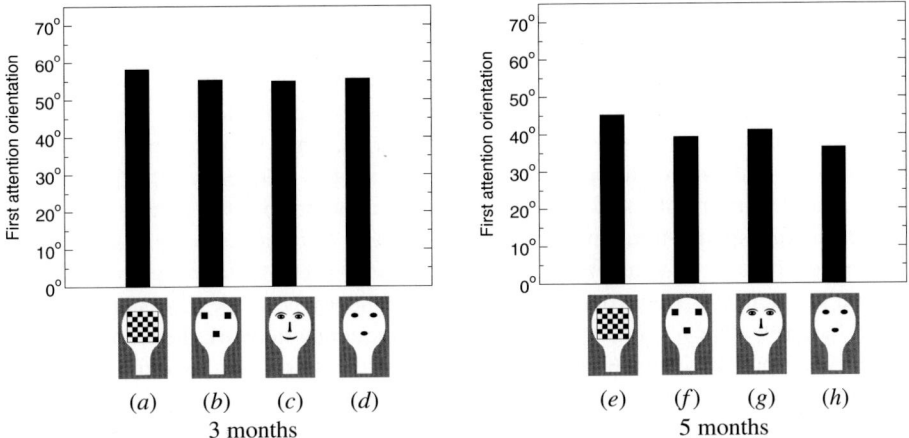

Fig. 10.3. Face preferences in young infants. In addition to newborns (Figure 10.2), Johnson and Morton (1991) also tested how infants at various postnatal ages up to 5 months respond to schematic patterns. They rotated the infant's chair toward the stimulus and measured the angle at which he or she first attended (or oriented) to it. Neither 3-month-old nor 5-month-old infants significantly preferred facelike schematic patterns (*b,c,d* and *f,g,h*) over the controls (*a* and *e*). Results at earlier ages were variable, depending on the testing method (e.g. whether the stimuli were presented in central or peripheral vision). These results suggest that early postnatal visual experience significantly shapes the infant face preferences. Replotted from Johnson and Morton (1991).

fixed set of predefined filters (Acerra, Burnod, and de Schonen 2002; Bartlett and Sejnowski 1997, 1998; Dailey and Cottrell 1999; Gray, Lawrence, Golomb, and Sejnowski 1995; Wallis 1994; Wallis and Rolls 1997; see Valentin, Abdi, O'Toole, and Cottrell 1994 for a review). Given the output of the filtering stage, these models show how face-selective neurons and responses can develop from training with real images. The HLISSOM model of face processing develops such neurons and responses as well. However, it is the first to use a self-organized V1 as the input stage, and the first to demonstrate that the same computational mechanism could be responsible for processing in both V1 and the higher face-processing area. HLISSOM thus unifies these high-level models with the V1 models discussed earlier.

Of the biological models, the Dailey and Cottrell (1999) and Acerra et al. (2002) models have goals most similar to those of this chapter. Acerra et al. (2002) simulated newborn face preferences, and their work will be reviewed in the next subsection. Dailey and Cottrell (1999) instead had a more general focus on whether face detection needs to be genetically encoded. They showed in an abstract model how specific face-selective regions can arise without genetically specifying the weights of each neuron. As was discussed in Section 2.1.3, some of the higher visual areas of the adult human visual system respond more strongly to faces than objects; others have the opposite preferences. Moreover, some of the face-selective areas have been shown to occupy the same region of the brain in different individuals (Kanwisher et al. 1997). This consistency suggests that those areas might be genetically specified for face processing.

To show that such explicit prespecification is not necessary, Dailey and Cottrell (1999) set up a pair of supervised networks that compete with each other to identify faces and to classify objects into categories. They provided one of the networks with real images filtered to preserve low-spatial-frequency information (i.e. slow changes in brightness across a scene), and another with the images filtered to preserve high-spatial-frequency information. These differences correspond to connecting each network to a subset of the neurons in V1, each with different preferred spatial frequencies. They found that the low-frequency network consistently developed face-selective responses, while the high-frequency network developed object-selective responses. Thus, they concluded that different areas may specialize for different tasks based on very simple, general differences in their connectivity, and that specific configuration of individual neurons need not be specified genetically to respond to faces.

Like the Dailey and Cottrell (1999) model, HLISSOM is based on the assumption that cortical neurons are not specifically prewired for face perception. To make detailed simulations practical, HLISSOM will model only a single high-level region, one that has sufficient spatial frequency information available to develop face-selective responses. Other regions presumably develop similarly, but become selective for objects or other image features instead.

10.1.3 Theoretical Models of Newborn Face Preferences

The computational models discussed in the previous section do not specifically explain why newborns should respond strongly to faces. Such explanations have all been conceptual, not computational (with the exception of Acerra et al. 2002). There are four main theories of this phenomenon: (1) the linear systems model, (2) sensory models (including the Acerra et al. (2002) computational model and the top-heavy conceptual model), (3) haptic models, and (4) multiple systems models. These theories will be reviewed below, showing how they compare to the pattern generation model, and arguing that it provides a simpler, more effective explanation.

Linear Systems Model

The linear systems model (LSM; Banks and Salapatek 1981; Kleiner 1993) is a straightforward and effective way of explaining a wide variety of newborn pattern preferences, and could easily be implemented as a computational model. Because it is general and simple, it constitutes a baseline model against which others can be compared. The LSM is based solely on the newborn's measured contrast sensitivity function (CSF). For a given spatial frequency, the value of the CSF will be high if the early visual pathways respond strongly to that size of pattern, and low otherwise. The newborn CSF is limited by the immature state of the eye and the early visual pathways, which makes low frequencies more visible than fine detail.

The LSM assumes that newborns pay attention to those patterns that give the largest response when convolved with the CSF. Low-contrast patterns and patterns with only very fine detail are only faintly visible, if at all, to newborns (Banks and Salapatek 1981). Conversely, faces might be preferred because they have strong spatial-frequency components in the ranges that are most visible to newborns.

However, studies have found that the LSM fails to account for the responses to facelike stimuli. For instance, some of the facelike patterns preferred by newborns have a lower amplitude spectrum in the visible range (and thus lower expected LSM response) than patterns that are less preferred (Johnson and Morton 1991). The LSM also predicts that the newborn will respond equally well to a schematic face regardless of its orientation, because the orientation does not affect the spatial frequency or the contrast. Instead, newborns prefer schematic facelike stimuli oriented right-side-up. Such a preference is found even when the inverted stimulus is a better match to the CSF (Valenza et al. 1996). Thus, the CSF alone does not explain face preferences, and a more complex model is required.

Acerra et al. Sensory Model

The LSM is a high-level abstraction of the properties of the early visual system. Sensory models extend the LSM to include additional constraints and circuitry, but without adding face-selective visual regions or systems. Acerra et al. (2002) recently developed such a computational model that can account for some of the face preferences found in the Valenza et al. (1996) study. Their model consists of a fixed Gabor-filter-based model of V1, plus a high-level sheet of neurons with modifiable connections. They model two conditions separately: newborn face preferences, and postnatal development by 4 months. The newborn model includes only V1, because they assume that the high-level sheet is not yet functional at birth.

Acerra et al. showed that the newborn model responds slightly more strongly to the upright schematic face pattern used by Valenza et al. (1996) than to the inverted one. This surprising result replicates the newborn face preferences found by Valenza et al. In the stimuli, only the internal facial features were inverted, not the entire pattern. In the upright case, the spacing is more regular between the internal features and the face outline (compare Figure 10.7*d* with 10.7*g*, top row). As a result, neurons whose RF lobes match the spacing respond more strongly, and the total response of all filters will be slightly higher for the facelike (upright) pattern than to the non-facelike (inverted) pattern.

However, the Acerra et al. model was not tested with patterns from other studies of newborn face preferences, such as Johnson et al. (1991). The facelike stimuli published by Johnson et al. (1991) do not have a regular spacing between the internal features and the outline, and it is unlikely that the model will replicate preferences for these patterns. Moreover, Johnson et al. used a white paddle against a light-colored ceiling, and so their face outlines would have a much lower contrast than the black-background patterns used by Valenza et al. (1996). Thus, although border effects may have contributed to the face preferences found by Valenza et al., they are unlikely to explain those measured by Johnson et al.

The Acerra et al. newborn model also does not explain newborn learning of faces, because their V1 model is fixed and the high-level area is assumed not to be functional at birth. Also importantly, the model was not tested with real images of faces, where the spacing of the internal features from the face outline varies widely depending on the way the hair falls. Because of these differences, we do not expect the Acerra et al. model to show a significantly higher response overall to photographs of real faces than to other similar images. The pattern-generation model will make the opposite prediction, and will explain how newborns can learn faces.

To explain learning of real faces in older infants, the Acerra et al. model relies on having face images strictly aligned in the input, having nothing but faces presented to the model (no objects, bodies, or backgrounds), and having the eyes in each face artificially boosted by a factor of 10 or 100 relative to the rest of the image. Because of these assumptions, it is difficult to evaluate how well their postnatal learning model corresponds to experimental data. In contrast, the HLISSOM model learns from faces presented at random locations on the retina, against natural image backgrounds, intermixed with images of other objects, and without special emphasis for faces relative to the other objects.

Top-Heavy Sensory Model

Simion et al. (2001) also presented a sensory model of newborn preferences, although their model is conceptual only. They observed that nearly all of the facelike schematic patterns that have been tested with newborns are top-heavy, i.e. they have a boundary with denser patterns in the upper than the lower half. They also ran behavioral experiments showing that newborns prefer several top-heavy (but not facelike) schematic patterns to similar but inverted patterns. Based on these results, they proposed that newborns prefer top-heavy patterns in general, and thus prefer facelike schematic patterns as a special case.

This hypothesis is compatible with most of the experimental data so far collected in newborns. However, facelike patterns have not yet been compared directly with other top-heavy patterns in newborn studies. Thus, it is not yet known whether newborns would prefer a facelike pattern to a similarly top-heavy but not facelike pattern. Future experimental tests with newborns can resolve this issue.

To be tested computationally, the top-heavy hypothesis would need to be made more explicit, with a specific mechanism for locating object boundaries and the relative locations of patterns within them. It would then be possible to test it with a variety of inputs, including photographs of real faces. Whereas the bulk of the current evidence suggests that newborns prefer face patterns in general, we expect that a computational test of the top-heavy model would find only a small preference (if any) for real faces, compared with many other common stimuli. Many real faces, such as those with beards, wide smiles, or wide-open mouths, are not necessarily top heavy, and would result in little or no response from the model.

This prediction is also supported by a systematic test of training pattern shapes with HLISSOM, presented in Section 10.2.6. Although many simple shapes including general top-heavy patterns result in weak face preferences, more facelike patterns are necessary to obtain selective responses that allow matching the behavior of the model with newborn data. Such patterns will therefore be used in training the HLISSOM model of newborn face preferences.

Haptic Hypothesis

Bushnell (1998) proposed an explanation very different from that of the sensory models: A newborn may recognize facelike stimuli as a result of prenatal experience with its own face, via manual exploration of its facial features. Some support for this position comes from findings that newborn and infant monkeys respond equally or more strongly to pictures of infant monkeys than to adults (Rodman 1994; Sackett 1966).

However, the process by which a newborn could make such specific connections between somatosensory and visual stimulation, prior to visual experience, is not clear. Moreover, the haptic explanation does not account for several aspects of newborn face preferences. For instance, premature babies develop face preferences at the same post-conception age regardless of the age at which they were born (Ferrari, Manzotti, Nalin, Benatti, Cavallo, Torricelli, and Cavazzutti 1986). Presumably, the patterns of hand and arm movements would differ between the intrauterine and external environments, and thus the haptic hypothesis would predict that gestation time should have been an important factor. Several authors have also pointed out strong similarities between newborn face preferences and imprinting in newly hatched chicks; chicks, of course, do not have hands with which to explore, yet develop a specific preference for stimuli that resemble a (chicken's) head and neck (Bolhuis 1999; Horn 1985). Some of these objections are overcome in a variant of the haptic hypothesis by Meltzoff and Moore (1993), who propose that direct proprioception of the infant's own facial muscles is responsible.

However, neither variant can account for evidence suggesting that newborns' preferences are specifically visual. For instance, newborns respond as well to patterns with a single dot in the nose and mouth area as to separate patterns for the nose and mouth (Johnson et al. 1991). This finding is easy to explain for visual images: In a blurred top-lit visual image, shadows under the nose blend together with the mouth into a single region. But the nose and mouth have opposite convexity, so it is difficult to see how they could be considered a single region for touch stimulation or proprioception. Similarly, newborns have so far only been found to prefer faces viewed from the front, and it is not clear why manual exploration or proprioception would favor that view in particular. Thus, in this chapter the newborn face preferences are assumed to be essentially visual.

Multiple-Systems Models

The most widely known conceptual model for newborn face preferences and later learning was proposed by Johnson and Morton (1991). Apart from HLISSOM simulations in this chapter that test some of its foundations, it has not yet been evaluated computationally.

Johnson and Morton proposed that infant face preferences are mediated by two hypothetical visual processing systems that they dubbed CONSPEC and CONLERN. CONSPEC is a fixed system controlling orienting to facelike patterns, assumed to be located in the subcortical superior colliculus–pulvinar pathway. Johnson and Morton proposed that a CONSPEC responding to three dark dots in a triangular configuration, one each for the eyes and one for the nose/mouth region, would account for the newborn face preferences (see Figures 10.5a and 10.6c for examples).

CONLERN is a separate plastic cortical system, presumably corresponding to the face-processing areas that have been found in adults and in infant monkeys (Kanwisher et al. 1997; Rodman 1994). The CONLERN system would assume control only after about 6 weeks of age, and would account for the face preferences seen in older infants. Eventually, as it learns from real faces, CONLERN would gradually stop responding to schematic faces, which would explain why face preferences can no longer be measured with static schematic patterns by 5 months (Johnson and Morton 1991).

The CONSPEC/CONLERN model is plausible, given that the superior colliculus is relatively mature in newborn monkeys and is involved in controlling attention and other functions (Wallace et al. 1997). Moreover, some neurons in the adult superior colliculus/pulvinar pathway are selective for faces (Morris, Ohman, and Dolan 1999), although such neurons have not been found in young animals. The model also helps explain why infants are less interested in faces in the periphery after 1 month: The preferences may change as the attentional control shifts to the not-quite-mature cortical system (Johnson et al. 1991; Johnson and Morton 1991).

However, subsequent studies showed that even newborns are capable of learning individual faces (Slater 1993; Slater and Kirby 1998). Thus, if there are two visual processing systems, either both are plastic or both are functioning at birth, and thus there is no a priori reason why a single face-selective visual system would be insufficient. On the other hand, de Schonen, Mancini, and Liegeois (1998) argue for *three* visual processing systems: a subcortical one responsible for facial feature preferences at birth, another one responsible for newborn learning (of objects and head/hair outlines; Slater 1993), and a cortical system responsible for older infant and adult learning of facial features. And Simion, Valenza, and Umiltà (1998a) proposed that face selectivity instead relies on multiple visual processing systems *within* the cortex, maturing first in the dorsal stream but later supplanted by the ventral stream (which is where most of the face-selective visual regions have been found in adult humans).

In contrast to the increasing complexity of these explanations, the HLISSOM model shows that a single general-purpose, plastic visual processing system is sufficient, if that system is first exposed to internally generated facelike patterns of neural activity. As reviewed in Section 2.3.4, PGO activity waves during REM sleep represent a likely candidate for such activity. If the PGO waves have the simple three-dot configuration illustrated in Figures 8.4c and 10.5a, they can explain the measured face-detection performance of human newborns.

The three-dot training patterns are similar to the three-dot preferences proposed by Johnson and Morton (1991). However, in their model the patterns were implemented as a hard-wired subcortical visual area responsible for orienting to faces. In HLISSOM, the areas receiving visual input learn from their input patterns, and only the pattern generator is assumed to be hard-wired. Both of these possible mechanisms require about the same amount of genetic specification. The crucial difference is that in the pattern generation approach the visual processing system can be arbitrarily complex because it learns the complexity from the input. In contrast, a hard-coded visual processing system like CONSPEC is limited by what can specifically be encoded in the genome. Such a system is a plausible model for subcortically mediated orienting at birth, but less plausible as a model of a cortical areas. In the future it may be possible to use imaging techniques to determine whether newborn face preferences are based on subcortical processes, or mediated by a hierarchy of cortical areas as they are in the adult.

Specifying only the training patterns also makes sense from an evolutionary perspective, because that way the patterns and the visual processing hardware can evolve independently. Separating these capabilities thus allows the visual system to become arbitrarily complex, while maintaining the same genetically specified function. Finally, assuming that only the pattern generator is hardwired explains how infants of all ages can learn faces.

In conclusion, previous face-processing models have not yet shown how newborns can have a significant preference for real faces at birth, or how newborns could learn from real faces. They have also not accounted for the full range of patterns with which newborns have been tested. Using the self-organized orientation map, the HLISSOM model of face detection will show how face-selective neurons can develop through internally generated activity, explaining face preferences at birth and later face learning. In the next two sections, the prenatal and postnatal phases of this process are each discussed in turn.

10.2 Prenatal Development

In this section, internally generated patterns are used to self-organize the full HLISSOM model prenatally, including both V1 and the FSA. Its performance on schematic facelike patterns is then shown to be remarkably similar to that of human infants. Importantly, the trained model responds strongly to real faces as well, but not to other naturally occurring objects. This behavior can be obtained with a variety of internally generated patterns that match the general outline of the human face.

10.2.1 Training Method

Previous developmental models have been tested only for small input areas such as those used in Chapters 4–9. In contrast, face detection requires processing head-sized stimuli at a distance of about 20 cm from the baby's eyes, filling a substantial portion of the visual field (about 45°). To model behavior at this scale, the orientation map simulation from Section 9.2 was expanded to a very large V1 area (approximately 1600 mm² in total) at a relatively low sampling density (approximately 50 neurons/mm²). The expansion was done using the scaling equations in Appendix A.2, which allow adjusting the simulation parameters algorithmically to obtain a simulation of a different size (the scaling equations will be described in detail in Chapter 15).

The cortical density was first reduced to the minimum value that would show an orientation map that matches animal maps (36×36), and the visual area was scaled to be just large enough to cover the visual images to be tested (288×288, i.e. width $\times 8$ and height $\times 8$). The FSA size was less crucial, and was set arbitrarily at 36×36. The FSA RF size was scaled to be large enough to span the central portion of a face input. The resulting network consisted of 438×438 retinal units, 220×220 PGO generator units, 204×204 ON-center LGN units, 204×204 OFF-center LGN units, 288×288 V1 units, and 36×36 FSA units, for a total of 408,000 distinct units. There were 80 million connections in total in the two cortical sheets, which required 300 MB of physical memory.

The RF centers of neurons in the FSA were mapped to the central 160×160 region of V1 such that even the units near the edge of the FSA had a complete set of afferent connections on V1, with no FSA RF extending over its edge. V1 was similarly mapped to the central 192×192 region of the LGN channels, and the LGN channels to the central 384×384 region of the retina and the central 192×192 region of the PGO generators. In the figures in this chapter, only the area mapped directly to V1 will be shown, to ensure that all plots have the same scale. The size of the input patterns on the retina was chosen to match the preferred spatial frequency of newborns, as cited by Valenza et al. (1996). Inputs were presented at the spatial scale where the frequency most visible to newborns produced the largest V1 response in the model. The rest of the simulation parameters are listed in Appendix C.3.

V1 was self-organized for 10,000 iterations with inputs consisting of 11 randomly located circular disks per iteration, each 50 units wide (Figure 10.4a). These simple patterns were used for clarity and simplicity, since the focus of these simulations is on the development of the high-level FSA region. If the V1 size were increased to provide more units for each location in the retina, the noisy disks patterns from the previous chapter could also have been used. The background activity level was 0.5, and the brightness of each disk relative to this surround (either +0.3 or −0.3) was chosen randomly. The borders of each disk were smoothed into the background level following a Gaussian width $\sigma_d = 1.5$.

The FSA was trained for 10,000 iterations based on the responses of the V1 network. Two triples of dark circular dots were used as input to V1, each arranged in a triangular facelike configuration (Figure 10.5a). As was discussed in Section 10.1.3, such patterns roughly correspond to the eye and nose/mouth areas of the face, as proposed by Johnson and Morton (1991). Each dot had a radius of 10 PGO units; the centers of the two top dots were separated by 50 units and they were located 54 units from the bottom dot. The dot was 0.3 units darker than the surround, which itself was 0.5 on a scale of 0 to 1. Each triple was placed at a random location each iteration, at least 118 PGO units away from the center of the other one to avoid overlap. The

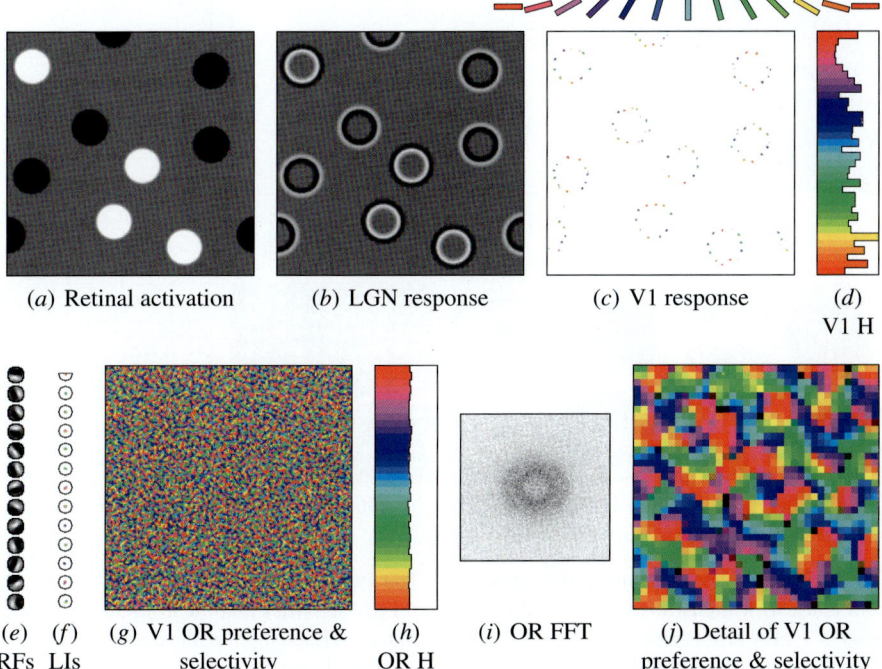

(a) Retinal activation   (b) LGN response   (c) V1 response   (d)
V1 H

(e) (f)   (g) V1 OR preference &   (h)   (i) OR FFT   (j) Detail of V1 OR
RFs LIs   selectivity   OR H   preference & selectivity

Fig. 10.4. Self-organization of the scaled-up orientation map. These figures show a scaled-up version of the "Disks" simulation from Figure 9.1, eight times wider and eight times taller. At this scale, each input includes multiple disks (a), the afferent weights of each neuron span only a small portion of the retina (drawn to scale in e) and the lateral weights only a small part of V1 (drawn to scale in f), and the orientation map has many more orientation patches (g). Its Fourier transform (i) is still ring-shaped and its OR histogram (h) flat. Zooming in on the central 36×36 portion of this 288×288 map, plot (j) also shows that the local structure and selectivity of the map are similar to those in Section 9.2. The map appears blockier because the neuron density was reduced to the smallest acceptable value so that the network would be practical to simulate. Plot (c) shows that the orientation preference of each neuron that responds to the input (plotted as in Figure 6.5) is still a good match to the orientation of the input at that retinal location, and the histogram of the responses is unbiased, although noisy (d). Thus, this network is a reasonable approximation to a large area of V1 and the retina.

angle of each triple was drawn randomly from a narrow ($\sigma = \pi/36$ radians) normal distribution around vertical.

Because the model is very large and expensive to simulate, V1 was trained first, followed by the FSA. This arrangement reduces computational cost without significantly affecting how face preferences develop, which is the focus of the next subsection.

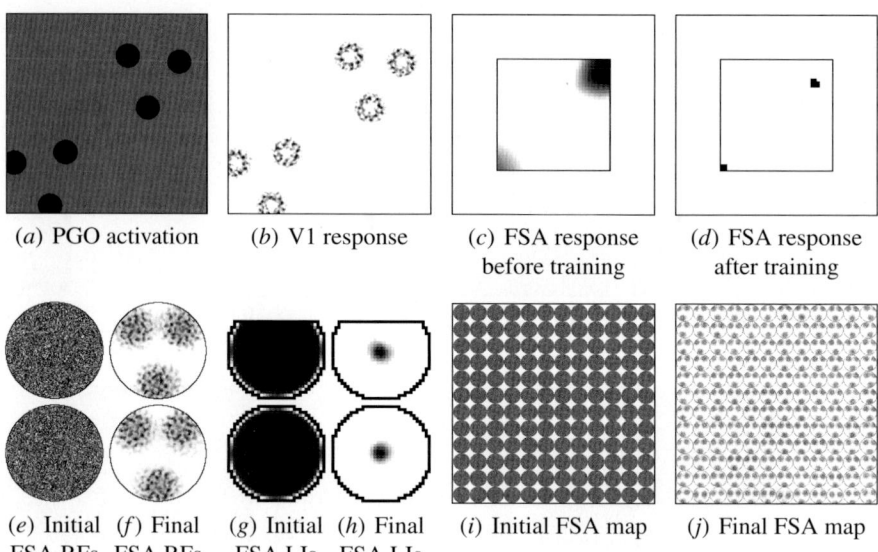

(*a*) PGO activation   (*b*) V1 response   (*c*) FSA response   (*d*) FSA response
before training   after training

(*e*) Initial (*f*) Final   (*g*) Initial (*h*) Final   (*i*) Initial FSA map   (*j*) Final FSA map
FSA RFs FSA RFs   FSA LIs FSA LIs

Fig. 10.5. Self-organization of the FSA map. The PGO activation is shown in gray scale from black to white (low to high), and the V1 and FSA activities and the afferent and lateral weights in gray scale from white to black (low to high). (*a*) Each input pattern consisted of two dark three-dot configurations with random nearly vertical orientations presented at random locations on the PGO sheet. (*b*) The V1 neurons compute their responses based on this input, relayed through the LGN. (*c*) FSA neurons initially respond to any activity in their receptive fields, but after training (*d*), only neurons with closely matching RFs respond. In the FSA plots, the inner square represents the FSA and is drawn to scale with the retina. The outer square is provided to help locate the FSA responses on the retina, as was done in Figures 4.4 and A.1(*a*). Through self-organization, the FSA neurons develop RFs selective for a range of V1 activity patterns like those resulting from the three-dot stimuli (*e* and *f*, drawn in the same scale as *b* for two sample neurons). The RFs are patchy because the weights target specific orientation patches in V1. This match between the FSA and the local self-organized pattern in V1 would be difficult to ensure without training on internally generated patterns. The FSA neurons also develop lateral inhibitory connections with a smooth Gaussian profile (*g* and *h*, drawn in the same scale as *c* and *d* for the two neurons in *e* and *f*). Plots (*i*) and (*j*) show the afferent weights for every third neuron in the FSA. All neurons develop roughly similar weight profiles, differing primarily by the position of their preferred stimuli on the retina and by the specific orientation patches targeted in V1. The largest differences between RFs are along the outside border, where the neurons are less selective for three-dot patterns. Overall, the FSA develops into a face detection map, signaling the location of facelike stimuli.

10.2.2 V1 and Face-Selective Area Organization

Through self-organization, the scaled-up map shown in Figure 10.4*g* emerged. The map has the same structural features and similar quantitative measures as the previous LISSOM OR maps (Chapters 5 and 9). However, as was seen in Figure 8.2, this scaled-up model can extract the salient local orientations even in large images, which has not yet been demonstrated for other models of V1 development.

After V1 had been trained, its weights were fixed and the FSA was allowed to activate and learn from the V1 responses to the three-dot patterns. The resulting face-selective map consists of an array of neurons that respond most strongly to patterns similar to the training patterns (Figure 10.5*j*). Despite the overall similarities between neurons, the individual weight patterns are unique because each neuron targets specific orientation patches in V1. Such complicated patterns would be difficult to specify and hardwire genetically, but they arise naturally from internally generated activity.

10.2.3 Testing Method

As the FSA training completed, the lower threshold of the sigmoid (i.e. θ_1 in Figure 4.5) was increased to ensure that only patterns that are a strong match to an FSA neuron's weights will activate it. An active neuron in the FSA thus indicates that there is a face at the corresponding location in the retina, which will be important for measuring the face-detection ability of the network.

The cortical maps were then tested on natural images and with the same schematic stimuli on which human newborns have been tested (Goren et al. 1975; Johnson and Morton 1991; Simion et al. 1998b; Valenza et al. 1996). For all such tests, the same parameter settings described in Appendix C.3 were used. Schematic images were scaled to a brightness range of 1.0 (i.e. the difference between the darkest and lightest pixels in the image). Natural images were scaled to a brightness range of 2.5, so that facial features in images with faces would have a contrast comparable to that of the schematic images. If both types of images were instead scaled to the same brightness range, the model would prefer the schematic faces over real ones. Using different scales is appropriate for these experiments because responses are compared only within groups of similar images, and the infant is assumed to adapt to each group. For simplicity, the model does not specifically include the mechanisms in the eye, LGN, and V1 that are responsible for such contrast adaptation.

The HLISSOM model provides detailed neural responses for each neural region. These responses constitute predictions for future electrophysiological and imaging measurements in animals and humans. However, the data currently available for infant face perception are behavioral. It consists of newborn attention preferences measured from visual tracking distance and looking time. Thus, validating the model on these data will require predicting a behavioral response based on the simulated neural responses.

As a general principle, newborns are assumed to pay attention to the stimulus whose overall neural response most clearly differs from those of typical stimuli. This idea can be quantified as

$$a(t) = \frac{F(t)}{\overline{F}} + \frac{V(t)}{\overline{V}} + \frac{L(t)}{\overline{L}}, \tag{10.1}$$

where $a(t)$ is the attention level at time t, and F, V, and L represent the FSA, V1, and LGN regions: $X(t)$ (either F, V, or L) is the total activity in region X at that time, while $\overline{X}$ is the average (or median) activity over the recent history. Because most stimuli activate the LGN and V1 but not the FSA, when a pattern evokes activity in the FSA the newborns would attend to it more strongly. Yet, stimuli evoking only V1 activity could still be preferred over facelike patterns if their V1 activity is much higher than typical.

Unfortunately, it is difficult to use such a formula to compare to newborn experiments, because the presentation order in those experiments is usually not known. As a result, the average or median value of patterns in recent history is not available. Furthermore, the numerical preference values computed in this way will differ depending on the specific set of patterns chosen, and thus will be different for each study.

Instead, a categorical approach inspired by Cohen (1998) will be used in the simulations below; it avoids these problems and leads to similar results. Specifically, when two stimuli both activate the model FSA, the one with the higher total FSA activation will be preferred. Similarly, with two stimuli activating only V1, the higher total V1 activation will be preferred. When one stimulus activates only V1 and another activates both V1 and the FSA, the pattern that produces FSA activity will be preferred, unless the V1 activity is much larger than for typical patterns. Using these guidelines, the computed model preferences can be validated against the newborn's looking preferences, to determine if the model shows the same behavior as the newborn.

10.2.4 Responses to Schematic Patterns

This section and the following one present the model's response to schematic patterns and real images after it has completed training on the internally generated patterns. The response to each schematic pattern is compared with behavioral results from infants, and the responses to real images constitute predictions for future experiments.

Figures 10.6 and 10.7 show that the model responses match the measured stimulus preferences of newborns remarkably well, with the same relative ranking in each case where infants have shown a significant preference between schematic patterns. These rankings are the main result from this simulation. Each category of schematic patterns is next analyzed in more detail, to understand what aspects of the model are responsible for the result.

Most non-facelike patterns activate only V1, and thus the preferences between those patterns are based only on the V1 activity values (Figures 10.6a,e–i and 10.7f–i). Patterns with numerous high-contrast edges have greater V1 response, which explains why newborns would prefer them. These preferences are in accord with the simple linear systems model (Section 10.1.3), because they are based only on the early visual processing.

Facelike schematic patterns activate the FSA, whether they are realistic or simply patterns of three dots (Figures 10.6b–d and 10.7a–d). The different FSA activation levels reflect the level of V1 response, not the precise shape of the pattern. Again the responses explain why newborns would prefer patterns like the three-square face over the three-oval face, which has shorter edges. The preferences between these patterns are also compatible with the LSM, because in each case the strength of the V1 response in the model matches newborn preferences.

The comparisons between facelike and non-facelike patterns show how HLISSOM predictions differ from the LSM. HLISSOM predicts that the patterns that activate the FSA would be preferred over those activating only V1, except when the V1 response is highly anomalous. Most V1 responses are very similar, and so the patterns with FSA activity (Figures 10.6b–d and 10.7a–d) should be preferred over most of the other patterns, as found in infants. However, the checkerboard pattern (Figure 10.6a) has nearly three times as much V1 activity as any other pattern with a similar background. Thus, the HLISSOM results explain why the checkerboard would be preferred over other patterns, even ones that activate the FSA.

Because the RFs in the model are only a rough match to the schematic patterns, the FSA can have spurious responses to patterns that to adults do not look like faces. For instance, the inverted three-dot pattern in Figure 10.7e activated the FSA. In this case, part of the square outline filled in for the missing third dot of an upright pattern. Figure 10.8 shows that such spurious responses should be expected with inverted patterns, even if neurons prefer upright patterns. This result may explain why some studies have not found a significant difference between upright and inverted three-dot patterns (e.g. Johnson and Morton 1991). Because of such effects, future experimental studies should include additional controls besides inverted three-dot patterns.

Interestingly, the model also showed a clear preference in one case where no significant preference was found in newborns (Simion et al. 1998a): for the upright three-dot pattern with no face outline (Figure 10.7a), over the similar but inverted pattern (Figure 10.7i). The V1 responses to both patterns are similar, but only the upright pattern has an FSA response, and thus the model predicts that the upright pattern would be preferred.

This potentially conflicting result may be due to postnatal learning rather than capabilities at birth. As will be shown in the next section, postnatal learning of face outlines can have a strong effect on FSA responses. The newborns in the Simion et al. study were already 1–6 days old, and Pascalis et al. (1995) showed that newborns within this age range have already learned some of the features of their mother's face outline. Thus, the pattern generation model predicts that if younger newborns are tested, they will prefer upright patterns even without a border. Alternatively, the border may satisfy some minimum requirement on size or complexity for patterns to attract a newborn's interest. For instance, the total V1 activity may need to be above a certain threshold for the newborn to pay any attention to a pattern. If so, the HLISSOM procedure for deriving a behavioral response from the model response (Section 10.2.3) would need to be modified to include this constraint.

Overall, these results provide strong computational support for the speculation of Johnson and Morton (1991) that the newborn could simply be responding to a three-dot facelike configuration, rather than performing sophisticated face detection. Internally generated patterns provide an account for how such "innate" machinery

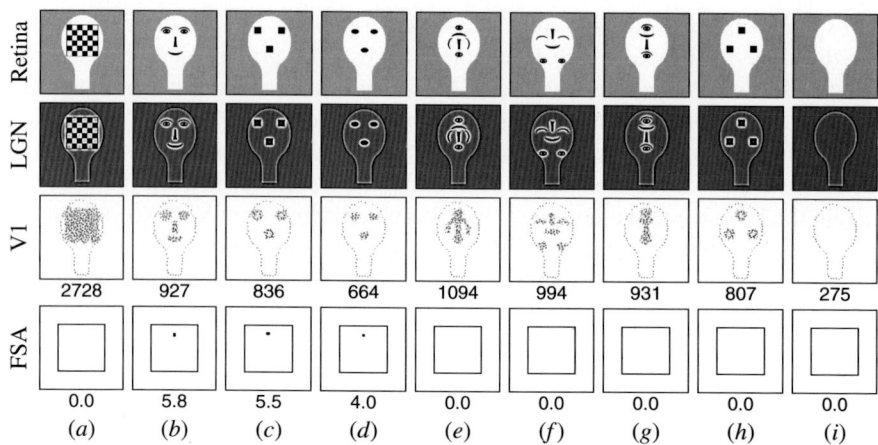

Fig. 10.6. Response to schematic images by Goren et al. (1975) and Johnson et al. (1991). The activations of the retina, LGN, V1, and FSA levels are shown using the plotting conventions from Figures 10.4 and 10.5. The top row shows a set of input images as they are drawn on the retina. These patterns were presented to newborn human infants on head-shaped paddles moving at a short distance (about 20 cm) from the eyes, against a light-colored ceiling. The newborn's preference was determined by measuring the average distance his or her eyes or head tracked each pattern, compared with other patterns. Below, $x>y$ indicates that image x was preferred over image y under those conditions. Goren et al. (1975) measured infants between 3 and 27 minutes after birth. They found that $b>f>i$ and $b>e>i$. Similarly, Johnson et al. (1991), in one experiment measuring within 1 hour after birth, found $b>e>i$. In another, measuring at an average of 43 minutes, they found $b>e$, and $b>h$. Finally, Johnson and Morton (1991), measuring newborns an average of 21 hours old, found that $a>(b,c,d)$, $c>d$, and $b>d$. The HLISSOM model has the same preference for each of these patterns, as shown in the images above. The second row shows the model LGN activations resulting from the patterns in the top row. The third row shows the V1 activations, with the numerical sum of the activities shown underneath. If only one unit were active at half strength, the sum would be 0.5; higher values indicate more activation. The bottom row displays the settled responses of the FSA, again with the numerical sum underneath. This sum represents the strength of the response of the model. The images are sorted left to right according to the preferences of the model. The strongest V1 response by nearly a factor of three is to the checkerboard pattern (a), which explains why the newborn would prefer that pattern over the others. The facelike patterns (b–d) are preferred over patterns (e–i) because of activation in the FSA. The details of the facelike patterns do not significantly affect the results — all of the facelike patterns (b–d) lead to FSA activation, generally in proportion to their V1 activation levels. The remaining patterns are ranked by their V1 activity alone, because they do not activate the FSA. In all conditions tested, the HLISSOM model shows behavior remarkably similar to that of the newborns, and provides a detailed computational explanation for why these behaviors occur. Reprinted from Bednar and Miikkulainen (2003a).

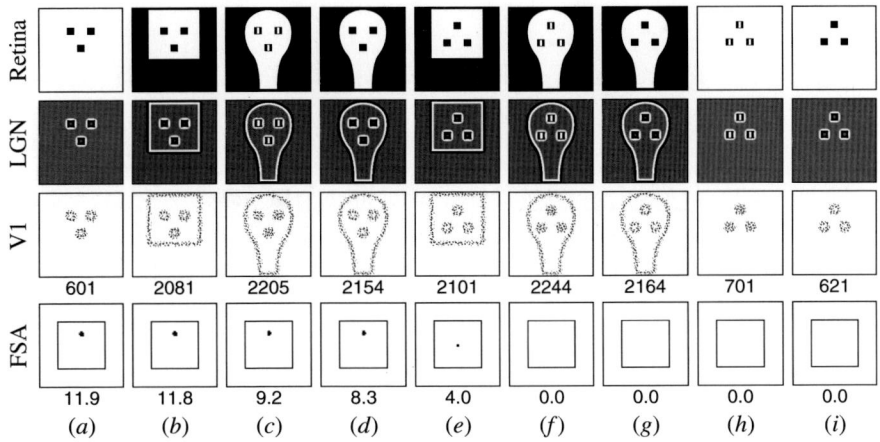

Fig. 10.7. Response to schematic images by Valenza et al. (1996) and Simion et al. (1998a).
Valenza et al. measured preference between static, projected versions of pairs of the schematic
images in the top row, using newborns ranging from 24 to 155 hours after birth. They found the
following preferences: $d>f$, $d>g$, $f>g$, and $h>i$. Simion et al. similarly found a preference for
$d>g$ and $b>e$. The LGN, V1, and FSA responses of the model to these images are displayed
here as in Figure 10.6, and are again sorted by the model's preference. In all cases where the
newborn preferred one pattern over another, so did the model. For instance, the model FSA
responds to the facelike pattern (d) but not to the inverted version (g). Patterns that closely
match the newborn's preferred spatial frequency (f and h) caused a greater V1 response than
their low-frequency versions (g and i). Some non-facelike patterns with high-contrast borders
can cause spurious FSA activation (e), because part of the border completes a three-dot pattern.
Such spurious responses did not affect the predicted preferences, because they are smaller than
the genuine responses (see Figure 10.8 for more details on how spurious responses typically
arise). Interestingly, Simion et al. found no preference between (a) and (i) in 1–6-day-old
infants. The model predicts that (a) would be preferred at birth, due to the FSA response, but
not by older infants who have learned face outlines postnatally. Reprinted from Bednar and
Miikkulainen (2003a).

can be constructed during prenatal development, within a system that can also learn
postnatally from visual experience.

10.2.5 Responses to Natural Images

Researchers testing newborns with schematic patterns often assume that the responses to schematics are representative of responses to real faces. However, no
experiment has yet tested that assumption by comparing real faces to similar but
non-facelike controls. Similarly, no previous computational model of newborn face
preferences has been tested with real images. HLISSOM makes such tests practical, providing an important way to determine whether the behavioral data based on
schematics can be used to predict preferences for real faces.

Face detection performance of the model was tested quantitatively using two image databases: a set of 150 images of 15 adult males without glasses, photographed

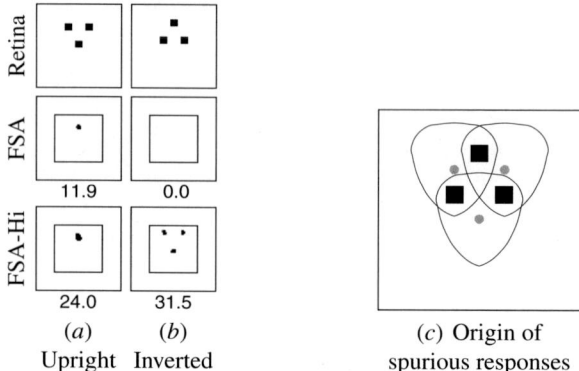

(a)   (b)   (c) Origin of
Upright   Inverted   spurious responses

Fig. 10.8. Spurious responses to the inverted three-dot pattern. Several studies have used an inverted three-dot pattern as a non-facelike control for an upright three-dot pattern with conflicting results (e.g. Johnson and Morton 1991; Simion et al. 1998a; Valenza et al. 1996). However, the results with HLISSOM show that this pattern does not make a good control, because of the many axes of symmetry of a three-dot pattern. The first two rows in (a) and (b) are reproduced from Figure 10.7a,i, and show that HLISSOM prefers the facelike upright pattern to the control. However, the preference is sensitive to the value of the FSA threshold θ_1 and the FSA input scale γ_A. For instance, if γ_A is increased by 30%, the model FSA responds more strongly to the inverted pattern (bottom row). The inverted pattern is not as good a match for any single neuron's weights, so the FSA activity spots are always smaller for the inverted pattern. However, with a high enough γ_A, the FSA responds in three different places (b) compared with only one for the upright pattern (a), and together the three small responses outweigh the single larger response. Plot (c) demonstrates how such spurious FSA responses arise in the model. These responses are shown superimposed on the retinal pattern as three small dots, and the outlines indicate the three-dot patterns that they represent. Each pattern shares two dots with the inverted input, shown as three black squares; these two shared dots are enough to activate the unit. In HLISSOM, γ_A is set to a value low enough to prevent such spurious responses, which ensures that FSA neurons respond only to patterns that are a good match to their (upright) RFs. For humans, the γ_A value represents the state of contrast adaptation at a given time, which varies depending on the recent history of patterns seen (Albrecht et al. 1984; Turrigiano 1999). Thus, these results suggest that infants will have no preference (or will prefer the inverted pattern) if they are tested on the high-contrast schematic patterns while being adapted to the lower contrast levels typical of the environment. Because such adaptation is difficult to control in practice, the inverted pattern is a problematic comparison pattern — negative results like those of Johnson and Morton (1991) may be due to temporary contrast adaptation instead of genuine, long-term pattern preferences.

at the same distance against blank backgrounds (Achermann 1995), and a set of 58 non-face images of various natural scenes (National Park Service 1995). The face image set contained two views of each person facing forward, upward, downward, left, and right (as shown in Figures 10.9a and 10.10f–i). Each natural scene was presented at six different size scales, for a total of 348 non-face presentations (examples shown in Figure 10.9f–i).

Fig. 10.9. Response to natural images. The top row shows a sample set of photographic images. The corresponding LGN, V1, and FSA responses are displayed as in Figures 10.6 and 10.7. The FSA is indeed activated at the correct location for most faces of the correct size and orientation (e.g. *a–d*), including 88% of those in the Achermann (1995) database. Just as importantly, the network is not activated for most natural scenes and man-made objects (*f–h*). In fact, the FSA responded to only 4.3% of 348 presentations of landscapes and other natural scenes from the National Park Service (1995). The spurious activations usually result from a V1 activation similar to that of a three-dot arrangement of contours (*d* and *i*), including related patterns such as dog and monkey faces (not shown). Response is low to images where hair or glasses obscure the borders of the eyes, nose, or mouth, and to front-lit downward-looking faces, which have low V1 activation from nose and mouth contours (*e*). The model predicts that newborns respond in the same way if tested. Credits: (*a*) copyright 1995 by University of Bern (Achermann 1995), (*b–e*) public domain; (*f–i*) copyright 1999–2001 by James A. Bednar. Reprinted from Bednar and Miikkulainen (2003a).

Overall, the HLISSOM performed very well as a face detection system: The FSA responded to 91% (137/150) of the face images, but to only 4.3% (15/348) of the natural scenes. Because the two sets of real images were not closely matched in terms of lighting, background, and distance, it is also important to analyze the actual response patterns to be sure that the percentage differences are genuine. Such an analysis indicates that the FSA responded with activation in the location corresponding to the center of the face in 88% (132/150) of the face images. It missed mostly faces where the hair obscured some of the borders between the eyes, nose, and mouth. At the same time, it generated spurious responses in 27% (40/150) of the face images, i.e. responses in locations other than the center of the face. Nearly half of the spurious responses were from the less-selective neurons that line the edge of the FSA (Figure 10.5*j*); these responses can be ignored because they would not occur in a model of the entire visual field. Most of the remaining spurious responses resulted from a genuine V1 eye or mouth response plus V1 responses to the hair or jaw outlines. In humans, such responses would actually direct attention to the general region of the face, and thus contribute to face preferences, although they would not

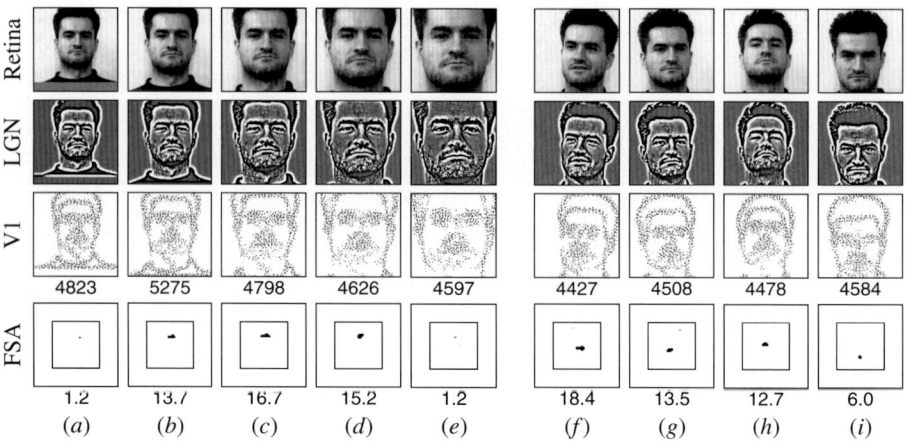

Fig. 10.10. Response variation with size and viewpoint. The three-dot training pattern of HLISSOM matches most closely a particular size and an upright frontal view. However, the model also responds to a range of other sizes (*a–e*) and viewpoints (*f–i*). The model activation is again displayed as in Figures 10.6–10.9. In the viewpoint experiment, the correct FSA location responded to 88% of the set of 150 images consisting equally of front, left, right, up, and down views. Most of these viewpoints result in similar responses, although 100% of the faces looking upward were detected correctly and only 80% of those looking downward were. Overall, HLISSOM predicts that newborns will respond to real faces even with moderate variation of sizes and viewpoints. Photographs copyright 1995 by University of Bern (Achermann 1995).

pinpoint the precise center. For the natural scenes, most of the responses were from the less-selective neurons along the edge of the FSA, and those responses can again be ignored. The remainder were in image regions that coincidentally had a triangular arrangement of three contour-rich areas, surrounded by smooth shading. Therefore, the HLISSOM performance was meaningful, demonstrating a clear difference between faces and non-faces, and making mistakes on inputs that are confusing for humans as well.

It is also important to evaluate how robustly HLISSOM performs when properties of the input are varied. The above dataset included mostly faces that were top-lit, which is the most common direction of light in the natural environment. As seen in Figure 10.9*b–e*, HLISSOM performs relatively well with other lighting conditions, having trouble only with frontal lighting. In such cases the nose and mouth contours are not as prominent, making them a poor match with the three-dot pattern. The response is also reliable across a range of size scales and viewpoints (Figure 10.10). Thus, training the model with a single type of pattern is sufficient for robust detection of real faces.

In summary, the FSA responds to most human faces of about the right size, illumination, and viewpoint, signaling their location in the visual field. It does not respond to most other stimuli, except when they contain accidental three-dot patterns.

The model predicts that when tested, human newborns will have a similar pattern of responses in the face-selective cortical regions.

10.2.6 Effect of Training-Pattern Shape

The preceding sections showed that a model trained on a triangular arrangement of three dots can account for face preferences at birth. This pattern was chosen based on Johnson and Morton's (1991) hypothesis that a hard-wired region responding to this pattern might explain newborn preferences. However, the actual shape of most internally generated activity in humans is unknown, and like retinal waves, the shape may not be very precisely controlled in general. Thus, it is important to test other possible training pattern shapes to see what range of patterns can produce similar results.

Accordingly, a series of simulations was run with a set of nine patterns chosen to be similar in overall size and shape to the three-dot patterns. These patterns are shown in the top row in Figure 10.11. Matching the size is crucial to make the networks comparable, because only a single-sized training pattern is used in each network. All training parameters were the same except for γ_F (the afferent scaling factor for the FSA), which was set manually so that each simulation would have the same amount of activity per iteration of training, despite differences in the patterns.

To make it possible to train and test so many networks, they were implemented in a reduced HLISSOM model that requires much less time and memory. This model consisted of a 24×24 FSA and did not include V1, allowing self-organization in 13 MB of memory (the rest of the simulation parameters are listed in Appendix C.2). Without V1, the model does not provide numerical preferences between the non-facelike patterns (i.e. those that activate V1 but not the FSA), but activity in the FSA allows facelike patterns to be distinguished from other types, which is sufficient to measure the face selectivity of each network.

Specifically, the face selectivity S_F was defined as the proportion of the average response $\overline{\eta_F}$ to a set of facelike patterns, out of the response to those patterns and the average responses $\overline{\eta_N}$ to a set of non-facelike control patterns:

$$ S_F = \frac{\overline{\eta_F}}{\overline{\eta_F} + \overline{\eta_N}}. \tag{10.2} $$

A value of 1.0 indicates that the network strongly prefers facelike patterns, i.e. of the patterns tested, only the facelike patterns caused any FSA response. Values less than 0.5 indicate that the network prefers non-facelike inputs.

The networks were tested both with schematic inputs (from Figure 10.6) and with real images (from Figure 10.13). To ensure that the comparisons between networks were fair, the sigmoid activity threshold (θ_l in Equation 4.4) was set to maximize the face selectivity of each network. That is, θ_l was set to the minimum value for each network at which the FSA would respond to every facelike pattern. The upper threshold was then set to $\theta_u = \theta_l + 0.48$, for consistency. If there was no response to any non-facelike pattern with these settings, S_F would be 1.0, the maximum. The θ_l parameter was set separately for schematics and real images.

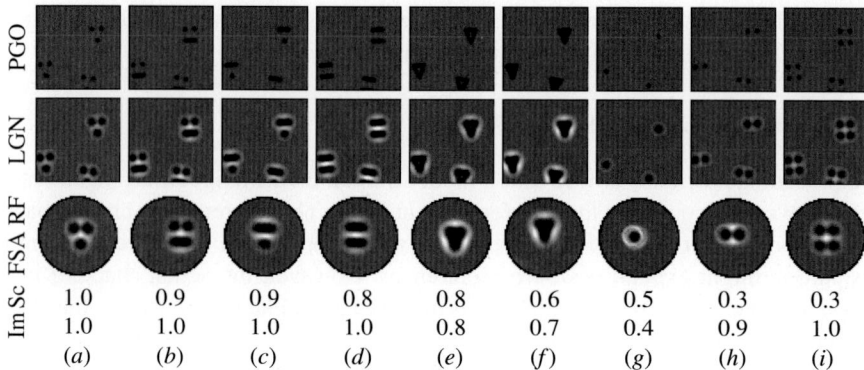

Fig. 10.11. Effect of training patterns on face preferences. Results are shown for nine matched face detection simulations (bypassing V1), each with a different set of input patterns. The top row displays examples of these patterns, drawn on the retina. The second row shows the LGN response to the retinal input, which forms the input to the FSA in these simulations. The third row plots a sample FSA receptive field after self-organization, visualized by subtracting the OFF weights from the ON; other neurons learned similar RFs. In each case the HLISSOM network learns FSA RFs similar to the LGN representation. These RFs are not patchy, because they no longer represent the patchy V1 activities. The two numerical rows quantify the face selectivity of each network. The row labeled "Sc" specifies the selectivity for facelike schematics (from Figure 10.6b–d) relative to non-facelike schematics (from Figure 10.6e–i). The row labeled "Im" lists the selectivity for the six face images from Figure 10.13 relative to the six comparable object images in the same figure. The different training patterns gave rise to different selectivities. Pattern (g) leads to equal responses for both facelike and non-facelike schematics (selectivity of 0.5), and (h) and (i) have a greater overall response to the *non*-facelike schematics (selectivity lower than 0.5). Thus, not all training patterns can explain preferences for schematic faces even if they match some parts of the face. Similarly, the single-dot pattern (g) has a selectivity below 0.5 for real faces, indicating a stronger response for the objects than for real faces. The other training patterns all have the same size as real faces, or match at least two parts of the face, and thus have selectivities larger than 0.5 for real faces. Overall, the shape of the training pattern is clearly important for face selectivity, both for schematics and real faces, but it need not be controlled very tightly to result in face-selective responses.

As Figure 10.11 shows, even with such a liberal definition of face selectivity, not all training patterns result in preferences for facelike schematics and real faces. Furthermore, several different patterns result in high face selectivity. As expected, the results vary most strongly for schematic test images (row "Sc"). The schematics are all closely matched for size, differing only by the patterns within the face, and thus the results depend strongly on the specific training pattern. Of those resulting in face selectivity, the training patterns in Figure 10.11a–e (three dots, dots and bars, bars, and open triangles) have nearly equivalent selectivity, although the three-dot pattern in Figure 10.11a has the highest.

The results were less variable on real test images (row "Im"), because real faces differ more from objects than the schematic faces differ from other schematic patterns. All training patterns that matched either the overall size of the test faces (Figure 10.11*a–f*), or at least the eye spacing (Figure 10.11*h,i*), led to face selectivity. In contrast, the single-dot pattern (Figure 10.11*g*) does not result in face preferences. Although it is a good match to the eyes and mouth regions in the real faces, it is also a good match to many features in object images.

These results show that if the training patterns have the right size, a weak selectivity for faces already develops. The shape of the pattern is also influential as well, as can be seen with schematic patterns. Of all training patterns tested, the three-dot pattern results in the highest selectivity, while a pattern matching a single low-level feature like eye size is not enough. The three-dot pattern is therefore a good default choice for pattern generation simulations, so that the predictions will most clearly differ from those of other models.

In conclusion, internally generated patterns explain how genetic influences can interact with general adaptation mechanisms to specify and develop newborn face-processing circuitry. The HLISSOM model of the visual system incorporates this idea, and is the first to self-organize both low-level and high-level cortical regions at the scale and detail needed to model such behavior realistically. The results match experimental data from newborns remarkably well, and for the first time demonstrate preferences for faces in real images. How these preferences change with experience of real faces will be discussed next.

10.3 Postnatal Development

In this section, the HLISSOM model will be extended to postnatal development. The prenatally trained system from the previous section is trained on real face images, and its face detection performance is compared with that of infants. It is found to learn faster and more robustly than a system that was not prenatally organized, display a similar decline to schematic inputs as infants do, and develop a mother preference like infants. Together, the simulations in this chapter show that prenatal and postnatal learning can explain much of the face preferences in young infants, and provide concrete predictions for future behavioral experiments.

10.3.1 Initial Trained and Naïve Networks

Psychophysical experiments on postnatal face detection have focused only on whether the infant prefers facelike over non-facelike patterns; the effect of spatial frequency (Figure 10.7) has not been measured. As a result, the reduced HLISSOM model (Section 10.2.6), which does not include V1 and is therefore not strongly sensitive to spatial frequency, is sufficient to model these experiments. The reduced model makes it practical to simulate much larger FSA RFs, which will be crucial for learning face outlines.

The final network trained with three-dot patterns in Section 10.2.6 formed the starting point for the postnatal learning phase. To determine whether the prenatal

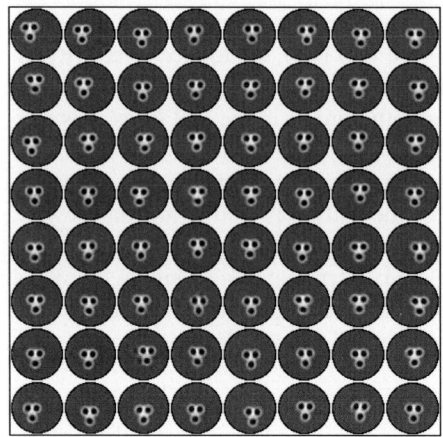 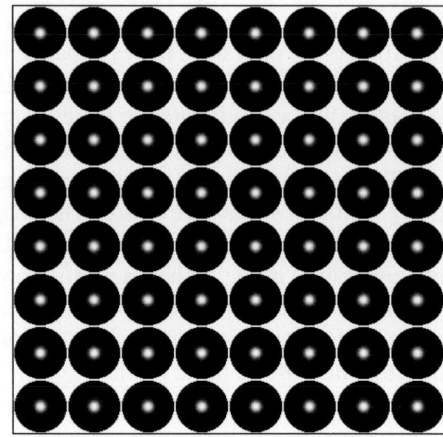

(a) Prenatally trained network (ON−OFF)   (b) Naïve network (ON and OFF)

Fig. 10.12. Initial afferent weights across prenatally trained and naïve FSA networks.
The RFs of every third neuron horizontally and vertically in each network are plotted. For the
prenatally trained network (a), the RFs were visualized by subtracting the OFF weights from
the ON (as in Figure 10.11). The RF patterns are roughly similar to faces, like the RFs of the
prenatal FSA trained with V1 input (Section 10.2). In contrast, the RFs of the naïve network (b)
were initially uniformly Gaussian; the ON and OFF weights were identical. These networks
form an unequal starting point for postnatal learning of faces, as will be shown in later figures.

patterns bias subsequent learning, a naïve network, which was not prenatally organized, was also tested. The goal is to determine whether neurons that are initially
face selective due to prenatal training will learn faces more robustly than neurons
that are initially unselective and learn only from the environment.

So that the naïve and prenatally organized networks would match on as many parameters as possible, the naïve network was constructed from the prenatally trained
network post hoc by explicitly resetting afferent receptive fields to their uniform-
Gaussian starting point. This procedure removed the prenatally developed face selectivity, but kept the lateral weights and all of the associated parameters the same.
The activation threshold θ_1 for the naïve FSA network was then adjusted so that both
networks would have similar activation levels in response to the training patterns;
otherwise the parameters were the same for each network. This procedure ensures
that the comparison between the two networks will be as fair as possible, because
the networks differ only by whether the neurons have face-selective weights at birth.
Figure 10.12 displays the state of each network just before postnatal learning.

10.3.2 Training and Testing Methods

The experiments in this section simulate gradual learning from repeated encounters
of specific individuals and objects against different backgrounds over the first few
months of life. Figure 10.13 shows the people and objects that were used and Figure 10.14 describes how the training images were generated. The prenatally trained

Fig. 10.13. Face and object images in postnatal training. Postnatal training inputs were formed by placing these face and object patterns in random locations in front of randomly chosen natural scenes (as shown in Figure 10.14). The faces are adapted from Rowley et al. 1998; the objects are from public domain clip art and image collections.

and naïve networks were each trained for 30,000 input presentations. The same random sequence of images was used in both cases, so that the influence of prenatal training on postnatal learning will be clear.

The RFs changed more in these simulations than in the postnatal learning experiments in Section 9.3, because the weights were initially concentrated only in the center (Figure 10.12) and needed to spread out over the whole receptive field. During this process, neural responses to typical training inputs varied significantly. To compensate for these changes, the sigmoid threshold θ_1 from Equation 4.4 was periodically adjusted for each network (as detailed in Appendix C.2). Without such compensation, the networks would eventually fail to respond to any training input as the weights become spread out over a large area and are normalized. Section 17.1.1 outlines how this process can be simplified by extending HLISSOM with automatic mechanisms for setting the threshold based on the recent history of inputs and responses, as found in many biological systems (Turrigiano 1999).

In the previous section, pattern preferences were measured in a fully organized network that had similar RFs across the visual field. With such a uniform architecture, image presentations generally result in similar responses at different retinal locations. In contrast, in this section the preferences will be measured periodically during early postnatal learning, before the network has become fully uniform. Therefore, input stimuli will be presented at 25 different retinal locations, the results will be averaged, and statistical significance of the difference between distributions will be computed. To match the typical analysis methods from the psychological experiments, all comparisons will use the Student's t-test, with the null hypothesis that the network responds equally strongly to both stimuli. As in psychological experiments, if $p \leq 0.05$ the difference is considered statistically significant.

The next three subsections will show that the prenatally organized network has a suitable bias for learning faces, its response declines over time similarly to infants, and it develops a mother preference like infants.

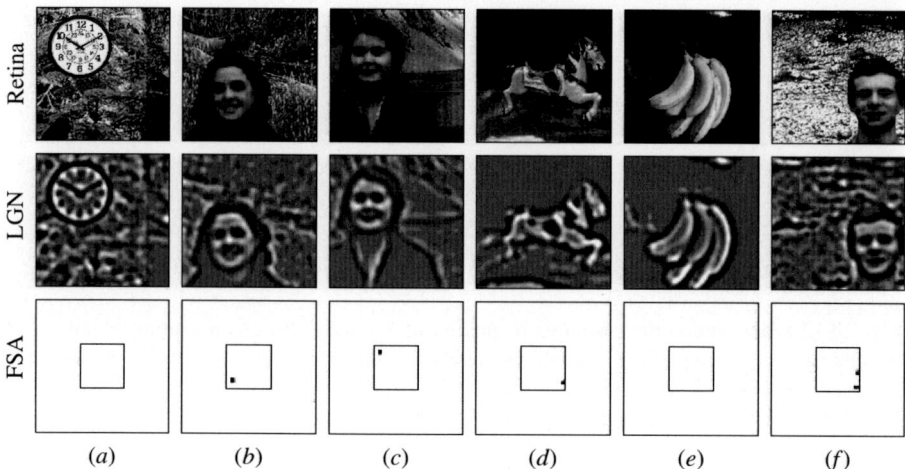

(a)   (b)   (c)   (d)   (e)   (f)

Fig. 10.14. Example postnatal training presentations. The top row shows six randomly generated images drawn on the retina during postnatal learning. Each image contains a foreground item chosen randomly from the images in Figure 10.13. The foreground item was overlaid onto a random portion of an image from a database of 58 natural scenes (National Park Service 1995), at a random location and at a nearly vertical orientation (drawn from a normal distribution around vertical, with $\sigma = \pi/36$ radians). The second row shows the LGN response to each of these sample patterns, and the bottom row the FSA response at the start of postnatal training. The FSA responds to groups of dark spots on the retina, such as the eyes and mouths in (b), (c), and (f) and the horse's dark markings in (d). Subsequent learning in the FSA will be driven by these patterns of activity. Because the prenatal training biases the activity patterns toward faces, postnatal self-organization will also be biased toward faces, as is shown in Figure 10.15.

10.3.3 Prenatally Established Bias for Learning Faces

Over the course of training, the RFs of both prenatally trained and naïve networks gradually learn to represent averages (i.e. prototypes) of faces and hair outlines (Figure 10.15). RFs in the prenatally trained network gradually become more face selective, and eventually nearly all neurons are highly selective. Postnatal self-organization in the naïve network is less regular, and the final result is less selective for faces.

For example, the postnatal network often develops neurons that respond strongly to the clock. The clock has a high-contrast border that is a reasonably close match to a face outline, and thus the same neurons tend to respond to both the clock and to real faces during training. However, the clock is a weak match to the three-dot training patterns of the prenatally trained network, and this network rarely develops clock-selective neurons. These results suggest that the prenatal training biases postnatal learning toward biologically relevant stimuli, i.e. faces.

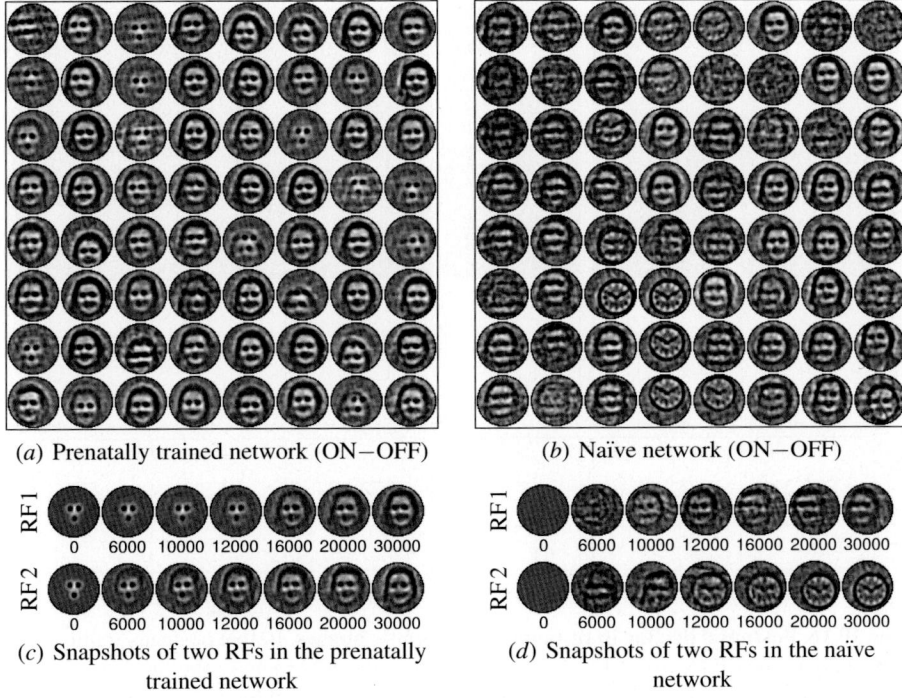

(a) Prenatally trained network (ON−OFF)

(b) Naïve network (ON−OFF)

RF1
0   6000 10000 12000 16000 20000 30000

RF1
0   6000 10000 12000 16000 20000 30000

RF2
0   6000 10000 12000 16000 20000 30000

RF2
0   6000 10000 12000 16000 20000 30000

(c) Snapshots of two RFs in the prenatally trained network

(d) Snapshots of two RFs in the naïve network

Fig. 10.15. Prenatally established bias for learning faces. Plots (a) and b show the RFs for every third neuron from the FSA array, visualized as in Figure 10.12a. As the prenatally trained network learns from real images, the RFs morph smoothly into face prototypes, i.e. representations of average facial features and hair outlines (c). By postnatal iteration 30,000, nearly all neurons have learned facelike RFs, with very little effect from the background patterns or non-face objects (a). Postnatal learning is less uniform for the naïve network, as can be seen in the RF snapshots in (d). In the end, many of the naïve neurons do learn facelike RFs, but others become selective for general texture patterns, and some become selective for objects like the clock (b). Overall, the prenatally trained network is biased toward learning faces, while the initially uniform network more faithfully represents the environment. Thus, prenatal learning can allow the genome to guide development in a biologically relevant direction.

10.3.4 Decline in Response to Schematics

Like human infants, the HLISSOM model gradually becomes less responsive to schematic patterns during early postnatal learning (Figure 10.16). This decline results from the normalization of the afferent weights (Equation 4.8). As the FSA neurons learn the hair and face outlines typically associated with real faces, the connections from the internal features become weaker. Unlike real faces, the facelike schematic patterns match only these internal features, not the outlines. As a result, the network responds gradually less strongly to schematic patterns as real faces are learned. Eventually the response drops below the fixed activation threshold (θ_1 in Equation 4.4) and at that point, the model no longer prefers facelike to non-facelike

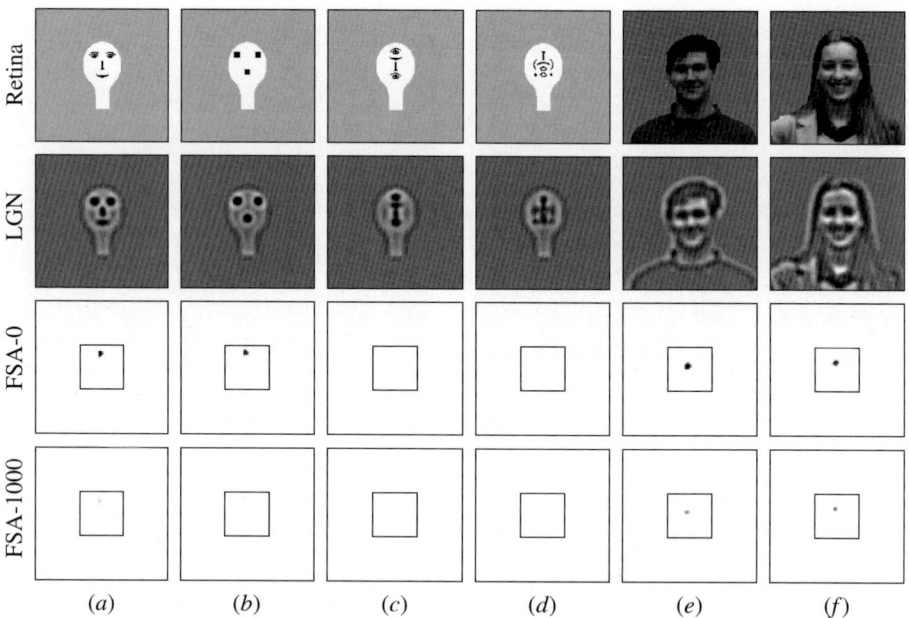

Fig. 10.16. Postnatal decline in response to schematic images. Before postnatal training, the prenatally trained FSA (row "FSA-0"), responds significantly more to the facelike stimulus (*a*) than to the three-dot stimulus (*b*; $p = 0.05$) or to the scrambled faces (*c,d*; $p < 10^{-8}$; Appendix C.2). These responses are similar to those found by Johnson and Morton (1991) in infants up to 1 month of age. In some of their experiments, no significant difference was found between (*a*) and (*b*), which is unsurprising given that they are only barely significantly different here. As the FSA neurons learn from real faces postnatally, they respond less and less to schematic faces. The bottom row shows the FSA response after 1000 postnatal iterations. The FSA now rarely responds to (*a*) and (*b*), and the average difference between them is no longer significant ($p = 0.25$). Thus, no preference would be expected for the facelike schematic after postnatal learning, which is exactly what Johnson and Morton (1991) found for older infants, i.e. 6 weeks to 5 months old. The response to real faces also decreases slightly through learning (*e,f*), because the newly learned average face and hair outline RFs are a weaker match to any particular face than were the original three-dot RFs. However, this decline is much smaller, because real faces are still more similar to each other than to the schematic faces. Thus, HLISSOM predicts that older infants will still show a face preference if tested with more-realistic stimuli, such as photographs.

schematics (because there is no FSA response, and V1 responses are similar). In a sense, the FSA has learned that real faces typically have both inner *and* outer features, and does not respond when either type of feature is absent or there is a poor match to real faces.

Yet, the FSA neurons continue to respond to real faces (as opposed to schematics) throughout postnatal learning (Figure 10.16*e,f*). Thus, the model provides a clear prediction that the decline in face preferences is limited to schematics, and that no decline will be found if infants are tested with sufficiently realistic face stimuli.

This prediction is an important departure from the CONSPEC/CONLERN model. In HLISSOM, an initially CONSPEC-like system is also like CONLERN, in that it will gradually learn from real faces. In contrast, in CONSPEC/CONLERN, CONLERN gradually matures and begins to inhibit CONSPEC, predicting that the decline also applies to real faces. These divergent predictions can be tested by presenting real faces to infants older than 1 month.

As was discussed in Section 10.1.1, the decline takes place at different times in the central vision and in the periphery. Such a difference can be due to gradual maturation of the fovea, as will be outlined in Section 10.4.

10.3.5 Mother Preferences

As was discussed in Section 10.1.1, infants a few days old prefer their mother's face over other similar faces (Bushnell 2001; Bushnell et al. 1989; Pascalis et al. 1995; Walton, Armstrong, and Bower 1997; Walton and Bower 1993). Similar behavior can be observed in the HLISSOM model. Designating one of the female images as the mother, it was presented in 25% of the postnatal learning iterations, corresponding to the estimated proportion of time the infant spends viewing the mother's face (Bushnell 2001). One of the other females with a similar face, designated as the stranger, was not presented at all during training. Over 500 training iterations, the FSA learned to respond to the mother significantly more strongly than to the stranger (Figure 10.17a,b).

Interestingly, the mother preference disappears when the hair outline is masked (Figure 10.17c,d), which is consistent with Pascalis et al.'s claim that newborns learn outlines only. However, Pascalis et al. (1995) did not test the crucial converse condition, i.e. whether newborns respond when the facial features are masked, leaving only the outlines. It turns out that HLISSOM does not respond to the head and hair outline alone either (Figure 10.17e,f). Thus, contra Pascalis et al. (1995), we cannot conclude that what has been learned "has to do with the outer rather than the inner features of the face."

In the model, the response declines with either type of masking because the model learns faces holistically, based on *all* facial features. As real faces are learned, the afferent weight normalization ensures that neurons respond only to patterns that are a good overall match to all of the weights, instead of matching only on a few features. Many authors have argued that adults also learn faces holistically (e.g. Farah et al. 1998). These results suggest that newborns may learn faces in the same way, and predict that newborns will not prefer their mother when her hair outline is visible but her facial features are masked. The time course of this behavior may to some extent depend on foveal maturation, as will be discussed in the next section.

10.4 Discussion

The HLISSOM simulations show that self-organization based on internally generated patterns and environmental inputs can together account for face detection in

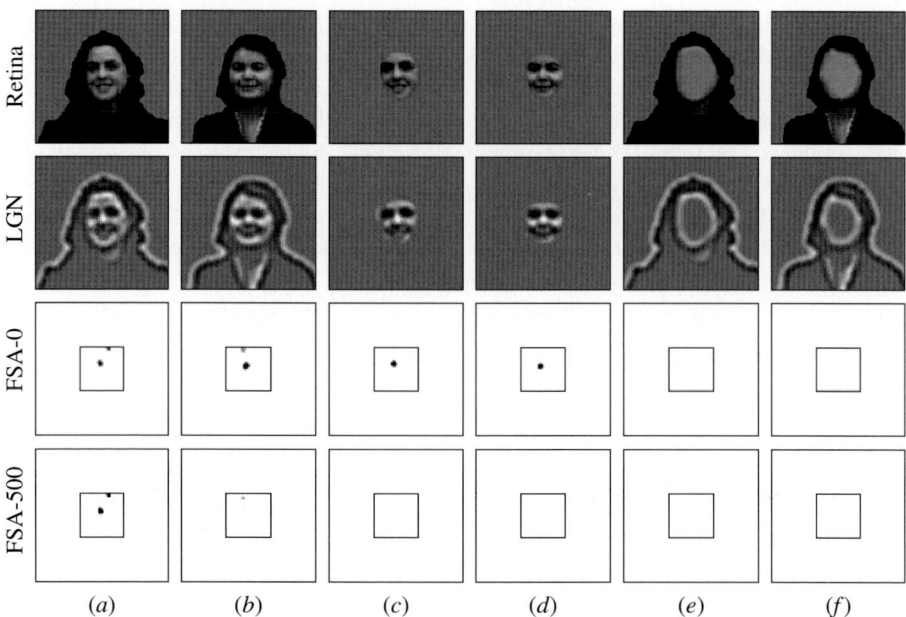

Fig. 10.17. Mother preferences based on both internal and external features. Initially, the prenatally trained FSA responds to both women well, with no significant difference ($p = 0.28$; plots *a,b* in the row labeled "FSA-0"). The response is primarily due to the internal facial features (*c,d*), although the hair and one of the eyes also align into a three-dot pattern in both figures, causing weak spurious activation (*a,b*). Subsequently, image (*a*), designated as the mother, was presented in 25% of the postnatal learning iterations, while image (*b*), the stranger, was not presented at all. After 500 iterations (bottom row), the response to the mother is significantly greater than to the stranger ($p = 0.001$). This result replicates the mother preference found by Pascalis et al. (1995) in infants 3–9 days old. The same results are found in the counterbalancing condition — when trained on face (*b*) as the mother, (*b*) becomes preferred ($p = 0.002$; not shown). After training with real faces, there is no longer any FSA response to the facial features alone (*c,d*), which replicates Pascalis et al.'s (1995) finding that newborns no longer preferred their mother when her face outline was covered. Importantly, no preference is found for the face outline alone either (*e,f*) suggesting that face learning in HLISSOM is holistic. This conclusion is contrary to Pascalis et al.'s (1995) conclusion but consistent with face learning in adults (Farah et al. 1998).

newborns and infants. This perspective leads to testable predictions and suggests future experiments. Importantly, several of HLISSOM's predictions differ from other models and theories, making it possible to distinguish between them in the future.

One easily tested prediction is that newborn face preferences should not depend on the precise shape of the face outline. The Acerra et al. (2002) model (Section 10.1.3) makes the opposite prediction, because in that model the preferences arise from precise spacing differences between the external border and the internal facial features. Results from the HLISSOM simulations also suggest that newborns will have a strong preference for real faces (e.g. in photographs), whereas the Acerra et al. model predicts only a weak preference for real faces, if any (Section 10.1.3).

Experimental evidence to date cannot yet decide between the predictions of these two models. For instance, Simion et al. (1998a) did not find a significant schematic face preference in newborns 1–6 days old without a contour surrounding the internal features, which is consistent with the Acerra et al. model. However, the same study concluded that the shape of the contour "did not seem to affect the preference" for the patterns, which would not be consistent with the Acerra et al. model. As discussed earlier, newborns may not require any external contour, as the pattern-generation model predicts, until they have had postnatal experience with faces. Future experiments with younger newborns should compare model and newborn preferences between schematic patterns with a variety of border shapes and spacings. These experiments will either show that the border shape is crucial, as predicted by Acerra et al., or that it is unimportant, as predicted by the pattern-generation model.

The predictions of the pattern-generation model also differ from those of the Simion et al. (2001) top-heavy model (Section 10.1.3). The top-heavy model predicts that any face-sized border that encloses objects denser at the top than the bottom will be preferred over similar schematic patterns. The pattern-generation model predicts instead that a pattern with three dots in the typical symmetrical arrangement is preferred over the same pattern with both eye dots pushed to one side, despite both patterns being equally top heavy. These two models represent very different explanations of the existing data, and thus testing such patterns should offer clear support for one model over the other.

On the other hand, many of the predictions of the fully trained pattern-generation model implemented in HLISSOM are similar to those of the CONSPEC model proposed by Johnson and Morton (1991). In fact, the reduced HLISSOM face preference network in Section 10.2.6 (which does not include V1) can be seen as the first CONSPEC system to be implemented computationally, along with a concrete proposal for how such a system could be constructed during prenatal development. The primary architectural difference between the trained HLISSOM network and CONSPEC/CONLERN is that in HLISSOM only neurons located in cortical visual areas respond selectively to faces in the visual field, whereas both the subcortical CONSPEC and the cortical CONLERN systems are face selective.

Whether newborn face detection is mediated cortically or subcortically has been debated extensively, yet no clear consensus has emerged from behavioral studies (Simion et al. 1998a). If future brain imaging studies do discover face-selective visual neurons in subcortical areas of newborns, HLISSOM will need to be modified to include such areas. Yet, the key principles would remain the same, because internally generated activity also shapes subcortical regions (Wong 1999). Thus, experimental tests of the pattern-generation model vs. CONSPEC should focus on how the initial system is constructed, and not where it is located.

Although the HLISSOM model is a good match with current experimental data, it could be extended to account for more variation in the input. For example, the current model is only able to detect facelike patterns at one particular spatial scale. Because all experimental data on face preferences in newborns come from experiments with life-sized input at a distance of around 20 cm, multiple sizes were not necessary to account for human data. It would be easy to extend the model to multiple face sizes (i.e. distances) by varying the spatial scale of the training patterns during self-organization (Sirosh and Miikkulainen 1996b). The FSA in such a simulation would need to be much larger to represent the different sizes, and the resulting patchy FSA responses would require more complex methods of analysis, but the resulting model should perform like the current model at the same scale. With the multi-scale model, it would be possible to make specific predictions about how human face detection varies over distance.

All HLISSOM experiments were based on upright training patterns, because the Simion et al. (1998a,b) studies suggest that the orientation of face patterns is important even in the first few days of life. In the areas that generate training patterns, such a bias might be due to anisotropic lateral connectivity, which would cause spontaneous patterns in one part of the visual field to suppress those below them (discussed further in Section 16.2.2). On the other hand, tests with the youngest infants (less than 1 day) have not yet found orientation biases (Johnson et al. 1991). Thus, the experimental data are also consistent with a model that assumes unoriented patterns prenatally, followed by rapid learning from upright faces. Such a model would be more complicated to simulate and describe than the one presented in this chapter, but could use a similar architecture and learning rules.

Another important aspect of postnatal learning that is currently not explicitly included in the HLISSOM model is that of fovea vs. periphery (this extension will be discussed in Section 17.2.10). Preference for schematic faces is not measurable in central vision until 2 months of age (Maurer and Barrera 1981), and is gone by 5 months (Johnson et al. 1991). This time course is delayed relative to peripheral vision, where preferences exist at birth but disappear by 2 months. As was reviewed in Section 10.1.3, Johnson and Morton (1991) propose two separate explanations for these phenomena. In the periphery, the preferences disappear because CONLERN matures and inhibits CONSPEC, whereas in central vision they disappear because CONLERN learns properties of real faces and no longer responds to static schematic patterns. HLISSOM instead suggests a unified explanation for both phenomena: A single learning system stops responding to schematic faces because it has learned from real faces. Why, then, would the time course differ between peripheral and central vision? As Johnson and Morton acknowledged, the retina changes significantly over the first few months. In particular, at birth the fovea is much less mature than the periphery, and may not even be functional yet (Abramov, Gordon, Hendrickson, Hainline, Dobson, and LaBossiere 1982; Kiorpes and Kiper 1996). As a result, schematic face preferences in central vision may be delayed. A single cortical learning system like HLISSOM is thus sufficient to account for the time course of both central and peripheral schematic face preferences.

Central and peripheral differences may also have a role in how mother preferences develop postnatally. In a recent study, Bartrip, Morton, and de Schonen (2001) found that infants 19–25 days old do not prefer their mothers significantly when either the internal features or the external features are covered. This result partially confirms the predictions of Section 10.3.5, although tests still need to be run with newborns only a few days old, like Pascalis et al. (1995) did. Interestingly, Bartrip et al. also found that infants 35–40 days old do prefer their mothers even when the external outline is covered. The gradual maturation of the fovea may again explain these later-developing capabilities. Unlike the periphery, the fovea contains many ganglion cells with small RFs, which connect to cortical cells with small RFs (Merigan and Maunsell 1993). These neurons can learn smaller regions of the mother's face, and their responses will allow the infant to recognize the mother even when other regions of the face are covered. In this way, simple documented changes in the retina can explain why mother preferences would differ over time in different parts of the visual field.

While the current HLISSOM simulations focus on how faces are detected, in the future the model could be used to study face recognition as well. These two tasks have very different requirements. In face detection, the system has to respond similarly to a wide range of different faces. This behavior is achieved in HLISSOM with low afferent learning rates: Each neuron develops preferences that match the long-term averages of faces. In contrast, in face recognition the responses to different individual faces have to differ significantly. Such behavior could be modeled in HLISSOM by including additional FSA-like regions with a higher learning rate: Different neurons would learn to prefer different faces. The final organization of such regions would not depend strongly on the prenatal training patterns, because the initial preferences would soon be overwritten with postnatal experiences. However, even for face recognition regions, prenatal training could speed up postnatal learning by ensuring that their initial state is close to patterns that will be experienced postnatally.

Because tests with human newborns are technically difficult and expensive to perform, understanding how face preferences develop can benefit from the study of model systems in other species. In particular, the phenomenon of chick imprinting has much in common with newborn and young infant face recognition (Bolhuis and Honey 1998; Horn 1985; Johnson and Morton 1991). Birds also exhibit REM sleep (Siegel 1999), and chicks have an "innate" preference for visual stimuli shaped like a head with a neck, on the day after hatching (Horn 1985). Interestingly, chicks do not significantly prefer such stimuli on the first day after hatching, and the preference does not depend on patterns experienced the first day. If such preferences arise from pattern generation, as in the HLISSOM model of face preferences, they may be due to patterns presented in REM sleep the first night. Such patterns may be triggered by the stress hormones that are released after hatching (as suggested by M. H. Johnson, personal communication, January 24th, 2002), but they would not take effect until the next REM sleep episode. Subsequent experimental studies of disrupting REM sleep in chicks can be conducted in conjunction with an HLISSOM-based model of chick imprinting. Such experiments would provide a concrete, practical test for high-level pattern generation as a general principle of development across species.

10.5 Conclusion

The HLISSOM face detection simulations show that internally generated patterns and a self-organizing system can together explain why newborns prefer facelike visual input, how neonatals learn faces, and how face detection develops in the longer term. The model also suggests why newborns prefer specific patterns, why the response to schematic faces decreases over time, and how mother preferences develop. Unlike in other models, the same principles apply to both central and peripheral vision in HLISSOM, and the results differ only because the fovea matures more slowly.

These explanations and simulation results lead to several concrete predictions for future infant experiments. Such experiments may eventually verify the underlying hypothesis that the genome steers development through internally generated patterns, allowing sophisticated abilities to be learned faster and more robustly.

PERCEPTUAL GROUPING

PGLISSOM: A Perceptual Grouping Model

Grouping of image elements into coherent objects is an intriguing, fundamental function of the visual cortex. Part IV of the book focuses on understanding this process, suggesting that self-organization of lateral connections plays a central role in it. In order to perform grouping, the LISSOM model is expanded in two ways. First, the flat two-dimensional map is extended into a two-layer structure, where long-range excitatory connections in the second layer perform binding and segmentation. Second, firing-rate units are extended into spiking units so that the system can represent temporal coding. The resulting model, PGLISSOM (perceptual grouping LISSOM; Choe 2001; Choe and Miikkulainen 2000, 2004), is used to demonstrate how V1 can perform grouping in schematic images. Simulating spiking and long-range excitation is computationally expensive, and the model can therefore include only the essential components. In particular, the high-level areas and subcortical areas (including the ON/OFF channels) of the HLISSOM experiments in Part III are not included in the simulations in Part IV.

In this chapter, the architecture and components of PGLISSOM are described in detail, showing how the network is initialized, activated, and trained. As a validation experiment, an orientation map is shown to self-organize like in firing-rate LISSOM models. In subsequent chapters in Part IV, PGLISSOM's temporal coding and self-organization processes are demonstrated and analyzed, and the model is shown to account for low-level perceptual grouping phenomena such as contour integration and certain illusory contours.

11.1 Motivation: Temporal Coding

Recently, considerable evidence has emerged suggesting that low-level perceptual grouping, such as integrating a sequence of line segments into a coherent contour, take place early in the visual system, most likely in V1 (Kapadia, Ito, Gilbert, and Westheimer 1995; Polat, Mizobe, Pettet, Kasamatsu, and Norcia 1998; Stettler et al. 2002). A map-level model such as LISSOM is ideal for testing this hypothesis computationally. However, the LISSOM models introduced in Parts II and III consist of firing-rate units representing cortical columns. As we saw in Section 2.4.1, superposition catastrophe will occur when activities of firing-rate units representing separate objects are combined. In order to study perceptual grouping in V1, firing-rate units must be replaced with spiking units.

A self-organizing map with spiking units can be interpreted to model cortex in two ways: (1) The spiking units can stand for individual, representative neurons in cortical columns, or (2) they can be interpreted as subgroups of neurons with oscillatory total activity. Current neuroscience data are not specific enough to favor either interpretation; however, it is easier to describe the model and its behavior in terms of neurons, so the first alternative is adopted in Part IV.

A map of spiking neurons can self-organize like a firing-rate neuron map, and it can segment simple objects (such as relatively small squares) through synchronization and desynchronization of spiking events (Choe and Miikkulainen 1997, 1998; Ruf and Schmitt 1998). The long-range inhibitory lateral interactions play a crucial role in both behaviors: They establish competition that drives self-organization, and they establish desynchronization that drives segmentation. It is not necessary to include long-range excitatory connections to achieve these behaviors.

However, to achieve simultaneous binding and segmentation of more complex objects such as long contours, long-range excitatory lateral connections need to be included in the model. Such specific excitation is necessary to overcome the segmentation that would otherwise result from the lateral inhibition, and the excitatory connections need to be long enough to bind together several contour elements. Simply making the short-range excitatory lateral connections longer in a spiking LISSOM map does not work. The neurons then respond to almost any input, and a map where the afferent receptive fields of all neurons look almost the same results. The two functions of perceptual grouping and self-organization therefore have conflicting requirements: One depends on having long and the other short excitatory lateral connections.

In order to account for perceptual grouping in a self-organizing visual cortex, the PGLISSOM model was developed. PGLISSOM includes both short-range and long-range excitatory lateral connections between spiking neurons. To prevent the long-range excitatory connections from interfering with the self-organizing process, PGLISSOM includes two layers (or maps). In the first map (SMAP), excitatory lateral connections are short range to allow self-organization to take place; in the second map (GMAP), they have a long range and implement perceptual grouping (Figure 11.1). The neurons in the corresponding locations in the two maps are linked with excitatory connections in both directions, allowing GMAP to self-organize properly, driven by the self-organization of SMAP.

In summary, PGLISSOM extends LISSOM in two important ways: Spiking neurons are included to represent grouping through temporal coding, and long-range excitatory lateral connections to coordinate synchronization for temporal coding. The model is motivated computationally, but the design is also a good match with the layered architecture found in the visual cortex, as will be discussed in Section 16.3.3. In the following sections, the components of the PGLISSOM model are described detail.

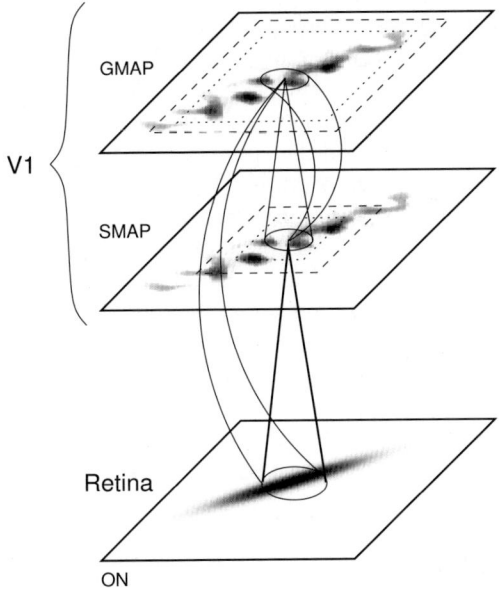

Fig. 11.1. Architecture of the PGLISSOM model. The cortical network consists of two layers (or maps). The lower map (SMAP) has short-range lateral excitation (dotted square) and long-range lateral inhibition (dashed square), and drives the self-organization of the model. In the upper map (GMAP), both excitation and inhibition have very long range, establishing perceptual grouping and segmentation. The two maps both receive afferent input directly from a model retina, representing the ON channel like the reduced LISSOM model (Figure 6.3). The neurons in the vertically corresponding locations on the two maps are connected via excitatory intracolumnar connections in both directions, tying such neurons together into a functional unit (i.e. a cortical column). All neurons are spiking neurons (Figure 11.2); their firing rate is visualized in gray-scale coding from white to black (low to high).

11.2 The Self-Organization and Grouping Architecture

The overall organization of the PGLISSOM model is shown in Figure 11.1. The model consists of two layers (or maps), one overlaid (or stacked) on top of the other. Both maps are based on the LISSOM model, but the extent of lateral connections in the two maps differs. To control input conditions better and to reduce computational cost (Section 11.4), the model will be trained with schematic images instead of natural images. The ON/OFF channels of the LGN can therefore be bypassed (as was discussed in Section 6.2): Both maps receive afferent input only through the ON channel, reduced into direct connections from the retinal receptors. The high-level areas and the subcortical areas of HLISSOM are not relevant for the grouping study either, and were omitted.

The lower map, SMAP, is similar to the LISSOM and HLISSOM cortical networks. Short-range excitatory connections establish a local neighborhood that enforces local correlation, and longer inhibitory connections establish competition that decorrelates more distant activities. Through these connections, SMAP drives self-organization in the model. In the upper map, GMAP, both the excitatory and inhibitory connections have long range. The excitatory connections form the basis for perceptual grouping: Through the self-organizing process, they learn to encode correlations in the input distribution, and the strength of the connections controls the degree of synchronization across the neurons. The inhibitory connections are broad and have a long range, causing two or more synchronized populations of neurons to desynchronize, thus establishing background inhibition for segmentation.

Neurons in the vertically corresponding locations in the two maps form a functional unit (cortical column), and they are connected to each other in both directions through excitatory intracolumnar connections. These connections influence the activity on the opposite map so that both self-organization and grouping behaviors are shared by both maps. This two-component model of the cortical column is called the SG model, corresponding to the SMAP and GMAP the columns form. The neurons are all spiking neurons, as will be described next.

11.3 Spiking Unit Model

A schematic diagram of the spiking neuron is shown in Figure 11.2. The model is based on the integrate-and-fire model discussed in Section 3.1.3 (Eckhorn et al. 1990; Gerstner 1998b; Reitboeck et al. 1993). Each neuron has three components: leaky synapses, weighted summation, and a spike generator. The synapses continuously calculate the decayed sum of incoming spikes over time. Four different kinds of input connections contribute to the weighted sum: afferent, excitatory lateral, inhibitory lateral, and intracolumnar connections (Figure 11.1). The activations of the different kinds of inputs are summed and compared with the dynamic threshold in the spike generator. Details of each component will be discussed below.

11.3.1 Leaky Synapse

Each connection in the model is a leaky integrator (Section 3.1.3), modeling the exponential decay of postsynaptic potential (PSP) in biological neurons. At each connection, an exponentially decayed sum of incoming spikes is maintained:

$$\tilde{\imath}(t) = \sum_{k=0}^{t} R(t-k)\, e^{-\lambda k}, \tag{11.1}$$

where $\tilde{\imath}(t)$ is the current decayed sum at time step t, $R(t-k)$ is the spike (either 0 or 1) received k time steps in the past, and λ is the decay rate. Different types of connections have separate decay rates: afferent connections (λ_A), excitatory lateral connections (λ_E), inhibitory lateral connections (λ_I), and intracolumnar connections (λ_C). The most recent input has the most influence on the activity, but past inputs also have some effect. As was discussed in Section 3.1.3, this sum can be defined recursively as

Afferent connections

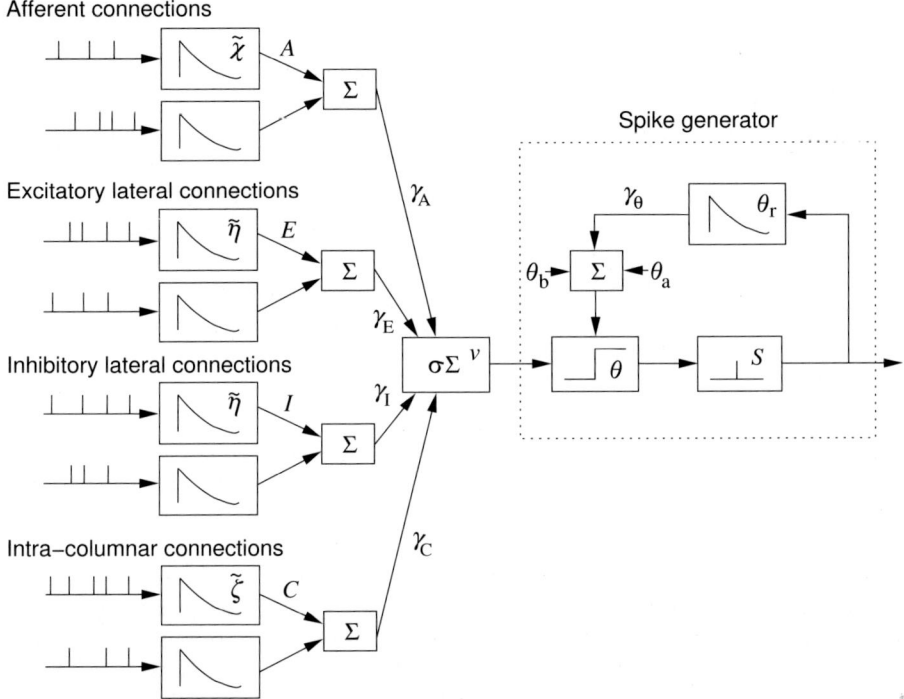

Fig. 11.2. The leaky integrator neuron model. Leaky integrators at each synapse perform decayed summation of incoming spikes, and the outgoing spikes are generated by comparing the weighted sum with the dynamic spiking threshold. Four types of inputs contribute to the activity: afferent, excitatory lateral, inhibitory lateral, and intracolumnar connections (Equation 11.3). The dynamic threshold consists of the base threshold θ_b, the absolute refractory contribution θ_a, and the relative refractory contribution θ_r (Equation 11.5). The base threshold is a fixed baseline value, and the absolute refractory term has a value of ∞ for a short time period immediately following an output spike. The relative refractory contribution is increased as output spikes are generated, and it decays to zero if the neuron stays silent.

$$\tilde{\imath}(t) = R(t) + \tilde{\imath}(t-1)\,e^{-\lambda}. \tag{11.2}$$

Such a formulation allows implementing the model efficiently, since the past spike values $R(t-k), k > 0$, do not need to be stored, nor do they have to be decayed repeatedly.

By adjusting the decay rate λ, the synapse can function as either a coincidence detector or as a temporal integrator. When the synaptic decay rate is high, the neuron can only activate when there is a sufficient number of inputs coming in from many synapses simultaneously. On the other hand, when the decay rate is low, the neuron accumulates the input. Thus, presynaptic neurons can have a lingering influence on the postsynaptic neuron. As will be shown in Section 12.2.1, by varying the decay rates for different types of connections, the relative time scales of the different connection types can be controlled, and desired synchronization behavior obtained.

11.3.2 Activation

The neuron receives incoming spikes through the afferent, lateral excitatory and in-hibitory, and intracolumnar input connections, and the decayed sums of the synapses are calculated according to Equation 11.2. These sums are then multiplied by the connection weights and summed to obtain the input activity (Figure 11.2). More specifically, the input activation $v_{ij}(t)$ to the spike generator of the cortical neuron at location (i, j) at time t consists of (1) the input from a fixed-size receptive field in the retina, centered at the location corresponding to the neuron's location in the cortical network (i.e. afferent input), (2) excitation and (3) inhibition from neurons around it in the same map, and (4) input from neurons in the same column in the other map (i.e. intracolumnar input):

$$v_{ij}(t) = \sigma \left(\gamma_A \sum_{xy} \tilde{\chi}_{xy}(t-1) A_{xy,ij} - \gamma_E \sum_{kl} \tilde{\eta}_{kl}(t-1) E_{kl,ij} \right.$$

$$\left. + \gamma_I \sum_{kl} \tilde{\eta}_{kl}(t-1) I_{kl,ij} + \gamma_C \sum_{cd} \tilde{\zeta}_{cd}(t-1) C_{cd,ij} \right), \quad (11.3)$$

where $\tilde{\chi}_{xy}(t-1)$ is the decayed sum of spikes from retinal receptor (x, y) during the previous time step, $A_{xy,ij}$ is the afferent connection weight from that receptor to cortical neuron (i, j), $\tilde{\eta}_{kl}(t-1)$ is the decayed sum of spikes from the map neuron (k, l), $E_{kl,ij}$ is the excitatory and $I_{kl,ij}$ the inhibitory lateral connection weight from that neuron, $\tilde{\zeta}_{cd}(t-1)$ is the decayed sum of spikes from neuron (c, d) in the other map in the same vertical column as (i, j), $C_{cd,ij}$ is the intracolumnar connection weight from that neuron, and γ_A, γ_E, γ_I, and γ_C are constant scaling factors. The function $\sigma(\cdot)$ is the piecewise linear approximation of the sigmoid (Equation 4.4), used to keep the input activity to the spike generator between 0.0 and 1.0.

The activation $v_{ij}(t)$ is then passed on to the spike generator, where comparison with the dynamic threshold is made. A spike is fired if the input activity exceeds the threshold:

$$S(t) = \begin{cases} 1 \text{ if } v_{ij}(t) > \theta(t), \\ 0 \text{ otherwise,} \end{cases} \quad (11.4)$$

where $S(t)$ represents the spiking output of the neuron over time. The dynamic threshold $\theta(t)$ determines how much activation is necessary to generate a spike. It depends on how much time has passed since the last firing event, as will be described next.

11.3.3 Threshold Mechanism

For a short period of time immediately after they have spiked, biological neurons cannot generate another spike. This short interval is called the refractory period and consists of two parts: (1) During the absolute refractory period, the neurons cannot fire no matter how large the input is, and (2) during the relative refractory period, neurons can only spike if they receive very strong excitatory input.

The dynamic threshold in the PGLISSOM spike generator implements a refractory period by providing a base threshold and raising the threshold dynamically, depending on the neuron's spike activity. The spike generator compares the input activity to the dynamic threshold and decides whether to fire (Figure 11.2). The threshold $\theta(t)$ is a sum of three terms:

$$\theta(t) = \theta_{\rm b} + \theta_{\rm a}(t) + \gamma_\theta \theta_{\rm r}(t), \qquad (11.5)$$

where $\theta_{\rm b}$ is the base threshold, $\theta_{\rm a}(t)$ implements the absolute refractory period during which the neuron cannot fire, $\theta_{\rm r}(t)$ implements the relative refractory period during which firing is possible but requires extensive input, and γ_θ is a scaling constant. More specifically, $\theta_{\rm a}(t) = \infty$ if the last spike was generated less than $t_{\rm r}$ time steps ago, otherwise $\theta_{\rm a}(t) = 0$. The relative refractory component $\theta_{\rm r}(t)$ is implemented as an exponentially decayed sum of the output spikes (Figure 11.2), i.e. a leaky integrator similar to the leaky synapses (Equation 11.2):

$$\theta_{\rm r}(t) = S(t) + \theta_{\rm r}(t-1)\, e^{-\lambda_\theta}, \qquad (11.6)$$

where λ_θ is the decay rate.

The usual integrate-and-fire model includes a similar dynamic threshold mechanism, but consists of $\theta_{\rm b}$ and $\theta_{\rm r}$ only (Eckhorn et al. 1990; Reitboeck et al. 1993). The absolute refractory period makes the model more realistic, but it also serves a computational purpose: It ensures that the neurons do not fire too rapidly. This property makes synchronization more robust against noise, as will be described in the next chapter.

The spikes generated this way are propagated through the connections and the average firing rates of the neurons in a small time window are gathered. Based on these firing rates, the connection weights are adapted. The details of the learning mechanism are explained next.

11.4 Learning

The PGLISSOM simulations begin with a network in an unorganized state: All connection weights in the network are initialized e.g. to uniformly random values between 0 and 1. The network is trained by presenting visual input, and adapting the connection weights according to the Hebbian learning rule. Instead of natural images as in Part III, schematic inputs will be used for the perceptual grouping experiments for two reasons: (1) Contour integration is a well-defined task with schematic inputs; it is possible to have tight control over stimulus configurations and therefore test behavior of the model systematically, as is done in psychological experiments on human subjects. (2) With schematic inputs, smaller networks can be used, making self-organization with spiking neurons computationally feasible.

The input to the network consists of oriented Gaussian bars similar to those in Section 5.3. To generate such an input pattern in a spiking model, the input neurons fire with different frequencies: At the center of the Gaussian, the spike rate is maximal, and the spike rate decreases gradually for neurons farther away from the center. At each input presentation, an input with a random orientation is placed at a random location in the retina, and the resulting retinal receptor activities are propagated through the afferent connections of the network. As in firing-rate LISSOM models, the input is kept constant while the cortical response settles through the lateral connections. The retinal receptors generate spikes at a constant rate and the cortical neurons generate and propagate spikes and adjust their dynamic thresholds according to Equations 11.1 through to 11.6. After a while, the neurons reach a stable rate of firing, and this rate is used to modify the weights.

The spiking rate $\eta(t)$ is calculated as a running average:

$$\eta(t) = \lambda_r \eta(t-1) + (1 - \lambda_r) S(t), \tag{11.7}$$

where λ_r is the retention rate, $\nu(t-1)$ is the previous spiking rate, and $S(t)$ is the output spike at time t (either 0 or 1). With this method, a short-term firing rate in a limited time window is calculated. For each input presentation, the average spiking rate of each neuron is calculated through several iterations. The weight modification then occurs as in firing-rate LISSOM models (Section 4.4.1). The afferent, lateral, and intracolumnar weights are modified according to the normalized Hebbian learning rule:

$$w'_{pq,ij} = \frac{w_{pq,ij} + \alpha X_{pq} \eta_{ij}}{\sum_{uv} (w_{pq,uv} + \alpha X_{pq} \eta_{uv})}, \tag{11.8}$$

where $w_{pq,ij}$ is the current connection weight (either A, E, I, or C) from neuron (p,q) to (i,j), $w'_{pq,ij}$ is the new weight to be used until the end of the next settling process, α is the learning rate (α_A for afferent, α_E for excitatory lateral, α_I for inhibitory lateral, and α_C for intracolumnar connections), and X_{pq} and η_{ij} are the spiking rates of presynaptic and postsynaptic neurons after settling. Note that in PGLISSOM normalization is done presynaptically instead of postsynaptically; this change does not affect self-organization but results in stronger grouping (Section 16.1.3; Choe, Miikkulainen, and Cormack 2000; Sakamoto 2004). As before, those connections that become near zero in the learning process are deleted, and the radius of the lateral excitation in SMAP is gradually reduced following a preset schedule.

This process of input presentation, activation, and weight adaptation is repeated for a large number of input patterns, and the neurons gradually become sensitive to particular orientations at particular locations. In this way, the network forms a global retinotopic orientation map with patchy lateral connections similar to that of the earlier LISSOM OR model. In PGLISSOM, this map will then synchronize and desynchronize the firing of neurons to indicate binding and segmentation of visual input into different objects.

11.5 Self-Organizing Process

After extending the LISSOM model with spiking neurons and excitatory lateral connections, a crucial question is whether the model is still capable of self-organization in the same way as the original model. Most importantly, do the lateral connections develop a characteristic patchy structure that could form the basis for perceptual grouping? This section will demonstrate that this is indeed the case: PGLISSOM self-organizes to form orientation maps and functionally specific lateral connection patterns.

11.5.1 Method

The SMAP consisted of 136×136 neurons, and GMAP of 54×54 neurons. GMAP was smaller to make simulations faster and to fit the model within the physical memory limit. The intracolumnar connections between SMAP and GMAP were proportional to scale, so that the relative locations of connected neurons in the two maps were the same. This connectivity scheme ensures that the global order of the two maps matches as they develop. The excitatory lateral connection radius in SMAP was initially seven and was gradually reduced to three. The lateral inhibitory radius was fixed at 10, i.e. smaller than in the LISSOM simulations, in order to reduce the computational requirements; reducing the radius further would not allow maps to self-organize properly. In GMAP, excitatory lateral connections had a radius of 40 and the inhibitory connections covered the whole map; such very long-range connections were necessary to achieve proper grouping behavior. Afferent connections from the retina had a radius of six in both maps (mapped according to Figure A.1), and intracolumnar connections a radius of two. The retina consisted of 46×46 receptors. As long as the relative sizes of the map, the retina, and the lateral connection radii are similar to these values, the maps will self-organize well. The rest of the parameter settings are specified in Appendix D.1.

The input consisted of single randomly located and oriented elongated Gaussians with major and minor axis lengths initially $\sigma_a = 3.9$ and $\sigma_b = 0.8$, and elongated to 6.7 and 0.7 after 1000 input presentations (of a total of 40,000 presentations). Such inputs result in sharp OR tuning and long lateral connections necessary for contour integration, as will be described in the next two subsections.

11.5.2 Receptive Fields and Orientation Maps

The self-organization process was very similar to that of the firing-rate LISSOM, and also matched experimental data well (as described in Sections 2.1, 5.1 and 5.3). As in earlier LISSOM simulations, the receptive fields are initially circular and have a random profile. They gradually become elongated and smoother as the training proceeds, and the neurons gradually develop orientation preferences.

The resulting receptive fields of SMAP and GMAP are shown in Figure 11.3a. The neurons belonging to the same column in SMAP and GMAP have highly similar orientation preferences. Most neurons are highly selective for orientation, and others

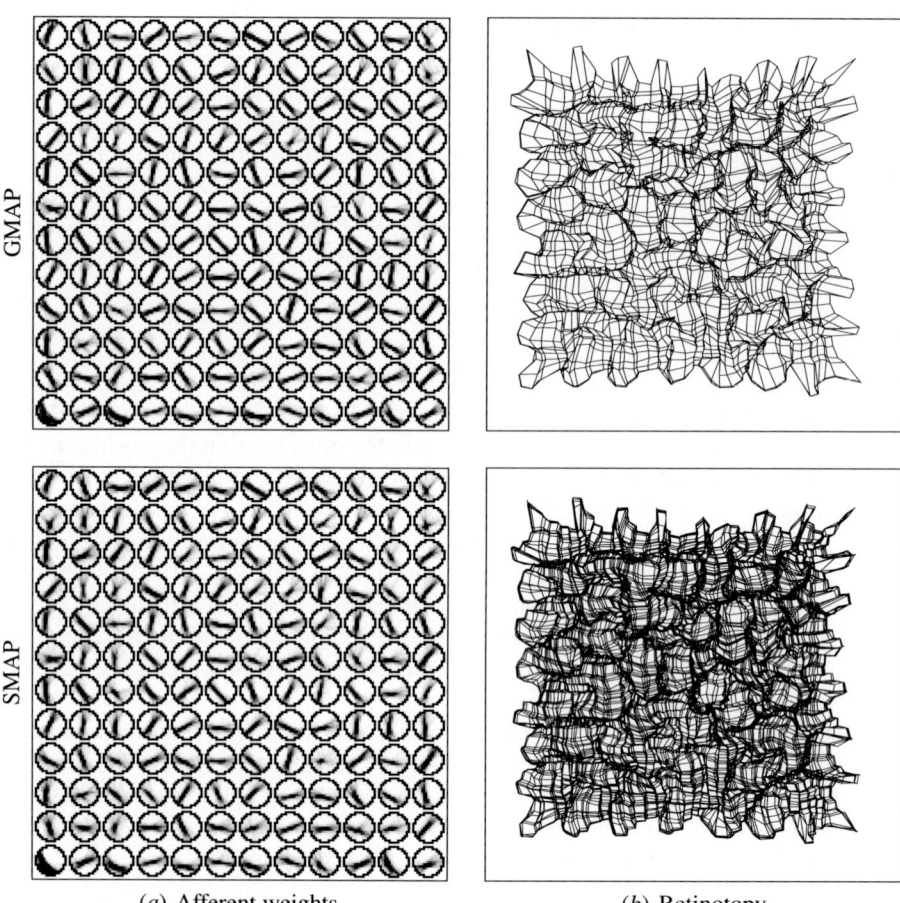

(*a*) Afferent weights   (*b*) Retinotopy

Fig. 11.3. Self-organized afferent weights and retinotopic organization. In (*a*), the afferent weight matrices of corresponding sample neurons in SMAP and GMAP are plotted (in gray scale as in Figure 6.4, organized as in Figure 5.8*a*): In SMAP, every fourth neuron horizontally and vertically is shown, and in GMAP, every tenth neuron. Both maps saw the same inputs during training, and due to the intracolumnar connections they developed matching orientation preferences. Since the ON/OFF channels in the LGN were bypassed in this simulation, the receptive fields are all unimodal. However, they display the same properties as the orientation model in Section 5.3: Most neurons are highly selective for orientation, and neurons near discontinuities are unselective. In (*b*), the center of gravity of the afferent weights of each neuron in the network are plotted as a grid in retinal space (as in Figure 5.11). Although the two maps differ in size, the overall organization closely matches: Neurons in the same cortical column receive input from the same locations in the retinal space. The overall organization of the map is an evenly spaced grid with local distortions, as observed in biology and in the LIS-SOM orientation map (Figure 5.11; Das and Gilbert 1997). The preferences are sharper and the distortions wider than in the LISSOM simulations because more elongated input patterns were used during training.

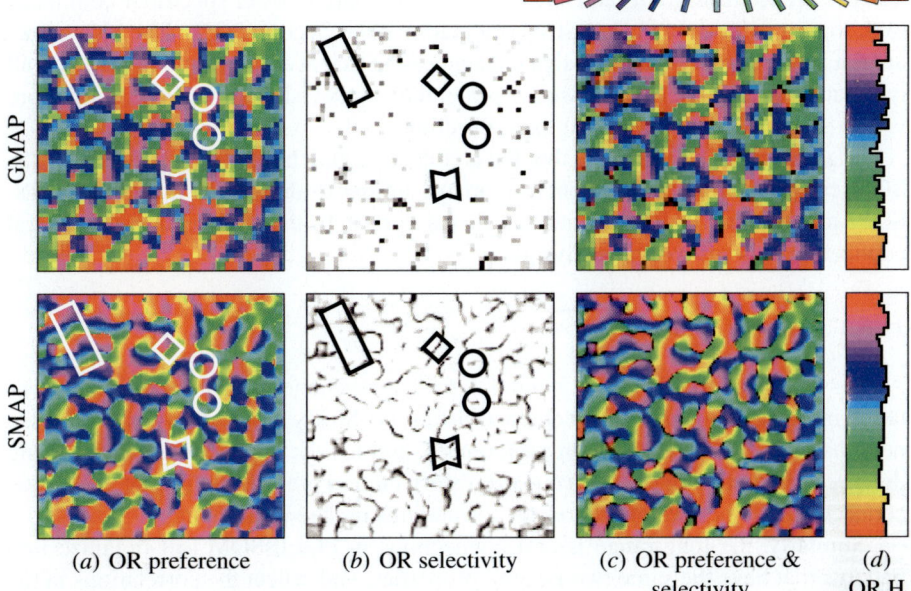

(a) OR preference   (b) OR selectivity   (c) OR preference &   (d)
                                              selectivity   OR H

Fig. 11.4. Self-organized orientation map. The orientation preference and selectivity of each neuron in SMAP and GMAP are plotted using the same conventions as in Figure 5.9. Because of the intracolumnar connections, the two maps develop similar organizations. As in LISSOM and in biological maps (Figures 2.5 and 5.9), the preferences change smoothly across the cortex, and exhibit features such as linear zones, pairs of pinwheels, saddle points, and fractures (outlined as in Figure 2.4). As in animal maps, the neurons at the pinwheel centers and fractures are unselective for orientation (these features are more prominent in the SMAP, which drives the self-organization). The orientation histograms are essentially flat and therefore free of artifacts. These plots show that realistic orientation maps can be formed with spiking neurons and with the SG model of cortical columns.

respond to a variety of orientations. The retinotopic organization in the two maps closely matches even though they had different numbers of neurons (Figure 11.3b). The receptive field centers are laid out in an evenly spaced grid that covers the whole retinal space, and the local distortions of this grid reflect the underlying orientation map.

To visualize the global order of the orientation maps, the orientation preferences and orientation selectivity of each neuron were measured (Figure 11.4). Despite the density difference, both maps have highly similar global order. The preferences change smoothly across the map, and contain features such as linear zones, matched pairs of pinwheels, saddle points, and fractures. Neurons at pinwheel centers and fractures are not very selective for orientation, similar to biological maps and the LISSOM orientation map (Sections 2.1.2 and 5.3.2).

In addition to the qualitative descriptions, quantitative measures introduced in Section 5.3.3 can be used to characterize the global properties of the maps. Orientation preference histograms and two-dimensional Fourier power spectra of both maps look similar to those of the LISSOM orientation map (Section 5.3.3): The histograms are flat and the spectra are ring-shaped, suggesting that at any location in the map, all orientation angles are equally well represented, as they should be. Such a structure is also very similar to biological maps (Section 5.1.1).

Together these results show that the architecture and the learning rules in PGLIS-SOM can develop realistic orientation maps, similar to those seen in previous LIS-SOM models, and in the mammalian visual cortex. In the next section, the lateral connection patterns will be analyzed.

11.5.3 Lateral Connections

As was discussed in Sections 2.2 and 5.1, the lateral connections in biological orientation maps have two prominent characteristics: (1) Strong connections exist between neurons with similar orientation preferences, and (2) the connections extend along the direction matching the source neuron's orientation preference (Figure 2.7). Such connections are believed to represent correlations in the visual input.

Similarly, the long-range lateral connections in PGLISSOM self-organize into patterns that have the same two general properties, and reflect the correlations in the input. The SMAP connections self-organize similarly to those in firing-rate LISSOM models; their properties were demonstrated in detail in Section 5.3.4. GMAP self-organization is different, however, since both excitatory and inhibitory connections have long range in GMAP, in order to establish segmentation and binding in the model.

Figure 11.5 plots the excitatory lateral connections of four sample neurons in the iso-orientation patches of the GMAP. In each case, the neuron is most strongly connected to others with similar orientation preferences. Thus, the figure shows qualitatively that the first property above holds in the model. The connections are also anisotropic, i.e. stretched along a particular direction with patches of connections found along this direction at roughly equal intervals. Such connection patterns are consistent with the second property exhibited in biological data.

As in the LISSOM models in the earlier chapters, such patterns emerge in PGLIS-SOM because the afferent and lateral connections adapt to encode the statistical structure in the training input. Since the training inputs are elongated Gaussian bars, the afferent connections form oriented receptive fields. Neurons with similar orientation preferences whose receptive fields are aligned along a straight axis will be activated simultaneously when a long input happens to fall upon those receptive fields. Due to the Hebbian learning process, such neuron pairs will develop strong lateral connections. Moreover, the connections are not strictly aligned along the axis, but there are also connections flanking the preferred axis. These flanks are larger farther from the source neuron, like a bowtie. The same pattern can be seen in biological data (Figure 2.7b).

This is an important observation, since the flanks allow grouping of not only straight contours, but also cocircular ones. Neurons not only respond to the optimal orientation at the optimal position, but also to slightly misoriented inputs (Fig-

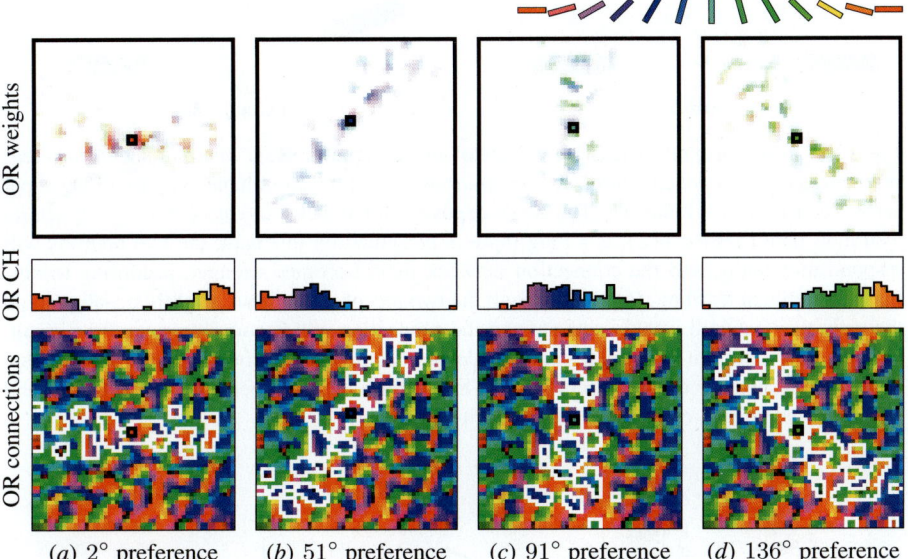

(a) 2° preference   (b) 51° preference   (c) 91° preference   (d) 136° preference

Fig. 11.5. Long-range lateral connections in GMAP. The long-range excitatory lateral connection patterns for four sample neurons in GMAP are shown on top, located in iso-orientation patches as shown in the map below. Similar plotting conventions are used as in Figure 5.12: The small black square identifies the neuron itself in both plots, and the white outline on the map indicates the extent of the lateral connections after self-organization and pruning; before self-organization the lateral connections covered the whole map, as shown by the black square outline on top. The color coding in the top plots represents the target neuron's orientation preference, selectivity, and connection strength, and the map below encodes orientation and selectivity. The histogram in the middle shows the distribution of the target neurons' orientation preferences. Each neuron is most strongly connected to its closest neighbors; the long-range connections are patchy and connect neurons with similar orientation preferences. They extend longer than those in Figure 5.12 because more elongated input patterns were used during self-organization. As in LISSOM, these connections extend along the orientation preference of the source neuron: (a) 2° red, (b) 51° purple, (c) 91° light blue, and (d) 136° light green. They are narrow around the neuron but wider farther away. As will be seen in Chapter 13, specific connection patterns like these are crucial for perceptual grouping such as contour integration.

ure 11.6). Thus, neurons with cocircular receptive fields can coactivate. When the two receptive fields can be connected with a straight path, but one or both are slightly misoriented from the axis of the path, they will both still be active. Their lateral connection will be strengthened, resulting in cocircular connection patterns.

The connection patterns in the model can also be measured quantitatively and compared with biological data. To measure the first property (i.e. that connections are stronger between neurons with similar orientation preferences), the percentage of GMAP excitatory lateral connections that connect receptive fields with varying orientation differences were calculated. The results are shown in Figure 11.7. The percentage of connections peaks at 0°, and rapidly decreases to zero as the orientation

(*a*) Collinear   (*b*) Cocircular

Fig. 11.6. Activating neurons with collinear and cocircular RFs. The plot shows two representative cases of coactivation, i.e. two neurons responding simultaneously to a long input across their receptive fields. (*a*) The two receptive fields (black bars) are precisely aligned on a straight path (dashed line). If a long input is presented on this path, the two neurons will respond maximally, and the connection between them becomes stronger, according to normalized Hebbian learning. (*b*) Even though the two receptive fields are slightly misaligned on the path, they can still weakly activate and the connection will become stronger. As a result, neurons that represent cocircular contours develop significant lateral connections, although not as strong as those that represent straight lines. Reprinted with permission from Choe and Miikkulainen (2004), copyright 2004 by Springer.

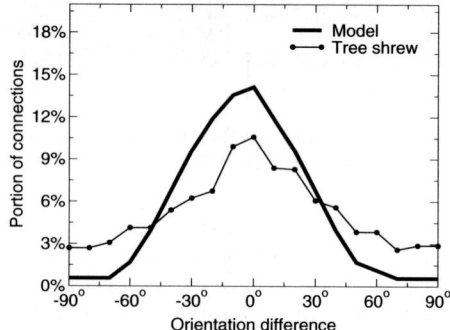

Fig. 11.7. Distribution of lateral connections in animals and in PGLISSOM. In the tree shrew V1, biocytin was injected in the cell body of seven different neurons, and the projections going to neurons of different orientation preferences were counted. The thin line with circles shows the median percentage of connections for each difference (adapted from Bosking et al. 1997). In the GMAP, the percentage of connections to neurons with different orientations were similarly counted (after pruning); the median over all neurons in the map is shown as the thick line. Both plots peak at 0°, and quickly fall off as the orientation differences become larger (this effect is slightly exaggerated in the model because in this experiment it was trained on straight elongated Gaussians only; cf. Section 13.4.2). In other words, strong excitatory lateral connections mostly link neurons with similar orientation tuning both in the model and in animals.

difference increases. In other words, strong excitatory lateral connections in GMAP are most likely to be found between neurons that have similar orientation preference. Such a result is consistent with experimental measurements (Figure 11.7), quantitatively verifying the first property.

This measure not only allows comparing the connectivity in the model and in biological data, but also suggests a possible functional role for the first property. As will be discussed in Section 13.1, contours are believed to be grouped through specific lateral interactions between contour elements (representing a local grouping function, or an association field). The measure presented above suggests that lateral connections could be implementing such a local grouping function.

The second property can also be measured quantitatively, by gathering statistics about the directions, angles, and distances of the source and target receptive fields. This study will be done in Section 13.2, showing that the second property holds quantitatively in PGLISSOM, and demonstrating that it implements a local grouping function. Edge-cooccurrence statistics in natural images and the connection statistics in the PGLISSOM model are shown to be strikingly similar, which makes the connectivity in PGLISSOM well suited for contour integration tasks.

11.6 Conclusion

To make it possible to understand self-organization of perceptual grouping phenomena, the PGLISSOM model expands the LISSOM framework inward in two ways. First, the single-unit model of the cortical column is extended to include two components: The S component takes part in self-organization (in SMAP), and the G component contributes to perceptual grouping (in GMAP). Second, the firing-rate neurons in LISSOM are replaced with spiking neurons, in order to represent grouping through temporal coding. The resulting PGLISSOM network self-organizes like the other LISSOM models, and the patterns of long-range excitation are appropriate for implementing perceptual grouping. In the following chapter, the synchronization behavior of the model is analyzed in detail. As a concrete example, its performance in contour integration tasks is then demonstrated in Chapter 13.

12

Temporal Coding

Experimental evidence reviewed in Section 2.4.2 suggests that temporally correlated activity may be the basis for binding and segmentation in perceptual grouping. In PGLISSOM such a temporal coding is generated by spiking neurons that synchronize their activities. It is important to understand how synchronization takes place in the model, mainly to gain insight into synchronization in biological networks, but also so that the PGLISSOM model as a whole can be tuned to function properly. In this chapter, the neuron model of PGLISSOM will be analyzed experimentally to find the conditions under which synchronization and desynchronization occur. Basic binding through synchronization and segmentation through desynchronization will be demonstrated first, followed by an analysis of how these processes are affected by the relative amounts of inhibition and excitation, spatial extent of the connections, synaptic decay rates, noise levels, neuron population sizes, and the length of the absolute refractory period. A special focus is on robustness against noise, which is crucial for these processes to function in biological networks.

12.1 Method

Synchronization is important for binding together populations of neurons that represent input features of the same coherent object. Desynchronization, on the other hand, signals that the input features belong to different objects. In this chapter, these processes will be illustrated using one-dimensional networks connected one-to-one to input and output. Such networks are sufficient for testing the various factors governing synchronization, and they are also easy to visualize in two dimensions because the activities can be plotted over time.

Unless stated otherwise, in all experiments the input neurons spike at every time step, the membrane potential of each neuron is initialized to uniformly random distribution within $[0..1]$, the afferent weights are fixed at 1.0, the lateral excitatory weights are fixed at $1.0/n_E$, and the lateral inhibitory weights are fixed at $1.0/n_I$, where n_E and n_I represent the number of excitatory and inhibitory lateral connections (the rest of the parameters are listed in Appendix D.3). In order to study synchronization behavior, the following parameters are systematically varied in the experiments: (1) lateral excitatory and inhibitory connection patterns and radii r_E and r_I, (2) their contributions γ_E and γ_I, (3) their synaptic decay rates λ_E and λ_I, (4) the size and pattern of the afferent input, (5) the degree of noise in the membrane potential, and (6) the duration t_r of the absolute refractory period.

For example, Figure 12.1 demonstrates synchronization behavior of a five-neuron network with full lateral connectivity under two conditions. In Figure 12.1a, all lateral connections were excitatory, whereas in Figure 12.1b they were inhibitory; all the other parameters were the same ($\gamma_E = 0.01, \gamma_I = 0.001, \lambda_E = 5.0, \lambda_I = 8.0$, full input activation, no noise in the membrane potential, no absolute refractory period). The simulation time is on the x-axis, and the membrane potential of each neuron is plotted over time in two ways: as a continuous y value in the top plot, and as a gray-scale value in the bottom plot. In Figure 12.1a, due to lateral excitation, the neurons reinforce each other's activity and soon start firing at the same time. In contrast, in Figure 12.1b, the lateral inhibition establishes competition between the neurons, and after a short time they each fire at different times.

Such complete and exclusive lateral excitation vs. inhibition results in extremely strong synchronization and desynchronization, and constitutes a good example of these behaviors. The PGLISSOM models include both excitation and inhibition, and the synchronization behavior depends on many other factors. In the following sections, these factors will be systematically varied, and gray-scale plots similar to those in Figure 12.1 will be used to illustrate the resulting behavior.

12.2 Binding Through Synchronization

In this section, two main factors that affect the quality of synchronized representations will be analyzed: the synaptic decay rate and the extent of lateral connections. Decay allows controlling how accurately the spiking events need to be timed. On the other hand, lateral connections are necessary to coordinate the firing of neurons, and their extent determines how large areas can be synchronized.

12.2.1 Effect of Synaptic Decay Rate

Previous models of spiking neurons have either adapted or selected the axonal delays to regulate synchronization behavior (Eurich, Pawelzik, Ernst, Thiel, Cowan, and Milton 2000; Gerstner 1998a; Horn and Opher 1998; Nischwitz and Glünder 1995; Tversky and Miikkulainen 2002). The biological basis for such delay tuning is unclear: Although e.g. axonal morphology (length, thickness, and myelination) can change over time (Eurich, Pawelzik, Ernst, Cowan, and Milton 1999), the fast and accurate delay tuning needed in the above models may not be easy to achieve in this way (Stevens, Tanner, and Fields 1998).

An alternative to delay adaptation is changing the decay rate of the PSP. Decay may be easier to alter in biological neurons since ion channels can be added

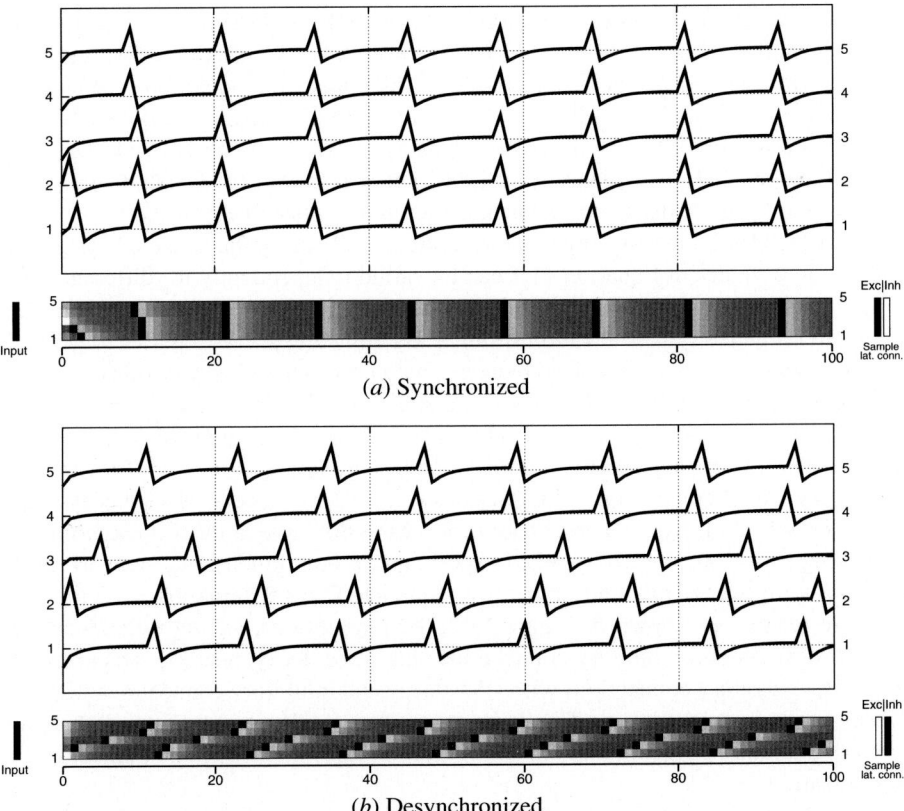

(a) Synchronized

(b) Desynchronized

Fig. 12.1. Synchronized and desynchronized modes of firing. A network of five neurons is connected only with excitatory lateral connections in (a) and only with inhibitory lateral connections in (b), and exhibits synchronized and desynchronized firing as a result. Simulation time is shown on the x-axis, and the membrane potential for each neuron is displayed in two ways: along the y-axis in a voltage trace plot on the top, and in gray scale from white to black (low to high) in the bottom. Each row in each plot represents a different neuron, with its index identified on both sides of the plot (1 to 5, from bottom to top). The black vertical bar on the left shows which neurons are activated by afferent input (black means on and white means off); in these two examples, all input neurons were activated. To the right of the scatterplot, two vertical bars illustrate the excitatory (left) and inhibitory (right) lateral connection ranges of one sample neuron in the network; all neurons had identical connections in these two examples. Black indicates that a connection to the neuron exists in that row, and white that it does not. The same plotting convention will be used throughout this chapter. With excitatory lateral connections in (a), all spikes (peaks) start to become vertically aligned around iteration 21, showing that all the neurons are firing at the same time. In contrast, with inhibitory lateral connections in (b), the neurons all fire at different times. An animated demo of these examples can be seen at http://computationalmaps.org.

or removed to tune the leakage of currents through the cell membrane. The number and distribution of ion channels can change through various mechanisms, including activity-dependent gene expression and activity-dependent modulation of assembled ion channels (Desai, Rutherford, and Turrigiano 1999; Nowak and Bullier 1997; see Abbott and Marder 1995 for a review). Synaptic decay has been utilized in computational models before (Eckhorn et al. 1990; Reitboeck et al. 1993), but the influence of different levels of decay on synchronization has not been fully tested.

PGLISSOM allows the effects of synaptic decay rates to be analyzed systematically. The λ values in Equation 11.1 can be varied independently for different types of connections (excitatory or inhibitory). In such simulations, the decay rate was found to influence synchronization strongly. By adjusting λ, it is possible to get both synchronized and desynchronized behavior with both types of connections.

Four separate experiments were conducted: (1) excitatory lateral connections with slow decay ($\lambda_E = 0.1$), (2) inhibitory lateral connections with slow decay ($\lambda_I = 0.1$), (3) excitatory lateral connections with fast decay ($\lambda_E = 1.0$), and (4) inhibitory lateral connections with fast decay ($\lambda_I = 1.0$). Except for the decay rate λ and the connection type, all other parameters were the same in the four experiments, including $\gamma_I = \gamma_E = 0.01$. In each experiment, a one-dimensional network of 30 neurons with full lateral connections was simulated for 500 iterations.

The results are shown in Figure 12.2. Two conditions, excitatory connections with fast decay and inhibitory connections with slow decay, result in synchrony. In contrast, excitatory connections with slow decay and inhibitory connections with fast decay result in desynchrony. This is an interesting result, since excitation does not always guarantee synchronization, and inhibition does not always guarantee desynchronization.

Nischwitz and Glünder (1995) showed that a similar result is obtained by varying the degree of delay among integrate-and-fire neurons connected via excitatory or inhibitory connections. A short delay with excitatory connections and long delay with inhibitory connections caused the neurons to synchronize, and in the opposite case to desynchronize. The current result indicates that synaptic decay adaptation, which appears more plausible than delay adaptation, can also control synchronization.

It is important to note that although synchronization can be achieved through slowly decaying inhibitory connections, excitatory connections are more likely to be responsible for coherent oscillation in the cortex. As will be discussed in Section 16.3.3, coherent oscillations have been found mostly in the superficial layers of the cortex, especially in layers 2/3. The long-range connections in those layers are mostly excitatory, suggesting that the synchronization is established through excitation.

In summary, synchronization can be regulated effectively by adjusting the synaptic decay rate. Because possible adjustment mechanisms are known to exist in biology, synaptic decay adaptation is an attractive alternative to models based on delay modulation.

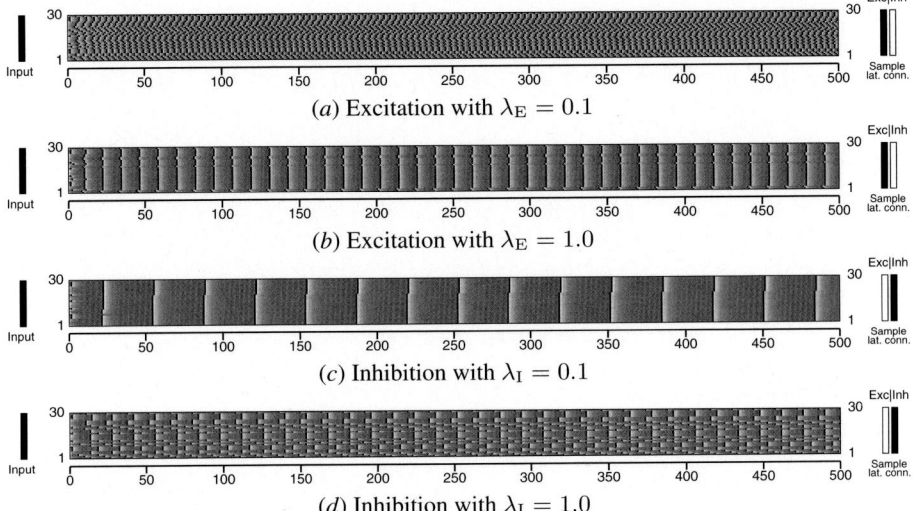

Fig. 12.2. **Effect of connection type and decay rate on synchronization.** Thirty neurons with full lateral connections, either excitatory or inhibitory, were simulated for 500 iterations (see Figure 12.1 for plotting conventions). Four experiments were conducted where the type of the lateral connections (excitatory or inhibitory) and the synaptic decay rates (λ) were altered. All other parameters were the same for all four cases. (*a*) Excitatory connections with slow decay ($\lambda_E = 0.1$) result in desynchronized activity. (*b*) Excitatory connections with fast decay ($\lambda_E = 1.0$) result in synchronized activity. (*c*) Inhibitory connections with slow decay ($\lambda_I = 0.1$) result in synchronized activity. (*d*) Inhibitory connections with fast decay ($\lambda_I = 1.0$) result in synchronized activity. Note that in the two synchronized cases (*b*) and (*c*), the firing rate is higher in (*b*): The input activity to the neuron $g(t)$ approaches the threshold faster because of the excitatory lateral input. The results show that synchronization behavior can vary greatly even for the same connection type if the synaptic decay rate differs.

12.2.2 Effect of Connection Range

In the second test, the goal was to determine whether local excitatory connections can synchronize a global population. Inhibitory lateral connections were excluded to simplify the simulations. Thirty neurons with varying degrees of excitatory lateral connection radii were simulated for 500 iterations. Five separate experiments were conducted with excitatory connection radii of 30, 10, 5, 2, and 0. Other simulation conditions were the same as in Section 12.1, except $\gamma_E = 0.01$ and $\lambda_E = 5.0$ so that the network would also synchronize under the smaller radii.

The results are shown in Figure 12.3. Global synchronization is achieved not only in the fully connected network as before (radius 30), but also in locally connected networks, down to a radius of 5. These results demonstrate that synchronization can propagate through locally connected neurons, which is consistent with other coherent oscillation models with local connections (Campbell et al. 1999; Terman and Wang 1995; Wang 1995, 1996). They show that synchronization may work as a basis for

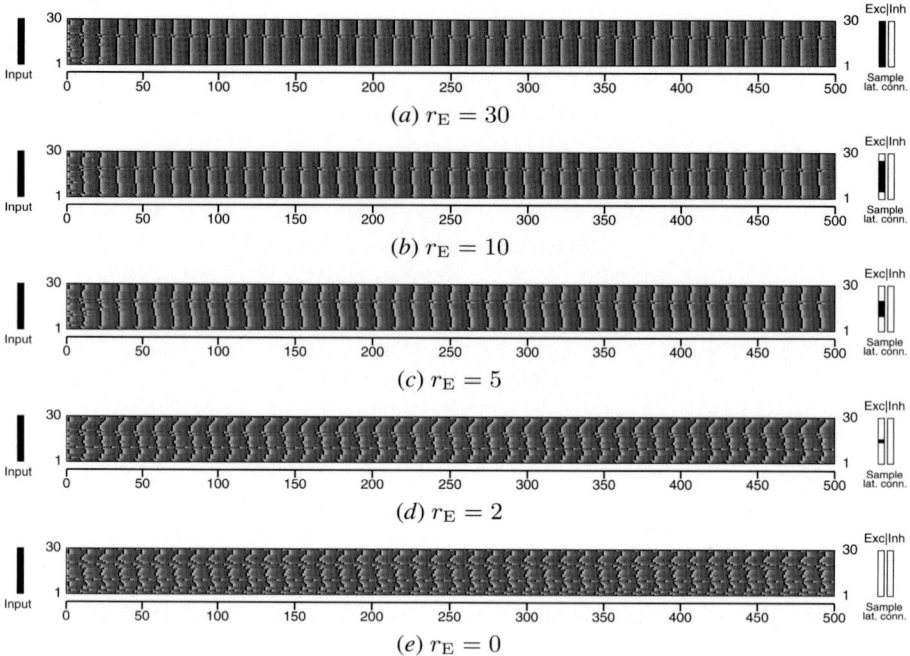

Fig. 12.3. Effect of excitatory connection range on synchronization. A network of 30 neurons with varying extent of lateral excitatory connections was simulated for 500 iterations.
Synchronization occurs through the excitatory connections even though the connections did
not cover the whole network. From (*a*) to (*e*), the lateral excitatory connection radius r_E was
reduced from 30 (i.e. full connectivity) to 10, 5, 2, and 0 (the bars at right depict the connections of the neuron in row 15). All other parameters were the same as before. Synchronization
starts to break once the radius reaches 2, but for a fairly local connection radius (e.g. 5), global
synchronization is maintained. As expected, with no excitatory connections (*e*), the initial random order of spikes is maintained throughout the simulation. Global synchrony can therefore
be established with local connections.

transitive grouping: If A and B are grouped together and B and C are grouped together, then A and C are perceptually grouped together (Geisler et al. 2001; Geisler
and Super 2000).

   In summary, fully connected networks synchronize well, but it is not necessary to
have full connectivity to achieve global synchrony. Global synchronization through
local connections in the PGLISSOM model may be a possible mechanism for transitive perceptual grouping.

12.3 Segmentation Through Desynchronization

Proper desynchronization is as important as synchronization, since it is the basis for segmentation. This section will show that inhibitory connections are necessary for segmentation, and that a small amount of noise is necessary for symmetry breaking.

12.3.1 Effect of Connection Types

As was seen in Section 12.2.1, excitatory and inhibitory connections have the opposite effect under the same decay rate. For perceptual grouping, both synchronization and desynchronization are necessary; such behavior may be efficiently achieved by utilizing both excitatory and inhibitory lateral connections. This hypothesis is tested in this section, verifying that including both types of connections indeed results in a desirable temporal representation for binding and segmentation.

As an abstraction of the grouping task and the underlying connectivity in the cortex, a one-dimensional network of 90 neurons was divided into two groups: Neurons [1..22] and [45..66] formed the first group, and neurons [23..44] and [67..90] the second group. Lateral excitatory connections were only allowed to connect neurons within the same group, whereas inhibitory connections connected the whole population.

Four separate experiments were conducted: one with both excitatory and inhibitory connections, another with excitatory connections only, the third with inhibitory connections only, and the fourth with no lateral connections at all. Other simulation conditions were the same as in Section 12.1, except $\gamma_E = 0.36$ and $\lambda_E = 5.0$ to compensate for the larger size of the network and the addition of inhibitory connections. Values of $\gamma_I = 0.42$ and $\lambda_I = 5.0$ were used for the inhibitory connections.

The results are shown in Figure 12.4. First, with both excitatory and inhibitory connections, neurons within the same group are synchronized, but across the groups where only inhibitory connections exist, desynchronization occurs (Figure 12.4a). Such temporal representation is well suited for perceptual grouping, since binding is signaled by synchrony and segmentation by desynchrony. Next, with only excitatory connections, segmentation does not occur (Figure 12.4b), and with only inhibitory connections, binding does not occur (Figure 12.4c). Finally, without any lateral connections, the neurons fire in different phases, determined by their randomly initialized membrane potential (Figure 12.4d).

In summary, binding and segmentation can be established in a network with both excitatory and inhibitory lateral connections. Omitting either kind of the lateral connections results in losing the ability to bind, segment, or both.

12.3.2 Effect of Noise

In previous sections, the initial membrane potential of each neuron was uniformly randomly initialized. Whether such initial perturbations are necessary will be tested

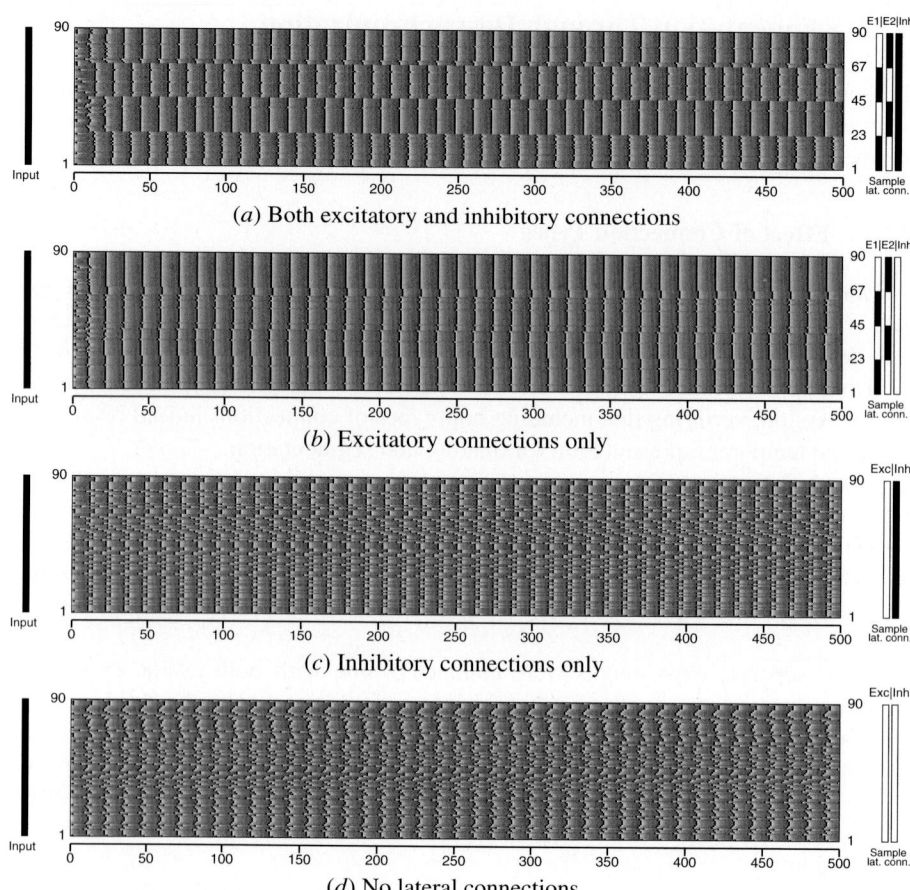

Fig. 12.4. Binding and segmentation with different connection types. A network of 90 neurons was divided into two groups and simulated for 500 iterations. Neurons [1..22] and [45..66] formed the first group (E1) and neurons [23..44] and [67..90] the second group (E2). All neurons in each group had the same lateral connections, shown at right. Excitatory lateral connections only linked neurons within the same group (E1 or E2), and inhibitory connections were global. (a) With both excitatory and inhibitory connections, the neurons within the same group are synchronized, while those in different groups are desynchronized. (b) With excitatory connections only, the neurons cannot desynchronize. (c) With inhibitory connections only, no coherently synchronized groups emerge. (d) When there are no lateral connections, neurons spike in different phases determined by their initial state. Both types of connections are therefore needed to establish simultaneous binding and segmentation.

in this section, analyzing the roles of initial and continual noise in symmetry breaking. The results are compared with the control case where the simulation is carried out without noise.

A network of 180 neurons with both excitatory and inhibitory lateral connections was simulated for 500 iterations. The network was divided into two groups as in the previous experiment (Section 12.3.1). Neurons [1..22], [45..66], [89..110], and [133..154] formed the first group, and neurons [23..44], [67..88], [111..132], and [155..180] the second group. Excitatory lateral connections only connected neurons within the same group, and their radius was limited to 90. The inhibitory connections were global.

Three separate experiments were conducted: with initial noise only, with continual noise only, and without noise. The parameters were the same in all three experiments: $\gamma_E = 0.48$, $\gamma_I = 0.42$, and $\lambda_E = 5.0$, $\lambda_I = 1.0$, and all other simulation conditions were the same as in Section 12.1.

The results are shown in Figure 12.5. With initial noise (i.e. random initial membrane potential), the neurons within the same group are synchronized while the two different groups are desynchronized (Figure 12.5a). Also, even if the neurons are initialized uniformly (at 1.0), when a small amount of noise (0.1%) is added to the membrane potential at each time step, the two groups will desynchronize (Figure 12.5b). However, without noise of any kind, symmetry is not broken and the two groups stay synchronized (Figure 12.5c). So, inhibitory connections alone are not sufficient for desynchronization. Cortical neurons actually operate in a noisy cellular environment, so including such noise in the model is realistic. It may also be desirable, in that it can make the behavior of the model more robust (Baldi and Meir 1990; Horn and Opher 1998; Terman and Wang 1995; Wang 1995).

In summary, a small amount of noise is needed for desynchronization; noise will therefore be used in the perceptual grouping experiments with PGLISSOM. The problem in biological systems, however, is not lack of noise, but that there may be too much of it. Next, how binding and segmentation can take place robustly under noisy conditions will be investigated.

12.4 Robustness Against Variation and Noise

The previous sections showed how the synaptic decay rate, the type and extent of the lateral connections, and the degree of noise can be controlled in the model to achieve synchronization and desynchronization for binding and segmentation. However, there are several factors that can possibly interfere with this process. For example, if the network is presented with different-size inputs simultaneously, the larger input could dominate the smaller input. If the level of noise is raised above a threshold, noise can dominate and coherent behavior may not be obtained. How robust the model is against such external factors will be analyzed in this section, and the components that contribute to its robustness will be identified.

12.4.1 Robustness Against Size Differences

One requirement for perceptual grouping is that input features should not be suppressed or promoted on the basis of size only, since smaller but complex input features in the scene can be equally important as large but simple features. Thus, a

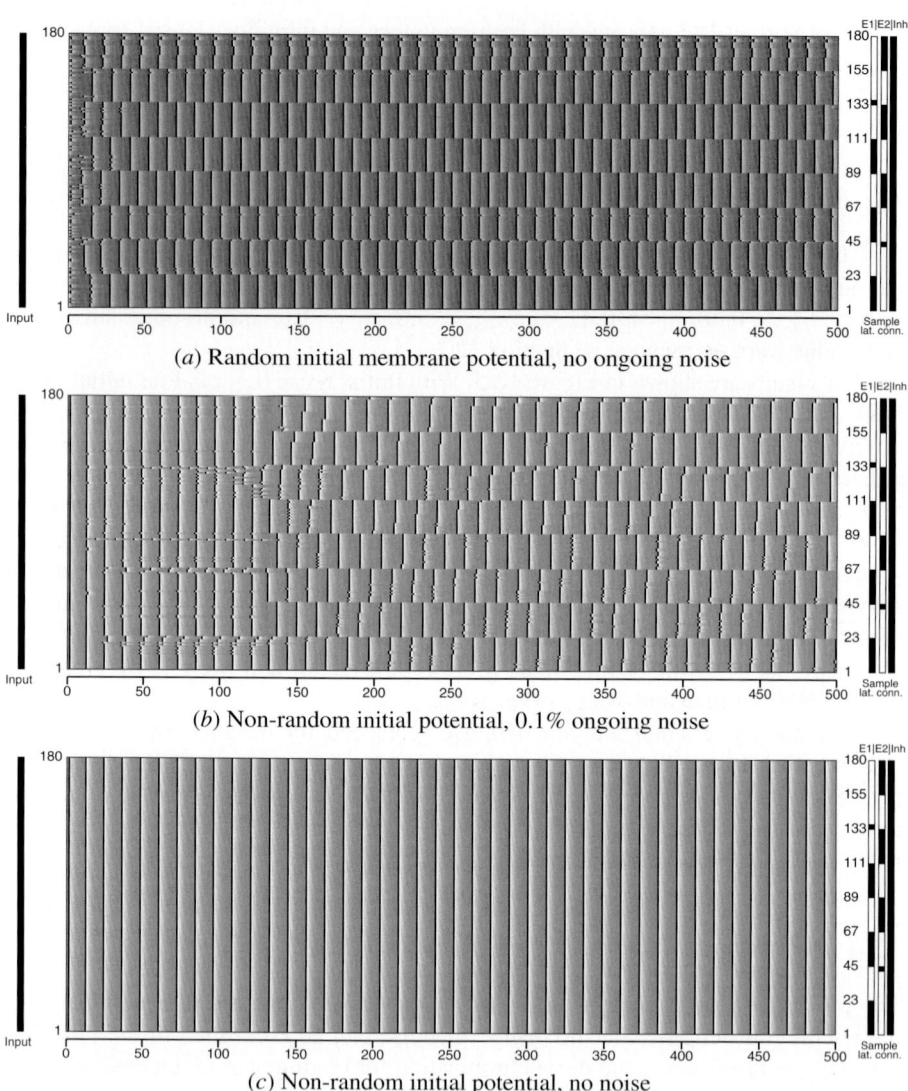

(a) Random initial membrane potential, no ongoing noise

(b) Non-random initial potential, 0.1% ongoing noise

(c) Non-random initial potential, no noise

Fig. 12.5. Effect of noise on desynchronization. A network of 180 neurons with both excitatory and inhibitory lateral connections was simulated for 500 iterations. The network was divided into two groups, as in the experiment of Figure 12.4. The first group (E1) consisted of neurons [1..22], [45..66], [89..110], and [133..154], and the second group (E2) of neurons [23..44], [67..88], [111..132], and [155..180]. The excitatory lateral connections were limited to the neurons in the same group within a radius of 90; the inhibitory connections were global. To illustrate, the plots at right show the lateral connections of neuron 45 in E1 and 132 in E2. (a) The membrane potential of each neuron was uniformly randomly initialized, and no noise was added afterward. The symmetry is broken and the two groups are separated as expected. (b) The membrane potentials initially were the same, but perturbed throughout the simulation by adding 0.1% of uniformly random noise. The neurons within the same group are synchronized at the same time as the two groups are desynchronized. (c) Without any noise (initial or continual), the symmetry was not broken and the entire network remained synchronized. A small amount of noise is therefore essential for proper desynchronization to occur.

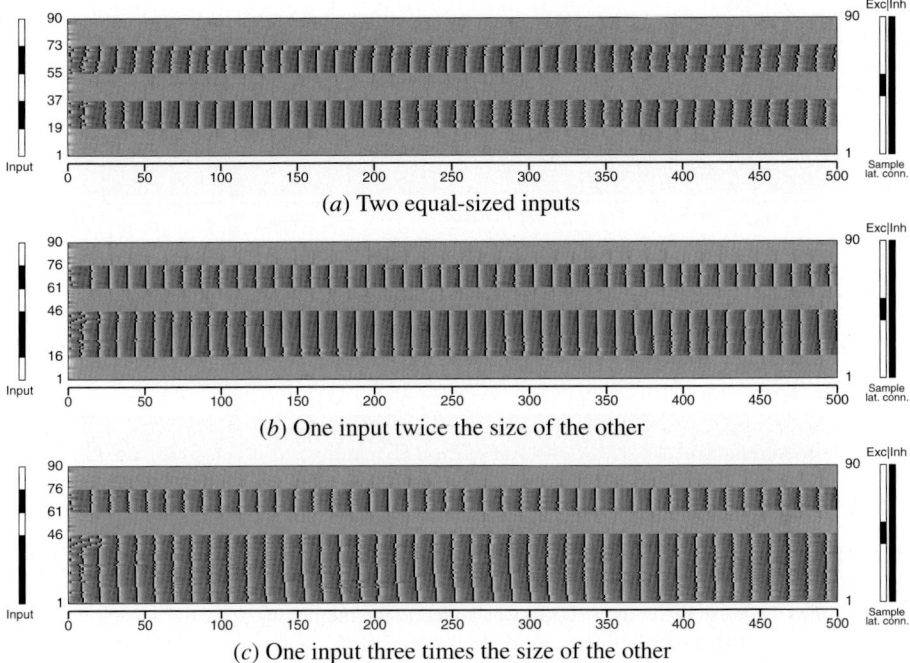

Fig. 12.6. **Effect of relative input size on synchronization.** A network of 90 neurons with both excitatory and inhibitory lateral connections was simulated for 500 iterations. The excitatory connection radius was 14 and inhibitory connections were global (as shown at right for neuron 45). The network was given two spatially separated inputs, and the size of the second input was varied. The rows (i.e. neurons) that received input are marked by black solid bars on the left. (a) The two inputs were the same size, activating neurons [19..36] and [55..72]. (b) One input was twice as long as the other input, activating neurons [16..45] vs. [61..75]. (c) One input was three times as long as the other input, activating neurons [1..45] vs. [61..75]. In all cases, the inputs are robustly bound and segmented, showing that the behavior is not affected by variation in the size of the input.

network of spiking neurons modeling such behavior should tolerate differences in input size.

To test if the PGLISSOM model is robust against such variation, a network of 90 neurons with both excitatory and inhibitory lateral connections was simulated for 500 iterations. The excitatory connection radius was 14 so that neurons representing different inputs were not connected, and inhibitory connections were global. Three separate experiments were conducted by presenting two inputs of relative sizes 1:1, 1:2, and 1:3 (Figure 12.6). The parameter values were the same in the three experiments: $\gamma_E = 0.7$, $\gamma_I = 0.6$, and $\lambda_E = 5.0$, $\lambda_I = 1.0$, to compensate for the larger number of excitatory connections compared with the previous experiments. Other simulation conditions were the same as in Section 12.1.

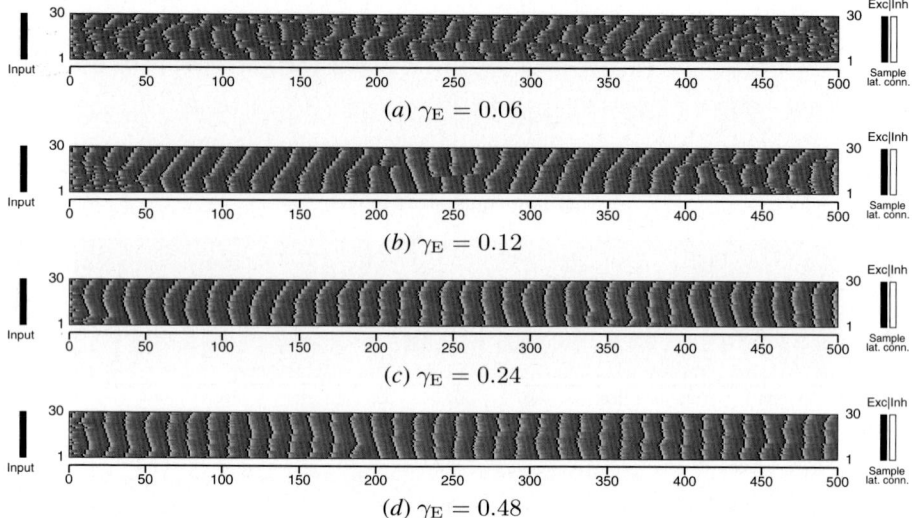

Fig. 12.7. Overcoming noise with strong excitation. A network of 30 neurons with global excitatory lateral connections was simulated for 500 iterations. A higher level of noise (1%, i.e. 10 times the noise in Section 12.3.2) was added to the membrane potential at each iteration. When the excitatory contribution is weak, as in (a) and (b), noise overwhelms it and causes the activities to desynchronize. However, as the excitatory contribution becomes stronger as in (c) and (d), it overcomes noise and achieves synchrony.

The results are shown in Figure 12.6. The two areas of the map representing the two objects are synchronized and desynchronized within and across the group, regardless of the input size. Note that these behaviors occur under identical parameter conditions, and thus demonstrate that the model is not affected by the size of the inputs alone.

12.4.2 Overcoming Noise with Strong Excitation

Cortical neurons operate in an inherently noisy environment. Noise can arise from several causes: For example, synaptic transmission may be unreliable or membrane potential may fluctuate. As was shown in Section 12.3.2, a small amount of noise is useful in desynchronizing separate representations. However, biological networks are likely to be very noisy, and a model should be robust against such high levels of noise as well.

A network of 30 neurons with global excitatory lateral connections was simulated for 500 iterations. Four separate experiments were conducted, increasing the excitatory contribution in four stages under a higher level of noise (1%, i.e. 10 times the noise in Section 12.3.2). The simulation conditions were the same as in Section 12.1, except $\lambda_E = 5.0$.

The results are shown in Figure 12.7. With a higher level of noise, weak excitatory connections cannot keep the neurons synchronized (Figure 12.7a,b), but as the

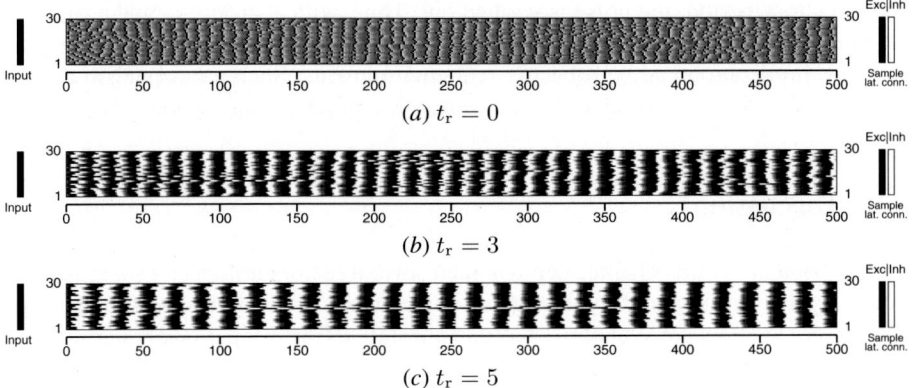

Fig. 12.8. Overcoming noise with a long refractory period. A network of 30 neurons with global excitatory lateral connections was simulated for 500 iterations. A significant level of noise (7%, 70 times the noise in Section 12.3.2) was added to the membrane potential at each iteration. Such high noise cannot be tolerated by just increasing the lateral excitatory contribution γ_E. However, increasing the absolute refractory period can make the model robust even in this case. (*a*) When there is no absolute refractory period ($t_r = 0$), the activities are random. (*b*) When the absolute refractory period lasts for three iterations, the activities start to synchronize loosely. (*c*) When it lasts for five iterations, the activities are strongly synchronized. With longer periods between firing, the noise is effectively washed out.

excitatory contribution γ_E is increased, the neurons start to synchronize. This result shows how a network of spiking neurons can robustly synchronize even in moderately noisy conditions: Strong excitation can be used to overcome the noise in such cases.

12.4.3 Overcoming Noise with a Long Refractory Period

Although increasing the excitatory contribution helps, there is a certain threshold where noise cannot be overcome this way. For example, 7% noise will break the synchronization behavior even with extremely strong excitatory connections because noise will dominate the spiking behavior of the network. However, it turns out longer refractory periods will make the network robust even in such cases.

A network of 30 neurons with global excitatory lateral connections was simulated for 500 iterations. Three separate experiments were conducted where the length of absolute refractory period was gradually increased. The simulation conditions were the same as in Section 12.1, except $\lambda_E = 5.0$ to make synchronization more robust.

The results are shown in Figure 12.8. Under significant noise (7%), the excitatory connections alone cannot keep the neurons synchronized (Figure 12.8*a*), but as the absolute refractory period is lengthened, the neurons start to synchronize again (Figure 12.8*b,c*). This result suggests that absolute refractory periods may have come to exist in biological neurons in part to overcome high levels of noise in the cortical environment. When the time interval during which the neuron can fire is smaller than the refractory period, the noise is washed out. Thus, with a strong γ_E and a long refractory period, the neuron can be made highly robust against noise, which suggests that synchronization can be robust in real environments. Such a mechanism can be seen as a way to increase reliability at the expense of processing speed.

12.5 Discussion

As experiments in this chapter demonstrate, several factors influence how firing becomes synchronized and desynchronized in a network of integrate-and-fire neurons. The observed effects of synaptic decay and absolute refractory period are particularly novel and potentially significant.

Synaptic decay has a similar effect on synchronization as the time it takes to integrate incoming PSPs: Both modulate the time it takes to reach the threshold. Integration time has been recognized earlier as a parameter that can alter synchronization behavior (Eurich et al. 1999, 2000). It is usually modeled by adding an additional delay to the spike arrival time. However, such an approach does not take into account that the PSP also decreases over time. The results in Section 12.2.1 suggest that decay plays a significant role in synchronization. Explicitly modeling decay therefore gives us a more accurate understanding of the mechanisms responsible for synchrony.

At this point, the effects of decay adaptation are computational predictions only; there is little biological evidence to support or falsify such processes. In the near future, it may be possible to verify whether the dendritic membrane potential can decay at different rates at different locations in biological neurons, and also whether there is such a difference between different types of synapses (e.g. glutaminergic vs. GABAergic synapse). If differences are found, they can be compared with the results presented in this chapter, allowing us to predict what role such different kinds of connections may play in modulating synchrony. An interesting further question is how the decay rate interacts with conduction delays. While it may be difficult to tune the delays in biological systems, it is possible that the decay rate adjusts to the delays so that robust synchronization behavior emerges under various conditions (Section 16.4.7).

Another novel result of this chapter is that a longer absolute refractory period can help overcome noise: Even in a highly noisy neural environment, synchronization can be achieved in this manner. From a computational point of view, such a mechanism can be seen as a way to increase reliability at the expense of processing speed: With a longer refractory period, firing rates will be lower and it will take longer to decode the information encoded in them (Section 16.4.7).

Together, the results in this chapter demonstrate qualitatively how the different factors affect behavior in networks of integrate-and-fire neurons. They serve as a practical guide that allows utilizing synchronization in large neural network systems, as will be done in the next chapter. In the future it may also be possible to carry the analysis a step further and develop a quantitative theory of how the different parameters modulate synchrony (similar to the analysis of connection types and delays by Nischwitz and Glünder 1995). For example, the PSP decay rate parameter λ or the length of the absolute refractory period can be continuously varied within a fixed range, and the degree of synchrony measured in each case. Such a study might lead to empirical equations that allow predicting the behavior of the network with different parameters. Like the equations that scale the model to different size networks (Chapter 15), such equations might also allow setting the parameters directly to obtain the desired behavior, eventually leading to a mathematical theory of synchronization.

12.6 Conclusion

In this chapter, a one-dimensional network of spiking neurons was systematically tested to determine how the different components of the PGLISSOM model contribute to synchronization and desynchronization of activity. The results show that increasing the decay rate can synchronize networks with excitatory connections and desynchronize networks with inhibitory connections, local excitatory connections can achieve long-range synchrony, and both excitatory and inhibitory connections are necessary for simultaneous binding and segmentation, and noise helps break the symmetry in such cases. The model was also shown to be robust against changes in input size, and against high levels of noise through strong excitation and long absolute refractory periods.

Understanding such qualitative and quantitative factors that affect synchronization allows us to predict how a network of spiking neurons behaves in a specific configuration, and provides a theory for the corresponding mechanisms in biology. These mechanisms play an important role in perceptual grouping tasks such as contour integration, which is the subject of the next chapter.

Understanding Perceptual Grouping: Contour Integration

Perceptual grouping is the process of identifying the constituents in the visual scene that together form a coherent object. Grouping takes place at several levels in the visual processing hierarchy, as was discussed in Section 1.3. Experiments with PGLISSOM focus on the early grouping task of contour integration, where there are plenty of neurobiological and psychophysical data to constrain, test, and validate the model. The hypothesis is that much of contour integration is performed in V1, based on mechanisms implemented in PGLISSOM.

The first section in this chapter defines and motivates the task and reviews psychophysical observations and computational models. The following sections demonstrate PGLISSOM in contour integration, including segmentation of multiple contours and completion of partial and illusory contours. Input-driven self-organization is also shown to possibly account for the differences in contour integration performance in different parts of the visual field.

13.1 Psychophysical and Computational Background

Contour integration is a well-defined task where performance can be readily measured both in humans and in computational models. It has therefore been extensively studied both in psychophysical experiments and in artificial neural networks. The underlying theory is remarkably clear and uniform across these studies and also extends to illusory contours. It is less clear, however, how contour integration circuitry can arise in early development and result in different performance in different parts of the visual field.

13.1.1 Psychophysical Data

A typical visual input for the contour integration task is shown again in Figure 13.1. The input consists of a series of short oriented edge segments (contour elements) aligned along a continuous path, embedded in a background of randomly oriented contour elements. The task is to identify the longest continuous contour in this scene.

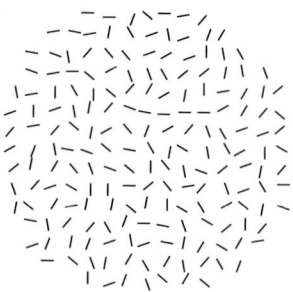

Fig. 13.1. Demonstration of contour integration. Look briefly at the circular area fill with line segments above; a continuous contour from top left to the right and slightly down should immediately emerge. This process is called contour integration: A good continuation of contour elements leads to a vivid perception of a single object. This function is believed to arise automatically in the orientation map of V1, mediated by lateral connections (Section 13.1.1).

Psychophysical experiments (Elder and Goldberg 2002; Field et al. 1993; Geisler et al. 2001; McIlhagga and Mullen 1996; Pettet et al. 1998; see Hess, Hayes, and Field 2004 for a review) suggest that there exists a highly specific pattern of interactions among the contour elements. Such interactions allow contour elements of certain positions and orientations to be more visible than in others. For example, Field et al. (1993) conducted a series of experiments where each subject was told to find a contour of similarly oriented Gabor patterns embedded among randomly oriented Gabor patterns. Several factors affected the performance. The relative orientation of successive contour elements (i.e. orientation jitter) along the longest contour (the path) was the most important such a factor. When the orientation of successive contour elements differed more, the performance degraded. Other factors, such as distance between elements and phase difference between successive Gabor patterns, also affected performance but to a lesser extent.

Based upon these results, Field et al. suggested that local interactions between contour elements follow specific rules and form the basis for contour integration in humans. In other words, these constraints form a local association field that governs how differently oriented contour elements should interact to form a coherent group (Figure 13.2). An association field can be described with two rules: (1) contour elements positioned on a smooth path (Figure 13.2a) and (2) contour elements aligned collinearly along the path (Figure 13.2b) are more likely to be perceived as belonging to the same contour. This idea can be formalized as a mathematical theory, and extended to account for non-uniform variance in the association profile as well (Ben-Shahar and Zucker 2004).

Pettet et al. (1998) further confirmed that lateral interactions between neighboring contour elements follow well-defined constraints. They compared the performance of human subjects with a model consisting of fixed lateral interaction constraints similar to the association field, and found that the model explained psychophysical data very well. In particular, it was consistent with the earlier result of Kovacs and Julesz (1993) showing that closed contours were easier to detect than

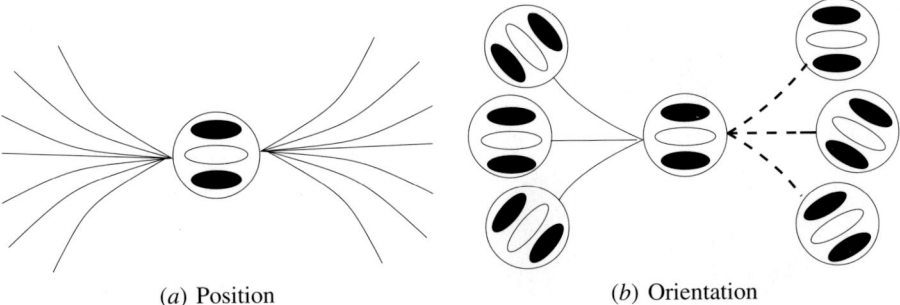

(a) Position   (b) Orientation

Fig. 13.2. Association fields for contour integration. The circular disks with black and white oriented bars represent Gabor wavelets, which is the typical input contour element used in the study by Field et al. (1993). Whether the elements interact depends on two factors: (a) The elements' positions have to be located on smooth (collinear or cocircular) contour, and (b) their orientations have to be parallel to the smooth contour. The solid lines indicate cases where integration occurs, and the dashed lines those where it does not. Adapted from Field et al. (1993).

open-ended contours. Pettet et al. reasoned that the missing lateral interaction between the two ends of the open-ended contour made it harder to perceive the contour, whereas reverberating lateral interaction along the closed loop made it easier.

Geisler et al. (2000, 2001) took a different approach in identifying the conditions that govern contour integration. Instead of proposing rules based on human performance, they extracted the rules from edge-cooccurrence statistics measured in natural images (Section 13.2.3). Edge-detected natural images were decomposed into outline figures, consisting of short oriented edges. The cooccurrence probability of each pair of edges belonging to the same physical contour in natural images was then calculated. Such edge-cooccurrence statistics (as also reported by Elder and Goldberg 2002; Krüger and Wörgötter 2002; Sigman, Cecchi, Gilbert, and Magnasco 2001) turned out to be very similar to the lateral interaction rules proposed by Field et al. and Pettet et al. Furthermore, Geisler et al. devised a method of extracting contours using these cooccurrence statistics: Two edges are grouped together if the probability of the edges occurring in the given configuration exceeds a threshold. Larger groups are then formed transitively: A and C were grouped together if A and B can be grouped together and B and C can be grouped together. Geisler et al. showed that this method of grouping accurately predicts human performance. Thus, they showed that the statistical structure in the environment closely corresponds to human perceptual grouping.

Moreover, Geisler et al. found that the statistical model explains more of the perceptual performance than previously believed, including the difference between open and closed contours. First, when the complexity of the contours and the eccentricity of the edge elements was carefully controlled, the advantage of closed contours disappeared in most cases. Where it still existed, it was perfectly predicted by the statistical model (Tversky, Geisler, and Perry 2004). Therefore, closed contours may

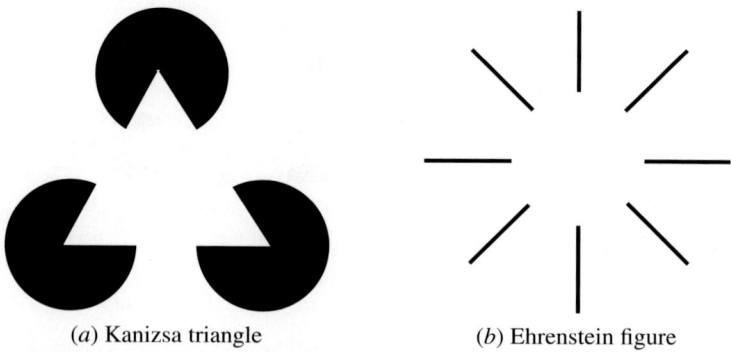

<table>
<tr><td>(a) Kanizsa triangle</td><td>(b) Ehrenstein figure</td></tr>
</table>

Fig. 13.3. Edge-induced vs. line-end-induced illusory contours. The figures illustrate the two mechanisms believed to underlie the process of forming illusory contours in the visual system: (a) The hovering white triangle is formed by edge inducers, i.e. by continuing the separate, collinearly aligned edges, and (b) the bright circle in the middle is composed by line-end inducers, i.e. by connecting the ends of the lines into a continuous circle.

be easier to perceive simply because each segment is a good continuation of a nearby segment in each direction, without any special reverberating mechanisms along the contour.

Contour integration takes place not only with disjoint line segments, but also with illusory contours, i.e. perceived boundaries where no luminance contrast actually exist. Illusory contours can be triggered by two types of stimulus configuration: edge inducers and line-end inducers. The Kanizsa triangle in Figure 13.3a is a representative of edge-induced illusory contours, where the contour forms collinearly with the inducing edges of the pacmen. The Ehrenstein pattern in Figure 13.3b is an example of line-end-induced illusory contours, where the boundary of the circle is orthogonal to the line ends near the center. Following the initial discovery by Schumann (1900), illusory contours were studied in depth by Ehrenstein (1941) and Kanizsa (1955). They have become an important subject in visual perception research as a test case in figure-ground separation, object recognition, and perceptual grouping in general (Lesher and Mingolla 1995; Petry and Meyer 1987).

Early on, there were two main theories of illusory contour perception: the bottom-up brightness theory and the top-down cognitive factor theory. Brightness theory maintained that illusory contours arise from a low-level mechanism that gives brightness to areas enclosed by illusory boundaries. On the other hand, cognitive theorists argued that illusory contours are purely a high-level cognitive phenomenon. However, more recent evidence suggests that neither of these theories can account for the full range of illusory contour phenomena. Illusory contours were discovered that arise from image configurations without subjective brightness, providing a counterexample to the brightness theory (Hoffman 1998; Kanizsa 1976; Parks 1980; Prazdny 1983). On the other hand, cells in V1 and V2 were found to respond to illusory contours, demonstrating that such contours can be processed early on, unlike what the cognitive theories suggested (Lee and Nguyen 2001; Peterhans, von der

Heydt, and Baumgartner 1986; Redies, Crook, and Creutzfeldt 1986; Sheth, Sharma, Rao, and Sur 1996; von der Heydt and Peterhans 1989). They arise at least partly based on the same mechanisms as ordinary contours, including lateral interactions in V1 and V2. Association fields and statistical models can potentially be used to explain much of illusory contour processing as well.

Another intriguing characteristic of contour integration is that it is stronger in certain parts of the visual field than in others. The two main divisions are: (1) The lower visual hemifield is better in illusory contour discrimination tasks than the upper visual hemifield (Rubin, Nakayama, and Shapley 1996), and (2) contour integration is stronger in the fovea than in the periphery (Hess and Dakin 1997; Nugent, Keswani, Woods, and Peli 2003).

Rubin et al. (1996) compared the performance of humans in discriminating the angle made by illusory contours in the lower vs. upper hemifield (see Figure 13.3a for an example stimulus). The pacman-like disks were rotated by small amounts so that the perceived square in the middle would look either thick (like a barrel) or thin (like an hour glass). Rubin et al. presented the inputs in either the lower or upper visual hemifields and measured the minimum amount of rotation (i.e. threshold) needed for the subject to reliably tell whether the input was thick or thin. The results showed that the threshold is much higher in the upper hemifield than in the lower hemifield, i.e. lower hemifield is more accurate in this task than the upper hemifield. Similar results are expected for contour integration tasks, although such experiments have not been carried out so far.

Along the same lines, Hess and Dakin (1997) and Nugent et al. (2003) investigated differences in contour integration performance in the fovea vs. the periphery. When line segments had a consistent phase (as in Figures 13.1 and 13.2), they found that, in the fovea, contour integration is accurate even for a relatively large orientation jitter, but quickly fails after a critical point. However, in the periphery, the accuracy decreases linearly as the orientation jitter increases. Hess and Dakin (1997) observed that such a linear decrease is predicted by a linear filter model (e.g. convolution with oriented Gabor filters), which does not require sophisticated lateral interactions. These results suggests that contour integration is a stronger process in the fovea than in the periphery.

How and why do such differences in performance occur? A likely explanation is that the cortical structures in these areas are different. Such differences could arise if the structures develop based on input-driven self-organization, and the areas receive different kinds of inputs during development.

The inputs can differ in two ways: due to passive environmental biases and due to active attentional biases, and there is evidence for both causes. Environmental biases seem to drive the upper vs. lower hemifield distinction. Because of gravity, objects tend to end up near the ground plane, making the input in the lower hemifield more frequent and complex. In animals with high dexterity such as monkeys, reaching for objects and manipulating them takes place mostly in the lower hemifield as well (Gibson 1950; Nakayama and Shimojo 1992; Previc 1990). Attentional biases, on the other hand, may be responsible for the differences between fovea vs. periphery. For example, Reinagel and Zador (1999) presented natural images to humans and gathered statistics about the locations in the image to which the human attended, by tracking eye movements. They found out that human gaze most often falls upon areas with high contrast and low pixel correlation. Since the attended areas mostly project to the fovea, the statistical properties will differ in the fovea vs. the periphery. Such evidence suggests that the input statistics can differ in different areas of the visual cortex, which may in turn lead to different development and different performance in these areas.

In summary, psychophysical results suggest that specific lateral interactions are crucial for contour integration. As we saw in the previous chapter, this is exactly what input-driven self-organization captures in the PGLISSOM model. Thus, computational models like PGLISSOM can show how these three observations are related: Human contour integration performance arises from specific neurobiological structures, derived from input statistics through self-organization. So far, PGLISSOM is the only model of contour integration that includes the self-organization component; the other models are reviewed next.

13.1.2 Computational Models

Several neural network models of contour integration have been developed, showing that specific lateral interactions are sufficient to account for the phenomena outlined above (Finkel and Edelman 1989; Grossberg and Mingolla 1985; Hugues, Guilleux, and Rochel 2002; Li 1998, 1999; Peterhans et al. 1986; Ullman 1976; VanRullen, Delorme, and Thorpe 2001; Wersing, Steil, and Ritter 2001; Yen and Finkel 1997, 1998). The models are able to detect and enhance smooth contours of oriented Gabor patterns embedded in a background of randomly oriented Gabor patterns (as in Figure 13.1), contours in natural images, and various illusory contours.

All these models use fixed formulas in determining the lateral interactions, and these interactions are similar to the association fields. For example in Yen and Finkel's (1997; 1998) coupled oscillator model, the units were connected with long-range lateral excitatory connections. The magnitude and time course of the synaptic interactions depended upon the position and orientation of the connected units. Excitatory connections were confined within two regions. One fanned out cocircularly around the preferred orientation axis of the central unit. The other extended out transaxially to a smaller area flanking the preferred orientation. Inhibitory connections linked to the rest of the surrounding neurons that did not receive excitatory connections. The connection strengths had a Gaussian profile, with the peak at the central unit. Through synchronization of the coupled oscillators, the model was able to predict human contour integration performance, showing that specific lateral interactions can be responsible for contour integration.

In Li's (1998; 1999) approach, the excitatory and inhibitory interactions were defined by fixed rules derived from specific constraints, as follows: (1) The system should not activate spontaneously, (2) neurons at a region border should respond strongly, and (3) the same neural circuit should enhance contours. Coupled oscillators were used to describe the dynamics of the orientation-selective cells, and

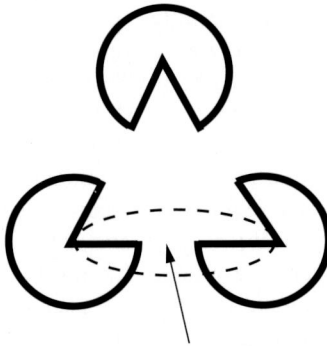

Fig. 13.4. Contour completion across edge inducers. After the retinal and thalamic preprocessing, the inputs received by the primary visual cortex are essentially edge-detected versions of the original images, as shown here for the Kanizsa triangle of Figure 13.3*a*. Each of the three sides (e.g. the one outlined with a dashed oval) has a gap (indicated by an arrow). The triangle boundary can be generated by filling in the gap through contour completion. Contour completion is therefore a possible mechanism underlying edge-induced illusory contours.

mean-field techniques and dynamic stability analysis were used to calculate the lateral connection strengths and the connectivity pattern according to these three constraints. The resulting lateral connection strengths were very similar to those of Yen and Finkel's, except there was no transaxial excitation; instead, areas flanking the center received specific inhibitory connections. The model also predicted contour integration performance well, and again showed that specific lateral connections can accurately predict human contour integration performance.

The first neural network model of illusory contours was based on edge inducers, and illusory contours were processed through contour completion (Ullman 1976). Figure 13.4 gives an example of this process. However, subsequent models were based on line-end inducers, and included edge inducers only as a special case. In these models, the corners where the edges meet (e.g. the throats of the pacmen in Figure 13.3*a*) and the tip of convex angles (e.g. the lips of the pacmen) constituted line ends, connected to obtain the illusory contour (Finkel and Edelman 1989; Peterhans et al. 1986). However, models strictly based on line-end inducers cannot account for psychophysical results where increasing the length of the inducing edge causes the illusory contour to become clearer (Shipley and Kellman 1992). Thus, neither model can account for both types of illusory contours.

For this reason, the model developed by Grossberg and Mingolla (1985) included both types of inducers. In the first stage, boundaries were formed orthogonally to the line-end inducers, and in the second stage collinearly to both line-end and edge inducers. In the second stage, the neurons had bipole (i.e. bowtie shaped) receptive fields similar to those found in V2 (von der Heydt and Peterhans 1989), combining the responses of two first-stage neurons. These neurons activated only when stimuli were present on both lobes of the receptive field, thereby binding the elements of the contour together. The neurons were connected into a systematic pattern by hand,

and through equilibrium values the resulting network successfully accounted for illusory contour perception where contour completion had an important role (Gove, Grossberg, and Mingolla 1993; Grossberg 1999; Grossberg et al. 1997; Grossberg and Williamson 2001; Ross, Grossberg, and Mingolla 2000).

Although the models described above have been successfully applied to explain experimental data, several important questions remain. Most importantly, how does the brain construct the circuitry required for contour integration and segmentation? As mentioned in Section 13.1.1, it is possible that statistical regularities in the visual environment drive a self-organizing process that results in such a circuitry. Demonstrating this process computationally is the main goal of this chapter.

Second, can the connections that implement contour integration produce illusory contours as a side effect? They are involved in binding line segments into a coherent representation across gaps, which could result in illusory contours in the extreme. This hypothesis is viable especially about edge-induced contours, which appear to occur early in the visual hierarchy. The hypothesis will be tested in this chapter using PGLISSOM as the computational platform; an extension to line-end-induced contours will be outlined in Section 17.2.12.

Third, can the adapting lateral interactions also account for the differences in human contour integration performance across the visual field? If the input distribution varies in the different parts of the visual field, the corresponding lateral connection patterns will be different, leading to different perceptual performance. This process can be modeled by the adaptive lateral connections in PGLISSOM, as will be done in the third main subsection of this chapter.

13.2 Contour Integration and Segmentation

This section will show how specific, self-organized lateral connections combined with synchronized activities can account for human contour integration performance. In addition, equally salient contours can be segmented via desynchronized activity, making PGLISSOM a unified model of binding and segmentation in the visual cortex.

13.2.1 Method

The contour integration experiments were run on the PGLISSOM network described in Section 11.5. The long-range lateral excitatory connections in the GMAP of this network are patchy, forming a substrate for binding. The inhibitory lateral connections are broad and have a long range, providing a baseline similar to global inhibition in other cortical models (Eckhorn et al. 1988; Terman and Wang 1995; von der Malsburg and Buhmann 1992; Wang 1995, 1996, 2000). Such inhibition allows input elements to be segmented by default, unless lateral excitation binds them together. To establish robust grouping in the self-organized network, the GMAP neurons were made more sensitive to excitation and inhibition, noise was added to their synapses,

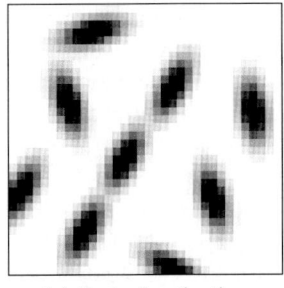

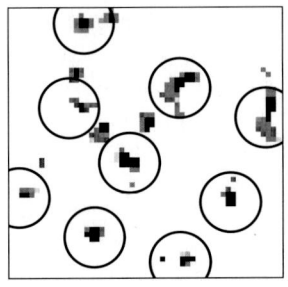

(a) Retinal activation   (b) GMAP response areas

Fig. 13.5. Measuring local response as multi-unit activity. (a) An example contour integration input, consisting of nine contour elements with different positions and orientations on the retina. Each element is an oriented Gaussian of length $\sigma_a = 1.9$ and width $\sigma_b = 1.2$. The firing rates of the retinal receptors are set according to these Gaussian values, plotted in gray scale from white to black (low to high). (b) The resulting activations of the GMAP neurons, measured as a leaky average firing rate with a 0.92 decay rate. The circles indicate areas where separate MUA values are measured. Each area is centered on a neuron whose receptive field is centered on one of the contour elements and whose orientation preference is the same as the element's orientation; due to local distortions in the retinotopic mapping, the circle's center in V1 is sometimes slightly displaced from the element's center on the retina. A few neurons outside the circles are also activated, driven by simultaneous input from two different contour elements. The circle radius is chosen such that the spurious activation is not included in the MUA measurements.

and the absolute refractory period was increased, as discussed in Chapter 12 (and described in detail in Appendix D.2).

In each experiment, two input examples with different arrangements of contour elements were tested in separate trials. Each contour element was an oriented Gaussian of length $\sigma_a = 1.9$ and width $\sigma_b = 1.2$, i.e. short enough to fit into a single receptive field (of radius 6). The patterns were generated to approximate contour integration inputs in human experiments as well as possible within the small model retina and V1 (Appendix D.2). The inputs were presented to the trained network for 500 iterations each, allowing the neurons to spike 50 times on average. Assuming that the firing rate is about 40 Hz (the γ-frequency band; Section 2.4.2), the simulations correspond approximately to 1.5 seconds in real time. The performance of the network was measured in GMAP, since it is the component that drives the grouping behavior in the model and therefore establishes its final output.

For each contour element, a distinct area in GMAP becomes activated (Figure 13.5). To measure the performance of the model, the degree of synchrony between these areas was calculated. The number of spikes in each area was counted at each time step. As briefly mentioned in Section 2.4.3, this quantity is called the multi-unit activity of the response, or MUA, and it measures the collective activity of a population of neurons (Eckhorn et al. 1988, 1990). MUAs over the 500 iterations, called MUA sequences, can be used to demonstrate qualitatively the process of contour integration for a given set of inputs.

In order to measure integration and segmentation performance quantitatively, the degree of synchronization between two areas can be measured using the linear correlation coefficient, or Pearson's r. That is, the product of the deviations from the mean of each MUA sequence is accumulated over time and normalized by the product of their variances:

$$r = \frac{\sum_i (X_i - \bar{X})(Y_i - \bar{Y})}{\sqrt{\sum_i (X_i - \bar{X})^2}\sqrt{\sum_i (Y_i - \bar{Y})^2}}, \tag{13.1}$$

where X_i and Y_i are the MUA values at time i for the two areas representing the two different objects in the scene, and $\bar{X}$ and $\bar{Y}$ are the mean MUA values of each sequence. The correlation coefficients between all possible pairs of MUAs are calculated: The higher the correlations, the more synchronized the areas are, thus representing a strong percept of a contour. The average correlation within the contour was compared with correlation between elements across contours and with background elements to establish how well the network integrates and segments the input elements.

The same method was used in all experiments in this chapter. Contour integration performance will be described first, followed by contour segmentation and contour completion. The effect of input distributions will be analyzed in the last section.

13.2.2 Contour Integration

As was discussed in Section 13.1.1, contour integration accuracy in humans is maximal when orientation jitter is $0°$, and the accuracy decreases as a function of increasing orientation jitter. To test whether PGLISSOM exhibits similar behavior, four contour integration experiments were carried out, with $0°$, $30°$, $50°$, and $70°$ of orientation jitter.

Figure 13.6 shows the MUA sequences for these four simulations. There are nine rows in each plot, corresponding to the nine contour elements in the input. The bottom three rows (rows 1 to 3) represent the three contour elements constituting the salient contour. How well these rows are synchronized compared with the rows representing the background elements determines how strongly the network perceives the contour. As expected, for $0°$ orientation jitter the three bottom MUA sequences are synchronized (Figure 13.6a), but as the orientation jitter increases (Figure 13.6b–d), the synchronized state is more difficult to maintain and the phases tend to shift back and forth.

This process can be quantified using the linear correlation coefficient r. The results are summarized in Figure 13.7, together with the human performance data from Geisler et al. (2001). The plot shows that at low orientation jitter, the model and humans both recognize the contours reliably, but as orientation jitter becomes larger, they both become less accurate in a similar manner. Correlation coefficients between MUA pairs corresponding to two background contour elements, or pairs between a background and a contour element in the salient contour, were usually less than 0.1 (i.e. not perceptually salient), except in rare cases where the jitter caused them to line up accidentally.

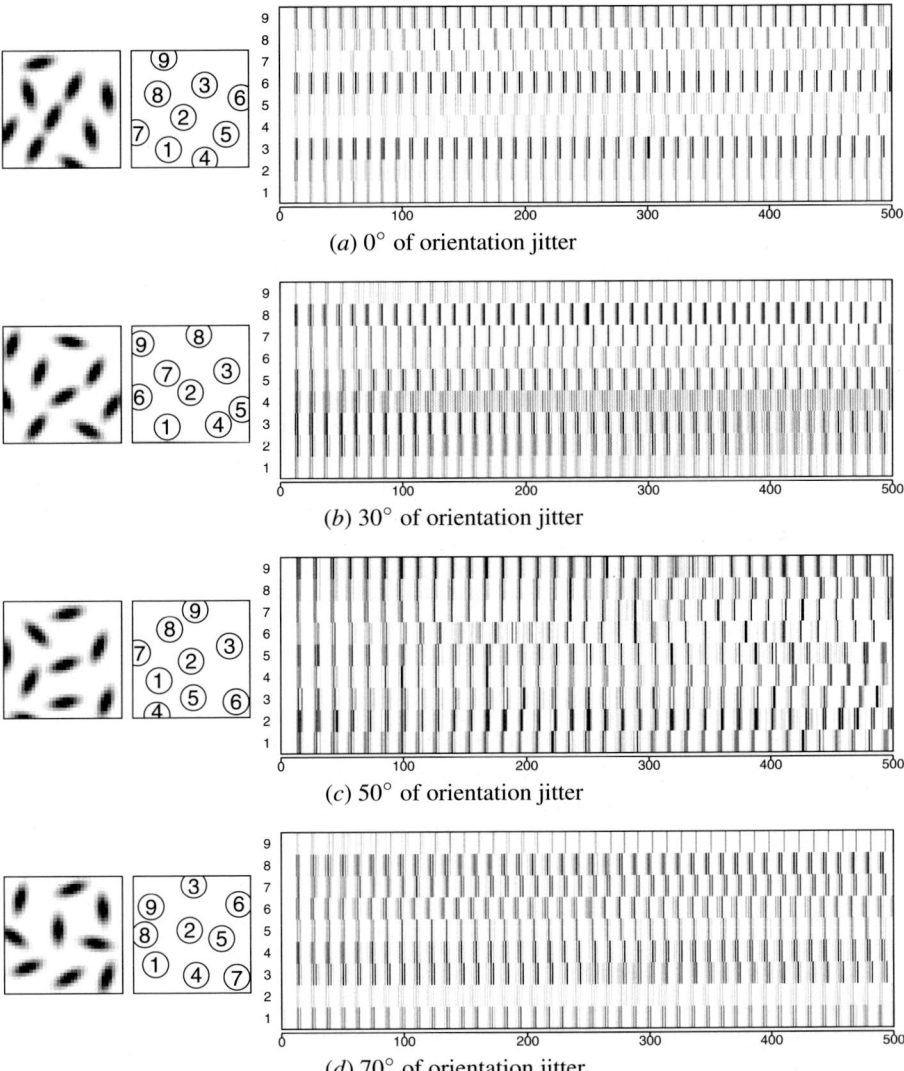

Fig. 13.6. Contour integration process with varying degrees of orientation jitter. In each subfigure, the input presented to the network is shown at left, the areas in GMAP where MUA was measured in the middle, and the resulting MUA plot at right. Each contour was composed of three contour elements, and embedded in a background of six randomly oriented elements. Each contour runs diagonally from lower left to top right with varying degrees of orientation jitter. The MUA of each area is plotted in gray scale from white (no neurons firing in the area at this time step) to black (all neurons firing). Time (i.e. simulation iteration) is on the x-axis and the y-axis consists of nine rows, each plotting the MUA of the area labeled with the row number. The three bottom rows (1 to 3) represent the MUAs of the salient contour, and the six top rows (4 to 9) the MUAs of the background elements. The contour is very strongly synchronized for $0°$ and $30°$ but relatively weakly synchronized for $50°$ and $70°$ of orientation jitter: The contours get harder to detect as the jitter increases. In all cases (a to d), the background MUAs are unsynchronized. A quantitative summary of these results is shown in Figure 13.7, and an animated demo can be seen at http://computationalmaps.org.

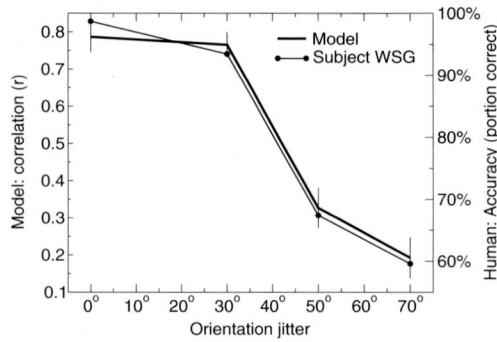

Fig. 13.7. Contour integration performance in humans and in PGLISSOM. The model's performance was measured as the average correlation coefficient between the MUA sequences in the salient contour, calculated over two trials, each with a different input example (left y axis). Human performance was measured as the percentage of correctly identified contours (right y axis; data by Geisler et al. 2001, root-mean-square (RMS) amplitude 12.5, fractal exponent 1.5, which is the closest match with the PGLISSOM input configuration). The x-axis is the orientation jitter in degrees, and the error bars indicate ± 1 SEM in the model (no error measures were published for the human data). In both humans and the model, contour integration is robust up to $30°$, but quickly breaks down as the orientation jitter increases (the difference between $30°$ and $50°$ is significant with $p < 10^{-4}$; the other differences are not significant with $p > 0.1$).

As described in Section 13.1.1, such contour integration performance is believed to depend on specific lateral connection patterns in the primary visual cortex. Next, the distributions of lateral connections in the model will be analyzed in order to demonstrate how they influence perceptual performance.

13.2.3 The Role of Lateral Connections

As was discussed in Section 11.5.3, lateral connections in PGLISSOM (as well as in the LISSOM orientation model) have two specific anatomical properties: (1) Strong connections exist between neurons with similar orientation preferences, and (2) the connections extend along the direction matching the source neuron's orientation preference. These properties allow the connections to encode specific local grouping functions, or association fields.

However, to understand the functional role of these connections in visual space (instead of cortical space), the relationships between the receptive fields of the connected neurons need to be examined. Which input features in a scene activate neurons that have strong lateral connections between them, and how strongly is a pair of input features bound together in the cortex through lateral connections? By comparing such connection statistics with human perceptual performance and natural scene statistics, it is possible to determine precisely what functional role the patchy lateral connections play in contour integration.

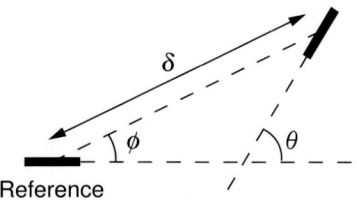

Fig. 13.8. Quantifying the spatial relationship between two receptive fields. For each pair of neurons connected with excitatory lateral connections, the afferent connection weights were examined to determine (1) the orientation preference of the neuron (shown as oriented bars), and (2) the location of the receptive field in retinal space (as the center of gravity of the afferent weight matrix). From these values, the direction ϕ, radial distance δ, and difference between orientation preferences θ between all pairs of neurons were calculated. Notice that these values define the spatial relationship between the two neurons in the retinal (or visual) space, not in the cortical space, and therefore allow comparing connectivity with edge-cooccurrence data. Such a comparison is presented in Figure 13.9. Adapted from Geisler et al. (2001).

Figure 13.8 illustrates the quantities that define the spatial relationship between a pair of receptive fields. These quantities were measured from all lateral excitatory connections in GMAP that remained after connection death. The results are summarized in Figure 13.9*b*. Two properties are evident in the plot: (1) The target receptive fields are most likely oriented along cocircular paths emanating from the center, and (2) the most likely target locations form a bowtie-shaped flank along the horizontal axis. These results show that neurons with receptive fields falling upon a common smooth contour are most likely to be connected with lateral excitatory connections. Such a pattern closely matches the association field proposed by Field et al. (1993; Figure 13.2), thus suggesting that perceptual grouping rules can be implemented as actual patterns of lateral connections in the brain.

In fact, such connection patterns predict the contour integration performance of the previous section very well. Since receptive fields aligned on an arc with smaller curvatures are more likely to be connected, inputs with smaller orientation jitter would be more strongly bound together than those with large orientation jitter. The model therefore offers an explanation for the observed performance in terms of specific neural structures.

Furthermore, these functional statistics in the model are similar to the local Bayesian edge-cooccurrence statistics in natural images (Geisler et al. 2001). Figure 13.9*a* summarizes the likelihood that a pair of edges under configuration (ϕ, θ, δ) fall upon a common physical contour, such as a tree trunk, boulder boundary, etc. Such natural contours are also found likely to follow cocircular paths. As demonstrated by Geisler et al. (2001) the edge-cooccurrence patterns accurately predict human contour integration performance, which also indirectly explains why PGLISSOM accurately predicts human contour integration performance: Both humans and PGLISSOM are biased toward integration of natural contours.

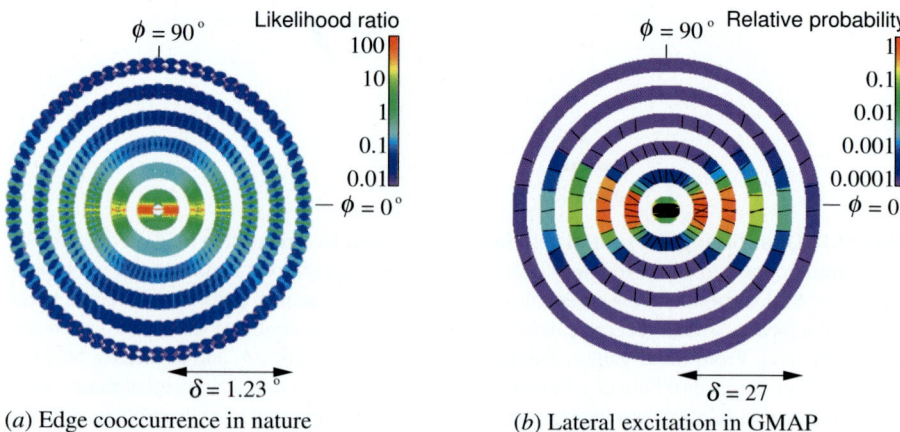

(a) Edge cooccurrence in nature   (b) Lateral excitation in GMAP

Fig. 13.9. Edge cooccurrence in nature and long-range lateral connections in PGLIS-SOM. The distributions of excitatory lateral connections in the model are compared with the edge-cooccurrence statistics in nature to see how well they match perceptual requirements. (a) The Bayesian edge-cooccurrence statistics in natural images (Geisler et al. 2001; reprinted with permission, copyright 2001 by Elsevier). Each location in polar coordinates (ϕ, δ) contains a small round disk, representing the likelihood ratios of all possible orientations θ at direction ϕ and distance δ by color coding; the θ with the highest ratio is shown in the foreground (θ, ϕ, and δ are defined as in Figure 13.8). Each likelihood ratio represents the conditional probability that a pair of edge elements in configuration (θ, ϕ, δ) belongs to the same physical contour vs. different physical contours in natural images. The conditional probabilities were determined through manual labeling of contours in real world images. The most likely elements are aligned along cocircular paths emanating from the center. (b) The distributions of θ, ϕ, and δ for the lateral excitatory connections in GMAP (Choe and Miikkulainen 2004; reprinted with permission, copyright 2004 by Springer). Each location (ϕ, δ) displays two values: (1) The color scale in the background shows the relative log-probability of finding a target receptive field at that location, and (2) the black oriented bars represent the most probable orientation θ of the target receptive field at that location (not plotted for the weakest connections). The figure shows that neurons with receptive fields aligned on a common smooth contour are most likely to be connected with lateral excitatory connections. This distribution corresponds closely to the edge cooccurrence patterns in nature, suggesting that the model is well suited for encoding grouping relations in natural images.

13.2.4 Contour Segmentation

Importantly, the synchronization process that establishes the contour percept can also separate different contours to different percepts. The same self-organized network with the same simulation parameters as in Section 13.2.1 was used for the contour segmentation experiment. Two contours and three background elements were presented as input, and the correlations between elements within and across the contours, between the contour and the background, and within the background were calculated. The MUA sequences of the nine areas are shown in gray-scale coding in Figure 13.10. The bottom three rows (1 to 3) correspond to the diagonal contour, the

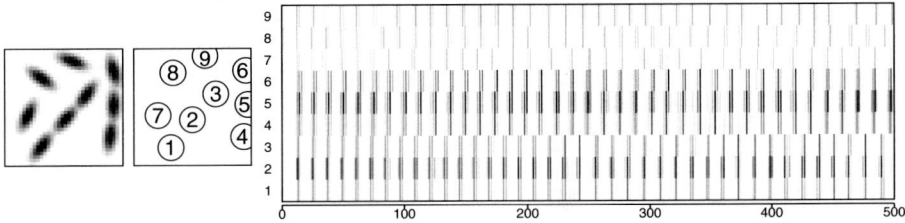

Fig. 13.10. Contour segmentation process. Input for the contour segmentation experiment consisted of two contours, diagonal and vertical, and three background elements. The same plotting conventions as in Figure 13.6 were used to illustrate the MUAs of the areas that responded to these inputs. The three bottom rows (1 to 3) correspond to the diagonal contour, the three middle rows (4 to 6) to the vertical contour, and the top three rows (7 to 9) to the background elements. The MUA sequences within each contour are synchronized. On the other hand, the MUA sequences of elements in different contours, of elements in the background, and of contour and background elements are desynchronized. In other words, the three areas representing the same contour fire together while the areas responding to the other contour and to the background are silent. Such an alternating activation of neuronal groups ensures that each coherent object is represented distinctly and not mixed with representations of other objects. An animated demo of this process can be seen at http://computationalmaps.org.

middle three rows (4 to 6) to the vertical contour, and the top three rows (7 to 9) to the background elements.

In the beginning, all areas are mostly synchronized, but as lateral interactions begin to take effect, the MUAs form two major groups firing in two alternating phases. The correlation coefficients of areas in the same contour are consistently high while those in different contours and in the background are low (Figure 13.11), signifying integration within each contour and segmentation across the contours. This result suggests that the same circuitry responsible for contour integration can also be responsible for segmentation between multiple contours.

PGLISSOM can segment up to about six contours this way. With more than six, representations for some objects will be synchronized instead of being desynchronized (a similar limitation was reported by Horn and Opher 1998 and Horn and Usher 1992). Over a longer period of observation, it may be possible to separate even more objects. Even if disjoint representations occasionally become synchronized, they do not stay in this state permanently. Synchrony is eventually broken, and another pair of representations that was previously desynchronized becomes synchronized. Therefore, even with a limited capacity for segmentation, a large number of objects can be segmented if the degree of synchrony is measured over a long period of time.

There is an interesting balance between segmentation and integration in the model. Segmentation cannot be made too strong, otherwise contour integration suffers. It turns out that with integration performance roughly comparable to that of humans, the system sometimes integrates when there is no contour. This behavior can explain how an interesting class of visual illusions, those based on edge-induced contour completion, may arise, as will be described next.

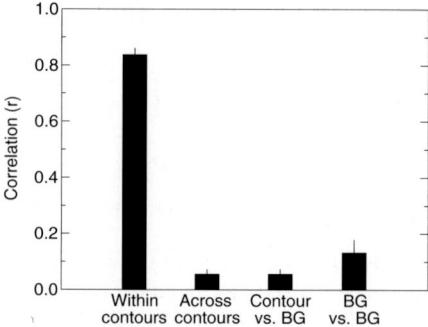

Fig. 13.11. Contour segmentation performance. The average correlation coefficients between two MUA sequences within the same contour, across different contours, between contour and background, and within the background are plotted, calculated over two trials. The error bars indicate ± 1 SEM. The MUA sequences within the same contour are highly correlated, whole those belonging to different contours or the background are not (the difference is significant with $p < 10^{-10}$). This result demonstrates quantitatively that neurons within each contour form a synchronized group, whereas neurons responding to different contours are desynchronized.

13.3 Contour Completion and Illusory Contours

As was discussed in Section 13.1.1, the same lateral interactions that implement contour integration could also underlie contour completion, i.e. filling in missing elements in a contour. Experiments with PGLISSOM strongly support this hypothesis. The model demonstrates that contour completion and the resulting illusory contours are a necessary side effect of the contour integration circuitry. In this section, the contour completion performance of PGLISSOM will be analyzed in detail, focusing on conditions under which completion occurs. This process gives rise to edge-induced illusory contours, and to a difference in detecting closed vs. open contours. These results suggest a possible mechanism for edge-induced illusory contours in V1. Similar mechanisms in V2 could be responsible for line-end-induced contours, as will be discussed in Section 17.2.12.

13.3.1 Method

The PGLISSOM network that was used to demonstrate contour integration and segmentation in Section 13.2 was tested in contour completion as well. The inputs included long contours with one element missing, and contours representing the edge-detected Kanizsa triangle (Figure 13.4). Because these inputs have more elements than those in Section 13.2, the elements tend to be closer than before. As a result, the radius of the MUA areas was reduced to five to avoid overlap.

As before, the network was activated for 500 iterations, and the MUA sequences for areas of GMAP representing the input contour elements and the gaps were measured. Each experiment consisted of two trials with the input positioned in a different location and orientation but with a similar structure.

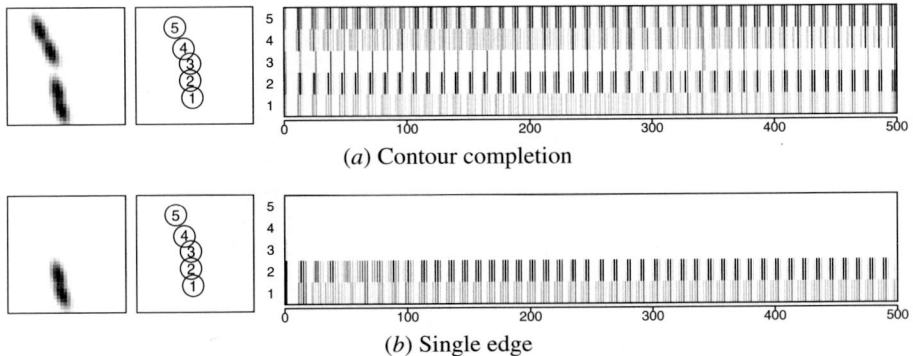

(a) Contour completion

(b) Single edge

Fig. 13.12. Contour completion process. (a) The four contour elements in the input with a gap in the middle correspond to one side in the edge-detected Kanizsa triangle (the dashed oval in Figure 13.4). In the MUA plot, the four contour elements are shown in the bottom and the top (rows 1–2 and 4–5) and the gap in the middle (row 3). Even though there were no inputs in the middle, the cortical area representing the gap is activated, and the activations are synchronized with the other four MUA sequences. This behavior indicates that contour completion occurred and the gap is perceived as an illusory edge. (b) In the second experiment, the input consisted of two contour elements from only one side of the gap. The MUA sequence for the gap is silent (row 3), indicating that contour completion did not occur. Thus, both sides of the gap need to be stimulated for the gap to be perceived as an edge.

13.3.2 Contour Completion

To test basic contour completion, PGLISSOM was presented with a straight contour with a gap in the middle as shown in Figure 13.12a. Such a contour represents one side of the edge-detected Kanizsa triangle in Figure 13.4. To make sure the contour elements on one side of the gap do not alone activate the gap, an input consisting of only half the contour was also presented to the network (Figure 13.12b). The prediction was that the network would fill in the gap in the first stimulus, but not in the second.

For the contour completion input (Figure 13.12a), there indeed is a significant MUA sequence for the gap (row 3), and it is synchronized with the rest of the sequences (rows 1–2 and 4–5): The gap is perceived as part of the contour. In contrast, with the single-edge input (Figure 13.12b), the MUA sequence representing the gap (row 3) is silent, while the rest of the MUA sequences (rows 1 and 2) are active and synchronized. Thus, both sides of the gap need to be stimulated for the gap to be perceived as an edge. The same self-organized circuitry in PGLISSOM that is responsible for contour integration can therefore account for contour completion as well. The contributions of the different kinds of connections to this process are analyzed next.

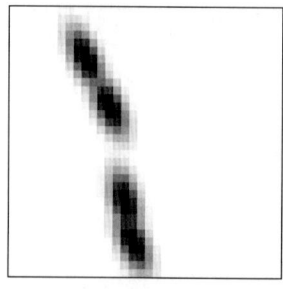

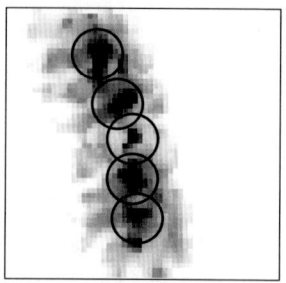

(*a*) Retinal activation   (*b*) Afferent GMAP input

Fig. 13.13. Afferent contribution in contour completion. The afferent contribution of the input in (*a*) to the GMAP activation is plotted in gray scale from white to black (low to high) in (*b*); the circles delineate the MUA areas as shown in Figure 13.12. The four areas corresponding to the four contour elements all receive strong afferent input. The center area, corresponding to the gap, receives weak afferent input, due to slight overlap with neighboring regions in the retina. However, as seen in Figures 13.14 and 13.15, it is not enough to activate its representation without a contribution from the lateral connections.

13.3.3 Afferent and Lateral Contributions

The filling in of gaps in the PGLISSOM model is to be expected, given that specific excitatory lateral connections project from the neighboring areas into the gap. However, it is also possible that afferent input is causing the completion. In animals and in the PGLISSOM model, receptive fields of neighboring areas in the cortex overlap. If the cortical area representing the gap receives enough afferent input from both sides around the gap, it can be activated the same way as the rest of the contour representations.

To check the amount of afferent input received by the gap, the net afferent contribution in GMAP was measured in the contour completion experiment (Figure 13.12*a*). A two-dimensional intensity plot (Figure 13.13) shows that the central area indeed receives some afferent input. Could such spurious afferent input be enough to activate the area representing the gap?

More generally, the question is whether the afferent contribution alone, or the lateral excitatory contribution alone, can cause the filling-in effect, or whether the phenomenon requires both kinds of contributions. To answer this question, two experiments were performed using the same method as in Section 13.3.2, with the single-contour input of Figure 13.12*a*. In the first experiment, the gap area received no afferent connections, and in the second there were no excitatory lateral connections.

The MUA sequences for the two experiments are shown in Figure 13.14. In both cases, the sequence representing the gap in the contour shows no activity at all, suggesting that contour completion did not occur in either case. For comparison, the average correlation coefficients in all three cases of lateral connectivity are shown in Figure 13.15. The correlation is high only when both afferent and lateral excitatory connections are included.

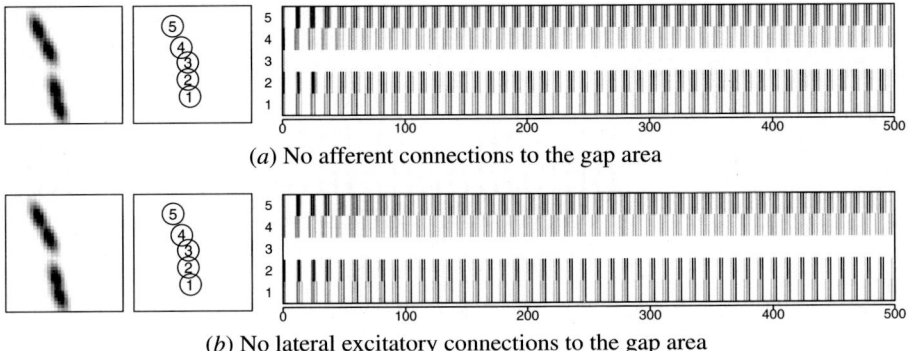

(a) No afferent connections to the gap area

(b) No lateral excitatory connections to the gap area

Fig. 13.14. Contour completion process with different kinds of connections. Networks without afferent connections to the gap area (a) and without lateral excitatory connections to this area (b) were tested in the contour completion task. In both cases, the MUA sequences for the four input contour elements (rows 1–2 and 4–5) are synchronized, whereas the sequences for the gap (row 3) are silent, suggesting that filling in did not occur. Contour completion therefore requires both kinds of connections.

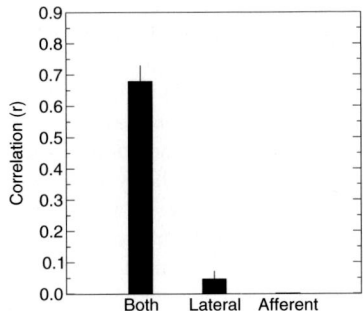

Fig. 13.15. Contour completion performance with different kinds of connections. The average correlation coefficients for the four MUA sequences representing the four input contour elements vs. the MUA sequence representing the gap are shown, calculated over two trials. Both afferent and excitatory connections are included in "Both". In "Lateral", the afferent connections are removed from the center, i.e. binding is due to excitatory lateral connections only. In "Afferent", the excitatory lateral connections are removed from the center, and binding is based on afferent connections only. The plot shows that both afferent and excitatory contributions are necessary for contour completion ($p < 10^{-7}$).

These results demonstrate that contour completion in PGLISSOM requires a contribution from both afferent and lateral excitatory connections. Such a condition can only occur when the input contour elements are aligned along a smooth path. The central receptive field is then partially activated by the input in the neighboring areas, and the cocircular projection of lateral connections amplify this activation above threshold. The next question is: Can this mechanism of contour completion be responsible for illusory contours as well?

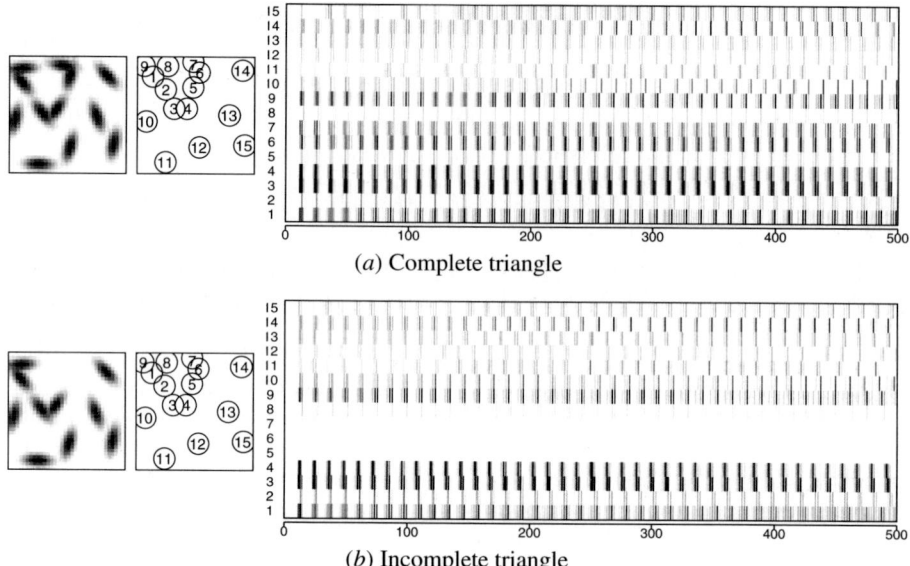

(a) Complete triangle

(b) Incomplete triangle

Fig. 13.16. Contour completion process in the illusory triangle. Each element in the triangle is identified by a number 1 through to 9 counterclockwise from the top left vertex, with 2, 5, and 8 denoting the gaps. (a) A complete triangle with gaps in the middle of each side approximates the central triangular part of the edge-detected Kanizsa triangle (Figure 13.4). The MUAs corresponding to gaps are all active and synchronized with the other inputs. Overall, the synchronization of all nine inputs means that the system is perceiving a single coherent object (as also demonstrated quantitatively in Figure 13.18). (b) When one vertex (elements 6 and 7) is removed, areas representing gaps 5 and 8 become almost silent: The perception of a triangle disappears, as it does in the incomplete Kanizsa triangle (Figure 13.17). An animated demo of these examples can be seen at http://computationalmaps.org.

13.3.4 Completion of Illusory Contours

To test the model in perceiving illusory contours, a simplified illusory triangle, embedded in a background of six randomly oriented edges, was presented to the network (Figure 13.16a). This triangle has gaps in each of the three sides, approximating the edge-detected Kanizsa triangle (Figure 13.4) as well as possible with the small model retina and V1. The network was also tested with one vertex of the triangle removed (Figure 13.16b) to see whether both sides of the gaps are necessary for the illusion to appear. Figure 13.17 shows the actual images corresponding to these inputs. Otherwise the same simulation method was used as in the single-gap experiment.

As expected, all gaps of the complete triangle are activated and synchronized with the neighboring contour elements (Figure 13.16a). In contrast, in the incomplete triangle (Figure 13.16b) only gap 2 (in the left edge) is filled; gaps 5 and 8 (on the sides) are not. These results are consistent with those of the single contour (Figure 13.12a). However, what makes this experiment particularly interesting is that the three sides of the triangle are also synchronized. The sides constitute three

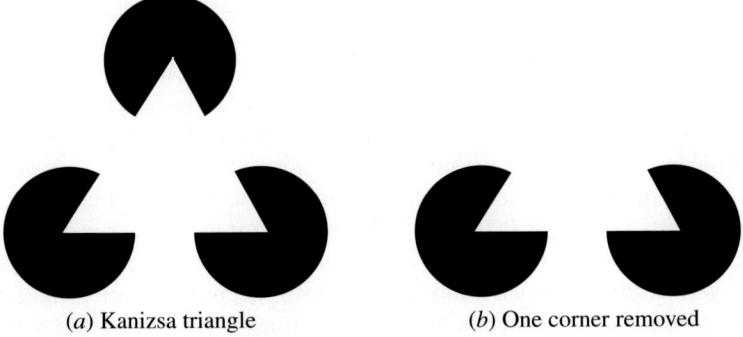

(*a*) Kanizsa triangle   (*b*) One corner removed

Fig. 13.17. Salience of complete vs. incomplete illusory triangles. The illusory object is vividly perceived in the complete Kanizsa triangle (*a*). However, when one corner is removed, this perception disappears (*b*).

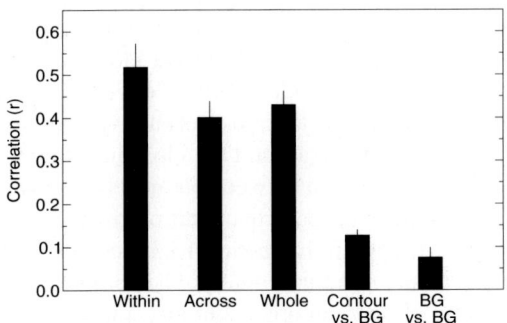

Fig. 13.18. Contour completion performance in the illusory triangle. Each side of the triangle is represented by a group of three MUA sequences, and constitutes a separate contour. The average correlation coefficients were calculated over two trials for MUA sequences representing two elements within the same side, across different sides, anywhere in the whole triangle, one in a contour and the other in the background, and within the background. The elements in each side are strongly synchronized, but so are elements across different sides and in the whole triangle (the differences between "Within", "Across", and "Whole" are not significant with $p > 0.1$). Furthermore, the elements in the triangle are significantly more synchronized than contour and background elements, and elements in the background ($p < 10^{-6}$). This result shows quantitatively that the three sides are perceived together as a single object.

independent contours with sharp angles between them, and based on the analysis in Section 13.2.4 would be expected to be desynchronized. However, as shown in Figure 13.18, all contour elements (within the same side, across different sides, and among the whole triangle) are highly correlated, suggesting that the network perceives only one object. How is such cross-contour synchronization possible?

At the vertices of the triangle, two contour elements with different orientation preference overlap. Since the afferent receptive fields in PGLISSOM are topologically organized, the two cortical areas responding to the two edges at the vertex are close by on the map. As shown in Figure 11.5, the excitatory connections not only connect to neurons with similar orientation preferences, but at a close range also to those with somewhat different orientation preferences. Thus, proximity of the inputs, as well as the good continuation of contours, determines the degree of synchronization. At the vertices, the two abutting inputs cause the corresponding cortical areas to synchronize, which in turn causes the three sides of the triangle to synchronize. As a result, the network represents the whole triangle as a coherent object.

These results show that PGLISSOM indeed performs contour completion, and also forms representations for whole objects. Inputs can be grouped through proximity as well as through good continuation. Such mechanisms arise automatically from the properties of afferent and lateral connections in the model, and may form a general principle for grouping (ordinary and illusory) in the visual system.

13.3.5 Salience of Closed Versus Open Contours

With a thorough understanding of how contour completion occurs in the model, let us now return to the psychophysical observation that closed contours are easier to detect than open contours (Kovacs and Julesz 1993; Pettet et al. 1998; Tversky et al. 2004). As discussed in Section 13.1.1, the most recent evidence suggests that there is no special reverberatory mechanism around the closed contour: The advantage is due to proximity and good continuation between elements (Tversky et al. 2004).

While it is difficult to replicate the control conditions in a small retina of PGLISSOM, the illusory triangle experiment in Section 13.3.2 can be used to test the fundamental principle of this theory computationally. The complete and incomplete triangles in Figure 13.16 form closed and open contours. Indeed, the perception of an illusory triangle breaks when one component is removed from the Kanizsa triangle (Figure 13.17).

To measure how salient the two objects are, the average correlation coefficients between the nine elements of the complete illusory triangle (elements 1 through to 9) and seven elements of the incomplete triangle (1–5 and 8–9) were calculated. The results are shown in Figure 13.19. The activities in the network for the closed contour are significantly more synchronized than those of the open contour, indicating that the closed contour is more salient.

The explanation for this effect in PGLISSOM is straightforward. Every part of the closed contour receives excitatory lateral contribution from *both* neighboring areas, and strong synchronization results along the contour. In contrast, at the two ends of an open contour the neurons only receive lateral excitation from one neighboring area, and the synchrony does not reach the same level of salience.

In this simple experiment, PGLISSOM has provided an independent computational confirmation of the current psychophysical theory on closed vs. open contours. The difference arises from local interactions based on proximity and good continuation, without a separate reverberatory mechanism. In the future, larger PGLISSOM models can be used to replicate the actual conditions in human experiments, leading to more detailed predictions and insights into this phenomenon.

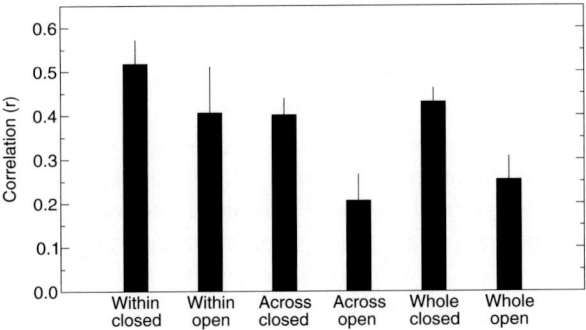

Fig. 13.19. Contour completion performance in closed vs. open contours. The average correlation coefficients between the MUA sequences in the complete and incomplete triangles (labeled "closed" and "open") of Figure 13.16 are shown, calculated over two trials and different groups of contour elements. Although the differences in correlation within each side are not significant ($p > 0.2$), the elements in the closed contour are significantly more correlated across the sides and within the whole object than in the open contour ($p < 0.004$), consistent with psychophysical results (Section 13.1.1). The correlations between contour and background elements and within the background were significantly weaker in both cases ($p < 0.03$), indicating that both contours were perceived as single objects. The PGLISSOM model therefore provides independent computational support to the theory that closed contours are salient because of proximity and good continuation rather than a special reverberatory mechanism.

13.4 Influence of Input Distribution on Anatomy and Performance

In Section 13.2 we saw how lateral connectivity plays a central role in contour integration in PGLISSOM. Because these connections are learned through input-driven self-organization, different input distributions during development result in different anatomy and performance. The self-organizing process can therefore potentially account for the observed differences in human visual performance across different parts of the visual field, such as upper vs. lower hemifield and fovea vs. periphery (Section 13.1.1). This hypothesis can be tested computationally by training PGLISSOM with inputs of varying frequency and complexity.

13.4.1 Method

The visual inputs that the cortex receives during training may vary in several ways. For example, inputs in the fovea and lower hemifield may be more frequent, shorter, sharper, curved, textured, or have higher contrast. As was discussed in Section 13.1.1, distributions of these features across the visual field have not yet been fully characterized, and it would be somewhat premature to test the model with a selection of such variations. However, at a more abstract level, two distinctly different dimensions of variation can be identified: (1) the amount of training each area receives, and (2) the complexity of the training inputs it sees. In this section, these dimensions, represented by input frequency and curvature, are varied systematically, leading to verifiable predictions about the resulting anatomy and performance.

In the frequency experiment, two PGLISSOM networks trained with different input presentation frequencies were compared. The first one was similar to the contour integration network described in Section 13.2, i.e. trained with single randomly located and oriented elongated Gaussians. The second one was otherwise the same, except every other input presentation was skipped during training. The simulation parameters (excitatory radius, learning rate, thresholds, and connection death) were adapted according to the same schedule as before, modeling maturation based on time and trophic factors (Sections 4.4.3 and 16.1.6). In other words, the second network received input half as frequently as the first network during its maturation.

In the curvature experiment, each training input consisted of three short elongated Gaussian bars that together formed a smooth contour, located and oriented randomly on the retina. The angles between the bars were changed to adjust the curvature of the input. For the first network, the training inputs had uniformly randomly distributed curvature in the range $[0°..25°]$, and for the second network in the range $[0°..10°]$.

Over the course of self-organization, the input Gaussians were slightly broader and became elongated slightly slower than those used in Section 11.5. As a result, the learning was slower but the resulting network performed more robustly on the wider variety of inputs (Appendix D.2). Other than the difference in input distributions, all PGLISSOM networks were trained as in Section 11.5. After training, the afferent and lateral connectivity patterns and contour integration performance of each network were measured and the differences analyzed as in Sections 11.5 and 13.2, as will be described next.

13.4.2 Differences in Connection Patterns

All four simulations resulted in a similar map of orientation preferences, matching the results of previous self-organization experiments (Section 11.5). Two interesting observations can be made based on orientation selectivity, as shown in the distributions in Figure 13.20. First, neurons in the more frequently stimulated network were more selective than those in the less frequently stimulated one (Figure 13.20a), suggesting that its initial responses are sparser but stronger for specific inputs. Second, the networks trained with different curvature are equally selective (Figure 13.20b), suggesting that any performance differences are likely to be due to lateral connections.

To uncover any differences between the resulting lateral connection patterns, the (ϕ, θ, δ) statistics were calculated on the four networks as in Section 13.2.3. In the frequency experiment, two major differences emerged: (1) The high probability areas extend out longer in the high-frequency network (Figure 13.21a) than in the low-frequency network (Figure 13.21b), i.e. the network with more frequent exposure to oriented edges can group together more distant inputs. (2) The most probable θ for a given (ϕ, δ) location tends to be cocircular in the high-frequency network, whereas

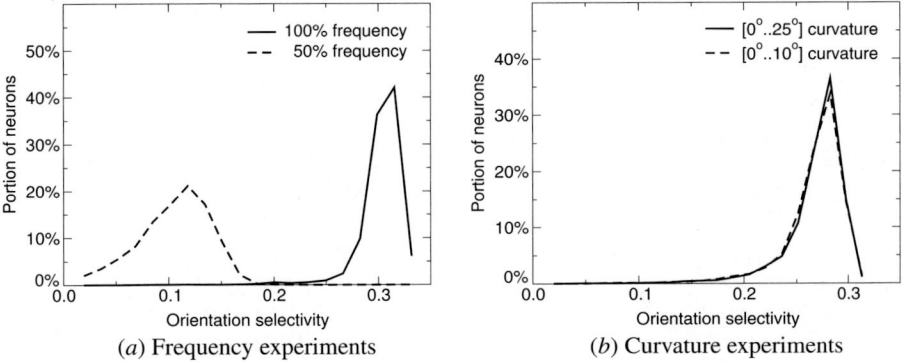

Fig. 13.20. Orientation selectivity in SMAP with different input distributions. For each of the four networks, the selectivity of neurons in the SMAP was measured (as described in Appendix G.1.3) and plotted as a histogram; GMAP selectivities were similar and are not shown. (*a*) The histogram for the 100% presentation frequency peaks at around 0.32, and that of the 50% frequency around 0.12, suggesting that the responses of the high-frequency network are sparser but stronger for specific inputs. (*b*) The histogram for the high curvature range $[0°..25°]$ and the low curvature range $[0°..10°]$ are almost identical. Given that the orientation preferences were also almost identical, any differences in their performance are likely to be due to the lateral connections.

in the low-frequency network it is more collinear (i.e. the black edges in the high probability areas are more parallel).

As we saw in Section 11.5.3, collinearity is the most prominent feature in the input, and is therefore learned more reliably. With extensive training, it is extended to large distances, as happened with the high-frequency network. Cocircularity develops more slowly than collinearity because the network responds less strongly in the cocircular arrangement. The high-frequency map had enough input presentations and was able to learn the secondary (cocircularity) property of the input as well.

In the curvature experiment, high probability areas (red and orange) along the horizontal axis are broader in the map trained with a broader range of curvatures (Figure 13.22*a*) compared with the one trained with a narrower range (Figure 13.22*b*). As expected, the input-driven self-organizing process has encoded the input distribution differences into the lateral connections. As a result, the map with exposure to higher curvature should be better at integrating cocircular contours.

In summary, differences in the input distribution, whether presentation frequency or complexity of inputs, result in specific, predictable differences in the afferent and lateral connection patterns. Such a difference in structure predicts that contour integration performance will also differ in these networks, as will be tested in the next section.

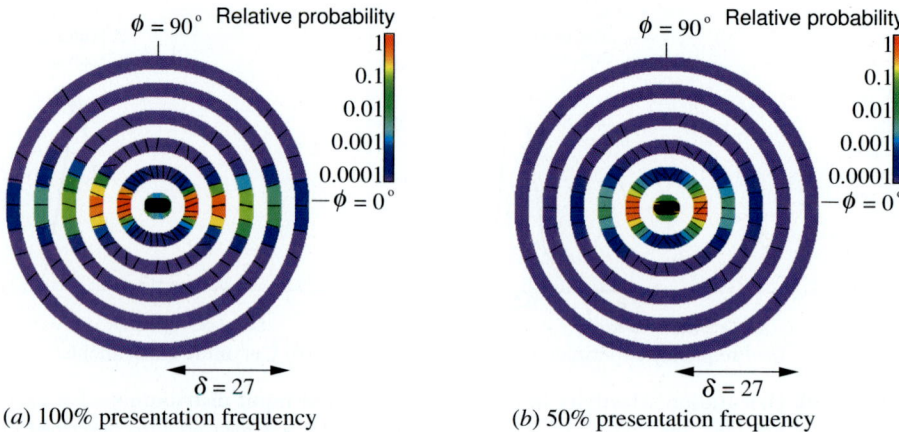

(a) 100% presentation frequency   (b) 50% presentation frequency

Fig. 13.21. Lateral excitatory connections in GMAP with different input frequencies. The connection probability distributions are displayed the same way as in Figure 13.9. As before, only GMAP is shown because it is responsible for contour integration in the model. The lateral connection profiles differ in two subtle ways: (1) The high probability areas (red and yellow) extend longer in the high-frequency map (a) than in the low-frequency map (b) (three vs. two rings of high probability). (2) The most probable θ (black oriented bars) are cocircular in (a), but mostly collinear in (b) (as seen e.g. in the second ring from the outside). These results predict that contours should be easier to detect in the high-frequency network.

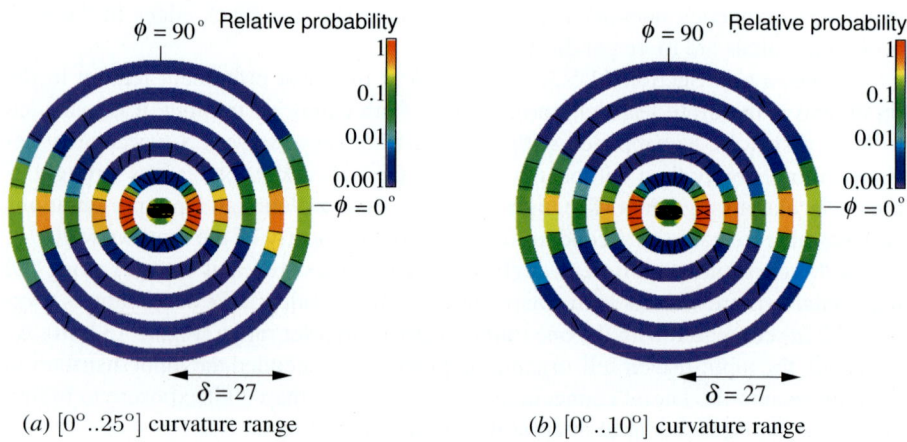

(a) $[0°..25°]$ curvature range   (b) $[0°..10°]$ curvature range

Fig. 13.22. Lateral excitatory connections in GMAP with different curvature ranges. The network trained with a broader range of curvatures (a) has broader areas of high probability connections (red and yellow) than the network trained with a narrower range (b). As a result, contours with more curvature and higher orientation jitter should be easier to detect in network (a) than in (b).

13.4.3 Differences in Contour Integration

For each of the four networks trained in Section 13.4.1, two contour integration experiments were performed, with orientation jitters $0°$ and $40°$. The $40°$ test case was chosen because contour integration performance in both humans and the model degrades most rapidly at around $40°$ (Figure 13.7), making the differences due the input distributions most clearly visible. The method described in Section 13.2.1 was used for all experiments. Figures 13.23 and 13.24 display the MUA sequences in each case. The MUAs are significantly more synchronized for the high-frequency network than for the low-frequency one when the orientation jitter is the same (compare Figure 13.23*a* vs. *b* and *c* vs. *d*). For the networks trained with different curvature range, the degree of synchrony was similar for $0°$ orientation jitter (Figure 13.24*a* vs. *b*), but the network trained with a broader range was significantly more synchronized in the $40°$ case (Figure 13.24*c* vs. *d*). The correlation coefficients between the MUA sequences confirm these observations (Figure 13.25). Frequency makes a difference both in perceiving collinear and cocircular contours, whereas curvature matters only with the cocircular ones.

Such performance differences are predicted by the afferent and lateral connection patterns described in the previous section. Each of the four networks has lateral connections that can group collinear contours, so any difference in performance with $0°$ orientation jitter must be due to the afferent connections. The neurons in the high-frequency network have more selective afferent connections, and therefore activate and synchronize more strongly for inputs that match their preferences. On the other hand, the afferent weights do not differ significantly in the curvature experiment, and neither does the performance of the two networks in the $0°$ case. In contrast, with $40°$ of jitter, the shape of the lateral connections makes a big difference. Each neighboring pair of contour elements is aligned on a cocircular path, and integration requires cocircular connections. Because the lateral connections in the high-frequency network and the high-curvature network are more cocircular, they can detect such contours with high orientation jitter much better than the low-frequency and low-curvature networks.

In summary, differences in the input distribution, even as simple as presentation frequency or curvature, can change how the maps are organized, which in turn can affect performance in contour integration. Such differences in structure and function are due to the input-driven nature of self-organization. This principle provides a possible developmental explanation for the differences in contour integration performance across different areas of the visual field found in psychophysical experiments.

13.5 Discussion

The results in this chapter suggest that contour integration, segmentation, and completion can be due to synchronization mediated by self-organized afferent and lateral connections, and may form a general principle for grouping (ordinary and illusory) in the visual system.

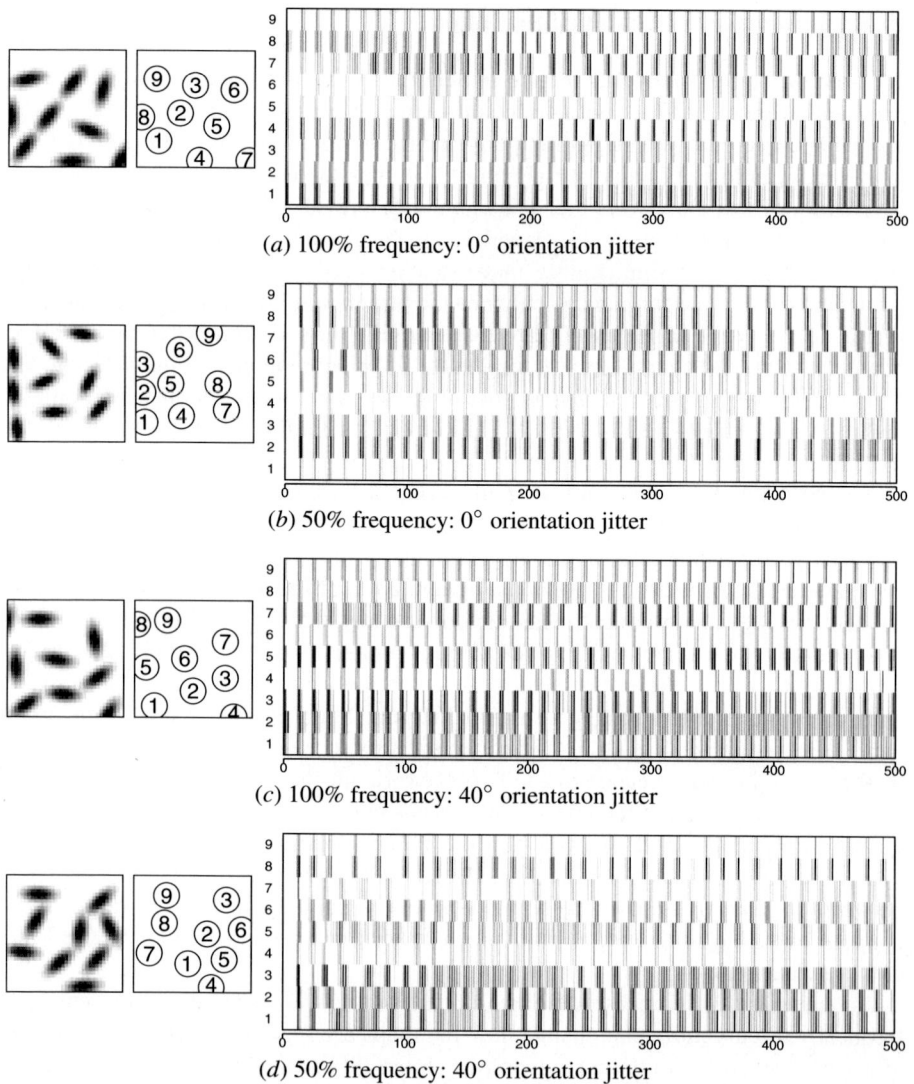

(a) 100% frequency: 0° orientation jitter

(b) 50% frequency: 0° orientation jitter

(c) 100% frequency: 40° orientation jitter

(d) 50% frequency: 40° orientation jitter

Fig. 13.23. Contour integration process with different input frequencies. In each MUA plot, the three bottom rows correspond to the MUA sequences for the three contour elements in the input and the rest correspond to background elements. For the same degree of orientation jitter (0° or 40°), the more frequently trained network is more strongly synchronized (*a* vs. *b*; *c* vs. *d*).

It may be possible to verify the synchronization hypothesis experimentally in the near future (Section 16.3.1). Meanwhile, the hypothesis is consistent with existing data on how temporal coding affects performance. Lee and Blake (2001) augmented the usual contour integration input with a temporal cues such as periodic flashing

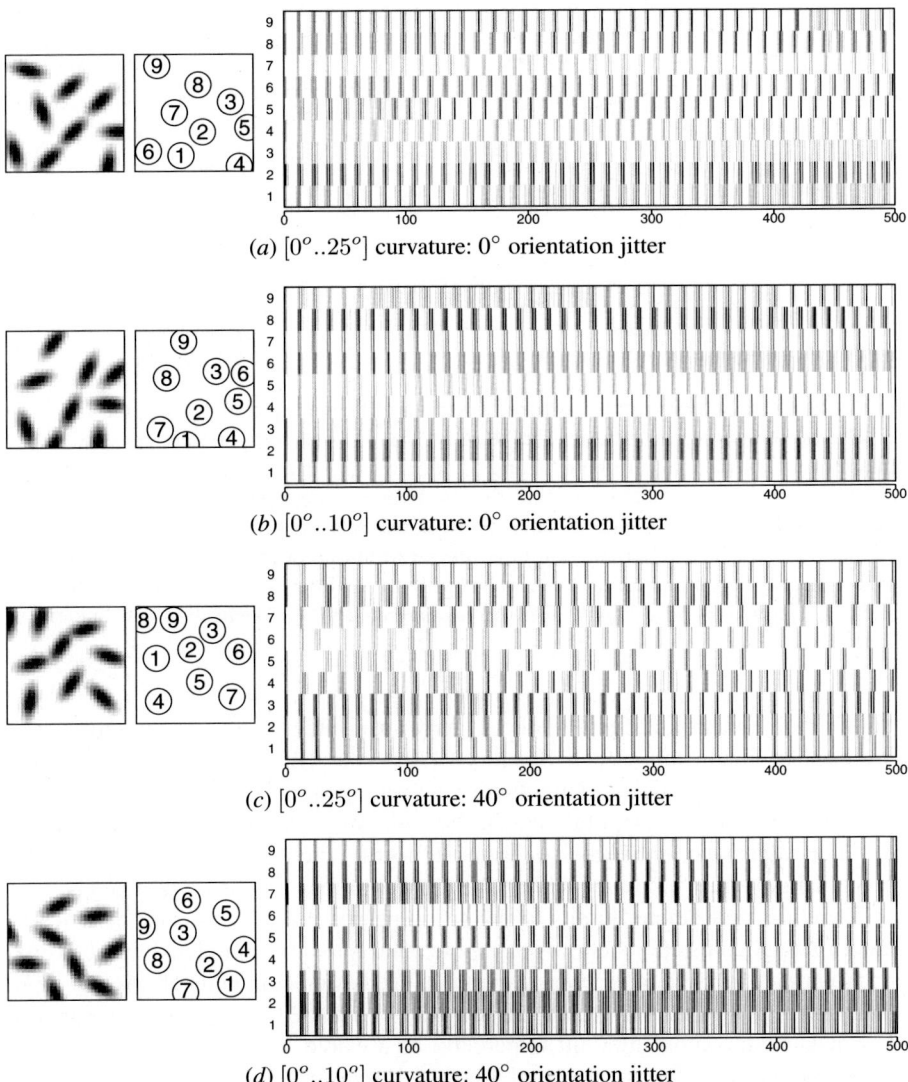

(a) [0°..25°] curvature: 0° orientation jitter

(b) [0°..10°] curvature: 0° orientation jitter

(c) [0°..25°] curvature: 40° orientation jitter

(d) [0°..10°] curvature: 40° orientation jitter

Fig. 13.24. Contour integration process with different curvature ranges. Both curvature networks show the same degree of synchrony for the 0° orientation jitter (*a* vs. *b*), but in the 40° case, the network trained with a broad range of curvatures becomes significantly more synchronized than the one trained with a narrow range (*c* vs. *d*). These observations and those from Figure 13.23 are confirmed quantitatively in Figure 13.25.

of contour elements. Strong spatial and temporal cues (such as smooth contours and synchronized flashing) resulted in accurate contour integration, as expected. However, when a weak spatial cue was combined with a weak temporal cue, the subjects performed better than expected. The two cues were not simply added together, but

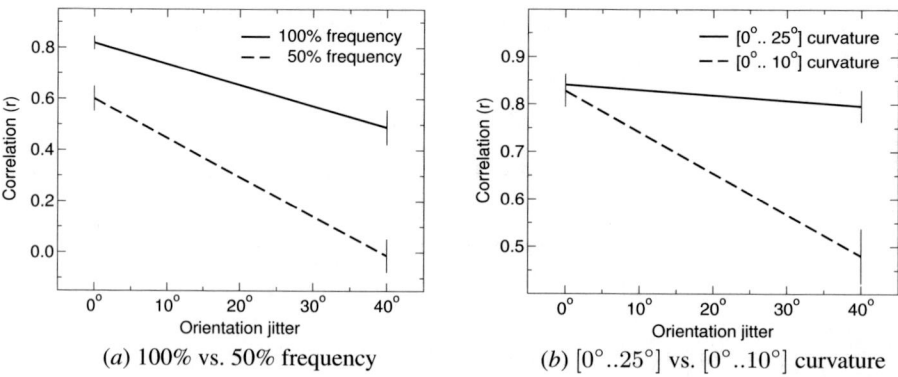

Fig. 13.25. Contour integration performance with different input distributions. The average correlation coefficients between the MUA sequences in each experiment are shown, calculated over two trials. (*a*) For both $0°$ and $40°$ orientation jitter, the high-frequency network was significantly more synchronized than the low-frequency network ($p < 0.003$). The difference is more pronounced in the $40°$ case, as predicted by the lateral connection distributions in Figure 13.21. (*b*) At $0°$ orientation jitter, the performance of broad and narrow curvature range networks is comparable ($p > 0.7$), but with $40°$ of jitter the broad curvature network performs significantly better ($p < 0.0009$), as predicted by the connection distributions in Figure 13.22.

interacted nonlinearly. The temporal cues may not even have to be continuously synchronized to obtain this effect: Beaudot (2002) showed that it is enough to have the contour become visible slightly before the background. A possible interpretation of these results is that temporal cues in the stimulus, such as synchrony and initial activation advantage, enhance synchrony in neural activity, which then allows the perceptual system to bind the individual elements of the contour more strongly.

There are well-defined limits to the grouping process as well. The specific excitatory lateral connections allow only those contours to be completed that fall on a cocircular path. On the other hand, the model needs small amounts of afferent input to fill in the gap, so that arbitrarily large gaps will not be filled in. High orientation jitter makes the contour difficult to perceive as a whole, as it does in humans. The raw segmentation ability of the model is also limited, and curiously similar to the limited number of short-term memory slots, usually quoted as 7 ± 2 (Miller 1956). It is difficult to say whether the memory and segmentation limits are related; however, similar limits for simultaneous representation have been observed on other temporal coding models (Horn and Opher 1998), and they seem to be a robust property of such systems.

The behavior of the model is primarily driven by the self-organized lateral connections. Their pattern matches edge distributions in natural images well, which results in good performance in contour integration. However, it is interesting to note that these connections were trained not with natural images, but with elongated Gaussian inputs. The gradual tapering of such patterns on both sides trains the connections to become cocircular. This result suggests that very simple visual inputs, such as those generated internally before birth (Section 2.3) could already prepare the animal for essential tasks in the actual visual environment. As demonstrated in Part III, training after birth with natural images will then further refine the circuitry for more accurate performance.

The PGLISSOM model can be extended in several ways to model a wider range of phenomena and to make it biologically more accurate. For example, the process of forming edge-induced illusory contours demonstrated in Section 13.3 could be extended to line-end-induced contours as well. As will be described in more detail in Section 17.2.12, the model could be extended with a V2 network, containing neurons with end-stopped receptive fields, and connected in a manner similar to GMAP in the current model. Synchronized activation in a group of such neurons in V2 would then be interpreted as a line-end-induced contour. In this manner, the same binding mechanism based on lateral interactions would account for both types of contours, providing an alternative to the bipole model discussed in Section 13.1.2.

The range of illusory contours could be expanded further by including feedback from higher levels of visual processing. For example, contours that establish an illusory object (such as the Kanizsa triangle of Figure 13.3a) cannot be explained entirely by low-level mechanisms (Hoffman 1998); they appear to be partly driven by object representations and cognitive factors as well. In fact, connections between lower and higher visual areas are reciprocal (Felleman and Van Essen 1991; Nelson 1995) and well suited for carrying out such computations. Extending PGLISSOM to include such high-level feedback is an interesting future research direction, as will be discussed in Sections 17.2.13 and 17.2.14.

The PGLISSOM model can also be extended in size. The retina and cortical maps are currently limited by the available computational resources, which makes it difficult to replicate the exact contour integration experiments done with humans, and especially those involving illusory contours and the perception of closed vs. open contours. The retina and the cortex would need to be an order of magnitude larger to approximate typical inputs consisting of about 200 line segments. Such simulations are currently not feasible, but the scaling techniques described in Chapter 15 and Section 17.2.9, coupled with the expected growth of computing power, should make them possible in a few years. While the current PGLISSOM model with small-scale inputs is a valid demonstration of the underlying processing principles, such larger-scale models would allow making detailed predictions that match actual psychophysical measurements. Such a large-scale model could also be trained with natural images, or with a combination of prenatal and postnatal inputs, further enhancing the realism of the model. It would then be possible to study new phenomena, such as interaction of multiple stimulus dimensions, as described in Section 17.2.8.

The model can also be extended with a more accurate representation of the input to the different visual areas. As was discussed in Section 13.1.1, contour integration is stronger in the fovea than in the periphery, and in the lower vs. upper hemifield. Section 13.4 demonstrated how such functional differences can result from different distributions of training inputs in these areas. To verify that such differences indeed exist, a method similar to that of Reinagel and Zador (1999) could be used: Input statistics from different parts of the visual field could be collected using eye-tracking devices while human subjects are freely browsing the environment. Such statistics will account for both environmental and attentional biases, thus accurately representing the input distributions in the different parts of the visual field. This information makes it possible to predict perceptual performance of the different cortical areas.

However, any such comparison would have to take into account the structural differences in the optics and the retina. For example, peripheral inputs tend to be more blurred, and there are far fewer photoreceptors in the periphery than in the fovea. As a result, small details that can easily be seen in the fovea may not be visible in the periphery. However, when inputs are larger, contour integration in the periphery could be similar to that in the fovea (as suggested by W. S. Geisler, personal communication, January 9th, 2004). Such structural factors should be taken into account when gathering input statistics and reasoning about the possible causes of functional divisions. The predictions of such an extended model can then be tested in developmental neurobiological experiments, by manipulating the visual input and measuring the resulting connectivity patterns and contour integration performance, as will be discussed in Section 16.4.8.

13.6 Conclusion

In this chapter, the self-organized afferent and lateral connections of the PGLISSOM model were shown to perform contour integration similarly to human subjects. The model shows how visual input statistics, lateral connection patterns, and perceptual performance are related. It suggests a concrete, testable explanation for how illusory contours arise as a side effect of normal performance and why performance in different parts of the visual field differs. Understanding these processes in the model allows ascribing function to low-level neural circuitry, and provides a foundation for building models of more complex visual tasks.

EVALUATION AND FUTURE DIRECTIONS

14

Computations in Visual Maps

So far we have seen how a wide variety of psychophysical and neurobiological phenomena can be explained by computations in a laterally connected, self-organized LISSOM network. In Part V, the LISSOM approach will be evaluated as a foundation for further research. In this chapter, the representations in LISSOM maps are analyzed experimentally, and shown to result in a sparse coding that reduces redundancies while preserving the most important features of the input. These representations serve as an efficient foundation for pattern recognition, as will be shown in an example application to handwritten digit recognition. In the next chapter, a method for scaling LISSOM to maps of different sizes, including the size of the entire visual cortex, is presented. The biological assumptions of the model are evaluated and predictions for future biological and psychological experiments are drawn from it in Chapter 16. Extensions of the model, future computational experiments, and new general research directions are proposed in Chapter 17. That chapter also briefly describes *Topographica*, the publicly available software package for simulating cortical maps, intended to support future computational research on computational maps in the cortex.

14.1 Visual Coding in the Cortex

How is visual information represented in the cortex? A number of researchers have proposed that the main goal of visual coding, besides representing the important features of the input, is to reduce redundancy (Atick 1992; Atick and Redlich 1990; Barlow 1985; Földiák 1990, 1991b; Rao and Ballard 1997; see Simoncelli and Olshausen 2001 for a review). The idea is related to methods used in compression of bitmap images, and the possible benefits are the same. Redundancy reduction could permit storing and transmitting the retinal image using fewer cells and connections, and as a result the visual cortex could process more visual information with limited resources.

The standard redundancy reduction methods aim at representing all likely inputs in a small number of coding units (e.g. neurons). Each image is coded into the activity of a small number of units, and the dimensionality of the representation is reduced. However, the cortex takes the opposite approach: A few million optic nerve fibers branch out to more than a hundred million cortical cells (Wandell 1995), and for each small region of the retina there is a large number of neurons with different feature preferences. Therefore, the visual input is expanded out and coded in the activity of a larger number of cells than in the retina. Coding in the cortex must therefore be based on an approach different from straightforward redundancy reduction.

Field (1994) suggested that the cortex might instead aim at representing the input with a minimum number of active units. For any given image, only a small subset of cortical units respond, with most neurons remaining inactive. For different images, different populations of cells are activated. Such sparse coding makes pattern recognition easier: Because each cell responds relatively rarely, it is easier to identify features. If a cell is active, it is possible to predict what inputs caused it to be active. Sparse coding also greatly reduces energy requirements, because spiking is metabolically expensive (Lennie 2003).

However, sparse coding by itself also conflicts with neurobiological evidence. Without redundancy reduction, at least the same number of cells would be active in V1 as in the retinal image, and all of this redundant activity would be carried through to the higher processing levels. The higher levels would then have to be at least as large as V1. However, the higher processing areas in the brain are invariably smaller than the primary visual cortex, and become smaller as one proceeds up the cortical hierarchy.

Taken together, however, sparse coding and redundancy reduction do constitute a strong, consistent hypothesis for the nature of the visual code. More specifically, the receptive field properties of V1 units produce a sparse representation of the input, because any given visual pattern matches only a small percentage of the neurons' RFs. Redundancy in this sparse response is then reduced by the lateral interactions within V1. As a result, an efficient, sparse coding of the input is formed, suitable for further processing by higher levels of the visual system.

This hypothesis rests crucially on the lateral interactions in V1. As proposed by Barlow (1972, 1989, 1990), lateral connections in V1 could suppress redundant activation by *decorrelating* the V1 responses. Such a process indeed takes place if neurons that respond to similar inputs are connected with inhibitory lateral connections. In such a case, the response of one neuron can be predicted based on the response of the other. Therefore, the activity of the second neuron is redundant, and a more efficient representation can be formed by eliminating the redundant response. Lateral inhibitory connections have exactly this effect: Whenever these neurons are active together, the inhibition tends to reduce their activation. Such decorrelation filters out the redundancies and concentrates the activity in independent feature-selective units.

The hypothesis is difficult to verify experimentally because it requires measuring activations of large numbers of neurons individually over very short time scales; such spatial and temporal resolution is not available with current imaging or recording techniques. However, computational models such as LISSOM are well suited for testing it. This section demonstrates that (1) LISSOM produces a sparse, decorrelated visual code, and (2) the specific self-organized lateral connections are crucial for this process. In Section 14.3, such coding will be shown to form an effective foundation for pattern recognition applications as well, using handwritten digit recognition as an example.

14.2 Visual Coding in LISSOM

In LISSOM, the self-organized connections store information about long-term activity correlations through Hebbian learning: the stronger the correlation between two cells, the stronger the connection between them. Because the long-range connections are inhibitory, they reduce the overall activation levels, while short-range lateral excitation locally amplifies the responses of active neurons. As will be shown in this section, this process makes the resulting responses sparse without losing information, i.e. by suppressing redundant activity.

Sparse responses can also be obtained through fixed, isotropic lateral interactions, like those used by nearly all previous computational models of the visual cortex. The connection strength between two neurons in such networks depends only on the distance between the neurons, not on their response properties. Although such interactions also reduce activity, the quality of the visual code is compromised, as will be demonstrated below.

14.2.1 Method

The experiments in this section were based on the reduced LISSOM version of the perceptual grouping network in the previous chapter. Because the analysis focuses on the overall activity patterns instead of grouping, firing-rate units were used instead of spiking neurons, and only the SMAP component of V1 was included in the simulations. (As was described in Section 11.2, SMAP determines the activity patterns in the model, whereas GMAP performs a grouping function among them.) To make input reconstruction practical, the retina was reduced to 36×36 receptors and the cortex to 48×48 units. Like the perceptual grouping network, the model was trained with long, oriented Gaussians (20,000 presentations of single Gaussians with axis lengths $\sigma_a = 30$ and $\sigma_b = 1.5$; Appendix F.1), and it developed a well-organized orientation map with long-range, patchy lateral connections (Figure 14.1).

Isotropically connected networks were constructed out of the self-organized networks by replacing the lateral inhibitory weights with isotropic weights. That way, all parameters and other components of the architecture were the same for both networks, making fair comparisons possible. A variety of isotropic profiles for the lateral connections were tested, including uniform disks, radial Gaussian distributions, and radial Cauchy distributions. The best performance was found using a sum of two Gaussians (SoG), chosen as a close match to the self-organized weight profiles (Figure 14.1b; Appendix F.1).

Sparseness of the cortical response was measured as the population kurtosis (i.e. the fourth statistical moment, or peakedness, of the neuronal activity distribution; Field 1994; Willmore and Tolhurst 2001). A small number of strongly responding

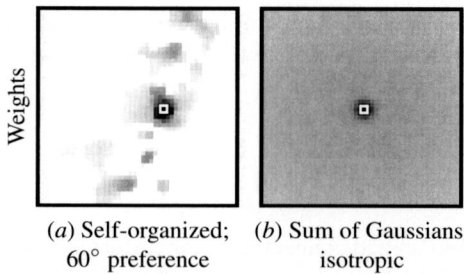

Weights

(a) Self-organized;   (b) Sum of Gaussians;
   60° preference   isotropic

Fig. 14.1. Self-organized vs. isotropic lateral connections. In (a), self-organized inhibitory lateral connection weights for a sample neuron in the LISSOM orientation map are plotted in gray scale from white to black (low to high); the small white square marks the neuron itself. In (b), the connections of a sample neuron in the network with isotropic long-range connections are shown. This network was constructed by adding two isotropic Gaussians: The smaller Gaussian was chosen as wide as the central peak in the self-organized weights, and the larger to extend as far as the longest self-organized lateral connections. Therefore, all neurons that are connected in the self-organized network are also connected in the sum-of-Gaussians network.

neurons, i.e. a sparse coding, results in high kurtosis, and a broad nonspecific pattern in low kurtosis. For each network, the average amount of kurtosis was measured for a set of 10,000 random input patterns, each with two long contours composed of two or three short, oriented Gaussians (Figure 14.2a).

Information content, on the other hand, can be measured in principle by reconstructing the input pattern from the cortical response. The idea is that accurate reconstruction is possible only if information about the input pattern is retained in the coding. A lossy coding, on the other hand, would result in incomplete reconstruction. If the networks were linear, it would be possible to perform the reconstruction simply by inverting the network, i.e. by backprojecting a set of V1 activity patterns through the afferent weights to produce a pattern on the retina. However, the neurons' activation functions are nonlinear, and their response depends on the nonlinear effect of the lateral connections. It is therefore not practical to reconstruct the input simply by mathematically inverting the network function.

However, an approximate inverse can be obtained by training a nonlinear neural network to compute it. One effective approach is to train a feedforward backpropagation neural network to map each V1 activity pattern to the retinal activity pattern that led to that V1 response. Such networks in general are effective in pattern recognition tasks, and also plausible as a model of how humans learn higher cognitive tasks (Bechtel and Abrahamsen 2002; Elman, Bates, Johnson, Karmiloff-Smith, Parisi, and Plunkett 1996; McClelland and Rogers 2003; Rumelhart et al. 1986; Sejnowski and Rosenberg 1987). One such reconstruction network was trained for the initial V1 response, another for the V1 response settled through self-organized lateral interactions, and a third for the V1 response settled through isotropic SoG lateral interactions. Each reconstruction network was trained and tested in a 10-fold cross-validation experiment with subsets of the same 10,000 retinal and V1 activity patterns that were used to measure kurtosis. The network and learning parameters were optimized to obtain the best performing network for each case (Appendix F.1). Any differences in reconstruction ability can then be attributed to the quality of the V1 representations.

In this manner, both how sparse the representations are and how well they represent information can be measured. In the next two subsections, representations formed with self-organized lateral connections, isotropic lateral connections, and without any lateral connections will be compared.

14.2.2 Sparse, Redundancy-Reduced Representations

The main effect of the self-organized lateral interactions is to make the cortical representation of the input sparser (compare Figure 14.2*b* and *d*). The recurrent excitation and inhibition focuses the activity to the neurons best tuned to the features of the input stimulus, producing a sharper response. The average kurtosis of the V1 activity patterns before settling was 35.4, which was more than doubled, to 85.7, by settling through the self-organized connections. The average total activity in the response reflected this change, reducing from 16.7 to 9.32 through the settling.

Importantly, the sparse coding is formed without losing information. As Figure 14.2*c,(e)* demonstrates, the retinal patterns can be reconstructed from the settled response just as well as from the initial response. In order to measure the reconstruction ability numerically, the percentage of output patterns that were identifiable, i.e. closest to the correct pattern in the test set, were counted. In both cases, 100% of test patterns (in a 10-fold cross-validation experiment) resulted in identifiable reconstructions.

Thus, the experiments with LISSOM provide computational evidence for the sparse coding with redundancy-reduction hypothesis. By decorrelating the V1 activity, the self-organized lateral connections form a sparse code without losing information.

14.2.3 The Role of Self-Organized Lateral Connections

Are self-organized lateral connections necessary to achieve sparse redundancy-reduced coding? It turns out that while the SoG network can indeed form a sparse code, it does so by reducing information instead of only redundancy.

In three control experiments, SoG networks were adjusted to perform sparse coding. First, the overall strength of the lateral inhibitory weights was set so that the Gaussian peak in the SoG was as high as the central peak of the self-organized weights (i.e. γ_I in Equation 4.7 was increased from 4 to 40 while γ_E remained at 0.9). The goal was to ensure that the SoG network included the lateral connections from the self-organized network, and differed from it by including additional connections as well. As a result, all activity in V1 was eliminated during settling. This result suggests that for the SoG network to form any visual coding at all, the individual long-range isotropic connections must be weaker than the individual self-organized connections. Consequently, any computation that the lateral connections perform,

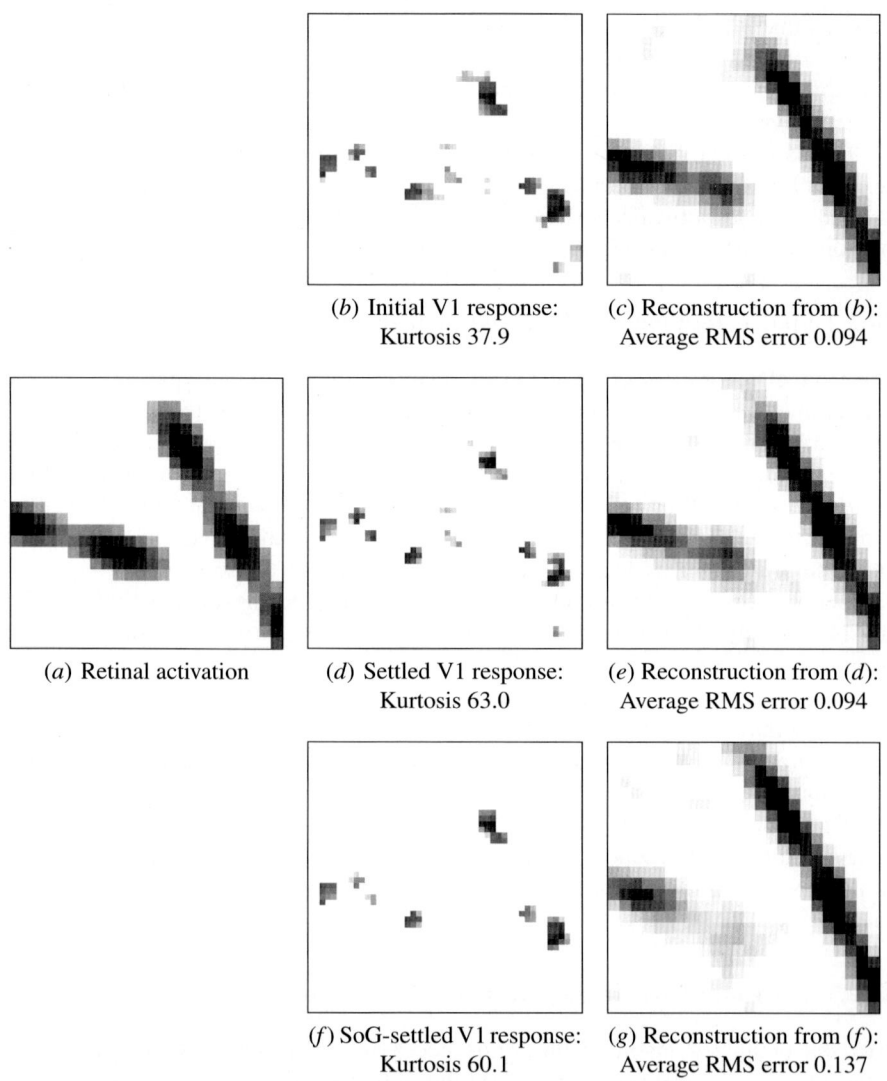

(*b*) Initial V1 response: Kurtosis 37.9

(*c*) Reconstruction from (*b*): Average RMS error 0.094

(*a*) Retinal activation

(*d*) Settled V1 response: Kurtosis 63.0

(*e*) Reconstruction from (*d*): Average RMS error 0.094

(*f*) SoG-settled V1 response: Kurtosis 60.1

(*g*) Reconstruction from (*f*): Average RMS error 0.137

Fig. 14.2. Sparse, redundancy-reduced coding with self-organized lateral connections. In (*a*), an example test input consisting of two multi-segment contours is shown. The V1 initially responds to this pattern with multiple large patches of activation (*b*), but lateral interactions focus the response into the most active neurons (*d*). This process results in a sparse code, as shown by the increased kurtosis values beneath each plot. The settling reduces redundancy but does not lose information; the input pattern can be reconstructed from both the initial response (*c*) and the settled response (*e*) equally well. When the lateral interactions are replaced with isotropic patterns, such as a sum of two Gaussians (*f*), a sparse code with a similar kurtosis results. However, crucial information about the input is lost in this process. All active neurons inhibit each other, and occasionally a crucial component of the representation is turned off. For example, the activity patch at the center represents the rightmost element of the left contour. It disappears in the settling process of the SoG network, and consequently the reconstruction image is missing this element as well (*g*). Self-organized patchy lateral connections are therefore crucial in forming a sparse redundancy-reduced coding of the visual input.

such as decorrelation or grouping, will perforce be weaker in SoG networks than in networks with self-organized lateral connections.

In the second experiment, the strength of the lateral inhibitory connections was reduced (to 11.4) until the average kurtosis of the responses reached 84.4, matching that of the self-organized network. However, the responses were now highly saturated, with average total activation of 27.1 compared with the 9.32 of the self-organized network. As a result, only 41% of the reconstructed input patterns were identifiable.

In the third experiment, therefore, the lateral inhibition and excitation were both adjusted simultaneously (γ_I to 14 and γ_E to 0.46) so that both the average kurtosis and the average total activation, at 82.3 and 9.16, were comparable to those of the self-organized network. Figure 14.2f shows a sample cortical response of this SoG network. Overall, it is very similar to that of the self-organized network: The experiment shows that it is possible for isotropic connections to achieve a sparse code similar to that of self-organized connections.

An important difference arises when the reconstruction is attempted based on the SoG representations. Whereas there was no loss of reconstruction ability in the self-organized case, the SoG network performs slightly but consistently worse: The reconstructed patterns are recognizable 99.0% of the time on average (the difference is significant with $p < 10^{-4}$). The reason is apparent in Figure 14.2f,g; with certain inputs, the settled SoG pattern is missing representations of parts of the input pattern, and as a result, those parts are also missing from the reconstructed pattern.

The loss of information in the SoG network primarily results from interactions between unrelated input components. As seen in Figure 14.2, even though the two contours have very different orientations, and thus are likely to be independent inputs instead of two parts of the same contour, their representations inhibit each other strongly in the SoG network. As a result, part of the V1 response disappears, allowing only one of the input contours to be reconstructed. In the self-organized case, even though the individual lateral inhibitory connections are stronger, they come from neurons that are often active together, reducing redundant activation only. The representations of unrelated contours do not inhibit each other, and both are retained in the settled response and in the reconstruction.

Similar but even worse results were observed for the other isotropic long-range connection patterns tested, including large Gaussians, Cauchy distributions, and uniform connections. If the isotropic connections were strong enough to provide a sparse code similar to that of the self-organized connections, they reduced the quality of the visual code. These results suggest that patchy, specific, self-organized connections are crucial for a sparse, redundancy-reduced visual code.

In conclusion, because the LISSOM model is computational, it allows testing hypotheses about visual coding in exact, quantitative terms. In doing so, self-organization is found to store long-range activity correlations between feature-selective cells in the lateral connections. During visual processing, this information is used to eliminate redundant information, and enhance the selectivity of cortical cells. As a result, the model establishes a sparse, redundancy-reduced coding of the visual input, which allows representing visual information efficiently with limited resources.

14.3 Visual Coding for High-Level Tasks

The sparse, redundancy-reduced coding is efficient, given that there is a limited number of neurons, and activation is expensive. Does it also provide an advantage in information processing? The example high-level application in this section suggests that it indeed does. Recognition of handwritten digits is easier when the visual input is represented on a LISSOM map, as opposed to the SOM version of the self-organizing map discussed in Section 3.4. The domain is first described below, followed by the recognition system architecture, and the results.

14.3.1 The Handwritten Digit Recognition Task

Handwritten digit recognition, a subtask of optical character recognition (OCR), is an important problem with many practical applications. Traditional approaches to this task include algorithmic and statistical methods, such as global transformation and series expansion, geometrical and topological feature extraction, and deriving features from the statistical distribution of points (Govindan and Shivaprasad 1990). More recently, neural networks have been successfully applied to this task as well, including general feedforward networks and dedicated and hybrid methods (Fukushima and Miyake 1982; Keeler and Rumelhart 1992; LeCun et al. 1995; Lee 1996; Lee and Lee 2000a,b; Martin, Rashid, Chapman, and Pittman 1993; Yaeger, Webb, and Lyon 1998. Digit recognition systems achieve 97.6–99.8% accuracy, rivaling estimated human performance at around 99.8% (LeCun et al. 1995).

In general character recognition, context information from the word and sentence may be available, and when recognition is done on-line, also information such as velocity, continuity of line segments, and the angle of motion. However, in its most basic and useful form, recognition is done off-line on isolated, normalized digits. Such raw bitmap images can be quite confusing, since many digits share similar features. For example, the digits 7 and 9, 4 and 9, 1 and 7, and 3 and 8 have large overlapping segments, and the distinct features are proportionally smaller than the overlapping ones. Although humans are good at paying attention to the distinct features in classifying digits, it is difficult to do so automatically.

For automatic recognition to be effective, it is necessary to form an internal representation that emphasizes the salient features of the input. Such representations must be highly separable, i.e. different for different digits, and easy to generalize, i.e. similar for the variations of the same digit. These requirements are difficult to achieve at the same time. If the representations are separated too far, generalization will often suffer. Good generalization, on the other hand, usually increases overlap between categories, degrading separation.

The hypothesis tested in the experiments that follow is that the sparse, redundancy-reduced representations in LISSOM are separable and generalizable, and thus constitute an effective foundation for pattern recognition. To test this hypothesis, LISSOM

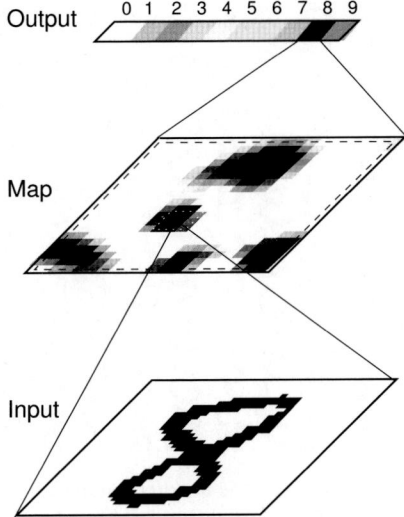

Fig. 14.3. Architecture of the handwritten digit recognition system. In the input, a normalized bitmap image of digit 8 is presented. The activation propagates through the afferent connections to the map, which is either a SOM or a LISSOM network; a LISSOM map is shown in this figure, together with an outline of the afferent (solid line), lateral inhibitory (dashed black line), and lateral excitatory (dotted white line) connections of one neuron. In LISSOM, the activity settles through the lateral connections into a stable activity pattern (Figure 14.8); in SOM, the response is due only to the afferent connections (Figure 14.7). This pattern is the internal representation of the input that is then recognized by the array of perceptrons at the output. In this case, the output unit representing 8 is correctly activated, with weak activations on other units representing similar digits such as 2, and 9. The gray scale from white to black represents activity from low to high at all levels.

will be used to form internal representations for handwritten digits that will then be categorized by a perceptron classifier. The performance of LISSOM representations in this task will be compared with those of the SOM self-organizing map network. As was discussed in Section 3.4, SOM is an abstract, computationally efficient model of cortical maps; however, because it does not have self-organizing lateral connections, the activation patterns on the map do not form a sparse, redundancy-reduced code. LISSOM representations lead to better recognition performance, thereby demonstrating that sparse redundancy-reduced visual coding provides an advantage for information processing.

14.3.2 Method

The recognition system consists of three levels (Figure 14.3): (1) an input sheet of 32×32 units, where the input digit is represented as a normalized bitmap; (2) a 20×20 unit LISSOM (or SOM) map, which is fully connected to the input sheet; and (3) an output array of 10 perceptron units, corresponding to digits 0 to 9, fully

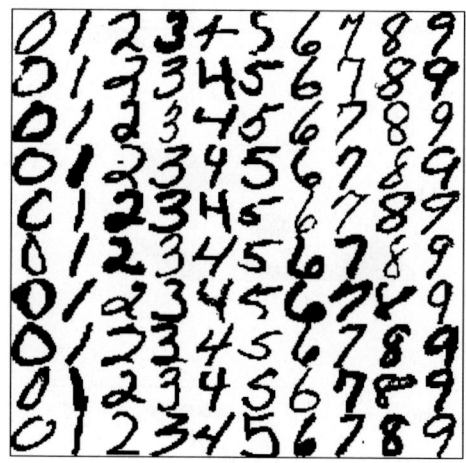

Fig. 14.4. Handwritten digit examples. One hundred samples from the NIST database 3 are shown demonstrating the variety of inputs in this task. Most of the digits are recognizable to a human observer; however, each digit occurs in different shapes, thicknesses and orientations, there is significant overlap between digits, and the classification is based on small, crucial differences. Such properties make automatic classification of handwritten digits very difficult.

connected to the map level. The map performs the feature analysis and decorrelation of the input, and the perceptrons perform the final recognition.

As training and testing data, the publicly available 2992 pattern subset of the National Institute of Standards and Technology (NIST) special database 3 (Garris 1992; Wilkinson, Garris, and Geist 1993) is used. These data contain samples from a large population of writers, coded into 32×32 bitmaps of single digits (Figure 14.4). The digits are first centered and scaled and the image is then normalized according to

$$\chi_{xy} = \frac{\chi_{xy_0}}{\sqrt{\sum_{uv} \chi_{uv_0}^2}}, \tag{14.1}$$

where χ_{xy_0} is the original input unit activity at location (x, y). Such normalization is useful because digit segments can vary in thickness. With normalization, the map activates approximately equally for both thick and thin digits.

The map activation and learning take place as described in Sections 3.4, 4.3, and 4.4, with three extensions that result in better performance in the character recognition domain and allow comparing the results of the two architectures more directly. First, in the SOM network, the Euclidean distance similarity measure is reversed and scaled so that the maximum response is 1 and minimum is 0, as in the LISSOM network:

$$\eta_{ij} = \frac{d_{\max} - \|\mathbf{X} - \mathbf{W}_{ij}\|}{d_{\max} - d_{\min}}, \tag{14.2}$$

where η_{ij} is the activity of map unit (i, j), $\mathbf{X}$ is the input vector and $\mathbf{W}_{ij}$ is the unit's weight vector, and $d_{\max}$ and $d_{\min}$ are the largest and smallest Euclidean distances between the input vector and the weight vectors on the map.

Second, in the LISSOM network, the RFs of all neurons cover the whole input image, for two reasons: (1) Such connectivity matches that of the SOM model, and (2) the resulting map representations do not have spatial organization by design; any such structure emerges from the visual coding, making the analysis and comparison easier.

Third, instead of keeping the total sum of the afferent weights constant in LISSOM, they are normalized so that the length of the weight vector remains the same:

$$A'_{xy,ij} = \frac{A_{xy,ij} + \alpha_A \chi_{xy} \eta_{ij}}{\sqrt{\sum_{uv} (A_{uv,ij} + \alpha_A \chi_{uv} \eta_{ij})^2}}, \tag{14.3}$$

where $A_{xy,ij}$ is the afferent weight between input unit (x, y) and map unit (i, j), χ_{xy} is the normalized activity of the input unit (x, y), α_A is the afferent learning rate. Since the input vectors are normalized to constant length, normalizing the weight vectors in the same way allows forming an accurate mapping of the input when the scalar product is used as the similarity measure (Section 14.4).

Although the LISSOM map can be organized starting from initially random afferent weights, a more controlled procedure was utilized in the character recognition experiments. A rough initial order was first developed in a SOM map, two copies were made of it, and their training was continued in two different ways: one as a LISSOM map, after normalizing the afferent weights and adding lateral weights, and the other as a SOM map. This procedure is useful because the resulting SOM and LISSOM maps are likely to have similar large-scale organization. The remaining differences reflect primarily the differences in visual coding, which makes comparisons easier.

The perceptrons receive the entire activation pattern on the map as their input. The activation for the perceptron unit ψ_k is calculated according to

$$\psi_k = \sum_{ij} \eta_{ij} P_{ij,k}, \tag{14.4}$$

where η_{ij} is the activity of map unit (i, j) and $P_{ij,k}$ is the connection weight between map unit (i, j) and perceptron k. The activity ψ_k represents the likelihood that the input belongs to category k. Thus, the digit represented by the perceptron with the largest activation is taken as the decision of the system. The perceptrons are trained with the delta rule (Haykin 1994; Widrow and Hoff 1960), by changing each weight proportionally to the map activity and the difference between the output and the target:

$$P'_{ij,k} = P_{ij,k} + \alpha_P (T_k - \psi_k) \eta_{ij}, \tag{14.5}$$

where α_P is the learning rate parameter and T_k is the target value ($T_k = 1$ if k is the correct digit, and zero otherwise).

Instead of perceptrons, other supervised classifiers such as backpropagation, radial basis function networks, or support vector machines (Ben-Hur, Horn, Siegelmann, and Vapnik 2001; Haykin 1994; Moody and Darken 1990; Rumelhart et al.

1986) could be trained to perform the recognition, and they would be likely to perform better. However, the goal of these experiments is not to engineer the best possible digit recognition system, but to demonstrate that some visual codes are easier to recognize than others. Because perceptrons are more sensitive to the separability of the input patterns than the other approaches, they should make such differences more clear in the performance of the system.

Digit recognition performance was measured in a 12-fold cross-validation experiment. In each split of the data into training and testing, the SOM and LISSOM maps were first organized with a training set, and the perceptron network was then trained with the responses of the final maps to the training set patterns: Different perceptron networks were trained with the SOM representations, the initial LISSOM representations, and the settled LISSOM representations. In addition, a fourth perceptron network was trained with the raw bitmap patterns in the training set. The performance of each of these networks was measured both with the training set and with the test set.

The simulation details are specified in Appendix F.2; the representations on the SOM and LISSOM maps and the recognition performance of the different networks is analyzed in the next two subsections.

14.3.3 Forming Map Representations

The final afferent weights for the SOM and LISSOM maps from one example split are shown in Figures 14.5 and 14.6. Even though they were trained from the same intermediate organization with the same sequence of inputs, SOM and LISSOM maps show different final organization. The SOM afferent weights are sharply tuned to the input patterns, and clear clusters for each digit 0 to 9 are found on the map. In contrast, the LISSOM afferent weights do not represent all digit categories equally. For example 2 and 5, which were somewhat less frequent in the dataset, are not represented distinctly anywhere, but appear only as a combination of digits 0 and 3.

Because of these differences in the afferent weights, the responses on the SOM map are continuous and diffuse, whereas LISSOM's initial responses are sparser, with contracted activity in several clusters (Figures 14.7b and 14.8b,c). The average kurtosis of these responses over all 2992 patterns was 0.84684, significantly higher than the 0.42988 of the SOM ($p < 10^{-5}$).

As expected, the strongest lateral connections in the LISSOM map link primarily to areas with similar afferent weights, i.e. those that respond to similar inputs (Figure 14.6). Their effect is to decorrelate and reduce redundant activation, forming a sparse response. The average kurtosis of the settled LISSOM activation was 2.2560, which is significantly higher than that of the initial responses ($p < 10^{14}$).

At first glance, the LISSOM map and its activity patterns seem less ordered than those of the SOM. However, the differences are mostly due to local vs. distributed style of representation, not regularity. Since the lateral connections in LISSOM link areas that respond to similar inputs, they implement more general neighborhoods than the two-dimensional local areas of the SOM. Representations far apart on the map can act as neighbors, and the responses are highly regular, even though they are less localized.

These distributed patterns act as attractors of the recurrent network, and they make the LISSOM representations more easily recognizable during performance. The settling process reduces the differences between similar patterns, by reducing redundant activation and focusing the activity at the usual distributed locations for that digit. In contrast, even though the initial responses to different digits may overlap significantly (Figure 14.8*b*), their differences become amplified during settling (Figure 14.8*c*). As a result, patterns within a category appear more similar and those across categories more different than they do in either the initial response or the SOM response, making classification of LISSOM representations easier.

In the following subsection, computational support for this informal interpretation will be provided by using perceptrons to measure the separability and generalizability of the map representations.

14.3.4 Recognizing Map Representations

Recognition performance of the perceptron serves as a measure of how good the map representations are. Performance on the training patterns can be used as a measure of separability of the patterns, i.e. how difficult the task is, and the performance on the test patterns can be used as a measure of how regular the representations are, i.e. how easy it is to generalize to new inputs.

Recognition performance based on settled LISSOM, initial LISSOM response, SOM, and raw input were measured and compared over the 12 splits of data into training and test sets. The settled LISSOM patterns turned out to be the easiest to learn (92.9%), followed by the initial LISSOM response (90.2%), SOM (88.7%), and raw input (84.8%). On the test sets, settled LISSOM patterns also performed best at 90.2%, followed by initial LISSOM responses (88.7%), SOM (85.9%), and far below, the raw input (54.4%). All of these differences are statistically significant ($p < 10^{-4}$).

The main conclusion from the digit recognition experiments is, therefore, that the sparse, redundancy-reduced internal representations provided by the LISSOM network are both most separable and easiest to generalize. In addition to being efficient, such representations provide a solid foundation for further stages of visual processing, such as pattern recognition.

14.4 Discussion

Comparing the kurtosis and reconstruction of the initial and settled LISSOM response demonstrates that self-organized long-range lateral connections are sufficient to form a sparse, redundancy-reduced visual code. The comparisons with isotropic long-range connection patterns suggest that they are also necessary: Whenever such patterns increased kurtosis, they also lost crucial information about the input. Note that this result applies to long-range connectivity only: Isotropic connections limited

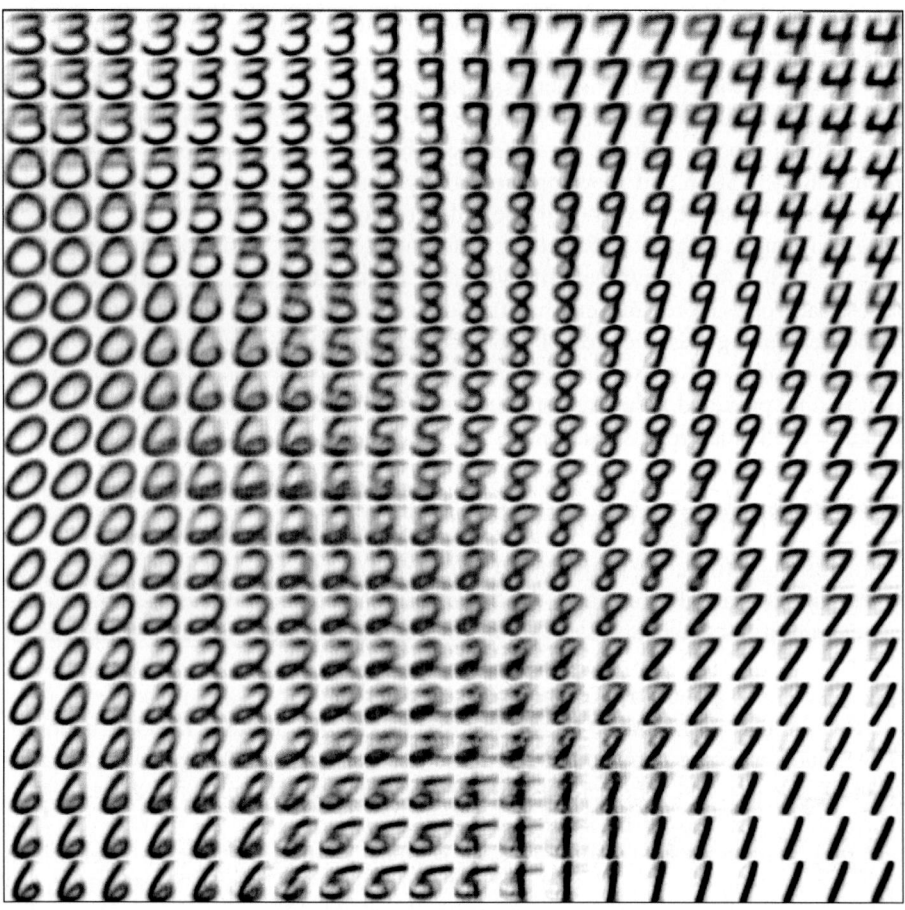

Fig. 14.5. Self-organized SOM afferent weights. The fuzzy digit-like images display the afferent weights for each unit in the 20 × 20 map, in gray scale from white to black (low to high). The SOM has a regular global organization with local clusters sensitive to each digit category. For example, the lower right corner is sensitive to inputs of digit 1, and this preference gradually changes to 7 and then to 4 along the right edge of the map.

to the range of the central Gaussian of the SoG, i.e. approximately the range of a single orientation patch in the OR map, would form a sparse response without significantly degrading the visual code. However, as was mentioned in Section 11.5.1, such a map would not self-organize or perform perceptual grouping properly, and would be an incomplete model of computations in V1.

The handwritten digit recognition experiments in turn demonstrate that the sparse, redundancy-reduced coding not only retains the salient features of the input, but is also particularly effective as input to later stages of the visual system. It is easier to recognize the input based on the LISSOM coding than on a comparable SOM coding which is not sparse and redundancy reduced.

Fig. 14.6. Self-organized LISSOM afferent and lateral weights. Compared with the SOM map in Figure 14.5, the afferent weights are less sharply tuned to individual digits and the clusters are more irregular and change more abruptly, resulting in more distributed responses (Figure 14.8). The black outline identifies the lateral inhibitory connection weights with above-average strength of the unit marked with the thick black square, which is part of the representation for digit 8. Inhibition goes to areas of similar functionality (i.e. areas sensitive to similar input), thereby decorrelating the map activity and forming a sparse representation of the input.

Such coding could be potentially useful in building artificial vision systems as well (Section 17.3.4). In such practical applications of the LISSOM model, it is important to make sure that the similarity between the input and the weight vectors is measured appropriately. Recall that the unit response in LISSOM is based on the weighted sum, i.e. the scalar product of the input and the weight vector, instead of Euclidean distance similarity measure as in SOM. The scalar product is biologically more realistic; however, it does not distinguish between differences in angle and length. In principle, both the input and the weight vectors should be normal-

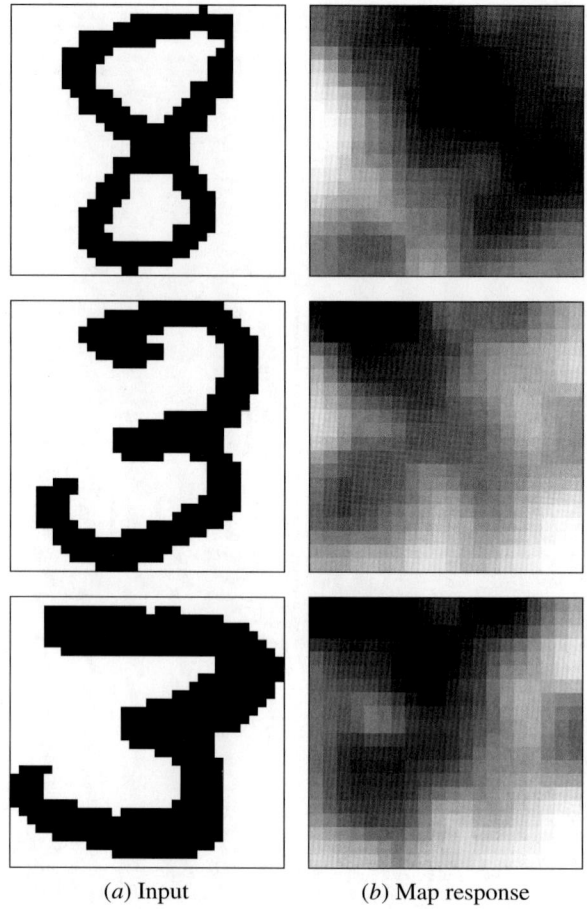

(a) Input   (b) Map response

Fig. 14.7. SOM activity patterns. Three samples of normalized input are shown in (a), and the response of SOM map to each input in (b). In each case, many units respond with similar activations, resulting in a broad and undifferentiated activity pattern over the map. Response patterns for different digits overlap significantly, making them difficult to classify.

ized to constant length, so that the similarity is measured only in terms of angles between vectors. In order to preserve the n-dimensional input distribution, a redundant $(n + 1)$th dimension must then be added to the input and weight vectors before normalization. The original dimensions are interpreted as angles and the $(n + 1)$th dimension represents the length of the vector, which is chosen the same for all inputs. After this transformation, the original input distribution becomes a submanifold of the $(n + 1)$-dimensional space. Since the dimensions are optimally chosen in the self-organizing process and the $(n + 1)$th input dimension is redundant, the map self-organizes to represent the original n-dimensional input distribution (Miikkulainen 1991; Sirosh and Miikkulainen 1994a).

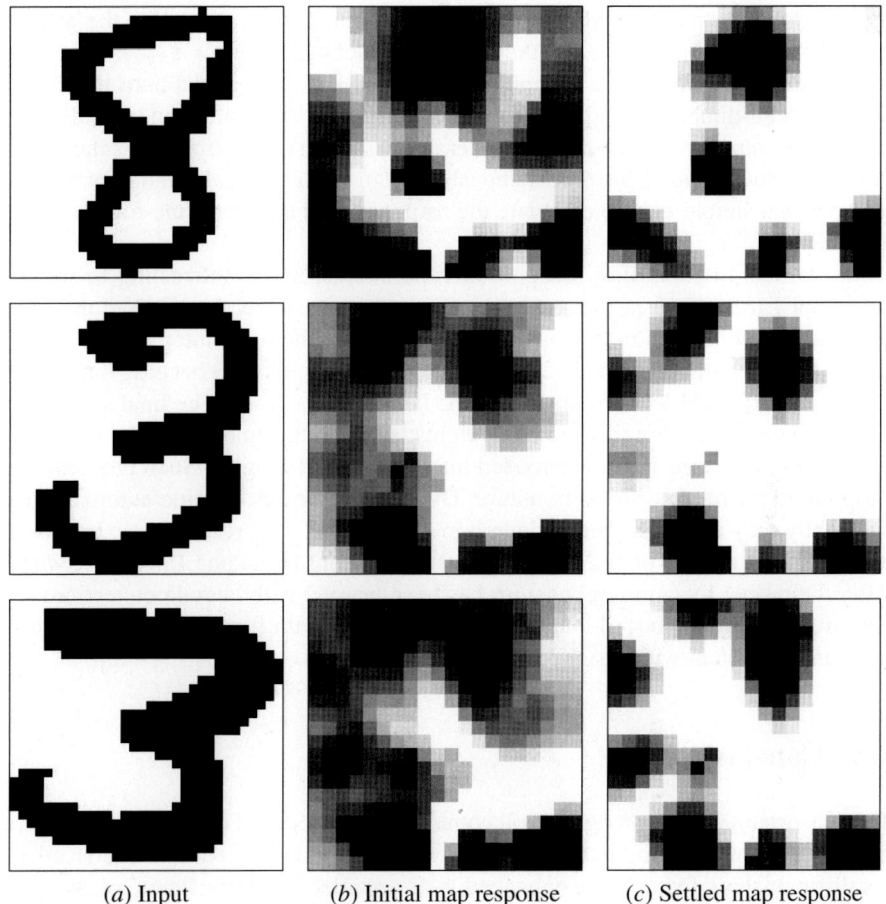

(a) Input   (b) Initial map response   (c) Settled map response

Fig. 14.8. LISSOM activity patterns. As in Figure 14.7, column (a) shows the normalized input; the LISSOM map activity before and after lateral interaction is shown in columns (b) and (c). The initial responses are sparser than in SOM, although the responses for different digits still overlap significantly. Settling through the lateral connections removes much of the redundant activation and focuses the response around the typical active regions for each digit. After settling, the patterns for the same digit have more overlap, and those for different digits less overlap than before settling, making the digits easier to recognize.

If the input vector lengths do not vary extensively, it may be possible to achieve robust self-organization simply by normalizing the afferent weight vectors to constant length. In some applications, like the digit recognition domain in this chapter, the input may also be normalized without losing crucial information, and the mapping will then be accurate. To a degree, such input normalization takes place in the ON/OFF channels, which respond mostly to edges in the input instead of constant activation (this effect is further enhanced by afferent normalization introduced in Section 8.2.3). In such cases, it is usually sufficient to maintain the total sum of the weights constant instead of length. Such normalization constrains the afferent weights of each neuron (i, j) to the hyperplane defined by $\sum_{xy} A_{xy,ij} = 1$, and self-organization produces a mapping of the input space projected onto this hyperplane. In high-dimensional spaces, the distortion due to this projection is small, especially when inputs are limited to a smaller dimensional submanifold, as is the case in the retina. Such a simplification is appealing from the biological standpoint because it provides a simple way to calculate the response, and a unified rule for modifying all synapses.

In the handwritten digit recognition experiments, it was interesting to see that the initial LISSOM responses were sparser and easier to recognize than the SOM responses, even though both maps were trained from the same intermediate organization with the same sequence of inputs. These differences result because the afferent weights in LISSOM learn to anticipate the lateral interactions. The final settled patterns are used to modify the afferent weights; as a result, some of the characteristics of the settled patterns become encoded into the afferent weights. Such processes are common in adapting systems in nature: Dynamic processes become automatic, and eventually may even become hardwired in the genome. This result is also promising from the point of view of building practical applications based on LISSOM networks. After the proper recognition behavior has been learned with lateral connections, the behavior can be transferred into the afferent weights with further training, resulting in a simpler system with faster performance.

14.5 Conclusion

The self-organized long-range lateral connections in LISSOM decorrelate the activation on the map, resulting in a sparse code where redundant activation is reduced and the distinguishing features are enhanced. Self-organized, specific lateral connections are necessary for such coding: When equally sparse representations are formed with isotropic lateral connections, information is lost. Such representations are efficient in that they allow more information to be represented in the same area of cortex, but they also provide an information processing advantage. They are more easily separable and generalizable, making further visual processing such as pattern recognition easier.

15

Scaling LISSOM simulations

Current computational models such as LISSOM can account for much of the structure of the visual cortex and how it develops, as well as many of its functional properties. However, other important phenomena, such as orientation interactions between spatially separated stimuli and long-range visual contour and object integration, have remained out of reach because they require too much computation time and memory to simulate. In this chapter, two interrelated techniques are presented for making detailed large-scale simulations practical. First, a set of linear scaling equations is derived that allows computing the appropriate parameter settings for a large-scale simulation given an equivalent small-scale simulation. Second, a method called GLISSOM is developed where the map is systematically grown based on the scaling equations, allowing the entire visual cortex to be simulated at the column level with desktop workstations. The scaling equations can also be used to quantify differences in biological systems, and to determine values for the model parameters to match measurements in particular biological species.

15.1 Parameter Scaling Approach

A given LISSOM simulation focuses on a particular area of the visual field with a particularly density of retinal receptors and V1 neurons. Modeling new phenomena often requires setting up a different area or density in the model. For example, a larger portion of the visual space, e.g. a larger part of V1 and the eye, may have to be simulated, or the area may have to be simulated at a finer resolution, or a species, individual, or brain area needs to be modeled that devotes more neurons or receptors to representing the same visual space.

Varying the area or density over a wide range can be difficult in a complex nonlinear system like LISSOM. Parameter settings that work well for one size are usually not appropriate for other sizes, and it is not always clear which parameters need to be adjusted. Fortunately, with LISSOM it is possible to derive a set of equations that allows computing the appropriate parameter values for each type of transformation directly. The equations treat the cortical network as a finite approximation of a continuous map, i.e. one that is composed of an infinite number of units (see Amari 1980; Fellenz and Taylor 2002; Roque Da Silva Filho 1992; Wu, Amari, and Nakahara 2002 for theoretical analyses of continuous maps). Under such an assumption, networks of different sizes represent coarser or denser approximations of the continuous map, and any given approximation can be transformed into another by conceptually reconstructing the continuous map and then resampling it. Given an existing retina and cortex, the scaling equations provide the parameter values needed to self-organize a functionally equivalent smaller or larger retina and cortex.

To be most useful, such scaling should result in equivalent maps at different sizes. Map organization must therefore not depend on the initial weights, because those will vary between different-size networks. As was shown in Section 8.4, the LISSOM algorithm has exactly this property: The organization is determined by the stream of input patterns, not by the initial weight values. In effect, the LISSOM scaling equations provide a set of parameters for a new simulation that, when run, will develop similar results to the existing simulation.

The equations for scaling area and density will be derived in the next section. The more minor LISSOM parameters can be scaled as well, using the methods described in Appendix A.2. Most of the simulations presented in this book were set up using these equations, and they were crucial for the large maps used in Section 10.2. In Section 15.3, these equations will be utilized systematically by developing an incremental scaling algorithm for self-organizing very large maps.

15.2 Scaling Equations

In this section, the method for scaling the visual area, retinal receptor density, and cortical density in LISSOM models is developed. The scaling equations are derived theoretically and verified to be effective experimentally, especially analyzing the limitations of the scaling approach. Although the method applies to all versions of LISSOM, the simulations are based on the reduced LISSOM orientation map without ON/OFF channels (Chapters 6, 7, and 14; Appendix B). This model is complex enough to demonstrate the power of scaling and simple enough to observe its effects clearly. The equations are derived for the central area of the retina with full representation in the cortex (Figure A.1), but they can also be extended to include the border area (as shown in Appendix A.2).

15.2.1 Scaling the Area

The simplest case of scaling consists of changing the area of the visual space simulated. The model can be developed quickly with a small area, then enlarged to eliminate border effects and to simulate the full area of a biological experiment. To change the area, both the V1 width N and the retina width R must be scaled by the same proportion k relative to their initial values N_o and R_o. (The ON and OFF channels of the LGN change just as the retina does; Appendix A.2.)

In such scaling, it is also necessary to ensure that the resulting network has the same amount of learning per neuron per iteration. Otherwise, self-organizing a larger network would take more training iterations; as a result, the final map would be different because nonlinear thresholding is performed at each iteration (Equation 4.6). To achieve the same amount of learning, the average activity per input receptor needs to remain constant. With discrete input patterns such as Gaussians, the number of patterns n_p per iteration must be scaled with the retinal area. With natural image inputs, it is sufficient to make sure that the images cover all of the full, larger retina; parameter n_p can be ignored. Assuming discrete inputs, the equations for scaling the area by a factor m are

$$N = mN_o, \quad R = mR_o, \quad n_p = m^2 n_{p_o}. \tag{15.1}$$

An example of such scaling with $m = 4$ is shown in Figure 15.1. The simulation details are described in Appendix B.3.

15.2.2 Scaling Retinal Density

Retinal density, i.e. the number of retinal units per degree of visual field, may have to be adjusted, e.g. to model species with larger eyes or parts of the eye that have more receptors per unit area. In practice, higher density means increasing retina size R while keeping the corresponding visual area the same. This type of change also allows the cortical magnification factor $N{:}R$, i.e. the ratio between the V1 and retina densities, to be matched with values measured in a particular species. The scaling equations allow R to be increased to any desired value, although in most cases (especially when modeling newborns) a low density suffices.

The parameter adjustments to change density are slightly more complicated than those for area. First, in order to avoid disrupting how the maps develop, the visual area processed by each neuron must be kept constant. More specifically, the ratio of the afferent connection radius and the retina width must be constant, i.e. r_A must scale with R.

Second, to make sure equivalent maps develop, the average total weight change per neuron per iteration must be remain the same in the original and the scaled network. When the connection radius increases, the total number of afferent connections per neuron increases dramatically. Because the learning rate α_A specifies the amount of change per connection and not per neuron (Equation 4.8), the learning rate must be adjusted to compensate; otherwise, a given input pattern would modify the weights of the scaled network more than those of the original network. The afferent learning rate α_A needs to be scaled inversely with the number of afferent connections to each neuron, which in the continuous plane corresponds to the area enclosed by the afferent radius. That is, α_A scales by the ratio $r_{A_o}^2/r_A^2$.

Third, because the average activity per iteration also affects self-organization, the size of the input features must also scale with R. For Gaussian inputs, the ratio between the width σ and R must be kept constant; other input types can be scaled similarly. Thus, the retinal density scaling equations are

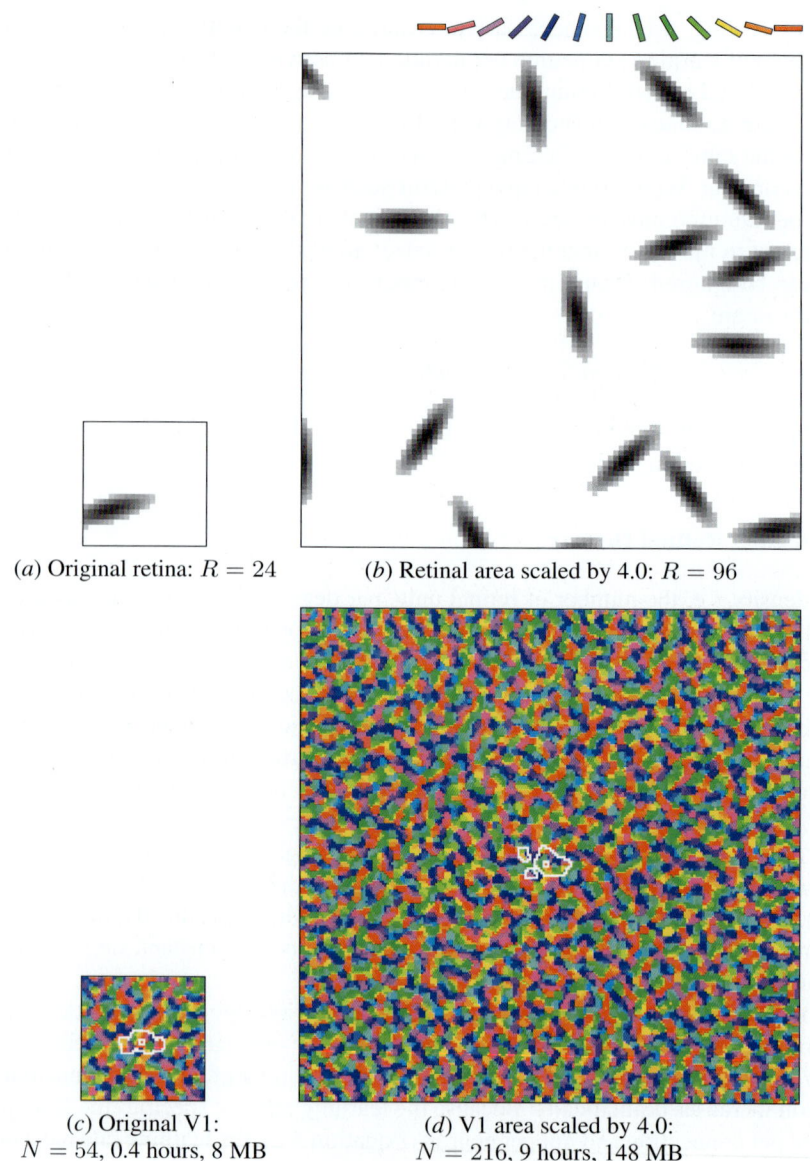

(a) Original retina: $R = 24$ (b) Retinal area scaled by 4.0: $R = 96$

(c) Original V1:   (d) V1 area scaled by 4.0:
$N = 54$, 0.4 hours, 8 MB $N = 216$, 9 hours, 148 MB

Fig. 15.1. Scaling retinal and cortical area. The small retina (a) and V1 (c) was scaled to a size 16 times larger (b,d) using Equation 15.1. To make it easier to compare map structure, especially in early iterations, the OR maps are plotted without selectivity in this chapter. The lateral inhibitory connections of one central neuron, marked with a small white square, are indicated in white outline. The simulation time and the number of connections scale approximately linearly with the area, and thus the larger network takes about 16 times more time and memory to simulate. For discrete input patterns like these oriented Gaussians, it is necessary to have more patterns to keep the total learning per neuron and per iteration constant. Because the inputs are generated randomly across the retina, each map sees a different stream of inputs, and so the patterns of orientation patches on the final maps differ. The area scaling equations are most useful for developing a model with a small area and then scaling it up to eliminate border effects and to simulate the full area of a corresponding biological preparation.

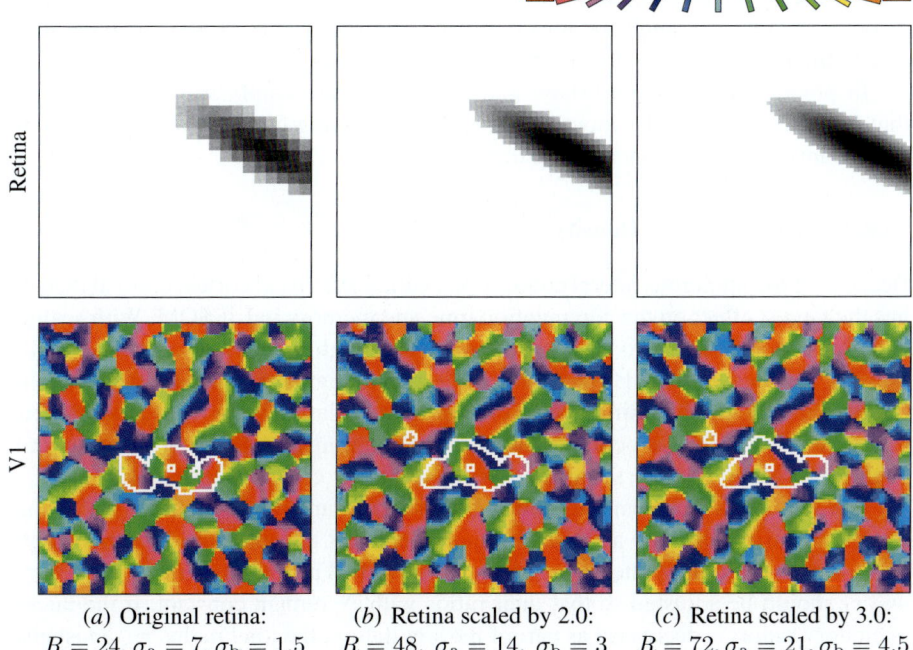

(*a*) Original retina: (*b*) Retina scaled by 2.0: (*c*) Retina scaled by 3.0:
$R = 24, \sigma_\mathrm{a} = 7, \sigma_\mathrm{b} = 1.5$ $R = 48, \sigma_\mathrm{a} = 14, \sigma_\mathrm{b} = 3$ $R = 72, \sigma_\mathrm{a} = 21, \sigma_\mathrm{b} = 4.5$

Fig. 15.2. Scaling retinal density. Each column shows a LISSOM orientation map from one of three matched 96×96 networks with retinas of different densities. The parameters for each network were calculated using Equation 15.2, and each network was then trained independently on the same random stream of input patterns. The size of the input pattern in retinal units grows as the retinal density is increased, but its size as a proportion of the retina remains constant. All of the resulting maps are similar as long as R is large enough to represent the input faithfully, with almost no change above $R = 48$. Thus, a low value can be used for R in practice. Such scaling of retinal density is useful for modeling species and areas with higher receptor resolution, and for matching the cortical magnification factor of a model to that of a particular species.

$$r_\mathrm{A} = \frac{R}{R_\mathrm{o}} r_\mathrm{Ao}, \qquad \alpha_\mathrm{A} = \frac{r_\mathrm{Ao}^2}{r_\mathrm{A}^2} \alpha_\mathrm{Ao}, \qquad \sigma_\mathrm{a} = \frac{R}{R_\mathrm{o}} \sigma_\mathrm{ao}, \qquad \sigma_\mathrm{b} = \frac{R}{R_\mathrm{o}} \sigma_\mathrm{bo}. \qquad (15.2)$$

Figure 15.2 demonstrates that these equations can be used to generate functionally equivalent orientation maps with different retinal receptor densities. The crucial parameters are those scaled by the retinal density equations, specifically r_A and σ. The value of R is unimportant as long as it is large enough to represent input patterns of width σ faithfully. The minimum such R can be computed based on the Nyquist theorem in digital signal processing theory (e.g. Cover and Thomas 1991): The sampling frequency, determined by R, has to be at least twice the spatial frequency of the input, which is determined by σ.

If ON and OFF channels are included in the model, σ is important only if it is larger than the centers of the LGN cells, i.e. only if the input is large enough so that it will be represented as a pattern at the LGN output. As a result, with natural images or other stimuli with high-frequency information (i.e. small σ), the receptive field size of the LGN cells may instead be the limiting factor.

In practice, these results show that a modeler can simply use the smallest R that faithfully samples the input patterns, thereby saving computation time without significantly affecting how the map develops.

15.2.3 Scaling Cortical Density

Because of the numerous lateral connections within the visual cortex, cortical density has the largest effect on the computation time and memory in LISSOM. With scaling equations, it is possible to develop the model through a series of computationally efficient low-density simulations, only scaling up to high density to see the details in the final model. Such scaling is also crucial for simulating maps with multiple feature dimensions, such as ocular dominance, orientation, and direction (Section 5.6), because such maps can be seen more clearly at higher cortical densities.

The equations for changing cortical density are analogous to those for retinal receptor density, with the additional requirement that the intracortical connectivity and associated learning rates must be scaled as well. The lateral connection radii r_E and r_I should be adjusted so that their ratios with N remain constant. If the lateral excitatory radius is decreased as part of the simulation, the final radius r_{Ef} must also be adjusted accordingly. Like α_A in the previous section, α_E and α_I must be scaled so that the average total weight change per neuron remains constant each iteration despite changes in the number of connections. Finally, the absolute weight level w_d below which lateral inhibitory connections are deleted must be scaled inversely to the total number of such connections, because normalization adjusts each weight inversely proportional to the number of connections (Equation 4.8). In the continuous plane, that number is the area enclosed by the lateral inhibitory radius. Thus, the cortical density scaling equations are

$$r_E = \frac{N}{N_o} r_{Eo}, \quad \alpha_E = \frac{r_{Eo}^2}{r_E^2} \alpha_{Eo}, \quad w_d = \frac{r_{Io}^2}{r_I^2} w_{do},$$

$$r_I = \frac{N}{N_o} r_{Io}, \quad \alpha_I = \frac{r_{Io}^2}{r_I^2} \alpha_{Io}. \tag{15.3}$$

Figure 15.3 shows how these equations can be used to generate closely matching orientation maps with different cortical densities. Larger maps are smoother and show more detail, but the overall structure is very similar.

The Nyquist theorem specifies theoretical limits on the minimum N necessary to faithfully represent a given orientation map pattern. In practice, however, the minimum excitatory radius r_{Ef} is the limiting parameter. For instance, the map pattern from Figure 15.3e can be reduced using image manipulation software to 18×18 without changing the global pattern of orientation patches. Yet, when simulated in LISSOM, the 36×36 (and to some extent, even the 48×48) map differs from the larger ones. These differences result from quantization effects on r_{Ef}. Because

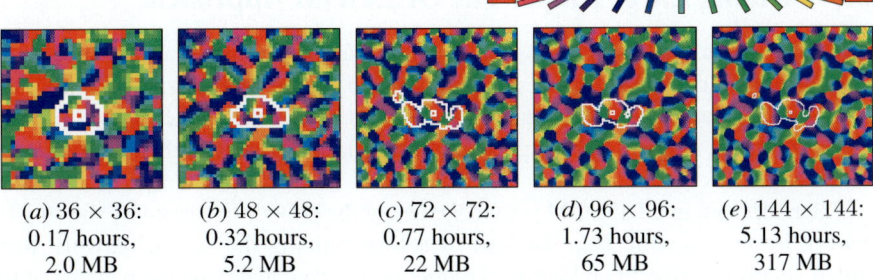

(a) 36 × 36: 0.17 hours, 2.0 MB   (b) 48 × 48: 0.32 hours, 5.2 MB   (c) 72 × 72: 0.77 hours, 22 MB   (d) 96 × 96: 1.73 hours, 65 MB   (e) 144 × 144: 5.13 hours, 317 MB

Fig. 15.3. Scaling cortical density. Five LISSOM orientation maps from networks with different densities are shown. The parameters for each network were first calculated using Equation 15.3, and each network was then trained independently on the same random stream of input patterns. The number of connections in these networks ranged from 2×10^6 to 3×10^8 (requiring 2 MB to 317 MB of memory), and the simulation time from 10 minutes to 5 hours. Despite this wide range of simulation scales, the final organized maps are both qualitatively and quantitatively similar, as long as their size is above a certain minimum (about 64×64 in this case). Larger networks take significantly more memory and simulation time, but offer greater detail and allow multiple dimensions such as orientation, ocular dominance, and direction selectivity to be represented simultaneously.

units are laid out on a rectangular grid, the smallest radius that includes at least one other neuron is 1.0. Yet, for small enough N, the scaled r_{Ef} will be less than 1.0. If such small radii are truncated to zero, the map will no longer have local topographic ordering, because there will be no local excitation between neurons. On the other hand, if the radius is held at 1.0 while the map continues to shrink, lateral excitation will take over a larger and larger portion of the map, making the orientation patches in the resulting map wider. Thus, in practice, N should not be reduced so far that $r_{Ef} < 1.0$.

Together, the area and density scaling equations allow essentially any size V1 and retina to be simulated without a search for the appropriate parameters. Given fixed resources, such as a computer of a certain speed with a certain amount of memory, they make it simple to trade off density for area, depending on the phenomena being studied. The equations are all linear, so they can also be applied together to change both area and density simultaneously. Such scaling makes it easy to utilize supercomputers for very large simulations. A small-scale simulation can be first developed with standard hardware, and then scaled up to study specific large-scale phenomena on a supercomputer. Scaling can also be done step by step while the network is self-organizing, thereby maximizing the size of networks that can be simulated, as will be described in the next section.

15.3 Forming Large Maps: The GLISSOM Approach

The scaling equations make it possible to determine the parameter settings necessary to perform a large-scale simulation. This approach can be generalized by applying it successively to larger and larger maps. This approach is called GLISSOM (growing LISSOM), and it allows scaling up LISSOM simulations much further, up to the size of the human V1.

The main idea in GLISSOM is to make use of the structure learned by the smaller network in each scaling step. Instead of self-organizing the scaled network from scratch, its initial afferent and lateral weights are interpolated from the weights of the smaller network. Such scaling allows neuron density to be increased while keeping the large-scale structural and functional properties constant, such as the organization of the orientation map. In essence, the large network is grown in place, thereby minimizing the computational resources required for the simulation.

GLISSOM is effective for two reasons. First, pruning-based self-organizing models such as LISSOM have peak computational and memory requirements at the beginning of training (Figure 15.4). At that time, all connections are active, none of the neurons are selective, and activity is spread over a wide area. As the neurons become selective and smaller regions of V1 are activated by a given input, simulation time decreases dramatically, because only the active neurons need to be simulated in a given iteration. GLISSOM takes advantage of this process by approximating the map with a very small network early in training, then gradually growing the map as selectivity and specific connectivity are established.

Second, self-organization in computational models, as well as in biology (Chapman et al. 1996), tends to proceed in a global-to-local fashion, with large-scale order established first, followed by more detailed local organization. Thus, small maps, which are faster to simulate and take less memory, can be employed first to establish global order, and large maps subsequently to achieve more detailed structure. In this manner, much larger networks can be simulated in a given computation time and in a given amount of memory.

Although the primary motivation for GLISSOM is computational, the scaling process is also well motivated biologically. It is an abstraction of how new neurons are integrated into an existing region during development. Recent experimental results suggest that new neurons continue to be added even in adulthood in many areas of primate cortex (Gould, Reeves, Graziano, and Gross 1999). Moreover, many of the neurons in the immature cortex (corresponding to GLISSOM's early stages) have not yet begun to make functional connections, having only recently migrated to their final positions (Purves 1988). Thus, the scale-up procedure in GLISSOM corresponds to the gradual process of incorporating those neurons into the partially organized map.

15.4 GLISSOM Scaling

GLISSOM is based on the cortical density scaling equations, with one significant extension: The initial weights of the scaled network are calculated from the exist-

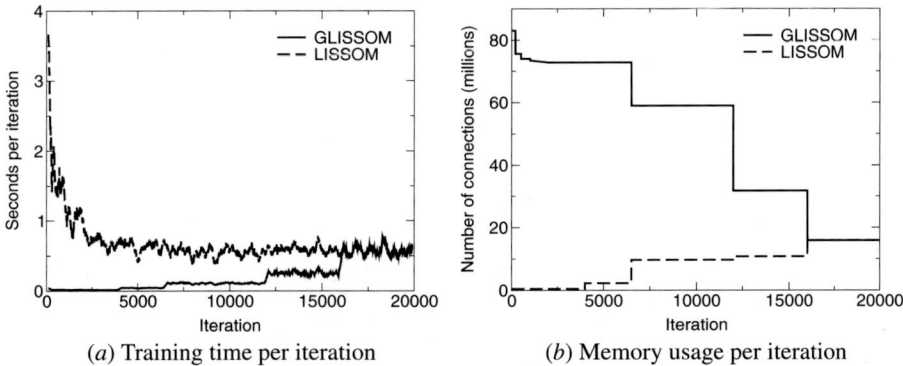

(a) Training time per iteration   (b) Memory usage per iteration

Fig. 15.4. Training time and memory usage in LISSOM vs. GLISSOM. Data are shown for a LISSOM network of 144×144 units and a GLISSOM network grown from 36×36 to 144×144 units as described in Section 15.4.2. (a) Each line shows a 20-point running average of the time spent in training for one iteration, with a data point measured every 10 iterations. Only training time is shown; times for initialization, plotting images, pruning, and scaling networks are not included. Computational requirements of LISSOM peak at the early iterations, falling as the excitatory radius (and thus the number of neurons activated by a given pattern) shrinks and as the neurons become more selective. In contrast, GLISSOM requires little computation time until the final iterations. Because the total training time is determined by the area under each curve, GLISSOM is much more efficient to train overall. (b) Each line shows the number of connections simulated at a given iteration. LISSOM's memory usage peaks at early iterations, decreasing at first in a series of small drops as the lateral excitatory radius shrinks, and then later in a few large drops as long-range inhibitory weights are pruned at iterations 6500, 12,000, and 16,000. Similar shrinking and pruning takes place in GLISSOM, while the network size is scaled up at iterations 4000, 6500, 12,000, and 16,000. Because the GLISSOM map starts out small, memory usage peaks much later, and remains bounded because connections are pruned as the network is grown. As a result, the peak number of connections (which determines the memory usage) in GLISSOM is as low as the smallest number of connections in LISSOM.

ing weights through interpolation. The interpolation equations are presented in this section, followed by experiments that demonstrate that the method is effective.

15.4.1 Weight Interpolation Algorithm

In order to perform interpolation, the original weight matrices are treated as discrete samples of a smooth, continuous function. Under such an interpretation, the underlying smooth function can be resampled at a higher density. The resampling is equivalent to the smooth bitmap scaling done by computer graphics programs (as will be shown in Figure 15.6). This type of scaling always increases the size of the network by at least one whole row or a column at once. However, unlike the growing SOM algorithms that add nodes to the original network (Bauer and Villman 1997; Blackmore and Miikkulainen 1995; Cho 1997; Fritzke 1994, 1995; Jockusch 1990; Rodriques and Almeida 1990; Suenaga and Ishikawa 2000), in GLISSOM the origi-

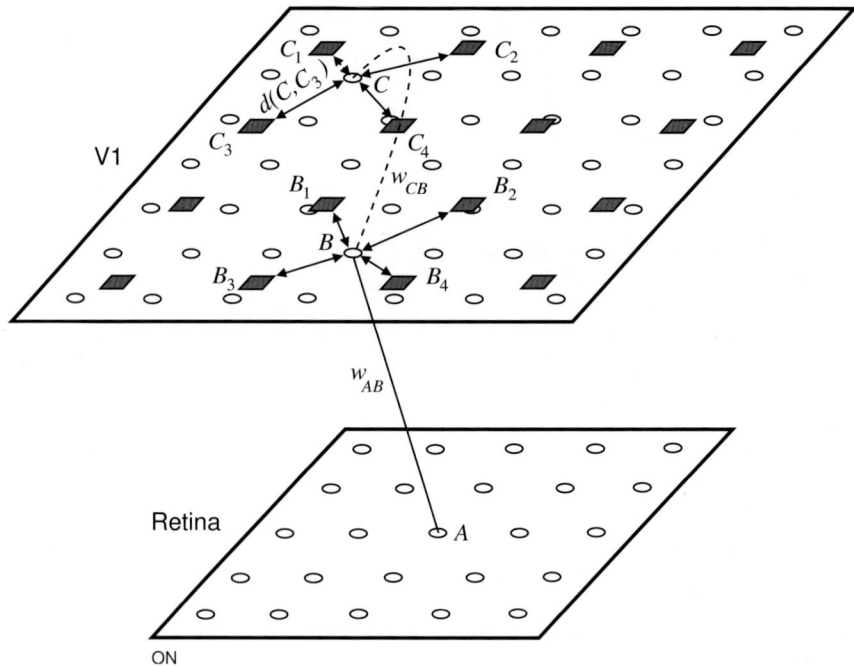

Fig. 15.5. Weight interpolation in GLISSOM. This example shows a V1 of size 4×4 being scaled to 7×7, with a fixed 8×8 retina. Both V1 networks are plotted in a continuous two-dimensional area representing the surface of the cortex. The squares in V1 represent neurons in the original network (i.e. before scaling) and circles represent neurons in the new, scaled network. A is a retinal receptor cell and B and C are neurons in the new network. Afferent connection strengths to neuron B in the new network are calculated based on the connection strengths of the ancestors of B, i.e. those neurons in the original network that surround the position of B (B_1, B_2, B_3, and B_4 in this case). The new afferent connection strength w_{AB} from receptor A to B is a normalized combination of the connection strengths w_{AB_i} from A to each ancestor B_i of B, weighted inversely by the distance $d(B, B_i)$ between B_i and B. Lateral connection strengths from C to B are calculated similarly, as a proximity-weighted combination of the connection strengths between the ancestors of those neurons. Thus, the connection strengths in the scaled network consist of proximity-weighted combinations of the connection strengths in the original network.

nal network is completely replaced by the scaled network. The interpolation process is similar to that used in continuous SOM algorithms (Campos and Carpenter 2000; Göppert and Rosenstiel 1997). However, whereas continuous SOM methods interpolate to approximate functions more accurately, in GLISSOM the result forms a starting point for further self-organization.

Let us first derive the interpolation procedure for the afferent connections. Assume the original and the scaled networks are overlaid uniformly on the same two-dimensional area, as shown in Figure 15.5. The afferent connection weight from retinal receptor A to neuron B in the scaled network is calculated based on the corresponding weights of B's original neighbors, i.e. the neurons B_i of the original network that surround B in its two-dimensional neighborhood h_B. These neurons are called B's ancestors. In the middle, each neuron has four ancestors; at the corners, each has only one, and along the edges, each has two. Each ancestor has an influence f ranging from 0 to 1 on the computed weights of B, determined by its proximity to B:

$$f_{BB_i} = 1.0 - \frac{d(B, B_i)}{d_{\max}}, \tag{15.4}$$

where $d(B, B_i)$ represents the Euclidean distance between B and its ancestor B_i in the two-dimensional area, and $d_{\max}$ is the maximum possible distance between B and any of its ancestors, i.e. the diagonal spacing between the ancestors. The afferent connection strength w_{AB} is then a normalized influence-weighted linear combination of the weights from A to B's ancestors:

$$w_{AB} = \frac{\sum\limits_{i \in h_B} w_{AB_i} f_{BB_i}}{\sum\limits_{i \in h_B} f_{BB_i}}, \tag{15.5}$$

where w_{AB_i} is the afferent connection weight from receptor A to the ith ancestor of B. Because receptive fields are limited in size, not all ancestors receive connections from that receptor; only those that do contribute to the sum.

The lateral connection strengths from neuron C to neuron B in the scaled map are computed analogously based on the connection strengths between the ancestors of C and the ancestors of B. The two kinds of lateral connections, excitatory and inhibitory, are computed separately through the same procedure. First, the contribution of C's ancestors to each B_i is calculated as

$$g_{CB_i} = \frac{\sum\limits_{j \in h_C} w_{C_j B_i} f_{CC_j}}{\sum\limits_{j \in h_C} f_{CC_j}}, \tag{15.6}$$

where $w_{C_j B_i}$ (either E or I) is the connection weight from the jth ancestor of C to the ith ancestor of B (if such a connection exists). The new lateral connection strength w_{CB} is then the influence-weighted sum of the contributions from all ancestors of B:

$$w_{CB} = \frac{\sum\limits_{i \in h_B} g_{CB_i} f_{BB_i}}{\sum\limits_{i \in h_B} f_{BB_i}}. \tag{15.7}$$

Because the neurons in the scaled network have more lateral connections than those in the original map, the new connections are usually pruned immediately during the scaling process: Each new connection is included in the scaled network only if it is

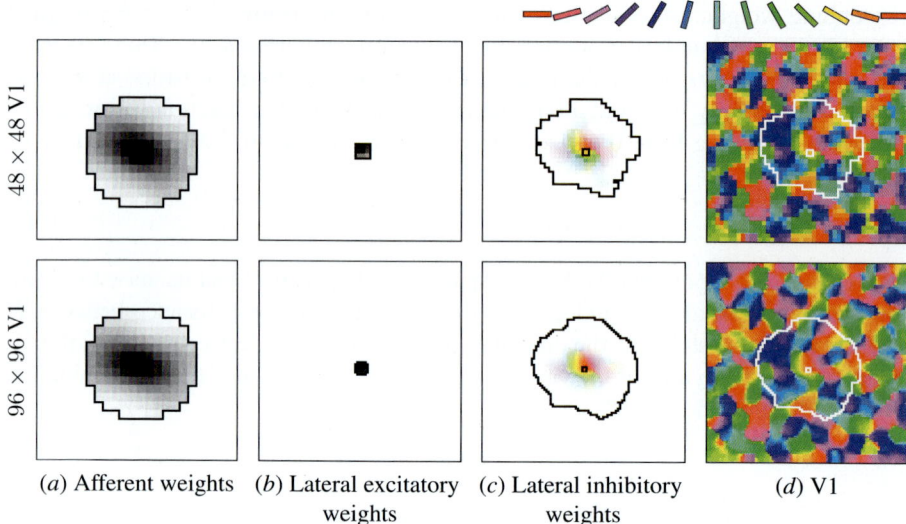

<table>
<tr><td></td><td>(a) Afferent weights</td><td>(b) Lateral excitatory
weights</td><td>(c) Lateral inhibitory
weights</td><td>(d) V1</td></tr>
</table>

Fig. 15.6. Scaling cortical density in GLISSOM. In a single GLISSOM cortical density scaling step, a 48×48 V1 (top row) is expanded into a 96×96 V1 (bottom row) at iteration 10,000 out of a total of 20,000. Smaller scaling steps are usually more effective, but the large step makes the changes more obvious. A set of weights for one neuron in each network is shown in (a–c). At this point in training, the afferent and lateral connection profiles are still only weakly oriented, and lateral connections have not been pruned extensively (the jagged black outline in (c) shows the current connectivity). The orientation map for each network is shown in (d), with the inhibitory weights of the sample neuron overlaid in white outline. The orientation map measured from the scaled map is identical to that of the 48×48 network, except that it has twice the resolution. This network can then self-organize at the new density to represent finer detail.

larger than the pruning threshold. This procedure makes sure the scaled networks require as little memory as possible.

Figure 15.6 shows an example scaling step for a partially organized orientation map and the weights of one neuron in it (to see the correspondence more clearly, the lateral connections were not pruned in this example). The larger map replicates the structures of the smaller one, and can be self-organized to represent further detail.

15.4.2 Method

The GLISSOM simulations were based on the same reduced LISSOM orientation model as the simulations in Section 15.2, with a 24×24 retina. The parameters of this model were adjusted with the density and area scaling equations to get the specific model for each of the comparisons.

Each GLISSOM simulation started with low cortical density. The scaling method was then used to increase the density gradually as the network self-organized. At the same time, the other parameters were adjusted according to the scaling equations to make sure the map stayed functionally the same. Similar scaling could be used to increase the retinal density during self-organization, but because retinal processing does not affect the computation and memory usage much, retinal density was not adapted in the simulations.

The precise scaling schedule is not crucial as long as the initial map is large enough so that the largest features of the final map can be represented approximately in the initial map. A linear increase from the initial size N_o to the final size N_f usually works well. If scaling is faster than linear, i.e. the simulation scales up quickly to large maps which are then self-organized for a long time, the final maps will be more refined but the simulation takes more time; conversely, slower than linear scaling results in faster simulation but less accurate maps. If the scaling steps are very large, the final map may have more distortions. On the other hand, many small steps incurs a significant overhead of having to organize the initial weights many times. Therefore, the most effective schedule usually consists of a few medium-size steps.

The scaling steps N were computed as

$$N = N_o + m(N_o - N_f), \tag{15.8}$$

where m is a constant whose values increase approximately linearly over the simulation. This equation allows specifying the scaling steps uniformly across experiments even with different network sizes. Unless stated otherwise, the simulations consisted of four steps, with $m = 0.20$ at iteration 4000, 0.47 at 6500, 0.67 at 12,000, and 1.0 at 16,000. The rest of the simulation details are described in Appendix B.3.

The GLISSOM maps formed in this manner were compared with LISSOM maps that were organized directly at the final size. The self-organization processes and the final maps are compared next.

15.4.3 Comparing LISSOM and GLISSOM Maps

The first result is that GLISSOM develops an orientation map in a similar process as a full-size LISSOM (Figure 15.7). Both networks pass through similar stages of intermediate order, while the GLISSOM map size gradually approaches that of the LISSOM map.

Second, as long as the initial GLISSOM map is sufficiently large to represent the global organization, GLISSOM results in an orientation preference map and weight patterns that are qualitatively and quantitatively equivalent to those of LISSOM (Figures 15.8 and 15.9).

Third, GLISSOM significantly reduces the overall computation time and memory usage (Figure 15.10). For example, for a final map with $N = 144$, LISSOM takes 5.1 hours for 20,000 training iterations, whereas GLISSOM finishes in 1.6 hours, yielding a speed-up ratio of 3.1. For the same simulation, LISSOM requires 317 MB of memory to store its connections, while GLISSOM requires only 60 MB, resulting in memory savings ratio of 5.2. Importantly, the speed-up and memory savings increase with larger networks, which means that GLISSOM can make simulation of very large networks practical.

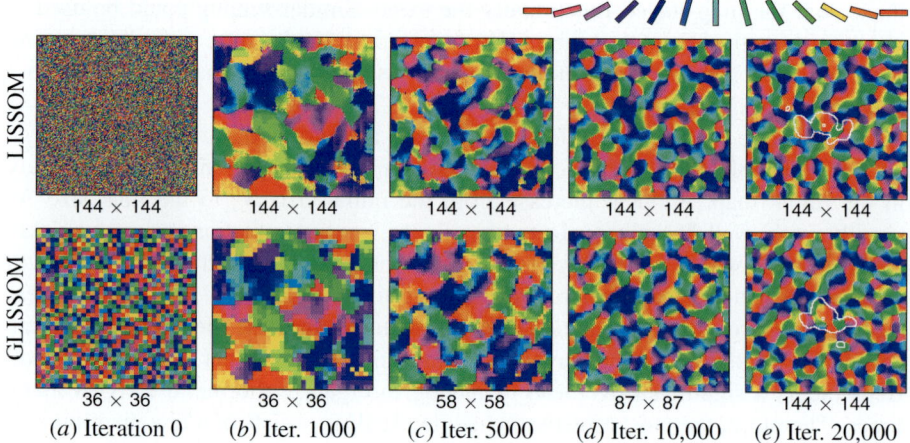

Fig. 15.7. Self-organization of LISSOM and GLISSOM orientation maps. The GLISSOM map is gradually scaled so that by the final iteration it has the same size as LISSOM. To make the scaling steps more obvious, this example is based on the smallest acceptable initial network; Figure 15.9 shows that results match even more closely for larger initial networks. At each iteration, the features that emerge in the GLISSOM map are similar to those of LISSOM except for discretization differences. An animated demo of these self-organization examples can be seen at http://computationalmaps.org.

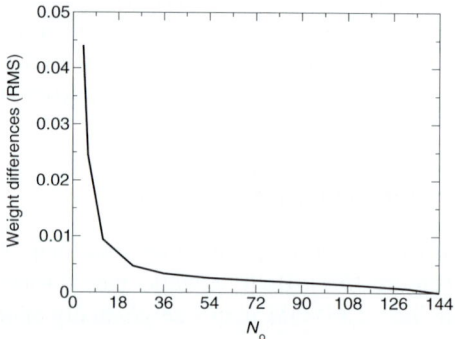

Fig. 15.8. Accuracy of the final GLISSOM map as a function of the initial network size. Each point shows the RMS difference between the final values of the corresponding weights of each neuron in two networks: a 144×144 LISSOM map, and a GLISSOM network with an initial size shown on the x-axis and a final size of 144×144. Both maps were trained on the same stream of oriented inputs. The GLISSOM maps starting at most as large as $N = 96$ were based on four scaling steps, whereas the three larger starting points included fewer steps: $N = 114$ had one step at iteration 6500, $N = 132$ had one step at iteration 1000, and there were no scaling steps for $N = 144$. Low values of RMS difference indicate that the corresponding neurons in each map developed very similar weight patterns. The RMS difference drops quickly as larger initial networks are employed, becoming negligible above 36×36. As was described in Section 15.2.3, this lower bound is determined by r_{Ef}, the minimum size of the excitatory radius.

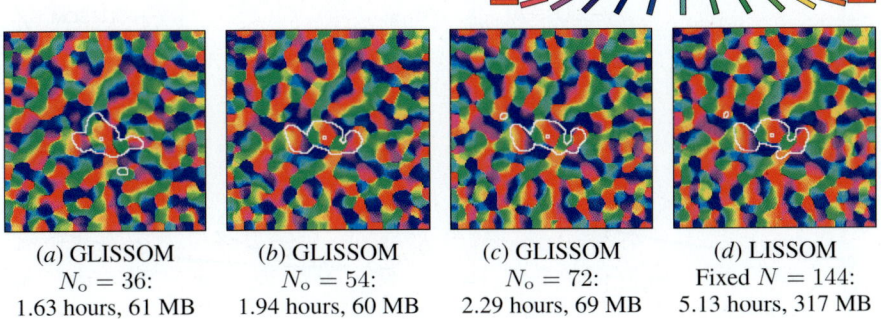

(*a*) GLISSOM	(*b*) GLISSOM	(*c*) GLISSOM	(*d*) LISSOM
$N_o = 36$:	$N_o = 54$:	$N_o = 72$:	Fixed $N = 144$:
1.63 hours, 61 MB	1.94 hours, 60 MB	2.29 hours, 69 MB	5.13 hours, 317 MB

Fig. 15.9. Orientation maps in LISSOM and GLISSOM. Above the minimum 36×36 initial network size, the final GLISSOM maps closely match those of LISSOM, yet take much less time and memory to simulate. Computation time increases smoothly as larger initial networks are used, allowing a tradeoff between accuracy and time. However, accurate maps are obtained substantially faster than with LISSOM. As long as the initial networks are small compared with the final maps, memory usage is bounded by the size of the final maps.

These results validate the hypothesis that a coarse approximation suffices for the early iterations in LISSOM. Early in training, only the large-scale organization of the map is important; using a smaller map for this stage does not significantly affect the final results. Once the large-scale structure settles, individual neurons become more selective and differentiate from their local neighbors; a denser map is required so that this detailed structure can develop. Thus, GLISSOM uses an appropriate map size for each stage in self-organization, in order to model development faithfully while saving simulation time and memory.

15.5 Scaling to Cortical Dimensions

The maps studied so far in this book represent only a small region of V1 and have a limited connectivity range. The results presented in this chapter make it possible to obtain a rough estimate of the resource requirements needed to approximate the full density, area, and connectivity of the visual cortex of a particular species. As discussed in the next section, with such a simulation it will be possible to study phenomena that require the entire visual field or the full cortical column density and connectivity. Calculating the full-scale parameter values is also useful because it can help tie the parameters of a small model to physical measurements. For instance, once the relevant scaling factors are calculated, the connection lengths, receptive field sizes, retinal area, and cortical area to be used in a model can all be derived from measurements in a biological preparation. Conversely, where such measurements are not available, GLISSOM parameter values that result in realistic behavior constitute predictions for future experiments.

In this section, the resource requirements and key LISSOM parameters are computed that make it possible to simulate the full human primary visual cortex at the

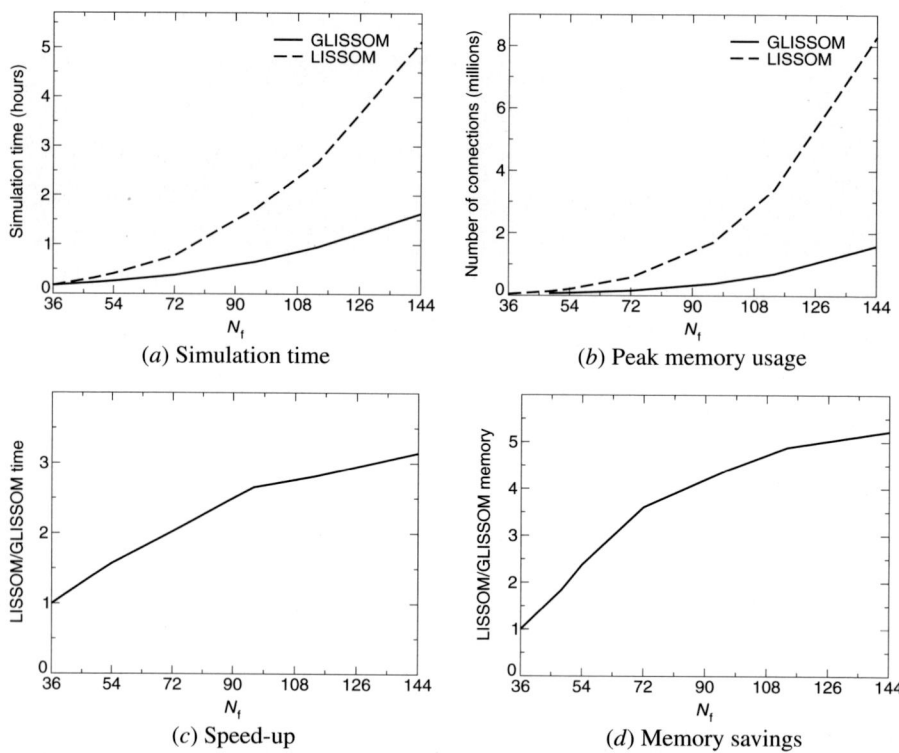

(*a*) Simulation time   (*b*) Peak memory usage

(*c*) Speed-up   (*d*) Memory savings

Fig. 15.10. Simulation time and memory usage in LISSOM vs. GLISSOM. Computational requirements of the two methods are shown as a function of the network size. In the LISSOM simulations the network had a fixed size N_f, as indicated on the x-axis; in GLISSOM the initial size was $N_o = 36$ and the final size N_f as indicated on the x-axis. Each point represents one simulation; the variance between multiple runs was negligible (less than 1% even with different input sequences and initial weights). (*a*) Simulation time includes training and all other computations such as plotting, orientation map measurement, and GLISSOM's scaling steps. The simulation times for LISSOM increase dramatically with larger networks, because larger networks have many more connections to process. In contrast, because GLISSOM includes fewer connections during most of the self-organizing process, its computation time increases only modestly for the same range of N_f. (*b*) Memory usage consist of the peak number of network connections required for the simulation; this peak determines the minimum physical memory needed when using an efficient sparse format for storing weights. LISSOM's memory usage increases very quickly as N_f is increased, whereas GLISSOM is able to keep the peak number low; much larger networks can be simulated on a given machine with GLISSOM than with LISSOM. (*c,d*) With larger final networks, GLISSOM results in greater speed-up and memory savings, measured as the ratio between LISSOM and GLISSOM simulation time and memory usage.

column level. These calculations apply to models with one computational unit per cortical column, and at most one long-range connection between units.

Most of the simulations in the preceding sections include a map whose global features match approximately a 5 mm × 5 mm = 25 mm^2 patch of macaque V1 (e.g. compare Figure 15.9 with Figure 9.4a). The full area of human V1 has been estimated at 2400 mm^2 (Wandell 1995), and so a full-size simulation would need to have an area about 100 times as large as the current simulations.

The full density of a column-level model of V1 can also be calculated. The total number of neurons in human V1 has been estimated at 1.5×10^8 (Wandell 1995). Each cortical unit in LISSOM represents one vertical column, and the number of neurons per vertical column in primate V1 has been estimated at 259 (Rockel, Hiorns, and Powell 1980). Thus, a full-density, full-area, column-level simulation of V1 would require about $1.5 \times 10^8/259 \approx 580,000$ column units in total, which corresponds to LISSOM parameter $N = \sqrt{580,000} \approx 761$.

More important than the number of units is the number of long-range lateral connections, because they determine the simulation time and memory requirements. Lateral connections in V1 can be as long as 8 mm (Gilbert et al. 1990), but the connections in the LISSOM models so far have been shorter in order to make them practical to simulate. For instance, scaling the parameters used in the previous sections to the full density would result in an inhibitory radius $r_I = 18$, but matching the full 8 mm connection length at full density would require $r_I = 8 \times 761/\sqrt{2400} \approx 124$. This larger radius requires about 45 times as much memory as $r_I = 18$, because the memory usage increase with the area enclosed by r_I. In the current LISSOM implementation, all of these possible connections must be stored in memory, so supporting such long connections would need enough memory for $761^2 \times (2 \times 124 + 1)^2 \approx 4 \times 10^{10}$ connections in V1. Thus, simulating the entire V1 at full density would require about $4 \times 4 \times 10^{10}/2^{30} \approx 150$ gigabytes of RAM (assuming 4 bytes per connection). Such simulations are currently possible only on large supercomputers.

In contrast, because all possible final connections do not need to be included in the initial network, GLISSOM can make use of a sparse lateral connection storage format that takes much less memory, and correspondingly less computation time. The memory required depends on the number of connections that remain active after self-organization, which in current GLISSOM simulations is about 15%. As the radius r_I increases, this percentage decreases quadratically, because long-range connections extend only along the preferred orientation of the neuron and not in all directions (Bosking et al. 1997; Sincich and Blasdel 2001). Thus, for the full-scale simulation, about $15\% \times 18^2/124^2 \approx 0.3\%$ of the connections would have to be included. Under these assumptions, the memory requirement reduces to approximately $0.003 \times 150 \times 1024 \approx 460$ MB. Thus, with GLISSOM it is possible to simulate the entire V1 at the level of laterally connected cortical columns on existing desktop workstations.

15.6 Discussion

The parameter scaling experiments in Section 15.2 showed that the LISSOM scaling approach is valid over a wide range of spatial scales. The GLISSOM experiments in Section 15.4 in turn showed that the equations can be used to reduce simulation time and memory requirements significantly and thereby make the study of large-scale phenomena tractable. The method is not specific to LISSOM; it should apply to most other models with specific intracortical connectivity, and it can be adapted to those with more abstract connectivity, such as a DoG interaction function. The growth process of GLISSOM should provide similar performance and memory benefits to most other densely connected models whose peak number of connections occurs early in training. Essentially, the GLISSOM method allows a fixed model to be turned into one that grows in place, by using scaling equations and an interpolation algorithm.

On the other hand, models that do not shrink an excitatory radius during self-organization, and therefore do not have a temporary period with widespread activation, benefit less from GLISSOM. For such models, it may be worthwhile to consider a related approach, whereby only the lateral connection density is gradually increased, instead of increasing the total number of neurons in the cortex. Such an approach would still keep the number of connections (and therefore the computational and memory requirements) low, while keeping the large-scale map features (such as the distance between orientation patches) constant over the course of self-organization.

The GLISSOM method is most effective when it can be initiated with very small maps. However, as was discussed in Section 15.2.3, the self-organizing process requires that the neighborhood radii are at least 1.0, even though the sampling limits imposed by the Nyquist theorem would allow smaller maps. One way to get around this limitation would be to approximate smaller radii with a technique similar to anti-aliasing in computer graphics. Before a weight value is used in Equation 4.7 at each iteration, it would be scaled by the proportion of its corresponding pixel's area that is included in the radius. Because the mask would only apply to small radii, the added computational overhead would not be large. This technique should permit smaller networks to be simulated faithfully even with a discrete grid.

Apart from their application to simulations, the parameter scaling equations provide insight into how structures in the visual cortex differ between individuals, between species, and during development. In essence, the equations predict how the biophysical correlates of the parameters differ between any two similar cortical regions that differ in size. The discrepancy between the actual parameter values and those predicted by the scaling equations can help explain why different brain regions, individuals and species will have different functions and performance levels.

For instance, Equation 15.3 and the simulation results suggest that learning rates per connection should scale with the total number of connections per neuron. Otherwise, neurons in a more densely connected brain area would have significantly more plasticity, which (to our knowledge) has not been demonstrated. Consequently, unless the number of synapses per neuron is constant, the learning rate must be regulated at the level of the whole neuron rather than being a property of individual synapses. This principle conflicts with assumptions implicit in most incremental Hebbian models that specify learning rates for individual connections directly. Future experimental work will be needed to determine whether such whole-neuron regulation of plasticity does occur, and if not, whether more densely connected regions also are more plastic.

Similarly, Equation 15.3 suggests that pruning is not based on an arbitrary fixed threshold, but depends on the total number of connections to a neuron. In the model, this behavior results from the divisive weight normalization, which ensures that increasing the number of connections makes each one weaker (as was discussed in Section 3.3, such normalization is consistent with recent biological results on neuronal regulation). If the pruning threshold were not normalized by the number of inputs, a fixed value that prunes e.g. 1% of the connections for a small cortex would prune all of the connections for a larger cortex. These findings provide independent computational and theoretical support for earlier experimental evidence that pruning is a competitive process, and not one based on a fixed threshold (Purves 1988).

The scaling equations are also an effective tool for making cross-species comparisons, particularly between species with different brain sizes. In effect, the equations specify the parameter values that a network *should* implement if it is to have similar behavior to a network of a different size. However, as pointed out by Kaas (2000), different species do *not* usually scale faithfully, probably due to geometrical, metabolic, and other restrictions. As a result, as V1 size increases, the lateral connection radii do not increase as specified in the cortical density scaling equations, and processing becomes more and more local. Kaas (2000) proposed that such limitations on connection length may explain why larger brains, such as human and macaque, are composed of so many visual areas, instead of just expanding the area of V1 to achieve the same functionality (see also Catania et al. 1999). The scaling equations in LISSOM provide a concrete platform on which to measure the tradeoffs between a small number of large visual areas and a large number of small, hierarchically connected visual areas.

15.7 Conclusion

The scaling equations and the GLISSOM method allow detailed laterally connected cortical models like LISSOM to be applied to much more complex, large-scale phenomena. Using GLISSOM, it should be possible to model all of V1 at the column level with desktop workstations. These methods also provide insight into how the cortical structures compare in brains that differ widely in size. Thus, the scaling equations and GLISSOM can help explain brain scaling in nature as well as provide a method for scaling up computational simulations of the brain.

16

Discussion: Biological Assumptions and Predictions

The experiments presented in this book provide computational support for the hypotheses presented in Chapter 1 about cortical structure, development, and function. To be well founded, a computational model should make only the assumptions that are necessary, and those assumptions should be compatible with biological evidence. Second, the model should suggest a realistic set of biological and psychological experiments that can verify or refute it. In this chapter, the assumptions underlying the self-organization, genetically driven development, and temporal coding in the LISSOM model are evaluated, and predictions are made based on the simulations. The next chapter focuses on computation, reviewing important new directions for future work.

16.1 Self-Organization

Many of the fundamental assumptions of the LISSOM model, such as the computation of the input activity as a weighted sum, the sigmoidal activation function, and Hebbian weight adaptation with normalization, are common to most neural network models. As was discussed in Chapter 3, their computational and biological validity has been examined in detail by other researchers. However, there are steps in the LISSOM self-organizing process that make it more complex than the usual abstract model of self-organizing maps. These are: (1) recurrent lateral interactions, (2) adapting lateral connections, and (3) independent multiplicative normalization for each connection type. The self-organizing process in LISSOM is also based on a number of assumptions that were made out of computational necessity and have not yet been fully characterized experimentally. Those are: (4) short-range excitation and long-range inhibition, (5) connection death, and (6) parameter adaptation.

In this section, these assumptions will be evaluated based on how biologically valid and crucial they are for the self-organization phenomena discussed in this book. Assumptions necessary for genetically driven development and for functional effects such as grouping will be discussed in later sections.

16.1.1 Recurrent Lateral Interactions

Perhaps the most important difference between LISSOM and other self-organizing map models is the settling of activity through recurrent lateral interaction. This process affects self-organization by modifying the activation patterns in two ways:

1. It concentrates activity in the maximally active regions of the network, suppressing activity elsewhere.
2. It decorrelates activation across the network through inhibitory lateral connections.

The first effect generalizes the winner-take-all process of SOM and other abstract models. Instead of finding one winner and adapting the neurons in a single continuous neighborhood around it, recurrent interaction selects a set of maximally active regions. This soft winner-take-all process is necessary for afferent connections to self-organize efficiently: Had there been no such process, all neurons would adapt for all inputs and the afferent weights would become the same for all neurons with the same anatomical RF. The explicit lateral interactions also eliminate the search mechanism for finding the winner which is used in SOM, and make it possible for all neurons to compute in parallel.

The second effect, decorrelation, is crucial for efficient coding of input. By decorrelating neural activity to the same extent that they are known to be correlated, the redundancy of the cortical activity is reduced most efficiently, as was discussed in Section 14.2. Without such recurrent lateral interactions, the neurons would act much like linear filters; with lateral interactions, the network activity is concentrated in a set of best-responding neurons. In this way, the input image is represented as a sparse coding of the primary visual features.

The cortical plasticity results of Chapter 6 also depend crucially on the lateral interactions. For example, the dynamic changes in the RF sizes of unstimulated neurons occur because there is less lateral inhibition from the surround. In the same way, the perceptual shift observed after the dynamic RF experiment results from reduced lateral inhibition. The extent of lateral excitation also determines the range over which cortical neurons can adapt and compensate for lesions.

In addition to self-organization, the recurrent lateral interactions affect the visual function of the map. They result in the tilt aftereffect discussed in Chapter 7, and possibly other aftereffects and illusions as well. They modulate synchronization across spatially separate regions, thus contributing to perceptual grouping as described in Part IV. Such functional aspects of recurrent lateral interactions will be discussed in Sections 16.3, 16.4, and 17.2.

16.1.2 Adapting Lateral Connections

As has been shown using the SOM and other self-organizing models, afferent receptive field structures such as those for ocular dominance and orientation can self-organize even with fixed lateral interactions. Similarly, receptive fields can be dynamic and cortex can reorganize after retinal lesions even in models with recurrent, but non-adapting lateral interactions. What role do the adapting lateral connections serve?

As was discussed in Section 14.2, self-organized inhibitory long-range lateral interactions are most important for eliminating redundant activity and coding visual input efficiently. Self-organization produces a variety of receptive fields for each retinal location, and the RFs are organized in a smoothly varying fashion across the cortex. Therefore, each input causes initial activity in many neurons, and most of this activity is redundant. An efficient coding can be achieved by retaining activity in only those units that are best tuned to the features of that retinal region. Such an encoding can be achieved by decorrelating activity through lateral connections, making the feature representations of the visual input across the cortex more independent. Therefore, lateral inhibition is necessary between the receptive fields, and its strength should be organized according to the correlations between them. This type of organization is what LISSOM achieves by adapting the inhibitory long-range lateral connections. Interestingly, perceptual phenomena such as the tilt aftereffect emerge as a side effect of this process.

On the other hand, adapting long-range lateral excitation is crucial for perceptual grouping. The correlations they learn implement the Gestalt principles that allow the network to decide which elements in the input should be bound together into a coherent object. Because maps are locally smooth, it is not as important to adapt the short-range excitatory connections, although such adaptation also helps improve the efficiency of coding. These connections cumulate the activity of nearby units, amplifying their responses. This process should also depend on activity correlations between neurons: Similar neurons should contribute more excitation than those that are dissimilar. If the lateral excitatory connections are also self-organized, the weighting of lateral activity will be matched to these correlations. Such correctly weighted excitation will produce appropriately sized activity bubbles, and minimum spurious activity.

Importantly, lateral connections adapt synergetically with the afferent connections. The afferent organization determines the initial pattern of activity, and the afferent and lateral organizations together determine the final pattern after settling. The settled patterns in turn determine the weight changes by the Hebbian rule. If one of these connection types were to be fixed while the other type develops, the resulting connection patterns would be different. Therefore, the afferent and lateral connections adapt together in LISSOM, and form matching structures.

In biology, it is not yet clear whether afferent and lateral connections develop in a similar synergetic fashion. However, experimental evidence suggests that they develop at least about the same time in mammals. In the cat visual cortex, for example, lateral connections proliferate exuberantly and rapidly elongate in the first postnatal week, but they do not grow very much afterward (Callaway and Katz 1990; Katz and Callaway 1992). After the first week, the connections slowly refine into clusters by synaptic elimination and reach an adult-like organization at the end of 6 weeks. Simultaneously, afferent connections organize into ocular dominance and orientation columns: Rough ocular dominance and orientation columns are visible from about 2 to 3 weeks after birth and are adult-like also at about 6 weeks. These observations suggest that in the cat neocortex, a rough lateral connection structure emerges first, bootstraps self-organization, and gradually gets refined into connections that selectively associate neurons with similar properties. However, establishing the details of this process will require additional experiments, both in cats and in other mammals such as primates.

16.1.3 Normalization of Connections

An important part of Hebbian learning is a regulatory process that keeps the connection weights from increasing without bounds (Section 3.3). In LISSOM, the different kinds of connections are assumed to be regulated independently and multiplicatively.

The different connection types must be adapted independently because they self-organize from different types of activity correlations. The afferent connections learn correlations between the cortex and the receptors, the short-range lateral excitatory connections learn correlations between near neighbors within the cortex, the long-range inhibitory connections learn redundancies between distant neurons, and the long-range excitatory connections (in PGLISSOM) learn correlations within coherent objects. If the connection weights are all normalized together, these different types of correlations will influence all the weights, and interfere with self-organization.

As was discussed in section 3.3, there are two common ways to normalize the synaptic weights in self-organizing models. In LISSOM, the total weight of each type of connection is kept constant multiplicatively: After the weights are adapted, each weight is scaled by the total weight. An alternative way would be to normalize subtractively: After the weights are adapted, the increase in total weight divided by the number of weights is subtracted from each weight (e.g. Goodhill 1993; Miller et al. 1989). Subtractive normalization would not work well in LISSOM. The reason is that it always results in some of the weights increasing to a maximum value and others decreasing to zero: No intermediate weight values develop (Miller and MacKay 1994). Therefore, lateral connections will not store the precise correlations between neurons, and afferent connections will not develop precise representations of the input features. Furthermore, for stability, synaptic weights become fixed once they reach their maximum values. Therefore, gradual reorganization such as that observed with retinal and cortical lesions cannot take place. Such representations would not be as useful in visual coding and processing as the precise, continuous weights obtained through multiplicative normalization.

The form of the multiplicative normalization used is not crucial: When the inputs are relatively regular and laid out on a retina, either the constant sum of weights or the constant vector length normalization can be used (Section 14.4). Self-organization also works similarly whether the normalization is done postsynaptically, i.e. over incoming connections as in the firing-rate LISSOM models (Section 4.4.1), or presynaptically over outgoing connections as was done in PGLISSOM (Section 11.4). As long as the normalization is done separately for each weight type, suitable parameters can be found for either case, and organized receptive fields and lateral interactions will develop.

However, the site of normalization is important for grouping: Presynaptic normalization makes it easier for the model to segment different objects. In this case, the postsynaptic cell receives inputs through weights that are each scaled differently, according to the outgoing weights of each presynaptic cell. Even relatively low activity can result in a large weight, and the postsynaptic cell can be more sensitive to small changes in the input. In segmentation tasks, small differences in the activation levels must be magnified, and presynaptic normalization makes this process easier.

In postsynaptic normalization, all incoming weights are scaled by the same value. The inputs are treated more equally than in presynaptic normalization, and the behavior of the neuron becomes slightly more stable. This property makes postsynaptic normalization preferable for models that do not include grouping. Biological data to date do not rule out either form of normalization, and they could even coexist. Computationally, connection weights could be modeled as a product of two factors, the postsynaptic and the presynaptic weight, each normalized separately (Leow 1994; Leow and Miikkulainen 1997). The different normalization processes could interact, and depending on the input, one or the other might dominate; they could also be specific to only the excitatory or inhibitory synapses. Future research, both experimental and computational, is necessary to verify the precise form of normalization in biological systems.

16.1.4 The Role of Excitatory and Inhibitory Lateral Connections

In order for a LISSOM network to self-organize, the net lateral interactions between strongly responding units must be inhibitory at long ranges and excitatory at short ranges. Such lateral interactions are essential for concentrated activity bubbles to form and for self-organization to take place (Section 4.2.3). They are also a key ingredient common to most self-organizing models (Section 3.4.1; Miller 1994; Miller et al. 1989; von der Malsburg 1973).

The original biological inspiration for such interactions comes from the neural architecture of the retina, where long-range inhibition is well established. In the retina, lateral inhibition enhances contrast, especially at edges and boundaries of objects. Such interactions have been shown to produce an efficient coding of the retinal image, decorrelating and reducing redundancies in the photoreceptor activities (Atick 1992; Atick and Redlich 1990). Numerous researchers have proposed that lateral inhibition is a general principle of perceptual systems, and may occur similarly in the cortex (e.g. Blakemore et al. 1970).

Measurements of the activity levels of strongly stimulated cortical neurons indeed support the idea of long-range lateral inhibition and local excitation in the cortex (Grinvald et al. 1994; Sceniak, Hawken, and Shapley 2001). For instance, Grinvald et al. (1994) performed optical imaging experiments visualizing large-scale cortical activity. The responses to two stimuli were compared: a surround stimulus consisting of a high-contrast grating with a square hole (or mask) at the center, and a center stimulus consisting of three small high-contrast bars that fit within the masked region. When the surround and center stimuli were presented together, the center region was substantially less active than when the center stimulus was presented alone, indicating that the surround was inhibiting the center area. Similarly, mapping excitatory and inhibitory regions using high-contrast sine gratings shows that surround influences tend to be excitatory locally but inhibitory at longer ranges (Sceniak et al. 2001).

However, the long-range interactions in the cortex are more complex than the above experiments might suggest. Anatomical surveys show that 80% of the synapses of long-range lateral connections connect directly between pyramidal cells, which are thought to make excitatory synapses only (Gilbert et al. 1990; Hirsch and Gilbert 1991; Kisvárday and Eysel 1992; McGuire et al. 1991). The other 20% of the connections target inhibitory interneurons, which in turn contact the pyramidal cells, and thus represent inhibitory connections. Even though the inhibitory connections are outnumbered, the net effect at the columnar level has been difficult to establish with anatomical studies. For instance, the interneurons often synapse at regions such as the soma, where their effects may be larger than those of excitatory neurons, which synapse farther out on the dendrites (Gilbert et al. 1990; McGuire et al. 1991). Thus, the known anatomy is compatible with both long-range excitation and long-range inhibition.

Electrophysiological evidence indicates that in fact the same connections can have either excitatory *or* inhibitory effects, depending on how strongly neurons are activated (Hirsch and Gilbert 1991; Weliky et al. 1995; see Angelucci, Levitt, and Lund 2002 for a review). The balance between these two types of connections depends on image contrast: The incoming lateral connections of a neuron have a mildly excitatory influence when the surrounding area is activated weakly (as it would be by a low-contrast stimulus) and a strongly inhibitory effect when the surround is activated strongly (as it would be by a high-contrast stimulus; Hirsch and Gilbert 1991; Weliky et al. 1995). Thus, for high contrast stimuli, as in the Grinvald et al. (1994) study, the interactions are usually inhibitory, even though the anatomical connections are primarily between excitatory neurons.

The details of the cortical circuit implementing this contrast dependence remain unclear. One early proposal was that the inhibitory interneurons are inherently more effective than the direct excitatory connections, but have a higher threshold for activation (Sillito 1979). At very low stimulus levels, the excitatory effects would predominate, but at high levels the inhibitory interneurons would become progressively more active and eventually would suppress the response of the target cell. More recently, Douglas, Koch, Mahowald, Martin, and Suarez (1995) proposed a detailed circuit based on recurrent short-range excitatory lateral connections. They showed how the inhibitory connections can dominate the response even though they are fewer in number. Simplified versions of such circuits have been modeled by Stemmler et al. (1995) and Somers et al. (1996). They propose that these complex connections make it easier to detect weak, large-area stimuli while suppressing spatially redundant activation for strong stimuli. Figure 16.1 shows one such circuit that could give rise to contrast-dependent effects.

The two-layer (SG) model of cortical columns in PGLISSOM can be seen as a column-level abstraction of such circuits. In PGLISSOM, SMAP has long-range inhibition and short-range excitation and drives the self-organizing process; in GMAP,

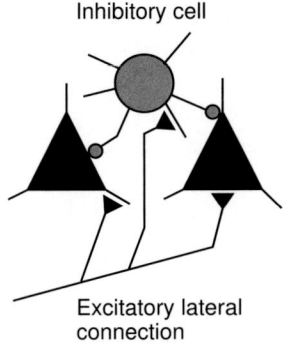

Fig. 16.1. Local microcircuit for lateral interactions. This circuit can potentially explain how lateral interactions can depend on the input contrast. A long-range lateral connection from an excitatory cell contacts two pyramidal excitatory cells (large black triangles) and one inhibitory cell (large circle). The inhibitory cell has a high threshold for activation, but strongly inhibits the pyramidal cells when activated. Weak excitation activates the pyramidal cells monosynaptically, and does not activate the inhibitory cell. However, strong excitation activates the inhibitory cell as well, causing a net inhibitory effect. In this manner, a single incoming excitatory long-range lateral connection could have inhibitory effects for strong stimuli (e.g. high-contrast patterns), and excitatory effects for weak stimuli. The SG model of cortical columns in PGLISSOM produces a similar effect, and can be seen as an abstraction of this circuitry at the columnar level. The excitatory synapses (shown as small triangles) adapt by Hebbian learning, but the inhibitory synapses (shown as small circles) are fixed in strength. Such learning can be approximated by direct Hebbian excitatory and inhibitory connections, as is done in PGLISSOM. Adapted from Weliky et al. (1995).

both connections have long range and implement grouping. When the combined effects of these interactions are measured on a cortical column in PGLISSOM, excitatory effects are found to dominate with low-contrast inputs, and inhibitory effects with high-contrast inputs, as they do in the cortex.

Importantly, self-organization is primarily driven by high-contrast inputs in PGLISSOM, and most likely in animals as well. Low-contrast patterns rarely cause a significant response because of the neurons' nonlinear activation function. The resulting synaptic changes are small and do not significantly affect the learning process. Thus, the simplifying assumption, common to all LISSOM models, that the long-range lateral interactions are primarily inhibitory during self-organization, is well founded. The GMAP layer can be omitted from models that do not focus on perceptual grouping; the remaining network includes short-range excitation and long-range inhibition, which is the necessary connectivity for proper self-organization to occur.

For computational convenience, the long-range inhibitory interactions are represented in all LISSOM models as direct connections instead of connections through interneurons (such as those in Figure 16.1). Because the interneurons can be brought to firing threshold rapidly and repeatedly without fatigue (Thomson and Deuchars 1994), they introduce only a small delay in the inhibitory process and can be approximated functionally by direct connections. Also, while there is no clear evidence for Hebbian strengthening of direct inhibitory synapses in the cortex, the inhibitory effects can be modified through Hebbian strengthening of excitatory synapses onto the inhibitory interneurons. Therefore, direct Hebbian learning is a valid abstraction of adapting lateral inhibition in the cortex, resulting in more parsimonious models with equivalent behavior.

16.1.5 Connection Death

An important component of self-organization in LISSOM is the pruning of unused lateral connections (Section 4.4.2). This process is useful computationally, but it is also well motivated biologically.

More than half of the long-range lateral connections in the neocortex are estimated to disappear during development (Callaway and Katz 1990; Katz and Callaway 1992; McCasland et al. 1992; Purves and Lichtman 1985). In the visual cortex, structured lateral connectivity emerges from an initially unstructured organization after axons projecting to incorrect targets die off (Callaway and Katz 1990). Which connections survive depends on how often they are active. The reason could be that synapses are nourished in proportion to their strength. Once formed, a weak synapse may survive only for a limited time without sufficient trophic factors.

The onset of connection death in LISSOM, t_d, models this survival time. Synapses whose strength falls below the survival threshold are not eliminated immediately, but only if they stay below the threshold until t_d. Even in prolonged self-organization, short-term fluctuations in synaptic strength will not cause inappropriate connection death in LISSOM. The connections are pruned at well-spaced intervals Δt_d, instead of eliminating them as soon as they become weak. As was seen in Section 14.2.3, the resulting patchy lateral connections are crucial in forming a sparse, redundancy-reduced visual code.

Connection death is also important for perceptual grouping. The long-range excitatory connections in PGLISSOM are pruned after training so that only the strong ones remain (Section 11.4). The resulting patchy connectivity represents activity correlations in the input, implementing the Gestalt principles that drive the grouping process. They also make it possible to adapt the lateral interactions dynamically during performance. Since the connectivity is patchy and stable, the strengths can be modulated at a fast time scale without changing their overall effect. Although not strictly necessary for grouping, such fast dynamic adaptation results in more robust synchronization (Baldi and Meir 1990; von der Malsburg 1981, 2003; Wang 1996).

An important side effect of connection death is that it limits how extensively the network can adapt to changes in internal and external inputs. For example, before the connections are pruned, the network can recover function even after relatively large cortical damage, but such plasticity is limited in the pruned adult network (Sections 6.4.2 and 16.4.4). Also, after the connections have been pruned to represent activity correlations in the input, if those correlations change, it will be difficult for the network to adapt, as it will be for animals (Sections 8.1 and 9.4.2). Connection death can therefore be seen as a process that makes the computational system more efficient, at the expense of the ability to adapt to changes.

16.1.6 Parameter Adaptation

As was discussed in Sections 4.2.3 and 4.4.3, consistent lateral inhibition is necessary for the self-organizing process, and gradually reducing the excitatory radius and gradually making the neurons more difficult to activate allows forming more regular maps. These mechanisms were included primarily for computational reasons, allowing the maps to self-organize even from very disordered starting points. Biological maps have more order initially, and thus may not require these processes.

However, biological counterparts do exist for the parameter adaptation processes in LISSOM. They represent maturation based on time and trophic factors, and can be used to establish a maturation schedule for LISSOM models independently of input-driven self-organization. Such maturation allows studying deprivation and critical periods, as reviewed in Section 2.1.4 and implemented in Sections 9.4 and 13.4.

For instance, several lines of evidence suggest that there is more net excitation during early development than later. First, immature neurons are connected by a network of excitatory gap junctions that are not seen in the adult (Sutor and Luhmann 1995). Second, cross-correlation studies in the primary visual cortex of the kitten showed that net lateral excitation extends to distances of 1 mm in the first 2 to 3 weeks (after compensating for cortical growth), and decreases to less than 400 μm by the seventh to ninth week (Hata et al. 1993). Third, direct studies of synaptic connections in the ferret visual cortex found that local excitatory synaptic connections increase rapidly in number and extent at the time of eye opening, and subsequently prune down to much more local connectivity (Dalva and Katz 1994). Thus, animal cortex may also have wider excitatory activations in early stages.

To fine tune the LISSOM map, the activation threshold for neurons is gradually raised so that neurons become more difficult to activate. Interestingly, cortical neurons also become harder to trigger electrically as they mature. Immature neurons have higher input resistances, longer time constants and more linear relationships between applied current and voltage than do mature cells (Prince and Huguenard 1988). Thus, older cells require more electrical stimulation to activate. These effects may be due to homeostatic plasticity processes, which tend to normalize the frequency of neuronal firing over time (Turrigiano 1999). That is, immature neurons have RFs that are not yet well developed and are not yet a good match to the statistics of visual scenes, and thus homeostatic mechanisms may lead them to fire more easily (for a given amount of electrical stimulation). Older neurons have well-tuned RFs, and can thus require a good match before responding. In LISSOM, these processes are approximated by gradually raising the sigmoid threshold. Extending LISSOM to include automatic mechanisms for regulating firing probability is discussed in Section 17.1.1.

For simplicity, LISSOM includes constant levels of inhibition throughout the simulation. However, the role of inhibition in animals is more complex. First, the neurotransmitter GABA has an excitatory effect on postsynaptic cells in early development, in contrast to its inhibitory effects in the adult. Second, direct electrical stimulation does not create inhibitory responses until about 10 days after birth in the rat (Sutor and Luhmann 1995). (Presumably, inhibition could be evoked before birth in animals such as monkeys with a longer gestation, but this possibility has not yet been studied.) Assuming that the homeostatic mechanisms mentioned above also apply to inhibition, it would be possible to extend LISSOM with an automatic mechanism for introducing inhibition. Once cells begin to activate regularly, a feedback mechanism could automatically increase inhibition to balance excitation. Such a mechanism would reflect the biological process, while allowing enough inhibition to initiate the self-organizing process.

16.2 Genetically Driven Development

In Part III, the hypothesis that input-driven self-organization is based on internally generated patterns as well as external visual inputs was tested in computational simulations. The first assumption was that simple patterns such as retinal waves could drive the early self-organization of V1. Second, higher levels could be similarly organized assuming more complex patterns could be generated in the brainstem as PGO waves and propagated during REM sleep. Third, pattern generation would have been discovered by evolution because it makes it easier to construct complex adaptive systems than hard wiring or general learning. These assumptions are discussed in more detail in this section.

16.2.1 Self-Organization of V1

The shape and distribution of internally generated activity patterns determine how the maps and connections develop in HLISSOM (Chapter 9). Although a variety of such patterns have been detected experimentally, retinal waves (Section 2.3.3) are currently the most likely cause for the early organization of V1. In general, such patterns have to satisfy four main requirements.

First, there needs to be a mechanism for generating internal patterns consistently while V1 develops. Such a mechanism has indeed recently been mapped out in ferrets (Butts, Feller, Shatz, and Rokhsar 1999; Feller 1999; Feller, Butts, Aaron, Rokhsar, and Shatz 1997): retinal waves emerge from the spontaneous behavior of neurons connected together by gap junctions. In essence, one neuron fires randomly, which excites its neighbors, and then regulatory mechanisms step in to keep the activity localized. The result is an activity spot that appears randomly, drifts, and disappears. It is likely that other pattern generation mechanisms will be found in other species once their developing sensory systems are studied in detail. A variety of such mechanisms are already known to exist in the motor systems of different vertebrate and invertebrate species (see Marder and Calabrese 1996 for a review).

Second, the internally generated activation needs to drive the activation of neurons in V1. Which sources of activity actually reach the developing V1 neurons is not yet known. However, the retinal wave patterns are known to occur before orientation maps and selectivity can be measured in V1 (see Issa et al. 1999; Wong 1999 for reviews), so they are correctly timed for this role. Further experiments will be needed to verify whether the retinal waves produce significant neural responses in V1 while orientation maps develop, or whether other sources of activity are more prominent at this time.

Third, the developing V1 needs to perceive the internally generated patterns as oriented. So far, no such patterns have been observed. For example, the retinal waves in the ferret are approximately as wide as the V1 receptive fields in the adult animal (Wong et al. 1993); if they were relayed directly to V1, they would activate all the inputs to many of the cortical cells, and would not appear oriented to most of them. However, as the simulations in Section 9.2 showed, the center–surround processing in the ON and OFF channels of the retinal ganglia and LGN could emphasize the edges of the retinal wave patterns enough to give them a distinct orientation. Alternatively, neurons in the LGN may respond only transiently, to the first appearance of activity in each part of the retinal wave, which again would make the patterns seen by V1 more like edges than like large activated areas. Although it is not yet known how the ganglia and the LGN respond to the retinal waves, in either of these cases the broad, internally generated patterns could drive the development of orientation maps.

Fourth, the internally generated patterns need to activate the ON and OFF channels differently. If the same patterns appear in both channels, V1 will not be able to learn the center–surround relationship between them, and will not be able to process natural image input. While the origins of retinal wave patterns are still not fully understood, they do result in different activations in the ON and OFF neurons in the retina (Myhr et al. 2001). It is possible that the activation is generated before it branches into ON and OFF channels, or else activity may be generated separately in each channel. In either case, such a difference should be enough to drive the development of V1 neurons, as was shown in Section 9.2.

Because retinal waves are consistent with these computational requirements and little is known about the properties of other spontaneous activity, they are currently the most likely candidate for prenatal self-organization of V1 orientation maps. Other sources of patterns could contribute to this process in addition or even instead of retinal waves, provided they satisfy the requirements above.

16.2.2 Self-Organization of Higher Levels

The face-selective area simulations in Chapter 10 rely on similar assumptions about the PGO waves (Section 2.3.4) as the V1 simulations do on retinal waves. Like retinal waves, PGO waves are not the only possible cause for prenatal self-organization, but they are the most likely cause for higher levels, given the computational requirements and our current understanding of internally generated patterns. These assumptions are evaluated in this section.

First, a neural mechanism must exist for generating spatial configurations of activity similar to the three-dot pattern (Section 10.2.6). Such activity might occur through a variation of the mechanisms involved in retinal waves. A relatively large circular area could be activated, which could then break up into individual activity bubbles through mutual inhibition.

In preliminary experiments using LISSOM it turned out relatively easy to generate such multi-spot activity using short-range excitation and long-range inhibition, starting from an initial larger set of activated neurons. Biasing the patterns to produce vertically oriented sets of spots, with more on top than below, is slightly more difficult, but can be achieved using a gradient of activation threshold or excitation strength across the dimension representing vertical in the cortex. Such a gradient makes it more likely for bubbles to form in the upper portion of an activated region than in the lower. Such mechanisms might be implemented in the developing brainstem, giving it the capability to generate spatial patterns that lead to newborn face preferences.

The second assumption is that such patterns can be propagated to the higher levels of the visual system during development. As described in Section 2.3.4, PGO waves generated during REM sleep are a good candidate for this process. However, there are a large number of theories for other functions of REM sleep (Harnad, Pace-Schott, Blagrove, and Solms 2003; Horne 1988; Moorcroft 1995; Rechtschaffen 1998), and it is important to evaluate whether pattern generation is consistent with them.

The most prominent theory is that REM sleep helps consolidate long-term episodic memories: Such memories are believed to be stored temporarily in the hippocampus as they are acquired, then rehearsed repeatedly during REM sleep (Alvarez and Squire 1994; Qin, McNaughton, Skaggs, and Barnes 1997; Sejnowski 1995; Shastri 2002; Smith 1996; Wilson and McNaughton 1994). Rehearsal helps store the memories permanently in the cortex through long-term learning that interleaves and integrates the new memories with existing ones. In this way an organism can quickly acquire new memories, without interference from older ones, and then store them more permanently through long-term learning. Evidence for the memory consolidation theory comes from several types of experiments and observations (Smith 1996). First, damage to the human or animal hippocampus often results in retrograde amnesia, i.e. forgetting of recent, but not distant, memories (although other functions may be affected as well, such as probabilistic category learning and spatial representation and navigation; Fuhs, Redish, and Touretzky 1998; Hasselmo, Bodelón, and Wyble 2002; Hopkins, Myers, Shohamy, Grossman, and Gluck 2003; O'Keefe and Burgess 1996; Touretzky 2002). Second, REM sleep deprivation can act much like retrograde amnesia. If an animal is deprived of REM sleep soon after training it on a new task, it performs poorly on that task later; deprivation before the task or long after the task does not have as large an effect. Third, when animals are trained on new tasks or exposed to novel environments, they spend more time in REM sleep. The conclusion is that new episodic memories depend on the hippocampus and REM sleep until the memories are consolidated into the cortex.

However, over the past few years, alternative explanations for each of these phenomena have emerged. For instance, retrograde amnesia does not necessarily indicate that memory is being consolidated (Moscovitch and Nadel 1998; Nadel and

Moscovitch 1997). Such amnesia could result from an impaired process of storing memories permanently in the hippocampus just as well as it can from disrupted consolidation in the cortex. Similarly, depriving subjects of REM sleep causes significant stress and changes the behavior of the animal, which in turn can make learning more difficult even if memory is consolidated normally (Horne 1988; Rotenberg 1992). Finally, REM sleep may in fact increase after learning and in novel environments in order to *counteract* excessive learning (Jouvet 1998, 1999).

It is possible that it has been difficult to establish a clear role for REM sleep because it may serve a variety of functions. For instance, REM sleep might be a way for the nervous system to rehearse important sensory patterns, replaying them to strengthen connections and responses. This mechanism could be used both for rehearsing memories, as in memory consolidation, and for training the system with genetically controlled activity patterns, like the three-dot patterns in Part III. Such training could be useful for constructing the system initially, and it could be crucial for integrating environment-driven learning with the genetic biases. While it is not yet possible to refute or verify this hypothesis experimentally, internally generated activity patterns resembling PGO waves are known to exist during sleep even before waking experience (Marks et al. 1995). The crucial assumption of the HLISSOM model is that higher cortical areas can learn from such patterns just as they do from externally evoked activity.

The third assumption is that the patterns propagated as PGO waves have the shape that can drive self-organization of higher levels, such as the three-dot configuration that results in prenatal face preferences in HLISSOM. At this point, PGO waves and other types of spontaneous activity have not been studied as thoroughly as retinal waves. In particular, it is not yet known whether they consist of large, coherent patches of activity, and whether those patches are organized in configurations such as three-dot patterns. Recent advances in imaging equipment may make measurements of this type possible in the near future (Rector et al. 1997), allowing testing these assumptions directly.

As internally generated patterns are measured in more detail in the future, candidates other than PGO waves may emerge; it may also be possible to identify other areas and behaviors where prenatal self-organization might play a role. However, based on our current biological and computational knowledge, the PGO waves are an explanation at least for the newborn preferences for faces.

16.2.3 Evolving Complex Systems

The HLISSOM simulations show how pattern generation can be an effective way to develop a functional system. The underlying assumption of the whole approach is that at some point during vertebrate evolution, pattern generation proved to be a more effective way to construct advanced perceptual and cognitive abilities than direct hard wiring and general-purpose learning. It is important to understand why, both to gain insight into the biological observations, and potentially to generate better artificial learning systems.

One important reason is that with pattern generation, the desired outcome can be encoded independently of the architecture that performs the task. For instance, it is possible to specify that neurons should respond to three-dot stimuli independently of the actual hardware that implements face processing. Using pattern generation, evolution could insert a measured amount of bias (Turney 1996) into the general-purpose learning system without changing the architecture of the learning system itself. The system can then learn to become specific for faces, using mechanisms like those explored by Dailey and Cottrell (1999), even without specific genetic control. In short, the divide-and-conquer strategy of pattern generation would allow such an architecture to evolve and develop independently of the function, which may have been crucial for the rapid evolution and development of complex adaptive systems.

In terms of information processing, pattern generation combined with self-organization may represent a general way to solve difficult problems like face detection and recognition. Rather than meticulously specifying the final, desired individual, the specification need only encode a process for constructing an individual through interaction with its environment (Section 2.3.2; see Elman et al. 1996; Marcus 2003; Mareschal, Johnson, Sirois, Spratling, Thomas, and Westermann 2005a,b for related and alternative explanations). The result can combine the full complexity of the environment with a priori information about the desired function of the system. In the future, this approach could be used for engineering complex artificial systems for real-world tasks, e.g. handwriting recognition, speech recognition, and language processing. This idea is explored in more detail in Section 17.3.5.

16.3 Temporal Coding

Like self-organization depends on adapting lateral connections and development depends on internally generated patterns, perceptual grouping in LISSOM depends on temporal coding through synchronization. The main assumptions are that neural synchrony drives perceptual performance, low-level temporal codes can be understood at a higher level, and synchronization can be implemented together with self-organization in different layers of the visual cortex.

16.3.1 Synchrony as a Perceptual Representation

The assumption that synchronous neural activity represents coherent percepts originates from two kinds of experiments: Either input properties and neural synchrony are compared in animals (Eckhorn 1999; Eckhorn et al. 1988; Gray 1999; Gray et al. 1989; Gray and Singer 1987; Singer 1993, 1999; Singer and Gray 1995), or psychophysical performance and timing between input features are compared in humans (Fahle 1993; Lee and Blake 2001; Leonards and Singer 1998; Leonards et al. 1996; Usher and Donnelly 1998). In the first case, whether the animal perceived the input as coherent is not known; in the second, whether the neurons fired in synchrony has not been observed. However, to verify the assumption, it would be necessary to observe both phenomena at the same time and establish a link between them.

Such an animal study might indeed be possible. Over the last decade, considerable evidence has emerged suggesting that under certain conditions, neural activity of the animal is faithfully represented in its observable behavior. For example, microstimulation of neurons in the monkey middle temporal area (MT, the visual motion center in the brain) causes a significant change in motion detection tasks (Salzman, Britten, and Newsome 1990). Also, the spike count of a single neuron in MT accurately predicts the behavior of the monkey in such tasks (Britten, Shalden, Newsome, and Movshon 1992; Celebrini and Newsome 1994).

In these studies on monkey MT, periodically modulated synchronization (or coherent oscillations) was not observed (Bair, Zohary, and Newsome 2001). However, it should be possible to apply the same methods to areas where such oscillations have been found, such as areas V1 and V2 in cats (Eckhorn et al. 1988; Gray et al. 1989; Gray and Singer 1987; Singer 1993). The animal could be trained to respond to the stimuli and act out the decision it made about the stimuli. The coherence in input features, the synchrony in neural firing, and the perceptual experience (manifested as behavioral performance) could all be measured and compared to verify explicitly that correlated firing of neurons represents perceptually salient events.

In order to account for graded responses in behavior, like those in the PGLISSOM grouping experiments of Chapter 13, it must be possible to observe different degrees of synchrony in the neural activity. In PGLISSOM, such a gradation is measured as correlation coefficients between the MUA sequences. The two sequences can match to a degree, and that degree can vary over time; the correlation of the two sequences over time corresponds to the degree of synchrony. Similar methods could be used in the experiments outlined above for comparing neural activity and behavior.

However, with graded synchrony, it will be difficult to interpret synchrony in more complex representations, like those constructed using the transitive grouping rule (Geisler et al. 2001; Geisler and Super 2000). For example, if representations (A, B) are synchronized to degree x, (B, C) to y, and (A, C) to z, how strong is the synchrony in this group, i.e. how strong is the sense of a coherent object? Should it be $\max(x, y, z)$, $\min(x, y, z)$, their average, or some other quantity? If $z, y < z$, should B be be interpreted as part of the representation of the object or not? Although it is possible to adopt various rules in computational models, such questions can be answered conclusively only through biological measurements of synchrony and performance, as outlined above.

16.3.2 Interpretation of Temporal Codes

The PGLISSOM approach assumes that synchronized activity is the perceptual representation employed by the brain, for the reasons outlined above. However, there are other forms of temporal information that neural systems could utilize as well, including time to first spike, inter-spike intervals, phase difference of spikes relative to background oscillation, variation of response amplitude over time, and degree of resonance in frequency-tuned response (Bruns, Eckhorn, Jokeit, and Ebner 2000; Eckhorn, Gail, Bruns, Gabriel, Al-Shaikhli, and Saam 2004; Freeman and Burke 2003; Izhikevich 2001; Kozma, Alvarado, Rogers, Lau, and Freeman 2001; Maass

1998; O'Keefe and Reece 1993; Oram, Wiener, Lestienne, and Richmond 1999). In a broader sense, the question is how the information in the spike sequence (or spike train) can be interpreted. Various techniques have been used in this task, including information theory, Bayesian inference, and other probabilistic methods (Agüera y Arcas and Fairhall 2003; Baldi 1998; Rieke, Warland, de Ruyter van Steveninck, and Bialek 1997). These methods work in different cases and lead to different results; however, there is little direct neurobiological evidence on what kinds of codes might actually be used by the brain, and for what purpose.

It might be possible to observe various temporal codes using these techniques in PGLISSOM, and it might be possible to adjust the model to express some of them in particular. Their possible causes and roles can be relatively easily assessed because the entire state of the model is known at all times and can be easily altered. Such computational studies may turn valuable in testing hypotheses about temporal codes, as well as in designing methods for measuring them.

Assuming that synchronized activity is used at the primary visual cortex to represent objects, how can the higher levels of the visual system respond to them? Somehow, such activity would have to be recognized as a coherent pattern, distinguishing groups of neurons that fire at different synchronized phases. One possibility is that different high-level neurons detect coincident low-level firing for different objects, and produce spatially separate rate codes as a result (Abeles 1982; Marsalek, Koch, and Maunsell 1997). Mechanisms such as depressing synapses, which synchronize synaptic transmission even when action potentials are only partially synchronized (Senn, Segev, and Tsodyks 1998; Tsodyks, Pawelzik, and Markram 1998), may make such coincidence detection robust in practice.

However, coincidence detection requires that a different neuron represents each possible binding of input features into a coherent object, which is unlikely to be a general solution to the interpretation problem. The alternative is that the high-level representations are distributed as well, each consisting of a collection of more abstract feature representations. These high-level patterns can then be synchronized together with their low-level constituents (von der Malsburg 1999).

Because high-level representations are difficult to identify and measure, it is difficult to distinguish between these alternatives experimentally. However, computational models can be used to study whether coincidence detection could be accurate or general enough, and what kind of connectivity would be required for synchronization at the high level. Such models might eventually lead to testable predictions about how the brain interprets temporal codes.

16.3.3 The Role of the Different Layers

The two-layer organization of PGLISSOM was developed because long-range lateral inhibition with short-range excitation was found computationally necessary for ordered self-organization (SMAP), and long-range inhibition and excitation for grouping (GMAP; Section 11.1). Such a design can also be interpreted in terms of the connectivity patterns in the cortex, leading to insights into their function.

As was discussed in Section 2.1.2, the visual cortex is often described in terms of six layers. These layers are connected in complex but systematic ways (Binzegger, Douglas, and Martin 2004; Çürüklü and Lansner 2003; Douglas and Martin 2004; Robert 1999). Afferent excitatory connections to neurons in layer 4 are often accompanied by inhibitory connections to their surrounding neurons, established through inhibitory interneurons in layer 6 (Ahmed, Anderson, Martin, and Charmaine 1997; Ferster and Lindström 1985; Grieve and Sillito 1995). The long-range inhibitory and short-range excitatory connections in SMAP can be interpreted as abstractions of this on-center off-surround effect, as well as the effect of direct inhibitory lateral connections of layers 3, 5, and 6 (Kisvárday, Kim, Eysel, and Bonhoeffer 1994; McDonald and Burkhalter 1993). On the other hand, the very long-range excitatory and inhibitory lateral connections in GMAP most naturally correspond to the long-range axonal projections in layer 2/3 (Gilbert and Wiesel 1989; Hirsch and Gilbert 1991; McGuire et al. 1991). As in the cortex, the inhibitory connections in the SMAP can be shorter than the connections in the GMAP. The third architectural component in PGLISSOM consists of the intracolumnar connections, corresponding to such known connections between layers 2/3, 4, 5 and 6 in the cortex (Callaway and Wiser 1996; Gilbert and Wiesel 1979).

Thus, the PGLISSOM architecture has a natural interpretation as a biological model. Consequently, PGLISSOM suggests that the specific anatomical arrangements of the visual cortex may be due to different functional requirements, as they are in PGLISSOM: The lower layers may contribute primarily to self-organization, and layer 2/3 mostly to perceptual grouping. As a matter of fact, layer 2/3 in the cortex contains fast-spiking cells known as chattering cells (Gray and McCormick 1996), postulated to contribute to coherent oscillations and suggesting that the long-range connections in layer 2/3 might play a central role in grouping (Grossberg 1999; Grossberg and Williamson 2001; Raizada and Grossberg 2001). Although the lateral excitatory and inhibitory connections appear to have opposite effects, they serve different purposes and operate through different mechanisms. While decorrelation takes place at the level of average activity, synchronization applies to temporal coding, and the system can therefore be both decorrelating and grouping at the same time. The PGLISSOM model further suggests that the lateral circuitry may adjust the balance of these two functions according to contrast (Section 16.1.4). With high-contrast inputs, which make grouping easier but strongly influence self-organization, the long-range interactions are mostly inhibitory; with low contrast these interactions are mostly excitatory, enhancing the grouping function.

In the future, the PGLISSOM model can be extended to model the biological circuitry in more detail, and some of its biological predictions can be verified, as will be discussed in Sections 16.4.8 and 17.1.3. In this way, models such as PGLISSOM can help explain the different functions found in the layered architecture of the visual cortex, and allow gaining insight into how the functional divisions may occur and how they interact.

16.4 Predictions

The main prediction of the LISSOM model is that the self-organization, pattern generation, and synchronization mechanisms described in this book are responsible for much of the structure, development, and function of the visual cortex. Because the model matches biological data well, many of its assumptions (outlined above) can also be interpreted as predictions. They are computationally advantageous and sometimes even necessary, suggesting that they might apply to biology as well.

The main advantage of a computational model is that its processes and mechanisms are completely observable at all times. Such observations lead to specific predictions and suggestions for future biological and psychological experiments. A number of such predictions are reviewed in this chapter, referring back to the individual chapters where the underlying mechanisms are discussed in more detail. Predictions about cortical structure are reviewed first, followed by those about development, perceptual grouping, and visual coding. The focus is on experimental (biological or psychological) studies that could be performed in the near future. Future computational work is reviewed in Chapter 17.

16.4.1 Cortical Organization

The LISSOM simulations in Chapter 5 show how multiple input feature dimensions and patchy lateral connections can self-organize at the same time based on a single input-driven Hebbian learning process. The model can be used to make several predictions about how the cortical organization depends on visual inputs.

Hubel and Wiesel (1959; 1965; 1974; Section 2.1.2) originally observed that input features such as ocularity and orientation seemed to be hierarchically organized in V1: For each retinal location there were two areas with different ocular dominance, further divided into orientation-selective patches. Further measurements have shown that neurons have more complex spatiotemporal receptive fields, and the features depend on each other more than originally thought. In many cases the emergent organization is well approximated by a hierarchy (e.g. Weliky et al. 1996), but e.g. the orientation maps may differ depending on the spatial frequency used in measuring them (Basole, White, and Fitzpatrick 2003; Issa et al. 2001). Thus, the hierarchical organization should be considered primarily as a way to describe the complex properties of neurons rather than an underlying principle of cortical organization.

Self-organizing models such as LISSOM can be instrumental in understanding how such complex cortical organization emerges. The V1 organization is not built in, but is constructed in a self-organizing process based on properties of visual inputs. What structures develop depends on how much each feature varies and how large the map is. The feature with the highest variance is represented as large map areas with uniform values, such as ocular dominance stripes. If the map has enough units, these areas are further divided into subareas with uniform values for the feature with the next largest variance, such as orientation preference. Eventually the features vary too little or there are too few units to continue the hierarchy, and the remaining features are represented simultaneously at the same level (Kohonen 1989; Miikkulainen 1993; Obermayer et al. 1990d).

How strongly expressed the hierarchy is depends on how different the variances of different features are (Section 5.5.3). An important prediction of self-organizing models is that different maps can be obtained by manipulating the variances. For example, if spatial frequency is artificially varied more and orientation less than in the normal visual environment, the visual cortex of a test animal should develop large spatial frequency stripes and orientation patches inside them. The model also predicts that in species with a larger V1, topographic organization would be visible for more visual features. However, as V1 gets larger to fit a wider variety of feature-selective cells, eventually it becomes infeasible to connect the entire map laterally, and the visual cortex is instead divided into several different visual areas (Section 15.6; Kaas 2000).

Such models also predict that the shapes of the observed structures, such as the periodicity of the preference patches, depend on correlations in the input patterns. For example, the stronger the correlations between the eyes, the narrower the ocular dominance stripes, and with perfect correlation they disappear entirely. Similar observations have already been made in cats, comparing normal and strabismic (uncorrelated) organizations (Section 5.1.2; Goodhill and Löwel 1995; Löwel 1994; Löwel and Singer 1992).

It should be possible to verify these predictions in biological experiments where the input variances and correlations in the visual environment are systematically varied and the resulting organization observed through optical imaging. If confirmed, they would suggest that the cortical organization is indeed constructed in a self-organizing process driven by the properties of the visual input.

16.4.2 Patterns of Lateral Connections

The LISSOM model predicts that lateral connection patterns follow the activity correlations set up by the organization of receptive fields. If the major activity correlations are organized according to ocular dominance, as in a strabismic animal, the patterns follow the organization of ocular dominance (Section 5.4.3). If the ocular dominance is a less important feature, as in normal animals, the lateral connections follow the organization of the orientation and direction maps (Section 5.6).

In the orientation map, the lateral connection patterns are determined not only by the orientation preference, but also orientation selectivity, i.e. by how tuned the neuron is to orientation. Highly selective cells have lateral connections that connect primarily to other highly tuned cells with similar orientation preference. When the connection patterns are mapped back to visual space, they appear elongated along the direction of orientation preference. This prediction has already been verified experimentally (Bosking et al. 1997; Sincich and Blasdel 2001). Other related predictions have not yet been tested. For instance, less selective cells in the model have lateral connections that are unspecific to orientation. The connections also follow the local organization of the orientation map: At pinwheel centers for example, the lateral connection patterns are unspecific to orientation, and more or less isotropic.

At discontinuities such as fractures, the patterns are elongated along the orientation preferences of the nearby cells. At saddles, they follow the orientations that make up the saddle, but also intermediate orientations that match the preferences in the middle of the saddle (Section 5.3.4).

Similarly, long-range connections for direction-selective cells in the LISSOM maps are specific for direction of motion, in addition to orientation. They extend along the preferred orientation, not direction, and avoid patches of orthogonal orientations and opposite directions. Preliminary experimental results in ferrets are consistent with this prediction (White, Bosking, Weliky, and Fitzpatrick 1996). Further, neurons at DR fractures connect with both directions and extend along their common orientation. At DR saddles, they connect with the directions of the saddle, and at DR pinwheels with all directions; these DR features can occur at a variety of OR map features. The ocular dominance preferences are overlaid with the OR and DR preferences: The neurons most selective for one eye prefer connections from that eye, but such preference is absolute only in strabismic animals.

As will be discussed in Section 17.2.1, the primary visual cortex also has an organization of spatial frequency and color (Issa et al. 1999, 2001; Landisman and Ts'o 2002a,b; Livingstone and Hubel 1984a). The lateral connections of cells with these feature preferences should be organized according to similar principles. The cells that are highly tuned to low spatial frequencies should preferentially connect to other cells with similar tuning. Color-selective cells typically occur near the center of ocular dominance columns in locations called blobs. The connections of cells at these locations should strongly prefer color-selective cells in other blobs.

A number of experiments can be performed on young animals to verify the above predictions. Two such experiments have previously shown that when kittens are deprived of visual input they develop shorter and less dense lateral connections in V1 (Callaway and Katz 1991), and that when they are made strabismic, these connections follow ocular dominance instead of orientation (Löwel and Singer 1992). Similarly, one would expect that if an animal is brought up in a visual environment with only diffuse light spots and no edges or lines, the lateral connections would not be elongated along the orientation axes of cells, but would be more isotropic. Further, it should be possible to devise conditions such that these patterns follow the organization of spatial frequency. If the diffuse light spots have varying sizes, but not edges and boundaries, orientation preferences of cortical cells should be weak. However, because the stimuli vary in size, spatial frequency should become the prominent component in the cortical representation. The lateral connections should then follow mainly the organization of spatial frequency, rather than the organization of the orientation map.

In essence, any manipulation of visual inputs that substantially changes the organization of feature-selective cells should alter the patterns of lateral connections to match the new activity correlations in the cortex. Such experiments would provide further evidence for the hypothesis that the lateral connections develop synergetically with the afferent connections, based on activity correlations in the cortex.

16.4.3 Tilt Aftereffects

The LISSOM model of tilt aftereffects (Chapter 7) makes specific predictions about the activation and adaptation processes underlying the effect. Assuming suitable protocols can be developed for measuring the tilt aftereffect in animals, these predictions could be verified in future animal experiments.

The LISSOM model predicts that the tilt aftereffect depends crucially on adapting lateral inhibition. This prediction could be verified by blocking intracortical inhibition mediated by $GABA_B$ receptors with bicuculline (Sillito 1979) and that mediated by $GABA_A$ receptors with phaclofen (Pfleger and Bonds 1995). The observed effect would then be due to adapting afferent connections only. According to LISSOM, its sign should be reversed and its magnitude should be significantly weaker (Figure 7.7).

The activity patterns during the tilt aftereffect could also be observed using optical imaging, directly verifying the model's predictions. Although the temporal resolution is currently not sufficient to measure such transient activity, certain key predictions should be testable already. For instance, if an indirect-effect test pattern is repeated regularly throughout adaptation, the LISSOM model predicts that the overall response to the test pattern will increase each time. Other models, such as fatigue or those relying on levels above V1 (described in Section 7.1.2), predict that the activity levels in V1 do not change or actually decrease instead.

If imaging techniques with sufficient temporal and spatial resolution become available in the future, plots like those in Section 7.4.2 could be computed for monkeys and compared with those of the model. After an orientation map is first measured on the animal, the response of the cortex would be measured for test patterns at orientations typical of direct and indirect effects. Next, the cortex would be adapted to a stimulus of a particular orientation. The test patterns would then be presented again, measuring the cortical response once more. When the earlier measurements are subtracted from the later measurements, there should be a net decrease in activity for orientation detectors near the orientation used during adaptation, a net increase in activity of those with more distant orientations, and no change for very distant orientations.

The required temporal resolution for such measurements is on the order of a few seconds to avoid significant adaptation to the test pattern, and the spatial resolution at the level of individual orientation patches, i.e. 0.1 mm. With such resolution, it should also be possible to calculate perceived orientations, as was done on the LISSOM model in Section 7.2.1. If those orientations were found to be within the range of the measured tilt aftereffect, the experiment would provide strong support for the lateral inhibition theory of direct and indirect tilt aftereffects.

16.4.4 Plasticity

If the same processes that self-organize the visual cortex during early life are assumed to operate in the adult, many observations of cortical plasticity, such as recovery after damage, can be explained computationally by the LISSOM model. The model also leads to several predictions verifiable in traditional plasticity experiments. These predictions are briefly reviewed below; the details are discussed Chapter 6.

A retinal scotoma changes the dynamic equilibrium of the cortex, and the cortical neurons shift their receptive fields and adjust their lateral connections as a result. The model predicts that if the scotoma is large enough so that the neurons responding to its center no longer receive afferent input, their receptive fields should remain unchanged; otherwise, the receptive fields should shift outward and the blind spot should disappear. Second, according to the model, the unstimulated neurons expand their receptive fields because they receive less inhibition from the surrounding neurons. If inhibition before and after the scotoma was blocked (e.g. by bicuculline), such an expansion should not occur. Third, the orientation map reorganizes to represent the boundaries of the scotoma, and these changes are integrated into a smoothly varying map structure.

Similar reorganization occurs after a cortical lesion. Because the neurons in the lesioned area no longer inhibit the neurons right outside, these surrounding neurons begin to respond to inputs to the lesioned area, and the loss of function is smaller than would be expected. The inhibitory connections between these surrounding neurons gradually increase, and their responses to such inputs become weaker, temporarily increasing the loss of function. Over a longer time however, the receptive fields of these neurons shift inward, regaining some of the lost representations again. The model predicts that neurons with orientation preferences perpendicular to the lesion boundary (and therefore most lateral connections from the lesioned area) shift the most, whereas those with parallel preferences are less affected. This prediction could be checked with standard recording and imaging techniques.

The model further predicts that the extent of surviving lateral excitation determines how much of the function is regained. If the neurons on the opposite sides of the damaged area are linked with excitatory connections, the RFs can reorganize as neighbors and take over the lost function completely. Therefore, in early development, when the lateral excitatory radius is large and the connections are less patchy (Callaway and Katz 1990; Dalva and Katz 1994), it should be possible to compensate for larger lesions. The prediction can be checked by inducing lesions at various stages of development of the animal and comparing the extent of reorganization. The model also suggests two mechanisms for speeding up recovery from such lesions. First, by blocking inhibition in the perilesion area after the lesion should eliminate the regressive phase, and allow the receptive fields to shift inward faster. Second, it should be possible to hasten the shift by stimulating the area extensively with inputs originally represented by the damaged area.

If successful, such experiments would provide convincing evidence that the cortex is a continuously adapting system in a dynamic equilibrium with internal and external input, as suggested by the LISSOM model.

16.4.5 Internal Pattern Generation

The main principle of the HLISSOM model is that the shape and distribution of internally generated activity patterns determine how the maps and connections in the visual cortex develop in early life. Further, the model predicts that differences in these patterns serve a purpose in information processing. If the internally generated patterns differ across species, they should have different cognitive and sensory-processing abilities as a result.

Internally generated activity has not yet been characterized in sufficient detail to constrain modeling efforts. However, such constraints could possibly be derived indirectly, based on computational grounds. For example, the results in Section 9.2 show that different types of internally generated patterns will result in different types of V1 receptive fields. Uniformly random noise tends to lead to four-lobed RFs, whereas disks alone result in two-lobed RFs. Although orientation maps have been measured in young animals, very little data are available about what receptive fields these neurons might have. In future experiments, both of these structures could be measured at the same time, so that it will be clear what types of RFs lead to orientation selectivity at different stages of map development. Such data should allow narrowing the range of possible models of orientation map development, e.g. by rejecting either those based on uniformly random noise or those based on retinal waves.

In humans, one of the most intriguing predictions is that the internally generated activity should have a specific spatial structure. In Section 10.2.6, a number of training patterns that matched the size of a face and several of its components were shown to result in a preference for faces, as seen in newborn humans. On the other hand, many other patterns did not, including single dots and pairs of dots. The model therefore predicts that in humans, at least some of the internal patterns belong in the first category. As was mentioned in Section 16.2.2, recent advances in imaging equipment may make such measurements possible in the near future (Rector et al. 1997).

Perhaps the most convincing test of the pattern generation idea would be to modify the internally generated activity in animals, measuring whether their brains develop differently in a systematic way. As was discussed in Section 2.3.3, some tests of this type have already been performed, measuring effects on the retina and the LGN. For instance, pharmacological agents can be used to make the retinal wave patterns larger, faster, and more frequent (Stellwagen and Shatz 2002). When the agent is applied to one eye, the LGN layer corresponding to that eye becomes larger. The effect is not simply due to increased metabolic activity or other nonspecific mechanisms, because no changes are seen when the agent is applied to both eyes at once. That is, the pattern of activity determines how the LGN develops, not simply the total amount of activity.

The effects of these manipulations have not yet been measured at the V1 level. HLISSOM predicts that faster waves would make neurons more direction selective, perhaps even resulting in a full direction map at eye opening (unlike in normal dark-reared animals; White and Fitzpatrick 2003). If the waves were made smaller than the RFs, the V1 neurons should become less selective for orientation, because they would not have seen large oriented edges during development. Once PGO wave patterns have been measured, similar modifications could be performed on them, leading to similar results. Each change in the pattern can be simulated in HLISSOM, generating a specific, testable prediction about how the change will affect V1.

In the long term, detailed analysis of the pattern-generating mechanisms in the brainstem could provide clues to how the pattern generators are specified in the genome. These studies would also suggest ways for representing the generator effectively in a computational evolutionary algorithm, which is currently an open question. Studies of genetics and phylogeny, perhaps paired with computational simulations, could then possibly determine whether pattern generation was crucial in producing complex organisms that could adapt to the environment postnatally. Such studies could be instrumental in understanding the evolutionary origins of complex abilities and adaptation in higher animals.

16.4.6 Face Processing

In Chapter 10, the pattern generation hypothesis was extended to explain how face preferences in human infants could arise, and how they could change during early life. The model leads to several concrete predictions that can be tested in psychological experiments with human infants.

For instance, HLISSOM suggests that the precise spacing of a face outline is not crucial for newborn face preferences. This prediction can be tested in human infants by presenting facelike patterns with a variety of outline shapes; the model predicts that the response will be similar regardless of the shape of the outline. Similarly, HLISSOM predicts that newborns will prefer facelike patterns to ones where the eyes have been shifted over to one side. This prediction contrasts the alternative "top-heavy" explanation for face preferences, and can be tested in similar experiments as the effect of outline.

A second set of predictions concerns the preferences that develop postnatally, with exposure to real faces. Pascalis et al. (1995) showed that covering the hair outline was sufficient to suppress infant's preference for his or her mother. HLISSOM predicts that covering the facial features will have a similar effect, i.e. that newborns learn faces holistically. This prediction can be tested using methods like those of Bartrip et al. (2001) and Pascalis et al. (1995), using human newborns. HLISSOM also predicts that older infants will continue to prefer realistic face stimuli, such as photographs or video of faces, even though they stop preferring schematic faces in the periphery by 2 months of age.

Nearly all of the proposed experiments can be run using the same techniques already used in previous experiments with human infants. They will help determine how specific the newborn face preferences are, and clarify what types of learning are involved in postnatal development of these preferences.

16.4.7 Synchronization

The PGLISSOM model shows that neural synchrony can serve as a foundation for perceptual representations in the visual cortex. It may be possible to verify this idea in future biological experiments, and also demonstrate how such representations are constructed and maintained.

First, the contour integration experiments in Chapter 13 suggest that how strongly and reliably the contour is perceived depends on how strongly synchronized the neural representations of the contour elements are. If synchronization were to be disrupted artificially, the perception of the contour should disappear. Such an experiment might be possible in the future using transcranial magnetic stimulation (TMS; Barker, Jalinous, and Freeston 1985; Hallett 2000; Lancaster, Narayana, Wenzel, Luckemeyer, Roby, and Fox 2004; Walsh and Cowey 2000). While the usual effect of TMS is to excite or inhibit an area of neurons temporarily, it might be possible to affect only their synchronization in three ways: (1) TMS could be applied to a related visual area, which could indirectly disrupt synchronization in the area of study; (2) with very low intensity, TMS could have a subthreshold excitatory effect, causing the target group of neurons to fire more rapidly and out of synchrony with other neurons; and (3) high-frequency repetitive TMS (1–60 Hz rTMS; George, Wassermann, Williams, Steppel, Pascual-Leone, Basser, Hallett, and Post 1996) at low intensity could be applied at the same or different phases at two separate sites, driving the firing of the neurons and inducing artificial synchrony or desynchrony. Applying such techniques or some combination of them while the subject is performing a perceptual grouping task should degrade his or her performance in the contour integration task without affecting the perception of the elements themselves.

Such methods could perhaps be even more illuminating and reliable if paired with simultaneous electro-encephalogram (EEG) recordings that measure the changes in brain activity resulting from the TMS (Ilmoniemi, Virtanen, Ruohonen, Karhu, Aronen, Näätänen, and Katila 1997). If TMS is found to result in a particular perceptual change and at the same time a characteristic change in the EEG, it might even be possible to develop a way to measure synchrony directly from the EEG. Such techniques would be highly useful in verifying whether synchronization indeed is the underlying representation for perceptual grouping.

Second, the results in Section 12.2.1 show that adapting the PSP decay rate is a possible way to modulate synchronization behavior in neurons. The effects of decay are similar to those of adjusting delay of signal propagation between neurons, which has been proposed before as a possible synchronization mechanism (Eurich et al. 2000; Gerstner 1998a; Horn and Opher 1998; Nischwitz and Glünder 1995; Tversky and Miikkulainen 2002). At this point there is no conclusive biological evidence to support either process. However, although axonal morphology can change over time and affect the delay, it would be difficult to make such changes fast and accurately enough (Eurich et al. 1999; Stevens et al. 1998). On the other hand, PSP decay may be easier to adjust, for example by controlling the properties of ion channels in the dendrites. It is therefore a good candidate for modulating synchrony.

Experimental techniques exist for measuring the various sources of temporal lag in the neuron (Nowak and Bullier 1997). Using similar techniques, it may be possible to verify whether the dendritic membrane potential decays at different rates at different locations in neurons, and also whether the rates depend on the synapse type (such as glutaminergic vs. GABAergic). If two neurons far apart are found to synchronize despite a long delay, the model predicts that their decay rates could be compensating for the delay. Further, using voltage or current clamping, it might be possible to control the decay of the PSP, and measure the effect on synchronization. In this way, it might be possible to verify that PSP decay rate could have a modulating effect.

Third, PGLISSOM simulations showed how a longer absolute refractory period could help synchronize neural activity even under noisy conditions (Section 12.4.3). Such an effect has not been demonstrated biologically, but it is nevertheless interesting to speculate how it could be utilized in biological systems. Given that homeostatic mechanisms have been found for regulating various other aspects of neuronal behavior (Turrigiano 1999), perhaps their refractory periods adapt as well. Synchronization would then be possible over different ranges of noise and variability. This hypothesis could be tested by changing the noise environment of a synchronized group of neurons and observing whether the refractory periods change to compensate.

Even if the refractory periods did not adapt dynamically, they could have been adapted through evolution. Assuming that synchronized activity will be found in different parts of the brain and in different species, the hypothesis could be tested by measuring the refractory periods of neurons located in different noise environments. Neurons operating under significant noise should have longer refractory periods than those where the signal is relatively free of noise. Further, it might be possible to strengthen synchronization artificially in such neuron populations. By artificially lowering the membrane potential immediately following a spike, the refractory period could be increased, which should result in more highly synchronized neuron activities in noisy conditions.

Fourth, fast adaptation of lateral excitatory connections in PGLISSOM promotes synchrony among the connected regions. Such adaptation has been proposed to be crucial for dynamic feature binding in general (Crick 1984; Sporns et al. 1991; von der Malsburg 1981, 2003; Wang 1996), and similar short-term plasticity has recently been observed in rat prefrontal cortex (Hempel, Hartman, Wang, Turrigiano, and Nelson 2000). However, it is not clear whether such plasticity depends on coinciding presynaptic and postsynaptic activity, and whether it indeed facilitates synchrony. To answer these questions, it will be necessary to measure plasticity and synchronization in the same experiment, perhaps by combining the techniques used by Hempel et al. (2000), Eckhorn et al. (1988), and Gray et al. (1989). Since short-term plasticity is believed to depend on Ca^{2+} channels (Fisher, Fisher, and Carew 1997; Hempel et al. 2000; Regehr, Delaney, and Tank 1994), it might also be possible to test the hypothesis directly by disabling short-term plasticity with Ca^{2+} antagonists such as nifedipine, or with Ca^{2+} channel blockers such as diltiazern (Jensen and Mody 2001). If synchronization becomes less robust, it suggests that fast adaptation indeed plays a role in establishing synchrony.

Such investigations of cellular mechanisms can lead to a deeper understanding of perceptual representations, including how neural synchrony occurs and how fine tuning of temporal behavior is possible.

16.4.8 Perceptual Grouping

The PGLISSOM model shows how input-driven self-organization can result in different perceptual grouping performance across the visual field (Section 13.4). This result leads to interesting predictions about the statistics of visual input and the anatomical foundations of perceptual grouping, as well as the role of internal pattern generation in the initial construction of the system. The assumption that grouping and self-organization are based on the lateral connections of neurons in different layers of the cortex may also be tested experimentally.

First, the PGLISSOM model can explain the observed performance differences assuming that the visual input statistics differ in specific ways across the visual field. This assumption should be verified in direct image measurements. Using an eye-tracking method similar to that of Reinagel and Zador (1999), statistics about edge frequency, curvature, and occlusion can be collected at different visual regions. To validate the model, edges and occlusions should be more frequent and curvature higher in the fovea vs. periphery and in the lower vs. upper hemifield. Further, the same method could be used to measure other input properties, such as color, contrast, and spatial frequency. The results should help characterize the structure of visual scenes and understand how visual attention biases the sensory statistics, forming a foundation for further computational modeling and psychophysical experiments of grouping performance.

Second, PGLISSOM predicts that if different areas of the visual cortex receive different inputs during development, they will develop different patterns of lateral connections. These patterns in turn cause perceptual performance to differ in corresponding areas of the visual field. Such predictions can be tested experimentally by manipulating the training inputs: Lateral connectivity can be measured in upper vs. lower hemifield and fovea vs. periphery to see whether there are any differences between these areas as the model suggests (Section 13.4.2); or, the cause of the differences can be experimentally tested by rearing animals in controlled visual environments where the input distributions in these areas are systematically altered.

For example, animals can be fitted with eye glasses that flip the input to the upper and lower hemifield. After the critical period, the connectivity patterns and contour integration in the lower and upper hemifield can be measured and compared with normally reared animals. If connectivity is genetically determined, there should be no noticeable difference to the control animals. In contrast, PGLISSOM predicts that the more cocircular connectivity and better contour integration would develop in the upper hemifield instead of the lower hemifield. Such an experiment would verify both that the contour detection depends on the connectivity and that the connectivity can be explained as an effect of input-driven self-organization. Finding such evidence would be an important step toward understanding how functional differences arise in the visual cortex.

Third, PGLISSOM can be brought together with internal pattern generation to suggest how prenatal and postnatal self-organization each contribute to constructing an effective contour integration circuitry. As was discussed in Section 13.2.2, essential for contour integration is that the lateral connections have a cocircular profile, matching the distribution of edges in natural images. Interestingly, PGLISSOM does not have to be trained with natural images to obtain such connectivity: It also develops with oriented Gaussian inputs. Such inputs might indeed be available during the prenatal development of the visual cortex: After the retinal waves are filtered by the LGN, they would appear as collections of smooth, oriented, and locally straight Gaussians to the developing V1 (Section 16.2.1). Although such Gaussians would be relatively short, they should be enough to develop a rudimentary contour integration ability prenatally. The circuitry would then be further refined postnatally with visual inputs, resulting in adult performance.

While it would be difficult to measure contour integration performance in newborns and in animals, it might be possible to measure synchrony in animals using multi-cellular recording techniques (Eckhorn et al. 1988; Gray et al. 1989). The prediction could then be verified by presenting contours to animals of different ages and comparing the extent of synchronized groups that form. Such an experiment would provide further evidence for prenatal self-organization, and lead to a precise understanding of how the ability to integrate contours develops.

Fourth, the model suggests that due to the opposite requirements for grouping and self-organization, functional differences might exist in the layered architecture of the visual cortex (Section 16.3.3). This prediction is difficult to verify with current experimental techniques, but it might become feasible in the future. For example, if deep or shallow layers could be selectively disabled, or the intracolumnar connections between the layers disrupted, their functions would become decoupled. Progress of development and the ability to form synchronized groups could then be measured separately in the deep layers and compared with that in the shallow layers. The PGLISSOM model predicts that the shallow layers would not properly self-organize and the deeper layers would not synchronize. Such an experiment would lead to significant insights into how cortical structures implement function.

16.4.9 Sparse Coding

One of the important predictions of the LISSOM model is that the inhibitory self-organized lateral connections decorrelate, reducing redundant activity and producing a sparse coding of the visual input (Sections 14.2 and 17.3.1). Such representations are efficient in energy and space, allowing more inputs to be represented with fixed resources, and also form an effective foundation for later stages of visual information processing.

In LISSOM, sparse representations emerge as a necessary component of the self-organizing process: The responses are focused into local neighborhoods so that distinct representations for the different parts of the input space can be formed. Indirect evidence for such focusing already exists from direct recordings of cortical activity. Pei, Vidyasagar, Volgushev, and Creutzfeldt (1994) found that the orientation selectivity determined from excitatory PSPs (and not action potentials) became increasingly sharper with time. Consistent with the LISSOM model, they independently attributed this phenomenon to intracortical excitation and inhibition.

Sparse coding implies that the information about a given visual image is represented by a small number of neurons, and conversely that each neuron conveys a significant amount of information about the input. The LISSOM network indeed focuses the initial activation in a few iterations to the best-responding neurons, and it is possible to identify what type of stimulus would have caused it to be active. Such a process is consistent with the response properties of neurons. The neurons divide up the task of representing the input space, and respond only to particular sharply defined regions of it. For a typical neuron in the primary visual cortex, a response of 10 action potentials following one brief stimulus presentation was found to be sufficient to classify the stimulus into a relatively small region in stimulus space, with a high degree of confidence (Geisler and Albrecht 1995).

With optical imaging techniques, it should also be possible to verify the sparse coding prediction directly. The time resolution of current imaging techniques (e.g. Senseman 1996) should be sufficient to observe how the sparseness of cortical activity changes in response to visual input, presumably due to the recurrent lateral interactions in the cortex. Sparseness can be quantitatively estimated by calculating the kurtosis of the activity (Barlow 1972; Field 1994), as was done with the LISSOM experiments of Chapter 14. For any given input, the kurtosis should increase with time as the activity becomes more focused and stable.

Any manipulation that affects the lateral interactions should affect the sparse code. Suppressing lateral inhibition, for example, would be expected to result in less sparse responses. It should be possible to apply an inhibitory blocker, such as bicuculline, and measure the kurtosis of cortical activity through optical imaging. If lateral inhibition plays a significant role in producing the sparse code, the less inhibited responses should be less sparse. Another experiment could focus on selectively preventing the long-range lateral connections from self-organizing during development. In such a case, the kurtosis of cortical activity should be much lower than in the normal case. As a consequence, neurons should also be far less selective for orientation and spatial frequency. The experiment would show that the activity coding in the cortex closely depends on the organization of lateral connections. The last step would be to verify that the sparse representations indeed constitute an advantage in information processing. Such experiments are difficult to design and conduct conclusively; however, if it is possible to prevent proper self-organization of lateral connections during development as outlined above, such an animal should have an impaired ability for visual pattern recognition due to a less efficient coding of visual inputs.

While sparse coding is currently a computational hypothesis consistent with biological data, the above experiments in the near future could establish it as a fundamental information processing principle in the visual cortex.

16.5 Conclusion

The LISSOM framework is based on a number of assumptions about the connectivity, adaptation, and temporal coding in the cortex. Although there is not yet sufficient evidence to verify these assumptions completely, they are plausible, and lead to testable predictions. The next chapter will discuss how LISSOM can be extended to explain new phenomena, and how it can serve as a starting point for new research directions.

Future Work: Computational Directions

The research discussed in this book can serve as a starting point for a wide variety of future computational investigations. Several such studies continue the research described in the individual chapters, and have already been discussed in those chapters. Others are new directions, and will be described in this chapter. First, the LISSOM architecture can be extended in several ways to match biological systems in more detail. Second, LISSOM models can be built to understand several new visual phenomena, both in V1 and in other visual areas. Third, LISSOM can be used as a starting point for new research directions, including modeling other cortical areas and building artificial vision and other information processing systems. The *Topographica* simulator, a publicly available computational tool for modeling cortical maps, is designed to make these future extensions practical to simulate and understand, as described in the end of this chapter.

17.1 Extensions to the LISSOM Mechanisms

The LISSOM equations and architecture are designed to account for a large fraction of the experimental results on the development and function of the primary visual cortex, while being as clear and parsimonious as possible. To more closely match specific experimental results, more complex mechanisms can be included in the model, as described in this section. These extensions should not change the fundamental principles and results of the model, only make it biologically less abstract.

17.1.1 Threshold Adaptation

One of the most important components of any detection or recognition system is the activation threshold. If set properly, the threshold allows the system to respond to appropriate inputs while ignoring low-level noise and other nonspecific activation. For instance, the threshold can allow an orientation-selective neuron to respond to stimuli near its preferred orientation, and not to other orientations. In terms of signal detection theory, the threshold is a balance between the false alarm rate and the false positive rate; in general this tradeoff is unavoidable (Egan 1975). Thus, it is important to set the threshold appropriately.

In LISSOM, the activation thresholds (θ_l and θ_u in Equation 4.4) are set by hand through trial and error. This procedure can be seen to correspond to evolutionary adaptation, which results in animals that are born with appropriate threshold values. However, biological systems make use of threshold adaptation as well. Homeostatic regulation mechanisms in general have recently been discovered in a variety of biological systems (see Turrigiano 1999 for a review). Some of these regulatory mechanisms are similar to the weight normalization already used in LISSOM (Equation 4.8). Others more directly adjust how excitable a neuron is, so that its responses will cover a useful range (i.e. neither always on nor always off). In particular, Azouz and Gray (2000) found the neuron's firing threshold to be proportional to the rate of depolarization in the PSP: When the PSP rises rapidly, neurons automatically raise their thresholds.

An automatic method of setting thresholds would also be computationally desirable for several reasons: (1) It is time consuming for the modeler to find an appropriate threshold value; (2) the threshold setting process is subjective, which often prevents rigorous comparison between different experimental conditions (e.g. to determine what features of the model are required for a certain behavior); (3) different threshold settings are needed even for different types of input patterns, depending on how strongly each activates the network (e.g. high-contrast schematic patterns vs. low-contrast real images); and (4) the optimal threshold value changes over the course of self-organization (because weights become concentrated into configurations that match typical inputs, increasing the likelihood of a response).

A first step in this direction is the mechanism used in the PGLISSOM simulations of Chapter 11, where the spiking threshold θ_b was set automatically based on input activity. A fixed percentage, usually 50 to 65%, of the maximum input activity over the map was set as the threshold at the beginning of each settling iteration (Appendix D). Another approach is to increase the threshold if the neuron has been highly active in the past, and decrease it if not (Burger and Lang 2001; Gorchetchnikov 2000; Horn and Usher 1989). With such mechanisms, the map responds to all inputs with roughly the same level of total activation. However, this level still needs to be adapted by hand during self-organization to make sure that the responses became gradually sparser. How to do it automatically, perhaps utilizing regulation mechanisms similar to those observed in biology, is currently an open question.

Adding automatic mechanisms for setting the thresholds would make the model significantly more complex mathematically. Even so, it would make the model biologically more realistic, and also much simpler to use in practice, particularly with real images.

17.1.2 Push–Pull Afferent Connections

Neurons in adult V1 retain their selectivity over wide variations in contrast; how strongly they respond depends primarily on how well the input matches their preferred features such as orientation, ocularity, and direction (Sclar and Freeman 1982).

As described in Section 8.2.3, HLISSOM includes a divisive normalization term to achieve such contrast invariance. While this method is simple and works well in most cases, it is an abstraction and should be replaced by biologically more accurate mechanisms.

One problem with divisive normalization is that it penalizes any activity in the anatomically circular receptive fields that does not match the neuron's weights. As a side-effect, the V1 responds less strongly to input where the stimuli are closely spaced. For example, the V1 network responds less to a high-frequency square-wave grating than to the same pattern with every other bar removed.

In future work, it may be possible remove this limitation by using a push–pull arrangement of weights rather than full-RF normalization (Ferster 1994; Hirsch, Gallagher, Alonso, and Martinez 1998b; Troyer, Krukowski, Priebe, and Miller 1998). With a push–pull RF, cortical neurons receive both excitatory and inhibitory afferent inputs from different parts of the retina, rather than the purely excitatory input, as in LISSOM and in most other models. One difficulty with push–pull weights is that the inhibitory weights need to connect the neuron with regions in the retina that are anti-correlated with it, and therefore such weights cannot be learned through Hebbian learning. Thus, either a new learning rule or a more complicated local circuit in the cortex will need to be developed so that push–pull weights can self-organize.

Such an extension should allow LISSOM to self-organize as before, but would represent the afferent circuitry more accurately, and also lead to reliable responses to a wider variety of input patterns.

17.1.3 Modeling Substructure Within Columns

LISSOM is a column-level model, and each unit in the model stands for the response patterns of a set of cells in a vertical column in the cortex. An important extension is to take more of the structure within the column into account, more precisely representing the fine-grained structure and processing that occurs at this level.

First, the responses recorded from LISSOM represent averages of multiple cells. These responses are a good match to data obtained with optical imaging techniques, which also measure averages over multiple nearby cells. For instance, LISSOM units in map regions near pinwheel centers and fractures in orientation maps tend to have lower orientation selectivity, just as in maps measured using optical imaging (Blasdel 1992b).

Interestingly, when the pinwheel neuron responses are measured using microelectrode recordings, they appear as selective as neurons in other parts of the map (Maldonado et al. 1997). However, the pinwheel centers do have a wider variety of orientation preferences in a small area. Thus, optical imaging techniques report lower selectivity probably because some neurons in that area respond to each of the different orientations.

In order to model the detailed behavior of individual neurons within pinwheel centers, LISSOM could be extended so that each unit in the current model is represented by a set of different units. Connectivity between each unit could be determined stochastically, so that each unit could function differently but the average response of all the units in the column would be similar to the current LISSOM model. Such a model is currently too expensive to simulate at the map level, but could become feasible in near future, especially through techniques outlined in Chapter 15. In this way, LISSOM could be extended to model low-level neural phenomena more accurately within the same basic framework.

Second, the circuitry and subfunctions in the column can be modeled in more detail. The SG model of the cortical column in PGLISSOM is already a step in this direction. Although it was motivated primarily on computational grounds (i.e. in order to implement both self-organization and grouping in the same map; Section 11.1), it has an intriguing biological implementation in terms of the layered structure of the cortex (Section 16.3.3). This interpretation can be expanded by implementing the circuitry in more detail. For example, the broad long-range inhibition in GMAP can be replaced by local inhibitory interneurons as outlined in Section 16.1.4, making it possible to determine precisely how layer 2/3 contributes to self-organization. Similarly, the connectivity within the column can be modeled in more detail, and the contribution of the deeper layers on synchronization analyzed.

Such more detailed models of cortical columns would allow understanding computations in maps more precisely, leading to predictions that can be verified with existing cellular recording techniques.

17.1.4 Phase-Invariant Responses

The behavior of cortical columns in the current LISSOM model is based on simple cells only, i.e. cells that respond most strongly when their preferred input is aligned with the ON and OFF subfields of their receptive field (Section 2.1.1). Such cells are thought to be the first in V1 to show orientation selectivity, but V1 also includes cells with more general responses (Hubel and Wiesel 1968). Termed complex cells, they respond to any input within their RF regardless of the alignment; in other words, their response is phase invariant. Such responses have been observed in the visual cortex, although the circuitry that gives rise to them is not well understood.

Most current models with phase-invariant responses are hierarchical: Complex cell behavior is obtained by pooling outputs from several simple cells (e.g. Hyvärinen and Hoyer 2001; Weber 2001). An alternative approach is to establish local recurrent connections within a single set of V1 neurons (Chance, Nelson, and Abbott 1999): Phase-invariant responses can then occur among the simple cells through recurrent excitation. It is not yet clear which approach is a closer match to how phase-invariant responses arise in V1.

The LISSOM model could be extended with additional sheets of neurons in V1 representing complex cells, or with a local circuit that pools the responses of simple cells into phase-invariant ones. Both of these extensions involve connecting neurons in a small local area, and assume that the area includes neurons that respond to different phases. Phase is indeed distributed randomly within a column and between nearby columns in animals (DeAngelis et al. 1999). The likely reason is that phase in the input may effectively be random over short time scales due to small eye movements known as microsaccades (Martinez-Conde, Macknik, and Hubel 2000).

Current LISSOM simulations tend to group RFs by phase similarity (in addition to similarity of orientation, ocular dominance, and direction selectivity), because neurons with similar phase preferences are activated together. For the phase-invariance extensions to work, LISSOM needs to be further augmented with a learning rule that associates stimuli over time, such as the trace learning rule (Földiák 1991a). In this variant of Hebbian learning, connections between neurons are strengthened if they respond soon after one another, instead of having to respond simultaneously. Based on microsaccade-like movements during training on visual images, the model should then develop phase-invariant responses and random phase distributions like those seen in animals. Such a model could be used to compare the two alternatives, and to draw predictions for future biological experiments.

17.1.5 Time-Lagged Activation

The direction map simulations in Sections 5.5 and 5.6 focused on how LGN cells with different lags can result in direction-selective responses in V1. However, any other source of different delays for signals reaching V1 neurons could also contribute to direction selectivity (Clifford and Ibbotson 2002). Since the biological mechanisms underlying such selectivity are not well understood, computational models could serve a pivotal role in evaluating the alternatives.

For example, different lags in the lateral connections in the cortical maps could be used to represent motion. If connections from nearby locations make synaptic connections on distal dendrites and connections from farther away on proximal dendrites, their effect would arrive at the soma at the same time. A coincidence detection mechanism could then detect these events and generate a spike, allowing the neuron to respond to moving inputs in a specific location, direction, and velocity. Alternatively, reverberating feedback loops (Amit 1994; Hebb 1949; Seung, Lee, Reis, and Tank 2000; Wang 2001) within V1 or between V1 and other areas could act as memory for previous inputs, providing information about past input patterns just as the lagged cells and connections do.

Future simulations can focus on where such lags might occur in different species, and how those differences can result in direction selectivity, leading to predictions for future biological experiments.

17.2 Modeling New Phenomena with LISSOM

In addition to the topics covered by current LISSOM simulations, the model can be used to understand a wide range of other visual phenomena. This section proposes a number of such studies, focusing on development, visual function, grouping, and scaling up to larger networks and to higher levels of visual processing. Each project is possible future work using the *Topographica* software described in Section 17.4.

17.2.1 Spatial Frequency, Color, and Disparity in V1

The LISSOM simulations in Chapter 5 focused on how orientation, ocular dominance, and direction maps develop in V1. However, the approach is very general and can be easily extended to include other dimensions of visual input, such as spatial frequency, color, and disparity. Maps for each of these dimensions can be developed by generating input that varies in these dimensions, self-organizing the model based on these inputs, and measuring the response properties of V1 neurons that result.

For instance, the current simulations are based on single-size ON and OFF cells (i.e. a single DoG center and surround radius), and thus include only a limited range of spatial frequencies. Spatial frequency maps can be simulated by including multiple sets of LGN cells, each with a different DoG size. The V1 network will organize into different groups preferring different spatial frequencies, which can then be compared against experimental spatial frequency maps (such as those observed by Issa et al. 2001).

Color maps can be developed in LISSOM by including separate groups of retinal and LGN neurons for the different colors. Each eye will be represented by three sheets of photoreceptors R, G, and B, corresponding to long, medium, and short wavelengths. One sheet of ON cells and another of OFF cells in the LGN will have center and surround RFs on all three photoreceptor sheets, and thus respond to differences in intensity. Eight other LGN sheets are connected to the photoreceptors in a manner that establishes four red/green opponent RF types (such as excitatory center on the red sheet and inhibitory surround on the green sheet), and four blue/yellow opponent RF types (such as excitatory center on the blue sheet and inhibitory surround on the red and green sheets). V1 receives input from all of these LGN cells, and should develop patches selective for colored areas of the input (e.g. regions with greater R activation than G activation). The model can be validated by comparing its color-selectivity structure to the color-selective areas found in biological V1 (Landisman and Ts'o 2002b). If the model is extended to include V2 (as discussed below), similar comparisons can be made with color maps in V2 (Conway 2003; Ts'o, Roe, and Gilbert 2001; Xiao, Wang, and Felleman 2003). It will also be interesting to determine whether the distribution of color representations in the model matches the statistical properties of color in natural images (Doi, Inui, Lee, Wachtler, and Sejnowski 2003; Lee, Wachtler, and Sejnowski 2002b), and whether lateral interactions contribute to constant perception of color under different lighting conditions (Barnard, Cardei, and Funt 2002; Brainard 2004).

Modeling disparity does not require additional LGN cells, but will require input patterns slightly offset in each eye, as they are in stereoscopic images. Through self-organization, such patterns will result in groups of cells in V1 that prefer different disparities, i.e. different distance between corresponding features. The model can again be validated by comparing with experimental results for disparity maps measured using optical imaging (such data are currently only available for V2; Ts'o et al. 2001).

Compared with orientation, ocularity, and direction, much less is known about how spatial frequency, color, and disparity are represented in the brain. Extending the model to these dimensions should lead to a number of specific, testable predictions, significantly advancing our understanding of how input features are represented in the visual cortex.

17.2.2 Differences between Species

The simulations in this book have drawn upon experimental data from multiple species, including human, monkey, cat, tree shrew, and ferret. This approach was necessary because most of the relevant experiments have so far been performed in only one species. For instance, only in the cat have lateral connections been measured in strabismic animals (Löwel 1994; Löwel and Singer 1992), and only in the ferret have direction maps been measured in V1 (Weliky et al. 1996). Because the primary visual cortex is remarkably similar across these species, pooling the experimental data in this way is generally valid. However, there are several differences between species as well; a computational model such as LISSOM can be instrumental in understanding which differences are significant and what their origins are.

Some of the main species-specific differences include: (1) Ocular dominance maps in the cat have a patchier, less stripe-like organization than in the monkey (Blasdel 1992a; Löwel 1994); (2) in the cat, V1 orientation maps have only a weak bias for horizontal and vertical orientations, unlike in the ferret (Müller et al. 2000); (3) orientation and ocular dominance patches are less likely to intersect at right angles in cat than in monkey (Müller et al. 2000; Obermayer and Blasdel 1993); and (4) in ferrets, some regions of the central visual field are entirely monocular, rather than binocular with alternating ocular dominance stripes as in other species (White, Bosking, Williams, and Fitzpatrick 1999).

There are a number of possible sources for such differences that could be modeled in LISSOM: (1) The shape of the head and the position of the eyes differ between species, which affects how correlated the patterns between the eyes are. Such differences in turn will change how the ocular dominance maps develop. (2) The anatomy and physiological properties of the retina differ between species. For instance, retinal ganglion cells in the rabbit are selective for motion direction, unlike in other species (see Clifford and Ibbotson 2002 for a review). (3) Whereas cats have time-lagged cells at the LGN level, similar cells have not yet been found in monkeys (Hubener et al. 1997; Löwel et al. 1988; Saul and Humphrey 1992). As a result, these species may represent time-varying input differently, which in turn may affect how the different features are organized in the cortex. (4) The various areas of the visual cortex, including V1, have significantly different sizes in species such as the ferret and the monkey, and the cortical area devoted to the corresponding visual area differs as a result (Kaas 2000). (5) Various developmental events (e.g. when spontaneous retinal activity stops and orientation maps emerge) take place at different times in different species (Blasdel, Obermayer, and Kiorpes 1995; Issa et al. 1999). (6) Internally generated activity patterns differ between species, potentially changing how the animal develops prenatally and how postnatal visual experience affects them (Jouvet 1998).

Species-specific differences can be modeled in LISSOM using different parameter values, demonstrating how self-organization depends on the specific input patterns seen by a developing visual area. Such hypotheses are difficult to test experimentally, but a computational model like LISSOM is ideal for the task: It is possible to set up hypothetical developmental scenarios and observe their outcome. In this way, it may be possible to determine which of the known anatomical and environmental differences could be responsible for the different maps and responses in different species.

17.2.3 Prenatal and Early Postnatal Development of V1

The simulations in Chapter 9 showed how orientation maps can be constructed in a self-organizing process that takes place both before and after birth. Once other feature dimensions have been simulated for a particular species (as proposed in Section 17.2.1), LISSOM can be used to construct a realistic and detailed model of how all the dimensions develop at once, based on internally generated activity and postnatal visual experience. A similar model can be built to understand how the ability to integrate contours could be constructed.

To allow for a detailed comparison, such studies need to focus on a single species. Currently, the most detailed data on the early development of maps are available for the ferret (although the cat is also a good candidate). The LISSOM model can be set up with parameters closely tied to measurements in ferrets, and the initial development of maps can then be simulated in detail. As mentioned in Section 2.1.4, experiments have shown that dark rearing, eyelid suturing, and modifying the visual environment can significantly change how maps develop in ferrets (Crair et al. 1998; Crowley and Katz 1999; Gödecke and Bonhoeffer 1996; Stellwagen and Shatz 2002; Weliky and Katz 1997), and it should be possible to replicate each of these experiments in the model.

Ocular dominance will be a particularly interesting test case, because OD maps have been found in animals before they have had any visual experience. Whether neural activity is required to develop them initially is currently controversial (Crowley and Katz 1999; Stellwagen and Shatz 2002); LISSOM simulations could help determine what types of activity are sufficient for this process. For instance, LISSOM simulations in Section 5.4.4 suggest that realistic adult maps require correlation between the two eyes, yet patterns like retinal waves are not correlated between the eyes. One possibility is that the OD map is constructed in two phases: prenatally with uncorrelated inputs (which leads to strabismic-like maps), and postnatally with correlated images that differ primarily in brightness. Simulations could demonstrate how an initial strabismic-like OD map changes into an adult-like OD map with visual experience, a hypothesis that could then be tested in future animal experiments. Alternatively, the developing cortex may receive simultaneous or alternating input from two sources, one uncorrelated (e.g. retinal waves) and one identical for both eyes (e.g. brainstem input during sleep). Simulations of this process should show adult-like maps at all stages of development, which again could be compared with animal measurements. The results of such comparisons would allow distinguishing between the two possible mechanisms of constructing the OD map.

The origin of direction selectivity is another interesting research issue because this property appears to develop differently from orientation and ocular dominance. Specifically, direction maps have not been detected in young ferrets raised in darkness, even though orientation and ocular dominance maps have been found robustly (White and Fitzpatrick 2003). Assuming that retinal waves result in orientation selectivity, perhaps the waves do not move fast enough or often enough to cause direction selectivity to emerge at the same time. Alternatively, perhaps the signals reaching V1 during early development do not have sufficiently different lag times, which again would prevent the direction selectivity from emerging. Through simulation studies, it should be possible to determine whether the amount of motion in retinal waves can lead to direction maps, or whether only orientation maps will develop. In the latter case, further simulations could verify whether direction selectivity can develop within an existing orientation and ocular dominance map based on postnatal training with moving natural images. Such simulations would result in predictions for future biological experiments, making it possible to determine how direction selectivity develops in the visual cortex.

Prenatal and postnatal simulations can also be set up to understand how contour integration circuitry is constructed. As was discussed in Section 16.4.8, Gaussian inputs such as LGN-filtered retinal waves should result in cocircular lateral connectivity patterns prenatally, which would allow the network to perform rudimentary contour integration. The lateral connections would be further refined through learning from visual inputs, eventually resulting in adult performance. Although it might be possible to verify this prediction experimentally already, further computational simulations would allow making the predictions much more detailed. PGLISSOM could be trained prenatally with Gaussians resembling input that the developing V1 receives, and postnatally with natural inputs (Section 17.2.8). Its ability to form synchronized representations for contours could then be tested at different stages of development, resulting in specific predictions for biological experiments.

Such computational studies would potentially allow accounting for all of the known data on how V1 develops in early life, and identifying specific gaps in our knowledge that can be addressed in further biological experiments.

17.2.4 Postnatal Internally Generated Patterns

When the V1 maps are constructed in two separate learning phases, prenatal and postnatal, the influence of the internally generated and environmentally driven stimuli can be clearly identified. Such a separation is a good model of spontaneous activity in the developing sensory areas, such as retinal waves, because the waves disappear at eye opening (Wong et al. 1993). But other activity, such as that during REM sleep, continues throughout development and adulthood (Callaway et al. 1987). These postnatal patterns suggest that pattern generation may also have a significant role beyond prenatal development.

Specifically, postnatal internally generated activity patterns may be interleaved with waking experience to ensure that postnatal development does not entirely overwrite the prenatal organization. Such postnatal patterns may explain why altered environments can only be learned partially (as found by Sengpiel et al. 1999), and why the animal spends so much time in REM sleep during the time when its neural structures are most plastic (Roffwarg et al. 1966). The postnatal patterns may help ensure that the visual system does not become too closely adapted to a particular environment (a phenomenon called "overtraining" in machine learning), which would limit its generality.

Such patterns would be needed only in systems that remain plastic in the adult, and they may provide a simple way to trade off between adaptability and genetically specified function in such systems. In future simulations, it should be possible to study how such interleaving interacts with experience. The results could be first validated with biological observations, such as those of Sengpiel et al. (1999), and then expanded to propose further experiments on how genetic bias is expressed in self-organizing systems.

17.2.5 Tilt Illusions

In Chapter 7, tilt aftereffects were shown to arise as interactions between subsequent visual patterns in the LISSOM model. Simultaneous inputs can also interact (as was demonstrated in a limited scale in Section 14.2.3), and cause distortions in perceived orientations. Such an effect, called the tilt illusion, is well documented psychophysically (Calvert and Harris 1988; Carpenter and Blakemore 1973; Gilbert and Wiesel 1990; O'Toole 1979; Smith and Over 1977; Wenderoth and Johnstone 1988; Westheimer 1990), but how it can arise from the two-dimensional spatial interactions in the cortex has not yet been demonstrated computationally.

In LISSOM, two stimuli should interact with each other as the lateral interactions settle, inhibiting neurons tuned to orientations between them. This effect should drive the two perceived orientations away from each other. Such an explanation was originally proposed by Carpenter and Blakemore (1973), and the principles have been demonstrated recently in an abstract model of orientation (Mundel, Dimitrov, and Cowan 1997).

With LISSOM, it should be possible to show how the tilt illusion depends on specific lateral connections, provided two extensions are made to the current simulations. First, because overlapping patterns could cause confounding effects, the inputs need to be separated spatially (as they are in psychophysical experiments). As a result, the radius of lateral inhibitory connections must be larger than that used in the tilt aftereffect simulations. Second, to self-organize such long connections, the training inputs would have to be correlated over a long range (as they are when the model is trained with natural images; Section 9.3.1). Spatially separated neurons will then develop lateral inhibitory connections, which causes the angle expansion. If it turns out that such connections would have to be longer than what can be simulated computationally, it may be possible to use shorter connections and more closely spaced test patterns by decoding the perceived orientations of overlapping lines using probabilistic methods (such as those of Zemel, Dayan, and Pouget 1998). In the extended model, the magnitude of the tilt illusion can be measured by computing the perceived orientations from each line alone, and comparing with the perceived orientation when both lines are presented at once.

Alternatively, it may be possible to test tilt illusions more economically using a combined orientation and ocular dominance simulation. In humans, when a different pattern is presented to the same location in each eye, they interact just as do two patterns presented to separate locations in one (or both) eyes (Carpenter and Blakemore 1973). Thus, it should be possible to test tilt illusions already in the network of Section 5.6.2, without first having to self-organize a model with a longer inhibitory radius.

Indirect tilt illusions similar to the indirect tilt aftereffect have been found in humans, and it might be possible to model them in LISSOM as well. Such an effect would arise if weakly activated units were facilitated by units at distant orientations; such facilitation could be mediated by lateral connections whose effective sign depends on local contrast, as it would in the extension to LISSOM proposed in Section 16.1.4. Implementing such extensions and observing their effects constitutes a most interesting direction of future work.

17.2.6 Other Visual Aftereffects

Many visual aftereffects similar to the tilt aftereffect are known to exist in biological vision. LISSOM could be extended to gain insight into these effects as well.

In addition to orientation, aftereffects of motion, spatial frequency, size, position, curvature, and color have been documented in humans (Barlow 1990; Howard and Templeton 1966; Schrater, Knill, and Simoncelli 2001; Wolfe 1984). For instance, a movement aftereffect known as the waterfall illusion can be induced by prolonged viewing of a moving stimulus: Stationary stimuli appear to be moving in the opposite direction (Kohn and Movshon 2003). Recent work also suggests that high-level tasks such as face perception have similar aftereffects (Leopold, O'Toole, Vetter, and Blanz 2001; Webster and MacLin 1999; Zhao and Chubb 2001). In all of these cases, the cortex adapts to a long-lasting stimulus, changing the perception of subsequent stimuli.

Using a LISSOM model that includes maps for the relevant features, it should be possible to demonstrate aftereffects for each of these dimensions. In each case, the effects would occur through short-term adaptation in specific lateral connections between feature-selective cells. For instance, presenting a continuously moving image to the direction map of Section 5.5 should result in a realistic movement aftereffect. Presenting single faces to the face-selective network of Section 10.3 should result in face-specific aftereffects.

Analogous aftereffects have also been found for other modalities, such as hearing, touch, muscle positioning, and posture (Howard and Templeton 1966). For instance, hearing a sound in one location can influence the perceived location of later sounds. That is, after adaptation, sounds presented in nearby locations appear to be farther away than they actually are, and the effect peaks at a certain distance, much like the direct tilt aftereffect. If development in these areas can be modeled with LISSOM (as is expected), aftereffects should also occur in such models. In this way, LISSOM could be used to provide a simple, unified explanation for a variety of perceptual aftereffect phenomena across modalities.

17.2.7 Hyperacuity

Like models of illusions and aftereffects, a LISSOM model of hyperacuity can provide useful information about how primary visual cortex adapts.

Performance in hyperacuity tasks, such as deciding whether two lines of same orientation are separated by a small perpendicular offset, improves with practice (Fahle, Edelman, and Poggio 1995; Weiss, Edelman, and Fahle 1993). The improvement occurs even without any feedback indicating whether each judgment is correct. The effect is specific to position and orientation, but transfers to some degree between eyes. This transfer is thought to indicate that at least some part of the effect arises in V1, because V1 is the first stage in the visual pathway where binocular inputs are combined.

Shiu and Pashler (1992) reported similar results for orientation discrimination tasks, although they found that the effect also depends on cognitive factors. Performance improved with practice only if the subjects were directed to pay attention to the orientation. However, the effect only occurred at the specific retinal location where the training examples had been presented, ruling out any deliberate cognitive strategy that the subject might have learned during the experiment. This result suggests that attentional mechanisms may activate circuitry in V1 (or other early visual areas) that regulates adaptation.

The LISSOM activation and learning mechanisms should be able to account for such basic psychophysical learning phenomena. The active units and lateral connections between them would adapt during repeated presentations. Over time, the area of the cortical map responding to those features would expand, allowing smaller differences to be represented and discriminated. However, the attentional effects might require an extension to high-level feedback, as discussed in Section 17.2.13. Such extended experiments might help clarify how and when adaptation occurs in early vision.

17.2.8 Grouping with Natural Input

Human contour integration performance depends on several stimulus dimensions in addition to orientation, including how random the background is, how jagged the path is, how much the elements of the contour are separated, and what spatial frequency, relative phase, color, and contrast the elements have (Field et al. 1993; Geisler et al. 2001; McIlhagga and Mullen 1996; Pettet et al. 1998). While human performance has been characterized in detail along most of these dimensions, their effect has not yet been analyzed computationally.

PGLISSOM can be trained with artificial input patterns where these dimensions are systematically varied. Alternatively, such training could be made more realistic by generating inputs based on known natural image statistics (as suggested by W. Geisler). Such inputs would need to be appropriately filtered based on known biological processes in the retina and LGN, because PGLISSOM is a model of V1 and assumes that such processes have already taken place. The model could then be used to predict human perceptual performance in more detail. It could also be used to identify the input statistics that are important for each stimulus dimension, by analyzing the lateral connection patterns and receptive field properties.

The next step would be to train the model with natural inputs such as moving natural images, as was done in Section 5.6.4. Currently such inputs cannot be used in PGLISSOM because the spiking network is computationally much more expensive to simulate than the firing-rate version. However, with the scaling-up techniques proposed in Section 17.2.9, such training should be possible. Natural images vary in all the stimulus dimensions described above, and should make PGLISSOM grouping performance sensitive to these dimensions.

Most interestingly, direction-selective cells that emerge from such training would allow grouping to take place based on motion, which is an important process in biological vision. Such a model would make it possible to compare the model directly with experimental observations on synchronization, which are usually based on moving inputs (Eckhorn et al. 1988; Gray and Singer 1987).

17.2.9 Scaling up to Large Networks

Developing techniques to simulate large networks accurately is a major issue in computational neuroscience research. Scaled-down versions of perceptual experiments had to be devised in this book in order to study them with the available model retina and cortex: For example, the tilt aftereffect inputs consisted of only a single line, instead of gratings used in human experiments; contour integration was performed over three to six elements in a background of zero to six elements, instead of 20 in 200. While the principles are the same and we believe the results are valid, such mismatch in scale makes it difficult to compare the results directly with human experiments. Also, large maps are necessary for many other important visual phenomena, such as visual attention, saccades between stimulus features, the interaction between the foveal and peripheral representations of the visual field, and the self-organization based on large-scale patterns of optic flow due to head movement. In order to understand them computationally, large parts of the visual cortex have to be simulated.

The parameter scaling equations and the GLISSOM map growing method introduced in Chapter 15 are the most promising avenue to date to scale up to large networks. As was discussed in that chapter, these techniques already allow simulating the entire V1 at the column level with desktop workstations. The techniques can be combined with parallel implementations of the LISSOM algorithm in order to simulate multiple visual areas or more-detailed column models (Chang and Chang 2002).

It may be possible to reduce the memory requirements of large simulations by modeling biological networks more directly. The lateral connection weights could be initialized based on known biological connectivity or connectivity derived from image statistics. Since such connectivity is usually sparse, a larger map can be constructed. Although the initial development of the map cannot be modeled in such networks, large-scale simulations can be run to test specific functional effects of biological connectivity patterns, and to test the components of the network in various realistic psychophysical tasks.

In order to deal with inputs as large as the entire visual scene, techniques could be developed for scanning the visual space sequentially with a small LISSOM network. If a model has only afferent connections, the input space can be partitioned into discrete grids the size of the LISSOM retina, and the responses for each grid location combined to form the global output. With lateral connections, the combination becomes more complicated because the lateral interactions between the different areas must be taken into account. As a first approximation, it may be sufficient to represent only the lateral connections of the LISSOM map to its eight neighboring locations in the grid. These same connections can then be used at all grid locations. In this manner, it may be possible to self-organize and run an arbitrarily large LISSOM map that is constructed on the fly from local components.

With the scale-up techniques, it should be possible to apply the LISSOM approach to many new visual phenomena in a realistic scale. Before its performance can be directly compared with that of humans, it will also be necessary to extend LISSOM to model foveated input, as will be described next.

17.2.10 Foveated Input and Eye Movements

Current LISSOM simulations are based on a uniform representation of the visual field, which is appropriate when modeling a small patch of the retina and the corresponding parts of the visual cortex. However, a number of visual phenomena depend on differences between central and peripheral visual processing (such as contour integration; Section 13.1.1). In the periphery, retinal ganglion cells are spaced much farther apart and have much larger receptive fields than in the fovea, and thus the mapping of visual space differs significantly between central and peripheral vision. As a result, object perception performance varies across the visual field (Levy, Hasson, Avidan, Hendler, and Malach 2001; Mäkelä, Näsänen, Rovamo, and Melmoth 2001; Strasburger and Rentschler 1996). For instance, faces in the periphery need to be both larger and have higher contrast to be recognized, compared with those in central vision.

To understand these experimental results, a large-scale version of LISSOM can be implemented that includes both central and peripheral processing. The architecture would be mostly the same as in the current model, perhaps scaled up with the techniques discussed in Section 17.2.9. In addition, a module would be included before the photoreceptors to perform a log-polar transformation of the visual image before computing the activation of the photoreceptors. This transformation could be adjusted over time to take into account that the fovea develops later than the rest of the retina (Abramov et al. 1982; Kiorpes and Kiper 1996). Such a transformation process would simulate the nonlinear distribution of retinal ganglion cells, without requiring changes to the LGN, V1, or FSA models.

Although modeling foveated input requires only these small changes to the LIS-SOM model of V1, to understand fully the effects of the fovea on self-organization and visual function, it may be necessary to include mechanisms for moving the direction of gaze.

In normal vision, the fovea is directed at several different visual targets each second, changing between targets with a quick saccade. The saccades are controlled by subcortical areas such as the superior colliculus and by high-level areas such as the frontal eye fields (see Bisley and Goldberg 2003 for a review). Including regions that generate eye movement would greatly complicate the model, but would also complete a loop between eye movement, retinal image, and subsequent eye movements. In the long run, such a model will be crucial for understanding how the visual system utilizes representations in the fovea and the periphery to make sense of the visual environment.

17.2.11 Scaling up to Cortical Hierarchy

In addition to scaling to larger maps, the visual cortex model can be scaled up vertically by including maps beyond V1 in the visual hierarchy, such as V2, V4, and MT. Such large-scale models will rely on detailed data now becoming available about the connectivity and functional properties of higher visual areas (e.g. Heeger, Boynton, Demb, Seidemann, and Newsome 1999; Kötter 2004; McCormick, Choe, Koh, Abbott, Keyser, Melek, Doddapaneni, and Mayerich 2004a; McCormick, Mayerich, Abbott, Gutierrez-Osuna, Keyser, Choe, Koh, and Busse 2004b; McGraw, Walsh, and Barrett 2004; Pinsk, Doniger, and Kastner 2004; Van Essen 2003; Wong and Koslow 2001). The ultimate goal would be to self-organize structures as complex and powerful as the primate visual system, with dozens of interacting visual areas allowing recognition of highly complex patterns. First steps toward this goal were presented in Section 10.2, where the high-level FSA was organized based on inputs from the V1 model.

Self-organization of hierarchical structures is a difficult unsupervised learning task in general (Becker 1992). The idea is to apply self-organization in multiple stages and to discover increasingly complex structures in the input. However, linear projections and nonlinear topographic mappings (such as the SOM) do not usually suffice: Each level will represent essentially the same information even if it is scaled or organized differently. In contrast, LISSOM includes several extensions that make it possible to discover high-level representations: (1) The afferent receptive fields are local, and higher levels receive information from broader areas than lower levels; (2) the activation function includes a threshold, ensuring that only the best-matching neurons respond; and (3) the lateral interactions decorrelate the representations, forming a sparse, redundancy-reduced code that makes recognition and classification at higher levels easier (as was shown in Section 14.3). The higher levels can then develop complex feature preferences like those found in higher levels of the visual cortex.

With a hierarchical model, it should be possible to show how high-level perceptual properties can develop. One important such property is translation-invariant and viewpoint-invariant responses. These invariances are crucial for large-scale object and face recognition, because large objects are usually not encountered at precisely the same retinal position and orientation each time. Lateral connections are believed to play a crucial role in this process (Edelman 1996; Marshall and Alley 1996; Wiskott and von der Malsburg 1996); however, existing models do not yet develop ordered maps for orientation or other low-level visual features, and do not explain how the lateral connectivity can develop along with the map (Bartlett and Sejnowski 1998; Földiák 1991a; Olshausen, Anderson, and Van Essen 1995, 1996; Stringer and Rolls 2002; Wallis and Rolls 1997). Instead, the models are based on hierarchically arranged sheets of neurons that respond to faces or specific objects over a wide range of positions or three-dimensional viewpoints.

Despite this difference in focus, the overall architectures of most transformation invariance models are similar to a hierarchical LISSOM model, and their mechanisms could be implemented in LISSOM. For instance, the VisNet family of models (Rolls and Milward 2000; Stringer and Rolls 2002; Wallis 1994; Wallis and Rolls 1997) achieves transformation invariance using the trace learning rule (Section 17.1.4; Földiák 1991a). Because a moving object will assume a number of different spatial positions and configurations over time, the trace learning rule ensures that responses to each of these views will become associated. At each subsequent hierarchical level, neurons will process larger areas of the visual field, leading to translation and viewpoint invariance at the highest levels (comparable to monkey inferotemporal cortex).

A hierarchical LISSOM model would provide a first unified account of how topographic maps can develop in a hierarchy of visual areas, how their function depends on self-organized lateral connections, and how high-level properties such as transformation invariance emerge in this process. In the following four subsections, opportunities for understanding several high-level perceptual phenomena with hierarchical LISSOM are reviewed.

17.2.12 Line-End-Induced Illusory Contours and Occluded Objects

PGLISSOM was used in Section 13.3.4 to show how edge-induced illusory contours could arise based on the same mechanisms as normal contour integration. An important future experiment with hierarchical LISSOM is to include end-stopped cells in the model, and demonstrate how line-end-induced illusory contours, and possibly occluded objects, could be detected in V2.

As was discussed in Section 17.1.4, the behavior of cortical columns in the current LISSOM models is based on simple cells, which are the first in V1 to show orientation selectivity. Such cells in the cortex usually respond more strongly when the input stimulus, such as an oriented line element, gets longer. In contrast, end-stopped cells also found in V1 respond best to an input with a particular length (Gilbert and Wiesel 1979), and they have been proposed to be responsible for both line-end-induced illusory contour completion and occluded object recognition (Finkel and Edelman 1989; Kellman, Yin, and Shapley 1998; Rensink and Enns 1998; Sajda and Finkel 1992; Weitzel, Kopecz, Spengler, Eckhorn, and Reitboeck 1997).

End-stopped cells have been found in layer 4 in cats. They are thought to arise when a layer-6 cell, which typically has a wider receptive field, inhibits (through an inhibitory interneuron) a layer-4 neuron with a smaller receptive field: Inputs that are exactly as long as the smaller RF excite such a layer-4 neuron the most (Bolz and Gilbert 1986; Gilbert 1994). The end-stopped cells further project to V2, and may form a basis for orientation columns for illusory contours that have been observed in V2 (Sheth et al. 1996).

In order to understand line-end-induced illusory contours and occluded object detection with PGLISSOM, it will first have to be extended with a set of neurons and intracolumnar circuitry in layers 4 and 6 that give rise to end-stopped receptive fields: These neurons in layer 6 have wider receptive fields and inhibit the neurons in layer 4, which have narrower receptive fields. During self-organization, these layer-4 neurons will develop into end-stopped cells and self-organize with the rest of the orientation map. A V2 map, receiving input from the end-stopped cells, will then self-organize based on the oriented end-stop activity, and form an orientation map of illusory contours in a process similar to how the V1 forms an orientation map of ordinary contours.

The model can be tested in illusory contour detection and in occluded object detection tasks. If successful, it would demonstrate how such behavior can arise from input-driven self-organization in V1 and V2. Further, since the V2 in such an extended model will behave as the V1 in the current LISSOM, tilt aftereffects should occur between illusory contours in much the same way as they do in ordinary contours in LISSOM (Chapter 7). Such aftereffects have indeed been found in psychophysical experiments (Berkley, Debruyn, and Orban 1993; Paradiso, Shimojo, and Nakayama 1989; van der Zwan and Wenderoth 1994, 1995); the model predicts that the same underlying mechanism is responsible for both of them.

In this manner, a hierarchical LISSOM model can be used to understand self-organization and function of V2 and higher levels of visual processing. An important further extension of the hierarchy is to include feedback from higher levels, as will be discussed next.

17.2.13 Feedback from Higher Levels

Current hierarchical LISSOM models, such as the HLISSOM network of Chapter 8 and those proposed so far in this chapter, are feedforward only: Activation propagates from the eye to LGN, to V1, and to higher levels, but not in the reverse direction. Including feedback connections is an important direction of future work that will allow us to model several new phenomena.

In the cortex, a large proportion of connections propagate in the reverse direction, connecting from higher levels to V1 and the LGN (see Gandhi, Heeger, and Boynton 1999; Lamme, Super, and Spekreijse 1998; Van Essen et al. 1992; White 1989 for reviews). The role of these feedback connections is not yet clear, but they may be involved in top-down pattern completion, attention, visual imagery, and large-scale object grouping. In many cases, they may achieve these effects by enhancing the existing lateral interactions (Freeman, Driver, Sagi, and Zhaoping 2003).

During self-organization, the feedback connections may also encourage different areas to develop synergetically, mediating competition and cooperation between multiple areas (Rolls 1990). Thus, over a large spatial scale, feedback connections may act like lateral connections within each area. The arrangement of maps into a hierarchy may even be primarily a means for making such large-scale connections more efficiently than could be achieved in a single large, laterally connected map (Kaas 2000).

As a first approach, feedback connections can be included in LISSOM just like afferent and lateral connections, as additional terms in each neuron's activation function. The principle is particularly clear if the activation function is written by indexing over receptive fields (as is done in Appendix A and implemented in *Topographica*):

$$
\eta_{ij}(t) = \sigma \left(\sum_\rho \gamma_\rho \sum_{kl \in \mathrm{RF}_\rho} X_{kl}(t-1) w_{kl,ij} \right),
\tag{17.1}
$$

where the index ρ indicates afferent, lateral, and feedback receptive fields (RF), $X_{kl}(t-1)$ is the activation of neuron (k,l) in that receptive field, and $w_{kl,ij}$ is the weight from that neuron to neuron (i,j). The sign of scaling factor γ_ρ is positive for afferent and lateral excitatory connections, and negative for lateral inhibitory connections. Although individual feedback connections are usually excitatory in the cortex (White 1989), they may have inhibitory effects for strongly activated neurons (like lateral connections do; Weliky et al. 1995). Both approaches can be tested in LISSOM by changing the sign of γ_ρ for the feedback connections. In future work, this approach can be further generalized by adding different delays for afferent, lateral, and feedback connections, thus replacing "1" in the equation with a specific d_ρ for each connection type. For instance, lateral connections have a higher latency, on average, than feedback connections (Nowak and Bullier 1997), and would thus have longer delays.

Using a hierarchical version of LISSOM with feedback, it should be possible to explain visual phenomena like pattern completion, where an object-selective neuron in a higher level can bias those neurons in a lower level that generally cause it to fire, thus completing missing or weak low-level features. Such feedback could also be used to account for a wider range of illusory contours, and for multi-modal integration, as will be discussed in the next two sections.

Feedback most likely plays a large role in perceptual systems, a role that has only recently began to be understood (Carpenter 2001; Dayan, Hinton, Neal, and Zemel 1995; Knoblauch and Palm 2003; Kosslyn and Sussman 1995; Murray, Schrater, and Kersten 2004; Pollen 1999; Schyns, Goldstone, and Thibaut 1998). Although much of the insight comes from computational experiments, it has been difficult to build large-scale computational models with self-organized hierarchy and feedback. The extensions to LISSOM proposed above, as well as their practical implementation

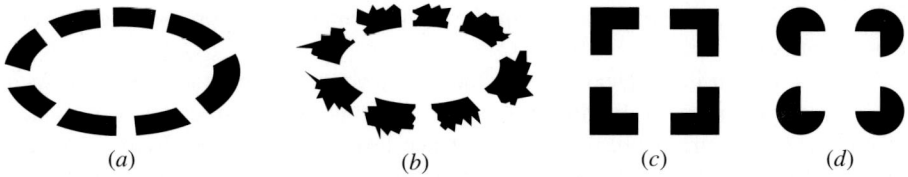

(a)   (b)   (c)   (d)

Fig. 17.1. High-level influence on illusory contour perception. Low-level mechanisms such as contour completion can account for some illusory contour perception phenomena. However, surrounding context can affect how salient the illusory contours are, suggesting that higher levels influence the low-level mechanisms. The patterns in (a) and (b) have the same oval outline in the middle, but only in (b) is the oval seen as a floating illusory object. Similarly, (c) and (d) have the same square outline in the middle, but the illusory square is prominent only in (d). In (a) and (c), the boundary elements are perceived as individual objects at the high level, suppressing the illusory effect. Adapted from Hoffman (1998).

provided by the *Topographica* simulator (Section 17.4), should make such models possible, making it possible to understand perceptual systems at a new level of detail.

17.2.14 High-Level Influence on Perceptual Grouping

Although synchronization in V1 can explain many perceptual grouping phenomena, feedback from higher levels of visual processing also has an effect in many cases. An interesting question is whether synchronized activities exist in high-level visual and cognitive areas of the brain, and how they might influence low-level perception and behavior.

In fact, correlated spiking has been found in the frontal cortex of awake monkeys. Such spiking forms synfire chains, where a population of neurons firing synchronously activates another population in a successive, feed-forward manner (Abeles 1991; Abeles, Bergman, Gat, Meilijson, Seidemann, Tishby, and Vaadia 1995; Abeles, Bergman, Margalit, and Vaadia 1993; Vaadia, Haalman, Abeles, Bergman, Prut, Slovin, and Aertsen 1995). Further, these chains correlate with the behavioral states of the animal: In a delayed response task, different chains were observed leading to different responses (Prut, Vaadia, Bergman, Haalman, Slovin, and Abeles 1998). Through backprojections, synfire chains could also affect synchrony in low-level areas. For example, Sillito, Jones, Gerstein, and West (1994) showed that to achieve synchrony, the LGN needs feedback from V1. Such observations suggest that synchronized firing indeed exists in higher cortical areas, and it can influence processing in lower levels.

Further evidence for this idea comes from certain complex illusory contour phenomena. For example, even though the central regions in Figure 17.1a and b are identical, the region is salient only in b (the same is true of c and d). This phenomenon suggests that all illusory contours are not based purely on bottom-up activation of V1 or V2 neurons; feedback from higher levels influences them as well. In Figure 17.1a and c, the parallel contours cause the boundary elements to be recognized as objects at a higher level, and feedback from their representations suppresses the illusory contour effect.

At this point it is not clear whether the high-level representations are local, such as columns on a map, or distributed, i.e. patterns of activation spread across an area of cortex. If they are local, they can influence lower levels simply by projecting a broad, diffuse set of connections back to the lower level; the neurons receiving these connections will then synchronize with the high-level representation. On the other hand, if the high-level representations are distributed, each neuron may project back just a small, focused set of connections; when the high-level representation synchronizes, the back projections will synchronize the corresponding low-level representation (von der Malsburg 1999).

Computational experiments with PGLISSOM can help distinguish between these alternatives. The model will first have to be extended with high-level object representations, such as localist and distributed representations for parallel lines. With each alternative, the contextual cues can be varied and the synchrony emerging at the lower level observed. Such a model can lead to a computational account on how high-level objects are represented in the cortex, and also how grouping is affected by feedback from higher levels.

Similar computational experiments with PGLISSOM can also be used to understand other phenomena of high-level influence on perceptual grouping. For example, which afterimages are observed after viewing patterns like those in Figure 17.1 (i.e. the boundary elements or the central illusory object) depends on the high-level context (Shimojo, Kamitani, and Nishida 2001); this phenomenon could be modeled with PGLISSOM that combines high-level feedback with short-term adaptation (such as that responsible for the tilt aftereffect; Chapter 7). Similarly, integration of inputs from different sensory modalities involves feedback from higher areas, and can be modeled with PGLISSOM as discussed in the next section.

17.2.15 Multi-Modal Integration

The different sensory modalities are known to interact in the brain: For example, auditory processing and visual perception influence each other (Churchland, Ramachandran, and Sejnowski 1994; McGurk and MacDonald 1976; Repp and Penel 2002; Stein, Meredith, Huneycutt, and McDade 1989), and so do touch and vision (Bach y Rita 1972, 2004; Zhou and Fuster 2000). Once LISSOM is extended to include a hierarchy of sensory areas and feedback from higher levels, it can be used to test hypotheses about how such interactions take place.

One well-documented multi-sensory phenomenon is the coupled development of auditory and sensory areas in the barn owl (Haessly, Sirosh, and Miikkulainen 1995; Knudsen and Knudsen 1985; Rosen, Rumelhart, and Knudsen 1995). The auditory spatial map in the inferior colliculus depends partially on visual input the animal receives during development. A hierarchical LISSOM network can be extended to model this phenomenon by including auditory and visual channels converging on a higher level map. The low-level maps learn to represent the auditory and visual space, and the high level learns associations between the two modalities. With back-projections, the high-level map then correlates the low-level maps as well, resulting in coupled development of the two modalities.

Another important issue is how the sensory information in different modalities is integrated during performance. One possibility is that a higher level area performs the integration (Section 17.2.14; de Sa 1994; de Sa and Ballard 1997). Another possibility is synchronization: When two representations in different modalities are synchronized, they are perceived as part of a single experience. Although coherent oscillations have been found in sensory areas other than vision (such as the olfactory bulb and the auditory system; Eeckman and Freeman 1990; Friedrich and Laurent 2001; Joliot et al. 1994), it is not yet known whether the different modalities are bound together through synchronization. Computational models such as PGLISSOM can be instrumental in testing this hypothesis.

Integration studies require simulating multi-modal brain areas, i.e. regions that receive strong input from multiple sensory modalities. One such candidate is the posterior part of the intraparietal area (PIP), where integration of tactile and visual information has been observed in fMRI studies (Saito, Okada, Morita, Yonekura, and Sadato 2003). A PGLISSOM model of this system would include sensory areas, a higher level area representing the PIP, and feedback mechanisms similar to those discussed in Section 17.2.14. The model would be first validated by matching its spike activity with the fMRI data. Different inputs would then be systematically presented and the resulting synchronization in the PIP and the low-level maps observed. In this way, it would be possible to predict what kinds of representations are activated and synchronized during multi-modal integration tasks. These predictions could then be verified in biological experiments, using e.g. the techniques proposed in Section 16.4.7.

In addition to high-level cortical areas like the PIP, subcortical areas such as the thalamus can contribute to multi-modal integration (Crabtree, Collingridge, and Isaac 1998; Crabtree and Isaac 2002; Hadjikhani and Roland 1998; see Calvert 2001; Sherman and Guillery 2001 for reviews). For example, Choe (2002, 2003a,b, 2004) recently showed how the recurrent activation between the thalamus and different cortical areas can establish analogical mappings between cortical representations (see Jani and Levine 2000; Kanerva 1998 for general neural mechanisms of analogy). Such an approach can be extended to establish mappings between sensory modalities, such as those between orthographic and phonetic representations of words and sharp visual edges and high-pitch tones. Corticocortical connections between different sensory areas initiate such mappings, and the thalamus selects the most appropriate ones among the resulting activity. If the PGLISSOM architecture were to be extended to include the thalamocortical loop, synchronization between different sensory modalities could be used to represent the analogy. Such a model would lead to concrete predictions about how information in different modalities is represented and associated.

A number of other psychophysical phenomena involving multi-modal integration have been described as well (Meredith and Stein 1986; Stein and Meredith 1993). Computational models such as PGLISSOM can serve an instrumental role in formulating hypotheses about their neurobiological foundations. Such models constitute a significant step toward understanding the multi-modal, multi-level, and recurrent nature of the perceptual system.

17.3 New Research Directions

The LISSOM framework was developed to understand the visual system in computational terms. However, experience with it can serve to motivate research in other areas, and even suggest entirely new research directions. A number of such ideas are reviewed in this section, including theoretical analysis of visual computations, training realistic natural and artificial vision systems, and constructing innate capabilities and complex systems through interactions between evolution and learning.

17.3.1 Theoretical Analysis of Visual Computations

Computational experiments with LISSOM demonstrate how the visual cortex can develop and function based on a number of biologically motivated computational principles. The next step in this direction is to analyze the model theoretically, determining what its computational goals are and which of its mechanisms are necessary to achieve them. The two main directions are defining an objective function for LISSOM self-organization, and characterizing the capabilities of temporal coding.

The ultimate goal in theoretical neuroscience is to understand why the cortical structures exist, i.e. what is their purpose and role in information processing (Arbib, Érdi, and Szentágothai 1997; Barlow 1994; Churchland and Sejnowski 1992; Dayan and Abbott 2001; Field 1994; Hecht-Nielsen 2002; Marr 1982; Rao, Olshausen, and Lewicki 2002). With LISSOM, a crucial question is: What is the goal of the self-organizing process? In Chapter 14, the self-organized LISSOM network was shown to form sparse representations by reducing redundancy in the input, and these representations were found to be effective in further stages of information processing. This observation suggests that the goal of self-organization is to form structures that allow representing the visual input in an optimal manner, given the biological constraints. What exactly is the objective function, and what are the constraints?

One possibility is that the process optimizes the ability to reconstruct the input, i.e. maximizes the information retained in visual cortex representations. Alternatively, these representations could be optimized for the needs of further stages in visual processing, such as pattern recognition. The process could be constrained by physical resources in the cortex, such as wiring length, i.e. the extent and total strength of the lateral connections, the total amount of activation in each cortical representation, or the smoothness and continuity of cortical activation patterns (Bell and Sejnowski 1997; Chklovskii, Schikorski, and Stevens 2002; Hochreiter and Schmidhuber 1999; Koulakov and Chklovskii 2001; Olshausen and Field 1997).

Such hypotheses can be verified by constructing a mathematical model with the proposed objective function and constraints, and showing that optimizing it results in processes and structures similar to those in the computational model (Wiskott and Sejnowski 1998). In fact, a settling and learning process very similar to that in LISSOM can be derived in this way by optimizing reconstruction with sparse representations (Olshausen 2003; Olshausen and Field 1997). The activity and the weights are optimized in two alternating phases using gradient ascent. First, given an input image and the weights, the most likely sparse network activity pattern is found. Second, given the input and the activity, the weights are found that allow reconstructing the input as accurately as possible. The activity update that results is similar to LISSOM settling and the weight update similar to Hebbian learning.

Using the same approach with further constraints (such as wiring length) and further objectives (such as pattern recognition accuracy), it may be possible to constrain the optimization process toward final structures similar to those found in LISSOM. Such a result would serve to identify a computational goal for the LISSOM process, and to the extent LISSOM accurately models the cortex, to biological self-organization as well.

In PGLISSOM, the self-organizing process is further extended with temporal dynamics. An important question is whether the computational goals that the system implements are also extended in this way, or whether spiking is simply a low-level implementation of rate coding. One way to analyze computational capacity of spiking networks is through Vapnik–Chervonenkis (VC) dimension analysis (Bartlett and Maass 2003; Valiant 1994; Vapnik and Chervonenkis 1971).

The VC dimension describes how complex classifications a given network can represent: the higher the VC dimension, the more complicated functions it can represent, but also the less likely it is to generalize well to novel inputs. Typically, the more hidden neurons the network has the higher its VC dimension. Zador and Pearlmutter (1996) first analyzed the VC dimension of leaky integrate-and-fire networks. Subsequently, Maass (1997) showed that networks with noisy spiking neurons require fewer neurons to perform the same function than rate-coding networks with linear threshold and sigmoidal activation functions, and are therefore more economical and likely to generalize better. It should be possible to use similar techniques in PGLISSOM, and perhaps demonstrate whether its self-organizing performance is in principle different from that of the rate-coded LISSOM.

In PGLISSOM, the mechanisms of Hebbian self-organization and grouping through synchronization are shown to operate in a single computational structure. Such a model can be used to motivate a theory of how these processes might interact. For example, it may be possible to include grouping as part of the objective function for self-organization, and demonstrate that the two-layer architecture results from combining these two functions optimally. In this manner, LISSOM can be used as a testbed for bringing together different theories of brain function, eventually resulting in a formal description of computations in the visual cortex.

17.3.2 Genetically Specified Pattern Associations

As was discussed in Section 16.2.3, internal pattern generation may be a way to encode species-specific knowledge into a self-organizing process. So far we have seen how such patterns could be useful in constructing the visual system, but the idea is more general. Internal pattern generation could play a role in establishing many other behaviors that have traditionally been considered innate, such as certain emotional responses.

As discussed in Section 17.2.4, internal patterns continue to be generated during REM sleep even in the adult, and they could affect other areas besides the visual system. For instance, the amygdala, an area primarily involved in processing emotion, is strongly activated during REM sleep (Maquet and Phillips 1998). This observation raises an intriguing possibility: Internally generated patterns may train the animal to associate a particular emotional valence (positive or negative) with a class of visual patterns (Bednar 2000).

Although this hypothesis is difficult to test directly, it is consistent with behavioral data. In a series of studies on rhesus monkeys raised alone (without other monkeys or mirrors), Sackett (1966, 1970) found behaviors suggesting that the monkeys had specific genetically determined preferences. For instance, beginning around 2–2.5 months of age, the monkeys became more playful when an image of an infant monkey was shown than when pictures of older monkeys were shown, and they would more often press a lever to continue the display of the infant image. At the same age, they became visibly disturbed and agitated when an image of a monkey with a threatening facial expression was shown, and were less likely to continue its display (Sackett 1966). Remarkably, these responses had developed without experience with either playful or threatening live monkeys. Also, the responses to the threatening images decreased with time and were no longer measurable by 3.5 months, suggesting that the original negative emotions associated with such images were overwritten by the experience where the threats never materialized.

Other associations in the same monkeys arose later in development, again without any environmental experience. These tests used live male and female monkeys visible in nearby cages, and measured which cage a test animal approached (Sackett 1970). When the ages of the visible monkeys matched the age of the test monkeys, deprived infants younger than 9 months approached both sexes equally. Older deprived infants preferred members of their own sex, and adult deprived monkeys preferred members of the opposite sex (Sackett 1970). When the visible age-matched monkeys were replaced with adults, deprived newborns strongly preferred females, but deprived males began to prefer males when they reached 48 months of age. Importantly, this change occurred at nearly the same age as in normal, non-deprived monkeys. Thus, positive or negative associations with visual stimuli can arise independently of visual experience in animals.

It may be possible to explain these behaviors through internal pattern generation in the HLISSOM framework. Specifically, a pair of high-level units representing positive and negative emotions can be added to the model, with connections to and from the highest cortical areas, and reciprocal inhibitory connections with each other. A set of internally generated patterns is then designed to represent the infant, male adult, and female adult faces. Through internal training, each type of pattern is associated with a positive or negative emotion, and the associations are tested using images of real faces. The model should produce the same emotional response to real faces as to their internally generated counterparts.

In this manner, genetically determined associations can be maintained for different classes of environmental stimuli throughout development, independently of environmental associations. The approach demonstrates how genetically supervised learning can occur in an otherwise unsupervised system.

17.3.3 Embodied, Situated Perception

The simulations discussed so far can all be performed off-line in experiments with a predetermined corpus of input patterns (e.g. a set of natural images or video). However, it can be very difficult to obtain a realistic, representative set of such patterns. The standard databases of e.g. face images include mostly canonical examples, instead of a realistic sampling of faces as seen by an infant. In order to train a model to perform like an infant, it will have to be trained with data seen by the infant. How can such data be obtained?

Einhauser, Kayser, Konig, and Kording (2002) recently took the first steps in collecting such realistic training data by attaching a video camera to an adult housecat's head, and collecting video while it freely roams the environment. Einhauser et al. argue that this approach is a good approximation of the visual experience of the adult cat, because cats tend to move their heads (and not just their eyes) when scanning a visual scene. However, the data may not be appropriate for modeling how the cat's visual system develops. The visual experience of a kitten may be quite different from that of an adult cat. For instance, very young kittens move around their environment much less than adult cats do, particularly during the earliest times after eye opening while their visual systems are first developing.

Moreover, it is impractical to extend this approach to human infants, or even to closely related primates. Apart from any ethical considerations, the video thus recorded would not necessarily be a good approximation of the infant's visual experience. In humans, eye movements can be significantly different from head movements: Humans often keep their heads still while scanning a scene or fixating on different parts of an object. Although eye-tracking equipment exists, it could not be attached to an infant or animal unobtrusively for several months.

Even if a realistic set of training data were obtained, video alone would not capture the crucial features of vision, particularly in the higher, cross-modal areas of the nervous system. Vision is an active process of interaction with the environment, driven by attention (Findlay and Gilchrist 2003; Gibson 1979; Goodale and Milner 1992; Milner and Goodale 1993; O'Regan and Noë 2001; see Findlay 1998 for a review). The video data can record the pattern of light falling on the eye, but it cannot capture the causal link between visual system activity, visual attention, and the eye movements that determine the specific images seen and thus future activity patterns. Given how difficult it is to replicate this feedback loop, it will be difficult to reach human levels of visual system performance by simply making training patterns more realistic.

The active vision approach directly addresses these issues. Vision is no longer limited to passive interpretation of visual input, but involves an active motor control process as well. This approach allows using resources efficiently, and makes many difficult computer vision problems easier to solve, such as inferring shape from shading, contours, and texture, and computing the optic flow (Aloimonos, Weiss, and Bandyopadhyay 1988; Bajcsy 1988; Ballard 1991; Ballard, Hayhoe, Pook, and Rao 1997; Blake and Yuille 1992; Harris and Jenkin 1998; Landy, Maloney, and Pavel 1995). However, in most active vision systems, the sensorimotor interaction is limited to the visual apparatus (e.g. rotation of stereo cameras), and the full range of movement in natural agents has not been fully utilized.

For these reasons, there is a growing sense that "embodied" or "situated" perception is the necessary next step toward understanding postnatal learning, as well as cognitive functions in general (Bailey, Feldman, Narayanan, and Lakoff 1997; Beer 2000; Brooks, Breazeal (Ferrell), Irie, Kemp, Marjanović, Scassellati, and Williamson 1998; Choe and Bhamidipati 2004; Clark 1999; Cohen and Beal 2000; Elliott and Shadbolt 2003; Goldstone 2003; Langley, Choi, and Shapiro 2004; Markman and Dietrich 2000; Pfeifer and Scheier 1997, 1998; Philipona, O'Regan, and Nadal 2003; Pylyshyn 2000; Regier 1996; Thompson and Varela 2001; Weng, McClelland, Pentland, Sporns, Stockman, Sur, and Thelen 2001; Ziemke 1999). For instance, it may be useful to study a relatively simple robot as an example of an individual interacting with its environment. Such a system may provide insights that could not be obtained from a purely computational study of human development, because it can preserve the causal link between eye, head, and body movements and the visual system activity that results.

Therefore, implementing models like LISSOM within situated devices such as simple robots is an important new direction in understanding how perceptual systems develop in animals.

17.3.4 Building Artificial Vision Systems

In addition to providing a precise understanding of the mechanisms underlying visual processing in the brain, computational models can serve as a foundation for artificial vision systems. Such systems have the advantage that they are likely to process visual information the same way humans do, which makes them appropriate for many practical applications.

First, LISSOM networks can be used to form efficient internal representations for pattern recognition applications. A method must be developed for automatically identifying active areas in the maps and assigning labels to neural populations that respond to particular stimulus features. This task is similar to the problem of interpreting the neural code (Section 16.3.2), and the same techniques can probably be used. One particularly elegant approach, demonstrated in Section 14.3, is to train another neural network to do the interpretation. In such a system, the feedback idea discussed in Section 17.2.13 can be useful in training the map. By adding backprojections from the interpretation network back to the map, a supervised process similar to learning vector quantization and top-down expectations could be implemented (Kohonen 1990; Xu and Oja 1990). The backprojections learn which units on the map are statistically most likely to represent the category; they can then activate the correct LISSOM units even for slightly unusual inputs, resulting in more robust recognition.

Second, a PGLISSOM network can serve as a first stage for object recognition and scene analysis systems, performing rudimentary segmentation and binding. Like contour integration, object binding and object segmentation are thought to depend on specific long-range lateral interactions (Gilbert, Das, Ito, Kapadia, and Westheimer 1996), so PGLISSOM is in principle an appropriate architecture for the task. With the scale-up approaches discussed in Sections 17.2.9 and 17.2.11, it should be possible to build networks large enough to cover the entire scene. The system would consist of multiple hierarchically organized PGLISSOM networks. At the lowest level, preliminary features such as contours would be detected, and at each successively higher level, the receptive fields cover more area in the visual space, eventually representing entire objects. At each level, synchrony would effectively represent coherent components and desynchrony would segment different components. A high-level recognition system could then operate on these representations to perform the actual object recognition and scene interpretation.

Including such postprocessing steps in LISSOM, and scaling it up to larger and hierarchical maps, should allow building robust pattern recognition and low-level vision systems up to the level of identifying coherent objects.

17.3.5 Constructing Complex Systems

The simulations on pattern-generator-driven self-organization demonstrate how genetic and environmental influences can interact in constructing a complex visual processing system (as was proposed in Section 2.3). The approach is more general, however, and can be seen as a general-purpose problem-solving approach that can be applied to a variety of fields, perhaps making it practical to develop much larger computing systems than we use today.

In the most straightforward approach, the pattern generator can be designed specifically for the task, as was done in Part III of this book. Such a generator allows the engineer to express a desired goal without having to hard-code it into a particular, inflexible architecture. In essence, the engineer will bias the learning system with generated patterns, allowing it to solve problems that would otherwise be difficult for learning systems. For example, simpler patterns can be learned before real data, thereby avoiding local minima in the search space of solutions (Elman et al. 1996; Gomez and Miikkulainen 1997; Nolfi and Parisi 1994; Tonkes, Blair, and Wiles 2000). Such bootstrapping may also allow the designer to avoid expensive and laborious manual collection and/or tagging of training datasets, as in tasks like handwriting recognition and face detection. For instance, a three-dot training pattern could be used to detect most faces, and only the patterns that were not detected would need to be tagged manually (Viola and Jones 2004).

More significantly, the pattern generator could be constructed automatically using evolutionary algorithms (EAs, such as genetic algorithms, evolution strategies, classifier systems, or genetic or evolutionary programming; Beyer and Schwefel 2002; Fogel 1999; Goldberg 1989; Holland 1975; Koza 1992; Mitchell 1996). In this approach, domain-specific knowledge necessary to design the generator by hand would not be needed. For instance, studying real faces may lead one to suggest that a three-dot pattern would be a good training pattern to bootstrap a face detector; however, often such knowledge can only be obtained through trial and error, and it would be better to have an algorithm to do it automatically. Indeed, the self-organizing system, the pattern generator, and the EA together can be considered a single general-purpose adaptive algorithm.

What benefits would such a system have over other adaptive systems, such as EAs or learning networks alone? Essentially, the combination of learning and evolution represents a balance between adaptation at different time scales (i.e. determines a proper tradeoff between bias and variance; Section 2.3.1). Short-term learning allows an individual network to become particularly well suited for the particular tasks on which it is tested. Long-term adaptation (i.e. selection by the EA) can ensure that short-term learning does not reduce generality. For instance, the EA can select training patterns to ensure that a system is able to handle events that occur rarely, yet are vitally important over the long term. For example, a computer vision system for detecting faults in manufactured devices can be trained both on the typical cases of correct devices, plus specifically generated examples of faults and defects. The EA can also select pattern generators that get the system "in the ballpark," to increase the chance that learning will succeed. Thus, by combining EAs and learning using pattern generators, it should be possible to evolve systems that perform better than using either approach alone.

A concrete first test of this idea could be devised in a pattern recognition domain such as identifying handwritten characters. Such a system has to be both flexible (to adapt to a particular person's handwriting) and general (to recognize writing by many people; Revow, Williams, and Hinton 1995). An EA could be used to design training patterns and a learning algorithm to adapt to those patterns and to real-world examples.

More specifically, the core of such a system would be a neural network that receives the handwritten digits as its input and produces the classification of each digit as its output. Three different ways of constructing such a network could be compared. First, the network weights could be evolved directly using an EA (Schaffer, Whitley, and Eshelman 1992; Stanley and Miikkulainen 2002; Yao 1999). Second, the network could be trained with a set of handwritten digit examples using a supervised learning algorithm such as backpropagation (Chauvin and Rumelhart 1995; Hecht-Nielsen 1989; Parker 1982; Rumelhart et al. 1986; Werbos 1974). Third, the network could be trained with some such examples, but also with inputs constructed by a pattern generator, which is evolved: The EA determines generator parameters such as the locations and sizes of a small set of two-dimensional Gaussians, and the generator is evaluated based on how well the network learns the task.

The expected outcome is that the direct EA would require a prohibitively large number of iterations, because it has to search in an extremely high-dimensional space of network weights. The environmentally driven learner, on the other hand, is likely to get stuck in a suboptimal local minimum, because it will start far from the desired solution, without any bias toward it. In contrast, the combined system should be able to discover a solution quickly because it only needs to evolve a small number of parameters of the generator. Its structure should also be simpler, determined largely by the pattern generator, which should allow it to generalize better to new data.

If successful, such an experiment would demonstrate how pattern generation can be applied to tasks that require both generality and flexibility, thus combining the benefits of evolution and learning. Such a system would be a significant step toward automatic construction of complex systems.

17.4 The *Topographica* Cortical Map Simulator

In this chapter, a number of computational research projects from low-level extensions of current models to new directions of long-term research have been proposed. As an integral component of the LISSOM project and an accompaniment to this book, *Topographica* is a software tool specifically designed to make such research possible. In this section, the design and implementation of *Topographica* is briefly reviewed and its use in modeling the visual cortex outlined; the software itself is available at http://topographica.org.

17.4.1 Overview

Future progress in understanding how the visual cortex develops and functions will require developing a large number of computational models. Building these models using existing simulation tools is time consuming, because they do not provide support for biologically realistic, densely interconnected topographic maps. Available biological neural simulators, such as NEURON (Hines and Carnevale 1997) and GENESIS (Bower and Beeman 1998), focus on detailed studies of individual neurons, or very small networks of them. Tools for simulating large populations of abstract units, such as PDP++ (O'Reilly and Munakata 2000) and Matlab (Trappenberg 2002), focus on cognitive science and engineering applications, rather than models of cortical areas. As a result, the current simulators do not provide support for constructing topographic map models, training them with perceptual input patterns, and analyzing their structure and function.

To fill this role, the *Topographica* simulator has been developed to make it practical to simulate large-scale, detailed models of topographic maps. *Topographica* is designed to complement the existing low-level and abstract simulators, focusing on biologically realistic networks of tens of thousands of neurons, forming topographic maps containing millions or tens of millions of connections. *Topographica* has been developed together with the LISSOM project, and it was used to implement the models discussed in this book; it is also well suited for the future research proposed in this chapter. However, *Topographica* is a simulator for topographic maps in the brain in general, not only those in the visual system, but in all domains where such maps occur.

In the subsections below, the design of *Topographica* is described, including the models and modeling approaches it supports, how the simulator is implemented, and how it can be used to advance the field of computational neuroscience.

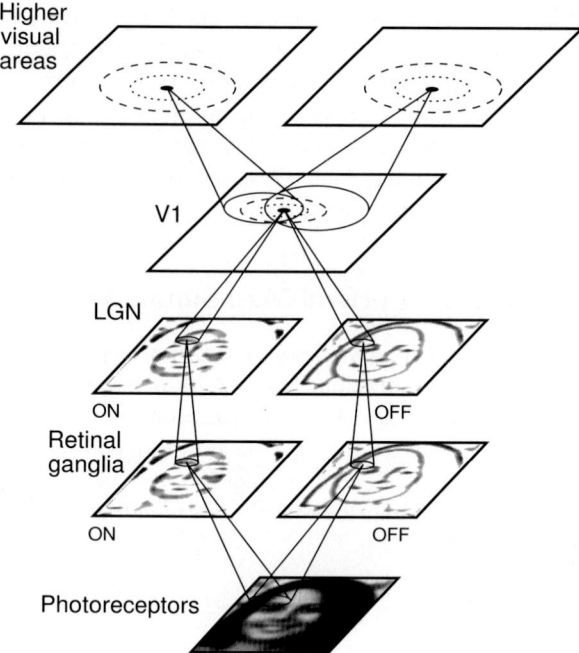

Fig. 17.2. Example *Topographica* **model.** In *Topographica*, models are composed of inter-connected sheets of neurons. Each visual area is represented by one or more sheets in this example model of the early visual system: For instance, the eye is represented by an array of photoreceptors plus two sheets representing retinal ganglion cells. Each of the sheets can be coarse or detailed, plastic or fixed, as needed for a particular study. Afferent and feedback connections link the different areas, and units within each area can be connected using lateral connections. The afferent connections are shown for one sample neuron in each sheet, and also the lateral connections for the sample neurons in V1 and higher areas. Similar models can be used for topographic maps in auditory, somatosensory, and motor cortex.

17.4.2 Scope and Design

The models supported by *Topographica* focus on topographic maps in any two-dimensional cortical or subcortical region, such as those in visual, auditory, so-matosensory, proprioceptive, and motor systems (Figure 17.2). Typically, models will include multiple regions, such as an auditory or visual processing pathway, and simulate a large enough area to allow the organization and function of each map to be studied. The external environment must also be simulated, including playback of e.g. visual images, audio recordings, and artificial test patterns. While current computational models usually include only a primary sensory area with a simplified version of an input pathway, larger scale models will be crucial in the future for understanding phenomena such as object perception, scene segmentation, speech processing, and motor control. *Topographica* is intended to support the development of such models.

To make it practical to model topographic maps at this large scale, the fundamental unit in the simulator is a two-dimensional sheet of neurons, rather than a neuron or a part of a neuron. Conceptually, a sheet is a continuous, two-dimensional area (as in the models of Amari 1980; Fellenz and Taylor 2002; Giese 1998; Roque Da Silva Filho 1992; Wu et al. 2002); this area is then approximated computationally by a finite array of neurons. This approach is crucial to the simulator design, because it allows user parameters, model specifications, and interfaces to be independent of the details of how each sheet is implemented.

As a result, the user can easily trade off between detailed simulations and computational requirements, depending on the specific phenomena under study. If enough computational power and experimental measurements are available, models can be simulated at full scale, with as many neurons and connections as in the animal system being studied. More typically, a less-dense approximation will be used, requiring only ordinary personal computers. Because the same model specifications and parameters can be used in each case, switching between levels of analysis does not require extensive parameter tuning or debugging, as would be required in neuron-level or engineering-oriented simulators. The continuous approach also makes it possible to calculate parameter values for a small simulated system based on experimental measurements. Such model specification and calibration is facilitated by the neuron property databases that have recently become available (e.g. Gardner 2004; Mirsky, Nadkarni, Healy, Miller, and Shepherd 1998; Mirsky et al. 1998).

For most simulations, the individual neuron models in the sheets can be implemented at a high level, consisting of single-compartment firing-rate or integrate-and-fire units. More detailed neuron models can also be used, e.g. if such detail is necessary to validate the model against experimental data, or if the specific phenomena require simulating them. Such models can be implemented using interfaces to existing low-level simulators, such as NEURON and GENESIS, aided by neuron-model synthesis techniques such as L-Neuron (Ascoli, Krichmar, Nasuto, and Senft 2001) and neuron model description languages such as NeuroML (Goddard, Hucka, Howell, Cornelis, Shankar, and Beeman 2001).

Connectivity between neurons can be established using prespecified profiles, such as the Gaussian and random connection distributions used for the initial conditions of the simulations in this book. Adult patterns of connectivity can also be specified based on measurements in particular species and systems, such as the area and connectivity data becoming available in various neuroinformatics knowledge bases (e.g. Arbib and Grethe 2001; Kötter 2004; Mazziotta et al. 2001; Mori et al. 2002; Van Essen 2003, 2004; Van Horn, Grafton, Rockmore, and Gazzaniga 2004). This experimental data can also be used to validate the self-organized structures that develop in the model.

17.4.3 Implementation

Topographica consists of a graphical user interface (GUI), a scripting language, and libraries of models, analysis routines, and visualization techniques. The model library consists of predefined types of sheets, connections, neuron models, and learning rules, and can be extended with user-defined components. These building blocks are combined into a model using the GUI or the scripting language.

The analysis and visualization libraries include statistical tests and plotting methods geared toward large two-dimensional areas. They also focus on data displays that can be compared with experimental results, such as optical imaging recordings, for validating models and for generating predictions. Figure 17.3 shows examples of such visualization types in a screenshot of *Topographica*.

To allow large models to be executed quickly, the numerically intensive portions of the simulator are implemented in C++. Equally important, however, is that prototyping be fast and flexible, and that new architectures and other extensions be easy to explore and test. Although C++ allows the fine control over machine resources that is necessary for peak performance, it is difficult to write, debug and maintain complex systems in C++.

To provide flexibility, the bulk of the simulator is implemented in the Python scripting language. Python is an interactive high-level language that allows rapid software development and interactive debugging, and includes a wide variety of software libraries for tasks such as data analysis, statistical measurements, and visualization. Unlike the script languages typically included in simulators, Python is a complete, well-defined, mature language with an independent user base. As a result, it enjoys strong support outside of the field of computational neuroscience, which provides greater flexibility for users and makes maintenance of the models and software easier.

17.4.4 Further Development

The goal of *Topographica* is to provide a tool for the research community that allows researchers to prototype and test ideas about topographic maps rapidly without the extensive software development that was previously necessary. In particular, using the tools provided by *Topographica*, it should be possible to answer many of the research questions posed in this chapter.

The simulator is designed throughout to be general and extensible, and the goal is to keep developing it further in parallel with research progress in the area. To facilitate such development, *Topographica* is an open source project, with binaries, source code, example models, and documentation freely available on the Internet at http://topographica.org. As in any open-source project, the users of the software will have a large effect on its future. Researchers can share code and models through an on-line repository, and the simulator itself will include user-contributed extensions. Thus, *Topographica* is intended to serve as a shared platform and a catalyst for future research on understanding how cortical maps develop and function.

17.5 Conclusion

In this chapter, a variety of ideas for future computational neuroscience research were discussed. Although they span a number of topics and disciplines, it is possible

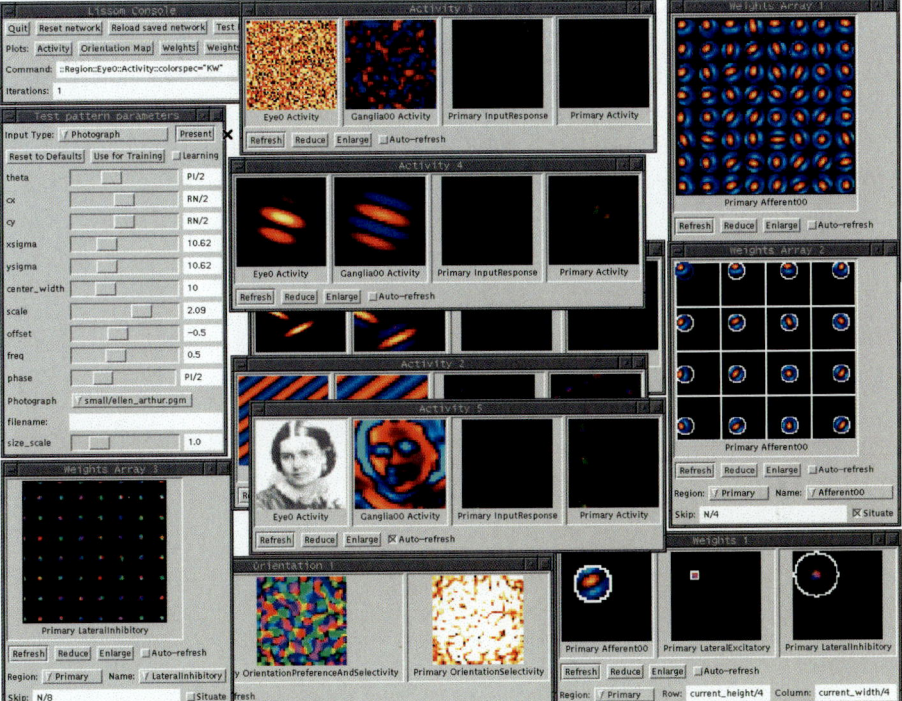

Fig. 17.3. Example *Topographica* screenshot. In this example session with *Topographica*, the user is studying the behavior of an orientation map in the primary visual cortex, using a model similar to the one depicted in Figure 17.2. The window at the bottom labeled "Orientation 1" shows the self-organized orientation map and the orientation selectivity in V1. The five windows labeled "Activity" show a sample visual image along with the responses of the retinal ganglion cells and V1 (labeled "Primary"; both the initial and the settled responses are shown). The input patterns were generated using the "Test pattern parameters" dialog at left. The window labeled "Weights 1" (lower right) shows the strengths of the connections to one neuron in V1. This neuron has afferent receptive fields in the ganglion cells and lateral receptive fields within V1. The afferent weights for 8×8 and 4×4 samplings of the V1 neurons are shown in the two "Weights Array" windows at right; most neurons are selective for Gabor-like patches of oriented lines. The inhibitory lateral connections for an 8×8 sampling of neurons are shown in the "Weights Array 3" window at lower left; neurons tend to receive connections from their immediate neighbors and from distant neurons of the same orientation. *Topographica* is designed to make this type of large-scale analysis of topographic maps practical, in addition to providing effective tools for constructing the models and their training and testing environments.

to see that many of the ideas interact. Models like LISSOM bring together several research areas and facilitate gaining a deep understanding about perceptual phenomena. Many of these ideas can be tested immediately, although some of them depend on large-scale models and high-performance computing. Importantly, at the current rate of technological progress, computers should be powerful enough to simulate the visual cortex at realistic detail within a decade. An appropriate goal for computational neuroscience is to produce models that can make use of that power. Such confluence is likely to lead to a fundamental understanding of perception and higher brain function, and result in novel algorithms for pattern recognition and artificial vision.

18

Conclusion

In the beginning of this book, three computational hypotheses about the development, structure, and function of the visual cortex were proposed:

1. Self-organization, plasticity and perceptual phenomena in the visual cortex are mediated by a single computational process based on recurrent lateral interactions between neurons and cooperatively adapting afferent and lateral connections.
2. A functioning sensory system can be constructed from a specification of a rough initial structure, internal training pattern generators, and self-organizing algorithm that learns from both internal and environmental inputs.
3. Perceptual grouping can be established through synchronized activity between neuronal groups, mediated by self-organized lateral connections.

In order to verify these hypotheses, a unified computational theory and a concrete simulation model called LISSOM was developed based on the known biological and psychological constraints, and a number of simulated experiments were performed with it. The results strongly support the hypotheses, matching biological and psychological data, and suggesting specific biological and psychophysical experiments for further verification. In this chapter, the main contributions of each main chapter are summarized, concluding with future prospects for computational understanding of the visual cortex.

18.1 Contributions

In Chapter 4, an algorithm called LISSOM that combines the self-organization of afferents and lateral connections was developed. The afferent weights of LISSOM develop like in other self-organizing algorithms and form nonlinear approximating surfaces for input distributions. The self-organizing lateral connections, an original contribution of the model, learn correlations in activity between the neurons, dynamically modulating the map response, which allows modeling several new phenomena. In addition, an LGN component in the model makes it possible to self-organize from moving natural inputs. Such a design is primarily biologically motivated, with the goal of establishing a computational interpretation for several experimental observations.

In Chapter 5, LISSOM was used to demonstrate how the observed patterns of orientation, ocular dominance, direction selectivity, and their combinations can arise from input-driven self-organization, together with patchy lateral connectivity. The maps were analyzed using the same techniques as those for experimental data, and shown to agree qualitatively and quantitatively. The model developed shaped receptive fields and maps with structures similar to those in the primary visual cortex. The patterns of lateral connections follow the organization of the map, matching the experimental data known to date. The model makes several predictions about the lateral connection patterns and interactions of the different features, much of which have not been studied experimentally nor computationally before.

Chapter 6 extended the input-driven self-organization to understanding cortical plasticity. The objective was to demonstrate that the self-organized network is in a dynamic equilibrium with the inputs and reorganizes like the cortex when the inputs are altered. No single model so far had accounted for both plasticity and development of the primary visual cortex: LISSOM constitutes such a unified model. Phenomena such as compensation for blind spots and dynamic receptive fields were shown to result from the rapid reorganization of afferent and lateral connection weights. Similar reorganization occurred after lesions in the cortical network. Based on the computational simulations, the model suggested techniques to hasten recovery following stroke and cortical surgery.

Chapter 7 showed how the same self-organizing processes can account for functional phenomena in the adult. The tilt aftereffect was studied in detail, and shown to result from increased inhibition during adaptation. The results from the model match human performance very well. The direct aftereffect with small angles was shown to take place as predicted, and the model suggested a novel explanation for the indirect effect: It arises indirectly as a result of weight normalization. The study demonstrated how a computational model can be used in lieu of a biological system to gain insight into the biological process.

In Chapter 8, the HLISSOM extension of LISSOM outward to subcortical and higher level visual areas was introduced. HLISSOM is the first model to show how genetic and environmental influences can interact in multiple cortical areas. This level of detail is crucial for validating the model on experimental data, and for making specific predictions for future experiments in biology and psychology. Also crucial is that the results of the self-organizing process depend on the stream of input patterns seen during development, not on the initial connection weight values. This result was demonstrated experimentally on the HLISSOM model.

In Chapter 9, V1 neurons in HLISSOM were shown to develop biologically realistic, multi-lobed receptive fields and patterned intracortical connections through unsupervised learning of internally generated and visually evoked activity. These neurons organized into biologically realistic topographic maps, matching those found at birth and in older animals. Postnatal experience gradually modified the orientation map into a precise match to the distribution of orientations present in the environment. This smooth transition has been measured in animals, but had not been demonstrated computationally before.

In Chapter 10, prenatal internally generated activity in HLISSOM was shown to result in a newborn visual system that prefers facelike patterns, and also detects faces in real images. This hypothesis follows naturally from experimental studies in early vision, but had not been previously proposed and tested. Further, HLISSOM was used to show how postnatal learning with real faces can explain how newborns learn to prefer their mothers, and why the preference for schematic faces disappears over time. The model suggests that the psychological studies claiming that newborns learn face outlines, and the studies claiming that responses to faces in the periphery decrease over the first month of age, should probably be reinterpreted. Instead, newborns may learn all parts of the face, and only responses to specific schematic stimuli may decline.

The LISSOM model was then expanded inward to PGLISSOM in Chapter 11, by replacing the firing-rate model of the neuron with a spiking model, and opening the cortical column to include an excitatory and an inhibitory component. The map still self-organizes as before, but it now can also implement perceptual grouping functions. PGLISSOM is the first model where these two process have been brought together, showing that they can coexist and both be due to adapting lateral interaction. If inhibition is strong enough, it will drive the self-organization of the whole system, and allow excitation to implement the grouping function in the time domain.

Conditions for synchronization in the model were studied in detail in Chapter 12. Synchronization can be robustly controlled in a network of spiking neurons, by adjusting the PSP decay rate and connection range. Since decay may be easier to regulate then delay, PGLISSOM suggests that it may be the mechanism used in biological systems to modulate synchronization. Such a network can be robust against noise, provided there is strong excitation and the refractory period is long enough. Thus, the model demonstrates that synchronization may be possible even in the noisy natural environment of the neuron, which has long been an open question.

Chapter 13 presented a series of experiments demonstrating how PGLISSOM can account for perceptual grouping phenomena. Contour integration performance in the model matches human performance, and contour segmentation is achieved simultaneously in the same network. The model predicts that differences in input distribution cause the network to develop different structure and functionality, as seen in the different areas of the vision system. The network also performs contour completion, thereby accounting for a class of illusory contours as well.

In Chapter 14, the representations of visual input formed in the LISSOM map were analyzed computationally. The self-organized inhibitory lateral connections decorrelate the activation on the map, resulting in a sparse, redundancy-reduced code that retains the original information well. Such coding allows representing more information with fixed resources, but it also provides an advantage for information processing: The patterns are more easily separable and generalizable, making further visual processing such as pattern recognition easier.

In order to make future research with larger models possible, in Chapter 15 a set of scaling equations was derived, showing how quantitatively equivalent maps can be developed over a wide range of simulation sizes. These equations are systematically utilized in a map growing method called GLISSOM, allowing the entire V1 be simulated at the column level with existing desktop computers.

Together, the results demonstrate a comprehensive approach to understanding the development and function of the visual cortex. They suggest that a simple but powerful set of self-organizing principles can account for a wide range of experimental results from animals and infants.

Perhaps the most important potential contribution of the LISSOM project, however, is to serve as a foundation and catalyst to further research in the area. As reviewed in Chapters 16 and 17, many future projects are possible based on the results presented in this book, some immediately, others in the near future. It is equally important to provide proper tools; the *Topographica* simulator reviewed in Section 17.4 is intended to serve that role, making it easy to initiate and carry out new research in computational modeling of the visual cortex.

18.2 Conclusion

The research reviewed in this book demonstrates how computational models can play a crucial role in understanding biological phenomena. In order to make a theory computational, it must be specified precisely and completely. It is then possible to test the theory as if it was the real system, in effect running simulated experiments that would be difficult to set up in biology. The results can be observed and analyzed exactly and completely, allowing insights that would otherwise not be possible. These insights must eventually be verified experimentally, but the experiments can be chosen more carefully if they are based on a solid computational theory. As our understanding of brain structures and mechanisms becomes more sophisticated, such computational models are likely to become an increasingly important part of neuroscience research.

LISSOM has already led to several insights and proposed experiments at the level of computations in cortical maps. It provides a framework for understanding the synergy of nature and nurture in development, the dynamic nature of a continuously adapting visual system, and the low-level automatic mechanisms of binding and segmentation. In the future, the same principles can be applied to understanding higher visual functions as well, and how the visual system is maintained, and how it can be repaired in case of damage.

The research in visual cortex is at an exciting stage. For the first time, we have the technology to look into the brain in enough detail to constrain computational models, and the computing power to build large models that help understand perceptual behavior. LISSOM and the *Topographica* software tool that accompanies it are intended to serve as a platform on which future research can be based, eventually aiming at complete understanding of computational maps in the cortex.

Appendices

A

LISSOM Simulation Specifications

Appendices A–F give the specifications and parameters for the models and computational experiments in this book. This appendix details the basic LISSOM model, starting with a generalized version of the activation equation that serves as a reference for how the activation parameters are used in practice. Later appendices describe the HLISSOM and PGLISSOM extensions, as well as the reduced LISSOM and SOM abstractions of self-organizing maps, and the experiments on sparse coding and pattern recognition. Appendix G then describes how the various map visualizations were calculated.

The specifications listed in these appendices can be used to reproduce the LISSOM simulation results on general-purpose simulation platforms, such as the *Topographica* simulator for cortical maps (Section 17.4). The executables and source code for this simulator are freely available at http://topographica.org; the site also contains implementations of a few LISSOM models as examples, including demos and animations illustrating how they work, specific instructions on how to run them, and how to modify them for other purposes.

A.1 Generalized Activation Equation

For clarity, the LISSOM activation equations in Chapter 4 were presented in their most concrete form, showing how activations are computed for a single retina, one pair of ON and OFF channels, and V1. This section generalizes those equations into a single equation applicable to all of the LISSOM simulations in this book, with an arbitrary number of input and LGN regions. Although more abstract, this unified version is concise and easily extensible to additional input dimensions and cortical areas in future work. This form is also the one implemented in the *Topographica* simulator, which makes it easy determine the simulator parameters from the values listed in this appendix.

In the general case, the activation of unit (i, j) in a LISSOM map at time t is

$$\eta_{ij}(t) = \sigma \left(\sum_{\rho} \gamma_{\rho} \sum_{kl \in \mathrm{RF}_{\rho}} X_{kl}(t-1) w_{kl,ij} \right), \qquad \text{(A.1)}$$

where the index ρ indicates a particular receptive field (RF; afferent, lateral, or feedback), $X_{kl}(t-1)$ is the activation of unit (k,l) in that receptive field, and $w_{kl,ij}$ is the weight from that unit to unit (i,j). The sign of scaling factor γ_{ρ} is positive for afferent and lateral excitatory connections, and negative for lateral inhibitory connections. This equation can also be extended to HLISSOM by including afferent normalization as in Equation 8.1, and to PGLISSOM by including spiking as in Equations 11.3–11.6.

As an example of how this equation is used, the combined orientation, ocular dominance, and direction simulation in Section 5.6.3 consisted of two eyes, 16 LGN regions, and V1. In V1, each neuron has 16 afferent RFs (four types of lag for both ON and OFF channels for the two eyes), and two lateral RFs (excitatory and inhibitory). Thus, ρ iterates over 18 RFs while the sum of the contributions from each RF is accumulated. A sigmoid is then applied to this entire sum to determine the actual response of the neuron. Other simulations have fewer RFs, but otherwise operate through the same steps.

For the first settling iteration, the lateral contributions are zero, because all units are initialized to zero at each input presentation. Thus, Equation A.1 reduces to the initial activation equation 4.6 for the first settling step. This equation also applies to LGN units: They have only had one RF so far in this book (i.e., a single eye), but multiple RFs can be included, e.g. if color inputs are to be processed (Section 17.2.1).

A.2 Default Parameters

All of the LISSOM simulations in this book were based on the same set of default parameters, with small modifications to these defaults as necessary to study different phenomena. The default model corresponds approximately to a 5 mm × 5 mm area of macaque V1; the V1 size was chosen to match the estimated number of columns in such an area and the other parameters were set to simulate it realistically. This model was introduced in Section 4.5 and used to organize an orientation map in Section 5.3. This section describes the default parameter values in detail, and later sections in this appendix show how each simulation differed from the defaults.

Although all of the parameters are listed here for completeness, most of these can be left unchanged or calculated from known values. Most of the rest can be set systematically and without an extensive search for the correct values. Each simulation has relatively few free parameters in practice, which makes it straightforward to use the model to simulate new phenomena.

Because a large number of closely related simulations will be covered by this appendix, the default parameter values are listed in a format that makes it convenient to calculate new values when some of the defaults are changed. That is, instead of listing numeric values, most parameters are shown as formulas derived from the

Parameter	Value	Description
N_{d_o}	142	Reference value of N_d, the cortical density
L_{d_o}	24	Reference value of L_d, the LGN density
R_{d_o}	24	Reference value of R_d, the retinal density
r_{A_o}	6.5	Reference value of r_A, the maximum radius of the afferent connections
r_{E_o}	19.5	Reference value of r_E, the maximum radius of the lateral excitatory connections
r_{I_o}	47.5	Reference value of r_I, the maximum radius of the lateral inhibitory connections
σ_{a_o}	7.0	Reference value of σ_a, the radius of the major axis of ellipsoidal Gaussian inputs
σ_{b_o}	1.5	Reference value of σ_b, the radius of the minor axis of ellipsoidal Gaussian inputs
t_{f_o}	20,000	Reference value of t_f, the number of training iterations
w_{d_o}	0.00005	Reference value of w_d, the lateral inhibitory connection death threshold

Table A.1. Parameters for the LISSOM reference simulation. These values define a reference simulation that serves as a basis for calculating the parameters for other simulations, as specified in Table A.2. The subscript "o" in each name stands for "original", as in the scaling equations in Section 15.2. These parameters have the same value in every simulation.

scaling equations in Section 15.2. These formulas differ slightly from the ones in Section 15.2 because they have been extended to support networks with an LGN, to add scaling for other additional parameters, and to make it easier to change the retina and cortex sizes.

The scaling equations require that one particular network size is used as a reference from which parameters for other sizes can be calculated; Table A.1 lists these reference values. Based on these values, the default LISSOM parameters are presented in Tables A.2 and A.3. The parameters in Table A.2 are constant for any particular simulation, and the parameters in Table A.3 vary systematically throughout each simulation. Figure A.1 illustrates how the parameters for retina, LGN and V1 sizes map to each other.

Sections A.3–A.9 explain how these three tables were used to compute the parameters for each different simulation (sometimes overriding the defaults). As an example, the parameters for the default simulation, i.e. the orientation map study presented in Section 5.3 can be computed by following Table A.2 line by line. The numerical value for each parameter is calculated by filling in the constants from Table A.1 and the previous lines. For instance, parameter N_d can be calculated as 142, L_d as 24, r_A as 6.5, and so on.

Most of the parameters in Table A.2 are temporary variables used only in later entries in this table and in Table A.3. They are introduced to make the notation easier to follow, and are not part of the LISSOM model itself. For the actual LISSOM parameters, the tables list the equation or section where each parameter is used. Once the LISSOM parameter values are obtained, the temporary values can be discarded.

Parameter	Value	Used in	Description
N_d	N_{d_o}	Table A.2	Cortical density, i.e. width and height of a unit area of cortex
L_d	L_{d_o}	Table A.2	LGN density, i.e. width and height of a unit area of the LGN (the area that projects to N_d)
R_d	R_{d_o}	Table A.2	Retinal density, i.e. width and height of a unit area of retina (the area that projects to L_d)
s_g	1.0	Table A.2	Global size scale of the model in area units N_d, L_d, and R_d
n_A	2	Table A.2	Number of afferent RFs per cortical unit (e.g. 1 ON and 1 OFF)
r_A	$\frac{L_d}{4} + 0.5$	Section 4.2.3	Maximum radius of the cortical afferent connections†
r_{Ei}	$\frac{N_d}{10}$	Section 4.2.3	Initial value for r_E, the maximum radius of the lateral excitatory connections, before shrinking†
r_{Ef}	$\max\left(2.5, \frac{N_d}{44}\right)$	Table A.3	Minimum final value of the r_E after shrinking†
r_I	$\frac{N_d}{4} - 1$	Section 4.2.3	Maximum radius of the lateral inhibitory connections†
s_w	$\frac{r_A}{r_{A_o}}$	Table A.2	Scale of r_A relative to the default
σ_A	$\frac{r_A}{1.3}$	Section 4.2.3	Radius of the initial Gaussian-shaped afferent connections†
σ_E	$0.78 r_{Ei}$	Section 4.2.3	Radius of the initial Gaussian lateral excitatory connections†
σ_I	$2.08 r_I$	Section 4.2.3	Radius of the initial Gaussian lateral inhibitory connections†
σ_c	$0.5 s_w \frac{R_d}{L_d}$	Equation 4.1	Radius of LGN DoG center Gaussian†
σ_s	$4\sigma_c$	Equation 4.1	Radius of LGN DoG surround Gaussian†
r_L	$4.7\sigma_s$	Section 4.2.2	Maximum radius of the LGN afferent connections†
N	$s_g N_d$	Section 4.2.1	Width and height of the cortex, in number of units
L	$s_g L_d + 2(r_A - 0.5)$	Section 4.2.1	Width and height of the LGN, in number of units
R	$s_g R_d + 2\frac{R_d}{L_d}(r_A - 0.5) + 2(r_L - 0.5)$	Section 4.2.1	Width and height of the retina, in number of units
s_r	$\left(\frac{L}{L_d + 2(r_A - 0.5)}\right)^2$	Table A.2	LGN area scale relative to the reference simulation†
σ_u	$3 s_w$	Equation 4.2	Radius of unoriented Gaussian inputs
σ_a	$\sigma_{a_o} s_w$	Equation 5.1	Radius of the major axis of ellipsoidal Gaussian inputs
σ_b	$\sigma_{b_o} s_w$	Equation 5.1	Radius of the minor axis of ellipsoidal Gaussian inputs
r_d	$25.0 s_w$	Equation 9.2	Radius of the full-brightness portion of disk-shaped patterns†
σ_d	$3.0 s_w$	Equation 9.2	Radius of the Gaussian falloff in brightness at the disk edge
s_b	1.0	Section 5.4.1	Brightness scale of the retina (contrast of fully bright stimulus)
o_b	0.5	Section 4.2.1	Brightness value of the background of the retina
R_p	L	Section 4.3.1	Width & height of the random scatter of discrete pattern centers
d_r	$2.2 r_A$	Section 4.3.1	Minimum separation between the centers of multiple inputs†
s_s	0.0	Section 5.4.4	Scale of the input pattern scatter from the calculated value
s_d	2.0	Table A.2	Input density scale (ratio between average cortical activity for one oriented Gaussian to the average for the actual pattern)
s_t	$\frac{1}{s_d}$	Tables A.2, A.3	Iteration scaling factor; can be adjusted to use fewer iterations if input patterns are more dense at each iteration, or vice versa
n_p	$\max(1, s_d s_r)$	Section 4.3.1	Number of discrete input patterns per iteration (e.g. Gaussians)

(Table continues on the next page)

(Table continued from the previous page)

Parameter	Value	Used in	Description
γ_A	1.0	Equation 4.5	Scaling factor for the afferent weights
γ_E	0.9	Equation 4.7	Scaling factor for the lateral excitatory weights
γ_I	0.9	Equation 4.7	Scaling factor for the lateral inhibitory weights
γ_L	$\frac{2.33}{s_b}$	Equation 4.3	Scaling factor for LGN's afferent weights†
γ_n	0.0	Equation 8.1	Strength of divisive gain control (only in HLISSOM)
t_{si}	9	Section 4.3.3	Initial value for t_s, the number of settling iterations
θ_{li}	0.083	Equation 4.4	Initial value for θ_l, the lower threshold of the sigmoid activation function
θ_{ui}	$\theta_{li} + 0.55$	Equation 4.4	Initial value for θ_u, the upper threshold of the sigmoid activation function†
t_f	$t_{f_o}s_t$	Section 4.4	Number of training iterations
α_{Ai}	$\frac{0.0070}{n_A s_t s_d}$	Equation 4.8	Initial value for α_A, the afferent learning rate†
α_{Ei}	$\frac{0.002 r_{E_o}^2}{s_t s_d r_E^2}$	Equation 4.8	Initial value for α_E, the lateral excitatory learning rate†
α_I	$\frac{0.00025 r_{I_o}^2}{s_t s_d r_I^2}$	Equation 4.8	Lateral inhibitory learning rate†
w_d	$2 w_{d_o} \frac{r_{I_o}^2}{r_I^2}$	Section 4.4.2	Lateral inhibitory connection death threshold
t_d	t_f	Section 4.4.2	Iteration at which inhibitory connections are first pruned

Table A.2. Defaults for constant parameters. This table specifies how the parameter values for the different simulations can be constructed, based on the reference values from Table A.1. These parameters have constant values in each simulation. Those with the subscript "i" represent the initial values for parameters that are changed over the simulation, as shown in Table A.3. The table is organized into sections including user-defined network size, connection radius, calculated network size, and input pattern parameters on the previous page, and connection strength, activation, and learning parameters on this page. The numerical values in formulas marked with a dagger (†) were determined empirically in earlier work (Bednar and Miikkulainen 2000b; Sirosh 1995). Parameters that are listed as being used in Table A.2 are temporary variables, introduced to make the notation easier to follow. Those listed as being used in various equations and sections are actual parameters of the LISSOM model itself.

A.3 Choosing Parameters for New Simulations

Despite the seemingly large number of parameters, few of them need to be adjusted when running a new simulation. The most commonly changed parameters are the cortical density N_d and area scale s_g, because these parameters directly determine the time and memory requirements of the simulations. The default N_d of 142 representing a 5 mm $\times$ 5 mm area is a good match to the density of columns in V1 orientation maps (Bednar, Kelkar, and Miikkulainen 2004), but in practice much smaller

Iteration	r_{E}	t_{s}	θ_{l}	θ_{u}	α_{A}	α_{E}
$0s_{\mathrm{t}}$	$\max(r_{\mathrm{Ef}}, r_{\mathrm{Ei}})$	t_{si}	θ_{li}	θ_{ui}	α_{Ai}	α_{Ei}
$200s_{\mathrm{t}}$	$\max(r_{\mathrm{Ef}}, 0.6000 r_{\mathrm{Ei}})$	t_{si}	$\theta_{\mathrm{li}} + 0.01$	$\theta_{\mathrm{ui}} + 0.01$	α_{Ai}	α_{Ei}
$500s_{\mathrm{t}}$	$\max(r_{\mathrm{Ef}}, 0.4200 r_{\mathrm{Ei}})$	t_{si}	$\theta_{\mathrm{li}} + 0.02$	$\theta_{\mathrm{ui}} + 0.02$	$\frac{50}{70}\alpha_{\mathrm{Ai}}$	$0.5\alpha_{\mathrm{Ei}}$
$1000s_{\mathrm{t}}$	$\max(r_{\mathrm{Ef}}, 0.3360 r_{\mathrm{Ei}})$	t_{si}	$\theta_{\mathrm{li}} + 0.05$	$\theta_{\mathrm{ui}} + 0.03$	$\frac{50}{70}\alpha_{\mathrm{Ai}}$	$0.5\alpha_{\mathrm{Ei}}$
$2000s_{\mathrm{t}}$	$\max(r_{\mathrm{Ef}}, 0.2688 r_{\mathrm{Ei}})$	$t_{\mathrm{si}} + 1$	$\theta_{\mathrm{li}} + 0.08$	$\theta_{\mathrm{ui}} + 0.05$	$\frac{40}{70}\alpha_{\mathrm{Ai}}$	$0.5\alpha_{\mathrm{Ei}}$
$3000s_{\mathrm{t}}$	$\max(r_{\mathrm{Ef}}, 0.2150 r_{\mathrm{Ei}})$	$t_{\mathrm{si}} + 1$	$\theta_{\mathrm{li}} + 0.10$	$\theta_{\mathrm{ui}} + 0.08$	$\frac{40}{70}\alpha_{\mathrm{Ai}}$	$0.5\alpha_{\mathrm{Ei}}$
$4000s_{\mathrm{t}}$	$\max(r_{\mathrm{Ef}}, 0.1290 r_{\mathrm{Ei}})$	$t_{\mathrm{si}} + 1$	$\theta_{\mathrm{li}} + 0.10$	$\theta_{\mathrm{ui}} + 0.11$	$\frac{30}{70}\alpha_{\mathrm{Ai}}$	$0.5\alpha_{\mathrm{Ei}}$
$5000s_{\mathrm{t}}$	$\max(r_{\mathrm{Ef}}, 0.0774 r_{\mathrm{Ei}})$	$t_{\mathrm{si}} + 2$	$\theta_{\mathrm{li}} + 0.11$	$\theta_{\mathrm{ui}} + 0.14$	$\frac{30}{70}\alpha_{\mathrm{Ai}}$	$0.5\alpha_{\mathrm{Ei}}$
$6500s_{\mathrm{t}}$	$\max(r_{\mathrm{Ef}}, 0.0464 r_{\mathrm{Ei}})$	$t_{\mathrm{si}} + 3$	$\theta_{\mathrm{li}} + 0.12$	$\theta_{\mathrm{ui}} + 0.17$	$\frac{30}{70}\alpha_{\mathrm{Ai}}$	$0.5\alpha_{\mathrm{Ei}}$
$8000s_{\mathrm{t}}$	$\max(r_{\mathrm{Ef}}, 0.0279 r_{\mathrm{Ei}})$	$t_{\mathrm{si}} + 4$	$\theta_{\mathrm{li}} + 0.13$	$\theta_{\mathrm{ui}} + 0.20$	$\frac{30}{70}\alpha_{\mathrm{Ai}}$	$0.5\alpha_{\mathrm{Ei}}$
$20000s_{\mathrm{t}}$	$\max(r_{\mathrm{Ef}}, 0.0167 r_{\mathrm{Ei}})$	$t_{\mathrm{si}} + 4$	$\theta_{\mathrm{li}} + 0.14$	$\theta_{\mathrm{ui}} + 0.23$	$\frac{15}{70}\alpha_{\mathrm{Ai}}$	$0.5\alpha_{\mathrm{Ei}}$

Table A.3. Default parameter change schedule. The values of these parameters at the beginning of simulation are given in Table A.2; this table describes how their values change at each subsequent iteration. The new values are calculated at the start of each listed iteration.

values also often work well (as shown in Section 15.2.3). The default s_{g} of 1.0 covers an area large enough to include several orientation patches in each direction, but more area is useful when processing larger images.

Apart from the simulation size parameters, most simulations differ primarily by the choice of input patterns. Starting from an existing simulation, usually only a few parameters need to be changed to obtain a similar simulation with a new set of patterns. If the new pattern is similar in overall shape, often all that is needed is to set the afferent input scale γ_{A} or the sigmoid threshold θ_{li} to a value that, on average, produces a similar level of cortical activity. Usually a quantitatively similar map results, as shown in the simulations with and without ON/OFF channels in Section 6.2.3.

For a large change in pattern shape or size, such as using natural images instead of Gaussian patterns, two parameters need to be adjusted. First, the input scale or threshold needs to be changed to get results as similar to the original working simulation as possible. Second, the input density scale s_{d} needs to be adjusted to compensate for the remaining differences in the amount of input per iteration. Of course, because the system is nonlinear, it is not always possible to compensate completely.

As an example, if Gaussian input patterns are replaced with large, sharp-edged squares, each input will produce multiple activity bubbles in V1 instead of one bubble. The input scale γ_{A} should be set to a value that results in bubbles about the same size as in the Gaussian simulation, and s_{d} should be set to the average number of bubbles per iteration in the new simulation. For input patterns with large, spread-out areas of activity, the lateral interaction strengths γ_{E} and γ_{I} can also be increased to ensure that distinct activity bubbles form. Other parameters do not usually need to be changed when changing the input patterns.

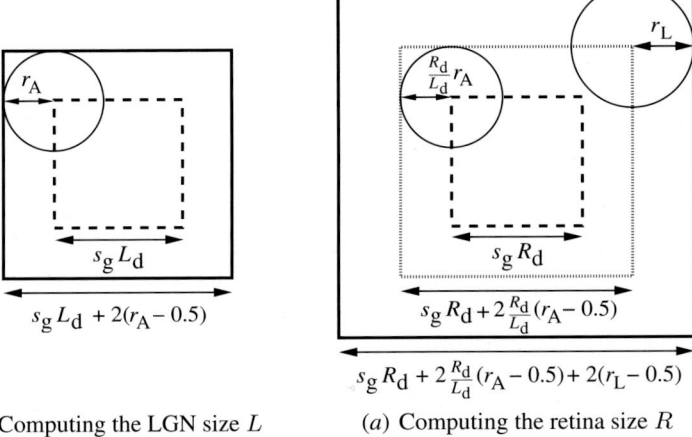

(a) Computing the LGN size L (a) Computing the retina size R

Fig. A.1. Mapping between neural sheets in LISSOM. In computing the LGN size L and the retina size R (Table A.2), a buffer area is added around the lower level sheet so that all neurons at the higher level have complete receptive fields. (a) In the mapping from LGN to V1, the outer square represents the LGN sheet, the dashed area maps point-for-point to V1, and the circle represents the receptive field of the top left V1 neuron. For instance, if $s_g = 8$ and $L_d = 24$, the dashed line encloses an area of 192×192 LGN units ($8 \times 24 = 192$). This area is extended on all sides by $r_A - 0.5$ units to make sure that all V1 neurons have complete receptive fields. Thus, the LGN contains 204×204 neurons in total ($192 + 2 \times 6 = 204$). (b) The mapping from retina to V1 is formed analogously, by extending the buffering down one more level. The outer square represents the retina, the dotted area maps point-for-point to the LGN, and the dashed area maps point-for-point to V1. The circle on the right shows the receptive field of the top right LGN neuron and the circle on the left represents the receptive field of the top left V1 neuron, with its radius expressed in retinal units (hence the factor R_d/L_d). For example, if $R_d = 48$, the dashed area is $s_g R_d = 8 \times 48 = 384$ retinal units wide and the dotted area $384 + 2 \times \frac{48}{24} \times 6 = 408$ retinal units wide. For an LGN radius of $r_L = 16.5$, the full retina therefore consists of 440×440 neurons ($384 + 2 \times \frac{48}{24} \times 6 + 2 \times 16 = 440$).

The connection weights in each RF are usually initialized to uniformly distributed positive random values; sometimes a Gaussian pattern is used instead to speed up the self-organization. The range $[0..1]$ was used for these values in all simulations; note however that the range does not matter because the weights in each group of RFs (afferent, lateral excitatory, and lateral inhibitory) are immediately normalized to sum to 1.0.

Sections A.4–A.9 describe the specific differences from the default parameters for each of the LISSOM simulations in this book. To determine the actual parameter values, one can begin with a copy of Tables A.2 and A.3, make the specified changes, and then calculate the new values for each parameter starting at the top of each table. For instance, for a simulation that changes θ_{li} to 0.05, parameter θ_{ui} would become 0.6 instead of 0.65 (Table A.2). The same method can be used to determine a consistent set of parameter values for any new simulations.

A.4 Retinotopic Maps

The retinotopic map simulations in Chapter 4 were based on the default values, except the training Gaussians were unoriented ($\sigma_a = \sigma_b = \sigma_u = 3$), the sigmoid threshold was lower ($\theta_{li} = 0.035$) so that the smaller patterns would produce as much cortical activity as oriented Gaussians do, the RF centers were randomly scattered by 5% in both the x and y directions, and the lateral connections were uniformly random instead of initialized to a Gaussian profile.

The default parameter values result in two Gaussian inputs per iteration. While in principle it would be possible to self-organize the network with patterns that consist of more than two, given that the Gaussians are relatively large compared with the retina, it would be difficult to distribute them uniformly on the retina while enforcing a minimum distance d_r between them. After two Gaussians have been placed randomly, the only remaining possible locations are often near the corners. The corners would therefore be trained more often than the other areas of the map, resulting in a distorted organization. More Gaussians could be used if a larger retinal and cortical area was simulated (i.e. with a larger s_g), or the Gaussians were narrower.

A.5 Orientation Maps

The LISSOM simulations with oriented Gaussians in Sections 5.3 and 6.2 (Figures 5.5–5.12, "Gaussians" in 5.13, "ON/OFF" in 6.4 and 6.5) were based on the default parameter values, except the inhibitory connection death threshold w_d was increased to $6w_{do}r_{Io}^2/r_I^2$ for historical reasons. These values were also used in the "Plus/Minus" simulation in Figure 5.13, except the sign of the input scale s_b was chosen randomly for each pattern. The other types of orientation map simulations presented in that figure are described in the following subsections.

A.5.1 Disks, Noisy Disks, and Noise

The simulations with circular disks ("Disks" in Figure 5.13) were based on the default LISSOM parameters, except only one disk was drawn per iteration ($n_p = 1$), the lateral interactions were stronger to allow long contours to be separated into distinct activity bubbles ($\gamma_I = 2.0$, $\gamma_E = 1.2$), and the LGN afferent scale was slightly stronger ($\gamma_L = 3/s_b$) because each activity bubble was slightly weaker than in the oriented Gaussian simulation.

The disk input patterns were fully activated in the circular region ($r_d = 25s_w$) around their centers, and the activation then fell off according to a Gaussian pattern ($\sigma_d = 3$). Each input center was separated far enough so that input patterns would never overlap ($d_r = 1.5r_d$). Because the disk stimuli are large compared with L_d, the area in which disk centers are chosen was increased so that even the neurons at the edges of the map are equally likely to receive input from all parts of the disks. Thus, the number of inputs per eye was corrected to reflect this larger area ($n_p = \max\left[1, \left(\frac{2r_d+L_d s_g}{2r_d+L_d}\right)^2\right]$).

The simulations with noisy circular disks ("Noisy disks" in Figure 5.13) were identical to the simulations with noiseless disks, except uniform random noise in the range $[-0.5..0.5]$ was added to the input pattern. The simulations with uniform random noise alone ("Noise" in Figure 5.13) were based on the noisy disk simulations, except no disks were drawn, the LGN parameters were adjusted to produce stronger LGN activations ($\sigma_c = 0.75$, $\sigma_s = 3\sigma_c$, and $\gamma_L = 5/s_b$), and the V1 parameters were changed to produced stronger V1 activations ($\gamma_E = 0.9$, $\gamma_I = 0.9$, $\gamma_A = 3$).

A.5.2 Natural Images

The orientation map simulations trained on natural images ("Nature" in Figure 5.13) were based on the default LISSOM parameters, except the retinal density was doubled to provide more image resolution ($R_d = 48$), only one image was drawn per iteration ($n_p = 1$), the input density scale was higher ($s_d = 4$) because each input resulted in about four activity bubbles on average, the number of iterations was fixed to the default value ($s_t = 0.5$) instead of being adjusted automatically for the input density scale, the LGN afferent scale was increased to produce activity for low-contrast inputs ($\gamma_L = 4.7/s_b$), and the sigmoid threshold was lower ($\theta_{li} = 0.076$) to allow responses to low-contrast stimuli.

The lateral interaction strengths and learning rates were also adjusted during training rather than being fixed. In the default simulation, the lateral inhibitory weights self-organize into only a few small regions, because Gaussian patterns have no long-range correlations. In contrast, natural images have significant long-range correlations, and inhibitory weights spread over a larger area. To keep the balance between excitatory and inhibitory lateral weights approximately constant, the lateral inhibitory strength was set to $\gamma_I = 1.75$ at first, increased to 2.2 at iteration $1000s_t$ and 2.6 at iteration $2000s_t$. The lateral inhibitory learning rate α_I was set to $\alpha_I = \dfrac{0.00005r_{Io}^2}{s_t s_d r_I^2}$ at first, increased to $\alpha_I = \dfrac{0.00010r_{Io}^2}{s_t s_d r_I^2}$ at iteration $1000s_t$, $\alpha_I = \dfrac{0.00015r_{Io}^2}{s_t s_d r_I^2}$ at iteration $2000s_t$, and $\alpha_I = \dfrac{0.00025r_{Io}^2}{s_t s_d r_I^2}$ at iteration $5000s_t$. With these modifications, the orientation map self-organizes robustly with natural image input.

A.6 Ocular Dominance Maps

The ocular dominance simulations in Section 5.4 (Figures 5.15–5.17, "Dimming" in Figure 5.19) were based on the default LISSOM parameters, except four LGN regions ($n_A = 4$) and two eyes were included, the sigmoid threshold was lower ($\theta_{li} = 0.035$) so that the smaller patterns would produce as much cortical activity as oriented Gaussians do, the training Gaussians were unoriented ($\sigma_u = 4.25$), and the input brightness scale s_b was randomly chosen from the range $[0..1]$ for the left eye and calculated as $1 - s_b$ for the right eye for each input.

The strabismic ocular dominance simulations (Figure 5.18, "Strabismic" in Figure 5.19) were otherwise similar, except the inputs were scattered randomly between the eyes ($s_s = 1.0$). The "Mild" ocular dominance simulations in Figure 5.19 were identical to "Strabismic", except the input scale was constant ($s_b = 1.0$), and the inputs were scattered slightly between the eyes ($s_s = 0.2$). The "Moderate" ocular dominance simulations were identical to "Mild", except the inputs were scattered more between the eyes ($s_s = 0.4$).

A.7 Direction Maps

The direction map simulations in Section 5.5 (Figures 5.21–5.24, "Speed 1" in Figure 5.25) were based on the default LISSOM parameters, except eight LGN regions were included ($n_A = 8$), only a single pattern was presented per iteration ($s_d = 1$) to avoid overlapping patterns that are moving in different directions, and the LGN afferent scale was slightly stronger ($\gamma_L = 2.38/s_b$) because the response is weaker when inputs are not perfectly aligned in each eye. Because $s_d = 1$, $s_t = 0.5$, which means that these simulations were run for 20,000 iterations with one pattern apiece rather than the default 10,000 with two patterns apiece. As long as input patterns can be placed randomly on the retina without overlap, and the responses are localized in V1 (as they usually are), using more patterns in fewer iterations is equivalent to using fewer patterns in more iterations.

The simulations with different speeds (Figure 5.25) were otherwise identical except the LGN afferent scale was adjusted for each speed to keep the average response comparable: $\gamma_L = 2.33/s_b$ for speed 0, $\gamma_L = 2.38/s_b$ for speed 1, $\gamma_L = 2.53/s_b$ for speed 2, and $\gamma_L = 2.80/s_b$ for speed 3.

A.8 Combined Orientation / Ocular Dominance Maps

The combined orientation and ocular dominance simulations in Section 5.6.2 (Figures 5.27 and 5.28) were based on the "Gaussians" orientation-only simulation, except four LGN regions ($n_A = 4$) and two eyes were included. For each input, the input scale s_b was chosen randomly for the left eye and calculated as $1 - s_b$ for the right eye.

A.9 Combined Orientation / Ocular Dominance / Direction Maps

In Section 5.6, the Gaussian-trained combined orientation, ocular dominance, and direction simulations (Figure 5.29, "Gaussians" in Figure 5.32) were based on the "Gaussians" orientation-only simulation, except 16 LGN regions ($n_A = 16$) and two eyes were included, only a single pattern was presented per iteration ($s_d = 1$) to avoid overlapping patterns that are moving in different directions, and the LGN afferent scale was adjusted to match the value computed for the direction-only map ($\gamma_L = 2.38/s_b$). For each input, the input scale s_b was chosen randomly for the left eye and calculated as $1 - s_b$ for the right eye. The speed of the moving patterns was 1, i.e. subsequent frames differed by 1.0 retinal units.

The combined simulation with noisy disks ("Noisy disks" in Figure 5.32), used the inputs from the "Noisy Disks" orientation-only simulation, drawn at speed 2. The parameters were the same as for the "Gaussians" combined OR/OD/DR simulation, except a higher input density scaling was used ($s_d = 2$) because the inputs more regularly activated the cortex, γ_I and α_I were adjusted as for the orientation-only simulation with natural images (Section A.5.2) to keep excitation and inhibition balanced, and the input scaling for the LGN was increased ($\gamma_L = 3.0/s_b$) to keep the LGN responses comparable.

The combined model trained with natural images (Figure 5.31, "Nature in Figure 5.32) was identical to the orientation-only natural image simulation in Section A.5.2 except 16 LGN regions ($n_A = 16$) and two eyes were included, the input density scale was increased without changing the number of iterations ($s_d = 2$ and $s_t = 1$), and the LGN afferent scale was adjusted to match the value computed for the direction-only map ($\gamma_L = 5.6/s_b$). For each input, the input scale s_b was chosen randomly for the left eye and calculated as $1 - s_b$ for the right eye. The speed of the moving patterns was 2.0.

B

Reduced LISSOM Simulation Specifications

As described in Section 6.2.3, as long as the inputs to the model consist of abstract patterns instead of natural images, simulations that utilize only the ON channel (or bypass the LGN entirely) lead to the same results as simulations that include both the ON and OFF channels of the LGN explicitly. Such reduced LISSOM networks allow demonstrating many phenomena efficiently and clearly, as was seen in Chapters 6, 7 and 11–15.

The reduced LISSOM simulations were based on the default LISSOM parameter values, except there was only one input sheet (instead of ON and OFF; thus $n_A = 1$). This sheet was mapped directly to the cortex like the LGN sheets (Figure A.1), and sized like the LGN sheets ($R = L$). The most important adjustment is to set the sigmoid thresholds so that the total cortical response is the same (on average) as with ON and OFF channels, to compensate for the higher average value of retinal activity compared with the LGN activity. In the reduced LISSOM simulations, the input threshold was therefore set to $\theta_{l i} = 0.1$. These values are taken as the default reduced LISSOM parameters, occasionally overridden in individual simulations as specified in the following sections and in Appendices D and F.

B.1 Plasticity

The retinal lesion experiments in Section 6.3 were run with the default reduced LIS-SOM parameters, except the inhibitory connection death threshold w_d was increased to $6 w_{d o} r_{I o}^2 / r_I^2$ for historical reasons, and the afferent learning rate α_A was increased to 0.003 after iteration 10,000; all other parameters remained the same as at the end of the self-organization. Faster learning makes the changes more visible, and also models the increased plasticity that might result during recovery from injury (Emery, Royo, Fischer, Saatman, and McIntosh 2003; Kaas 2001a,b). The cortical lesion experiments in Section 6.4 were based on the same parameters as the retinal lesion experiments, except the afferent learning rate remained at its default value even after the end of self-organization.

B.2 Tilt Aftereffect

The tilt aftereffect experiments in Chapter 7 were based on the default reduced LIS-SOM parameters, except for historical reasons the cortical density was slightly larger ($N_d = 192$), the final excitatory radius r_{Ef} was reduced to 1.5, the inhibitory connection death threshold w_d was decreased to 0.00005, and only a single pattern was presented per iteration ($s_d = 1$). Also for historical reasons, the input patterns were blurred with a uniform 3×3 convolution kernel before they were presented to the network.

For the average tilt aftereffect plots in Figures 7.5 and 7.6, each input line was presented at nine different locations chosen from the nodes of a regular 3×3 grid centered in the retina, with each grid step four retinal units wide.

B.3 Scaling

The scaling equation and GLISSOM simulations in Chapter 15 were based on the tilt aftereffect parameters (Section B.2), scaled as described in each section of Chapter 15. For instance, the area scaling simulation (Figure 15.1) compares two simulations with $L_d = 24$ and $N_d = 44$, differing by their area scales ($s_g = 1$ and $s_g = 4$).

The scaling simulations were run on a 600 MHz Intel Pentium III Linux machine with 1024 MB of RAM. The timing results are user CPU times reported by the GNU "time" command; the CPU time is essentially the same as the elapsed wallclock time because the CPU utilization was always over 99%.

C

HLISSOM Simulation Specifications

The HLISSOM model extends basic LISSOM by including afferent normalization (Equation 8.1) that allows processing natural images, as well as a PGO sheet and a face-selective area. The PGO sheet is connected to the LGN and V1 is connected to the FSA so that each unit has a full receptive field (Figure A.1). The simulation parameters are based on the default LISSOM parameters of Appendix A, except as specified in the subsections that follow.

C.1 V1 Only

The V1-only simulations showing the effect of γ_n (Figures 8.2 and 8.3) were based on disk-shaped input patterns. The parameters were the same as in the "Disks" simulation in Section A.5.1, except the area was very large ($s_g = 8$) to allow large, detailed retinal stimuli to be tested, the cortical density was very low ($N_d = 24$) to reduce memory and computational requirements, and the LGN radius was slightly smaller ($\sigma_c = 0.4$) for historical reasons. During self-organization, γ_n was zero; after self-organization, that parameter was adjusted as shown in the figure.

The small-scale V1-only HLISSOM simulations in Figure 8.5 and in Chapter 9 were based on the default LISSOM simulation parameters of Appendix A; in particular, afferent normalization was not used ($\gamma_n = 0$). In the input and weight stream simulations (Figure 8.5) the cortical density was lower ($N_d = 64$) so that the random initial weights would be visible.

The V1 simulation parameters were always evolved on a schedule for a 10,000-iteration run; in the prenatal simulations of Figure 9.1, the training was simply interrupted at 1000. The prenatal "Disks" and "Noisy Disks" simulations were based on the same parameters as the full-length simulations described in Section A.5.1, up to iteration 1000. The prenatal "Noise" simulation was identical to "Noisy disks", except no disks were drawn.

The "Nature" simulation in Figure 9.6 was based on the "Images" simulation described in Section A.5.2, except the inhibitory and excitatory strengths were fixed ($\gamma_I = 2.0$ and $\gamma_E = 1.2$), and the input density scale was higher ($s_d = 8$) without

Parameter	Value	Parameter	Value
N_d	24	σ_s	$1.6\sigma_c$
R_d	48	d_r	$20^{\frac{6.5s_w}{9.5}}$
r_A	$\frac{L_d}{0.96} + 0.5$	s_d	1.5
r_{Ei}	$\frac{N_d}{6}$	s_t	0.5
r_I	$\frac{N_d}{2.4}$	n_p	$\max(1, 0.5 s_d s_r)$
θ_{1i}	0.1	γ_A	$\frac{1.07}{n_A}$
s_w	$\frac{9.5}{r_{A_o}}$	γ_L	$\frac{10.2}{s_b}$
σ_A	$\frac{9.5}{1.3}$	w_d	$6w_{do}\frac{r_{I_o}^2}{r_I^2}$
σ_c	$0.75 s_w$		

Table C.1. Defaults for FSA simulations. FSA simulations were based on the defaults from Table A.2, modified as shown in this table. Some of these defaults are overridden in individual FSA simulations, as described in the text.

changing the time scale ($s_t = 0.5$). These same parameters were used in all postnatal simulations, but starting at 1000 after the prenatal training described above.

The simulations with prenatal training followed by natural images (Figures 9.3–9.6) were based on the prenatal training parameters specified until 1000, followed by 9000 iterations with the "Nature" parameters.

C.2 Face-Selective Area Only

The FSA (like any cortical area) has its own independent set of parameters. To keep the notation simple, the same parameter names are used for the FSA as for V1, and each area is described separately in this appendix. For example, when discussing the FSA parameters, N_d refers to the size of the FSA network and L_d to the size of the V1 network.

The default FSA parameter values are listed in Table C.1, overriding those in Table A.2. The values for σ_A, s_w, and d_r were determined empirically in earlier work (Bednar and Miikkulainen 2000a). The afferent radius r_A was significantly increased compared to the V1 simulations to allow whole faces and face outlines to be learned. The other parameters were adjusted accordingly; for example, to allow the initial activity bubbles to remain similar to those in networks with a smaller afferent radius, the afferent weights were initialized with a fixed-width Gaussian instead of random noise.

The prenatal phase of the FSA-only simulations in Section 10.3 (Figure 10.12) was identical to the default simulation in Table C.1.

The simulations with the different face training pattern types (Figure 10.11) were otherwise identical to the default, except γ_A was varied as shown in Table C.2 to ensure that the average FSA activity was the same for each pattern. These values

Figure	Pattern	γ_A		Figure	Pattern	γ_A
10.11a		1.070		10.11f		0.490
10.11b		0.697		10.11g		1.983
10.11c		0.697		10.11h		1.416
10.11d		0.550		10.11i		0.948
10.11e		0.577				

Table C.2. Parameters for different types of face training patterns. Each of the simulations in the subfigures of Figure 10.11 was based on the same parameters, except γ_A was adjusted by hand until the average activity resulting from each pattern was similar. The resulting γ_A is shown for each pattern in this table.

Iteration	θ_l (prenatally trained)	θ_l (naïve)	α_A
$20000s_t$	$\theta_{1i} + 0.070$	$\frac{10}{70}\alpha_{Ai}$	$\theta_{1i} + 0.120$
$20400s_t$	$\theta_{1i} + 0.070$	$\frac{10}{70}\alpha_{Ai}$	$\theta_{1i} + 0.120$
$22000s_t$	$\theta_{1i} + 0.070$	$\frac{9}{70}\alpha_{Ai}$	$\theta_{1i} + 0.120$
$24000s_t$	$\theta_{1i} + 0.070$	$\frac{8}{70}\alpha_{Ai}$	$\theta_{1i} + 0.090$
$28000s_t$	$\theta_{1i} + 0.070$	$\frac{7}{70}\alpha_{Ai}$	$\theta_{1i} + 0.070$
$32000s_t$	$\theta_{1i} + 0.070$	$\frac{7}{70}\alpha_{Ai}$	$\theta_{1i} + 0.070$
$36000s_t$	$\theta_{1i} + 0.080$	$\frac{6}{70}\alpha_{Ai}$	$\theta_{1i} + 0.050$
$40000s_t$	$\theta_{1i} + 0.090$	$\frac{6}{70}\alpha_{Ai}$	$\theta_{1i} + 0.045$

Table C.3. Parameter change schedule for postnatal FSA simulations. The FSA simulations in Section 10.3 continued past 20,000 iterations to model postnatal learning; the formulas above describe how the parameter values were obtained for these additional iterations. Together, Tables A.3 and C.3 specify the parameter change schedule for the entire FSA-only simulations.

were determined by presenting a set of random inputs while adjusting γ_A until the sum of the cortical response was the same as for the three-dot pattern (Figure 10.11a).

The postnatal phase of the FSA-only simulations in Section 10.3 (Figure 10.15) continued with the parameters that were in effect at the end of prenatal training, except α_A was reduced further as shown in Table C.3, the sigmoid range $\theta_u - \theta_l$ was reduced to 0.48, and the sigmoid's lower threshold θ_l was set separately for each network as described in Section 10.3 and shown in Table C.3.

	V1 settings					FSA settings				
Iteration	θ_1	γ_A	γ_E	γ_I	γ_n	θ_1	γ_A	γ_E	γ_I	γ_n
$28000s_t$	$\theta_{1i} + 0.140$	1.90	0.9	0.9	0	$\theta_{1i} + 0.130$	3.0	0.9	0.9	0
$30000s_t$	$\theta_{1i} + 0.267$	2.70	1.0	1.1	1	$\theta_{1i} + 0.250$	5.0	0.5	0.7	2
$35000s_t$	$\theta_{1i} + 0.317$	2.90	1.1	1.3	2	$\theta_{1i} + 0.400$	9.0	0.4	0.6	5
$40000s_t$	$\theta_{1i} + 0.417$	3.25	1.2	1.4	4	$\theta_{1i} + 0.710$	10.6	0.4	0.6	9

Table C.4. Parameter change schedule for combined V1 and FSA simulations. During iterations 20,000–28,000s_t, the network with both V1 and FSA (Section 10.2) was trained using the schedule for iterations 0–8000s_t in Table A.3. Beyond 28000s_t, it was trained as shown in this table. Because each cortical area has an independent set of parameters, the values for V1 and the FSA are listed separately.

In the statistical significance tests discussed in Figures 10.16 and 10.17, each input image was presented at 16 different locations chosen from the nodes of a regular 4×4 grid centered on the retina, with each grid step four retinal units wide. The example images shown in these figures are located at the center of the retina.

C.3 Combined V1 and Face-Selective Area

In the simulations of Section 10.2, V1 and FSA were combined into a single model. The retina, LGN, and V1 parameters were identical to the "Disks" simulation in Section A.5.1, except V1 density was very low ($N_d = 24$) to reduce memory and computational requirements, the V1 area was very large ($s_g = 8$) to allow large retinal stimuli to be tested, the LGN radius was slightly smaller ($\sigma_c = 0.4$) to match earlier simulations, and the simulation continued past 20,000s_t. The FSA parameters were identical to the default FSA-only parameters (Section C.2), except the FSA was slightly more dense ($N_d = 36/0.94$), the input region was significantly more dense ($L_d = 170$) because the FSA connects to V1 and not to LGN, the area scale corresponded to the full area of V1 ($s_g = 0.94$), the afferent radius was smaller ($r_A = 0.375L_d + 0.5$) and the weights initially random because only small training patterns were used, the afferent scale was larger ($\gamma_A = 3$) to compensate for the patchy activity in the input region (V1), and the parameters were changed on the schedule listed in Table C.4. The FSA also had only one input (V1) instead of two (the ON and OFF LGN channels), i.e. $n_A = 1$ for the FSA.

Because FSA training followed that of V1, V1 training iteration 20,000s_t was treated as if it was iteration zero in FSA training, and thus the parameters e.g. for iteration 28000s_t were determined like parameters for iteration 8000s_t of the default FSA simulation. After iteration 28000s_t, the V1 and FSA parameters followed the schedule shown in Table C.4. The V1 parameters were chosen to ensure that the network responded well to large natural images. First, the value of γ_n was gradually increased from zero to make the responses less dependent on image contrast. As a result, similar shapes that have different contrasts in different areas of an image would lead to comparable V1 responses. The other parameters were adjusted to compensate for the effect of γ_n, ensuring that V1 responses to the highest-contrast patterns were near 1.0, and lower contrasts resulted in lower V1 responses. The parameters of the FSA were chosen similarly, except the FSA sigmoid threshold θ_1 was also gradually increased so that the response would be nearly binary. In this way, the FSA response was used to decide whether a facelike pattern had occurred in the input, as described in Section 10.2.

D

PGLISSOM Simulation Specifications

The PGLISSOM simulations focused on demonstrating valid self-organization of the orientation map, and on grouping visual features. These simulations were based on the reduced LISSOM specifications of Appendix B, with slightly different parameter values to reduce the required computational resources.

D.1 Self-Organization

The self-organization simulations in Section 11.5 form the baseline for all PGLIS-SOM simulations. The parameter values for this simulation are specified in Table D.1 and the schedule for parameter adaptation over the course of self-organization in Table D.2. The parameter values were found to be effective by running several experiments, and small changes to them did not affect the global behavior of the model.

All weights were initialized with uniformly random values distributed within $[0..1]$; the afferent connections were mapped to the retina so that each neuron had a complete receptive field (Figure A.1). GMAP was smaller than SMAP to make simulations faster and more compact. The γ_C was lower in SMAP than in GMAP so that activity caused by high excitation in GMAP would not interfere with self-organization in SMAP, and to allow the self-organized global order of SMAP to be transferred to GMAP. The value γ_E was lower and γ_I higher in GMAP than in SMAP to prevent the map from becoming too active.

The inhibitory connections in GMAP did not adapt ($\alpha_{IG} = 0$); the initial broad connectivity remains to provide background inhibition, as explained in Section 11.2. The afferent and intracolumnar learning rates α_A and α_C in both maps were decreased over time, so that the order in the map could gradually start stabilizing. The long-range lateral inhibitory connections in SMAP and long-range excitatory lateral connections in GMAP were pruned on the same schedule, specified by w_d and t_d.

The base thresholds θ_b in both maps were set to θ_{b0} when the input was first presented, the activation-based values in Table D.1 were then computed, and the base thresholds were fixed to those values for the remaining settling iterations. As a result, the network would not become too active or completely silent for any of the training

Parameter	Value	Parameter	Value
N_S	136	γ_{ES}	0.8
N_G	54	γ_{EG}	0.2
L	46	γ_{IS}	0.9
r_A	6	γ_{IG}	2.5
r_{ESi}	7	t_f	40000
r_{EG}	40	n_p	1
r_{IS}	10	α_{Ai}	0.012
r_{IG}	54	α_E	0.008
σ_{ai}	3.9	α_{IS}	0.008
σ_{bi}	0.8	α_{IG}	0.0
θ_l	0.01	w_d	0.001
θ_u	1.3	t_d	40000
γ_A	1.1		

Parameter	Value	Used in	Description
r_C	2	Section 11.2	Maximum radius of the intracolumnar connections
θ_{b0}	0.05	Equation 11.5	Base threshold at the beginning of settling
γ_{bSi}	0.5	Table D.2	Initial value for γ_{bS}, SMAP base threshold scaling factor
γ_{bGi}	0.5	Table D.2	Initial value for γ_{bG}, GMAP base threshold scaling factor
θ_{bS}	$\gamma_{bS} \max_{ij} v_{ij}(t)$	Equation 11.5	SMAP base threshold
θ_{bG}	$\gamma_{bG} \max_{ij} v_{ij}(t)$	Equation 11.5	GMAP base threshold
γ_{CS}	0.5	Equation 11.3	Scaling factor for the SMAP intracolumnar weights
γ_{CG}	0.9	Equation 11.3	Scaling factor for the GMAP intracolumnar weights
γ_θ	0.4	Equation 11.5	Scaling factor for the relative refractory period
λ_E	3.0	Section 11.3.1	Excitatory synaptic decay rate
λ_I	0.5	Section 11.3.1	Inhibitory synaptic decay rate
λ_C	1.0	Section 11.3.1	Intracolumnar synaptic decay rate
λ_θ	0.5	Equation 11.6	Refractory period decay rate
λ_r	0.92	Equation 11.7	Spiking rate decay rate
t_r	0	Section 11.3.3	Length of the absolute refractory period
t_w	15	Section 11.4	Length of time window for computing average spiking rate
α_{Ci}	0.012	Equation 11.8	Initial value for γ_C, the intracolumnar learning rate

Table D.1. Defaults for PGLISSOM simulations. The subscript "S" identifies parameters of the SMAP, and "G" those of the GMAP; parameters without these subscripts had the same values for both maps. Although many of the parameters are the same as in firing-rate LISSOM models (top half), their default values are slightly different to take into account the two-map architecture and the spiking units. A number of new parameters related to intracolumnar connections and spiking are also introduced (bottom half). Some of these defaults are overridden in the synchronization and grouping simulations, as described in the text.

Iteration	r_{ES}	γ_{bS}	γ_{bG}	α_A	α_C	σ_a	σ_b
$0s_t$	r_{ESi}	γ_{bSi}	γ_{bGi}	α_{Ai}	α_{Ci}	σ_{ai}	σ_{bi}
$500s_t$	$0.57r_{ESi}$	γ_{bSi}	γ_{bGi}	α_{Ai}	α_{Ci}	σ_{ai}	σ_{bi}
$1000s_t$	$0.429r_{ESi}$	γ_{bSi}	γ_{bGi}	α_{Ai}	α_{Ci}	$1.718\sigma_{ai}$	$0.875\sigma_{bi}$
$5000s_t$	$0.429r_{ESi}$	γ_{bSi}	$1.3\gamma_{bGi}$	$0.667\alpha_{Ai}$	$0.667\alpha_{Ci}$	$1.718\sigma_{ai}$	$0.875\sigma_{bi}$
$15000s_t$	$0.429r_{ESi}$	$1.15\gamma_{bSi}$	$1.3\gamma_{bGi}$	$0.667\alpha_{Ai}$	$0.667\alpha_{Ci}$	$1.718\sigma_{ai}$	$0.875\sigma_{bi}$
$40000s_t$	$0.429r_{ESi}$	$1.15\gamma_{bSi}$	$1.3\gamma_{bGi}$	$0.667\alpha_{Ai}$	$0.667\alpha_{Ci}$	$1.718\sigma_{ai}$	$0.875\sigma_{bi}$

Table D.2. Parameter change schedule for PGLISSOM simulations. Starting with the initial values given in Table D.1, these parameters were adapted at each iteration as shown in the table.

inputs. The scaling factors γ_{bS} and γ_{bS} were later adjusted as shown in Table D.2 so that the activities could gradually become sparser. This method has a similar effect as adapting the sigmoid activation function (described in Section 4.4.3). To speed up self-organization, t_r was set to zero so that the neurons could fire as rapidly as possible.

The input consisted of single randomly located and oriented elongated Gaussians. Over time, these Gaussians were made longer than in previous LISSOM simulations, so that sharper orientation tuning and longer lateral connections could develop, improving contour integration in the model (self-organizing very long-range lateral connections is computationally expensive and was therefore avoided in previous simulations). Continuous input values were used to approximate spiking input, making the simulations more efficient without discernible effect on how the model behaves. The training took about 30 hours and 178 MB of memory on a 1 GHz AMD Athlon Linux machine.

D.2 Grouping

After self-organization, several parameters were adjusted slightly in order to make grouping more robust in Chapter 13. First, the excitatory learning rate α_E in GMAP was set to 0.1. Although not strictly necessary for grouping, such fast adaptation of lateral excitatory connections allows the network to quickly adjust the weights remaining after connection death to a level that allows robust synchronization (von der Malsburg 1981, 2003; Wang 1996). It does not affect the patchy structure of the lateral connections nor the organization of the map. Second, in GMAP γ_E was increased to 0.8, γ_I increased to 5.0, and 4% noise was added to the membrane potential of all GMAP neurons, as described in Sections 12.3.2 and 12.4.2. Third, the absolute refractory period t_r was increased to 4.0. After self-organization, fast simulation is not critical, but it is important to have a high enough temporal resolution of activity so that multiple groups of neurons can desynchronize at the same time. Higher t_r results in such higher resolution; it also makes synchronization more robust, as discussed in Section 12.4.3.

Each contour integration test pattern was generated to approximate the patterns used with human subjects (such as that in Figure 13.1) as well as possible within the small model retina and cortex. The input patterns consisted of oriented Gaussians with $\sigma_a = 1.87$ and $\sigma_b = 1.22$. The elements of each contour were placed on the retina at approximately collinear locations (cocircular in the curvature experiment of Section 13.4) with a minimum separation of $0.5\sigma_a^2$ and an orientation that corresponded to the desired degree of jitter. In addition, the neurons in SMAP and GMAP had to be highly selective for that input in that location. The background elements were then placed on the retina at random locations and orientations with the same distance and selectivity constraints until no more possible such locations existed.

In the input distribution experiments in Section 13.4, slightly broader Gaussians were used to train the networks, and the Gaussians became elongated slightly slower. Initially, $\sigma_a = 3.9$ and $\sigma_b = 1.1$, and they were increased to $\sigma_a = 7.1$ and $\sigma_b = 0.9$ at iteration 10,000. As a result, self-organization was slower; on the other hand, the connectivity patterns remained broader, allowing the networks to perform robustly on the wider variety of inputs.

D.3 Synchronization

The synchronization and desynchronization simulations in Chapter 12 were all based on a simplified PGLISSOM network where synchronization behavior could be clearly demonstrated. The network consisted of a one-dimensional array of neurons, connected one-to-one to input and output. The input neurons spiked at every time step, and the membrane potential of each neuron was initialized to uniformly random distribution within $[0..1]$. The afferent weights were fixed at 1.0, the lateral excitatory weights at $1.0/n_E$ (where n_E is the number of incoming excitatory lateral connections of the neuron), and the lateral inhibitory weights at $1.0/n_I$ (where n_I is the number of incoming inhibitory lateral connections). The afferent contribution was set to $\gamma_A = 0.63$ and decay rate to $\lambda_A = 1.0$, spike generator decay rate was $\lambda_\theta = 0.05$, sigmoid threshold and ceiling $\theta_l = 0.0$, $\theta_u = 3.0$, relative refractory threshold proportion $\gamma_\theta = 0.65$, and base threshold $\theta_b = 0.1$.

The remaining parameters were systematically varied in the experiments: (1) lateral excitatory and inhibitory connection patterns and radii r_E and r_I, (2) their contributions to the neuron activation γ_E and γ_I, (3) their synaptic decay rates λ_E and λ_I, (4) the size and pattern of the afferent input, (5) the degree of noise in the membrane potential, and (6) the duration t_r of the absolute refractory period.

E

SOM Simulation Specifications

In order to establish a baseline for comparison with LISSOM, the simulations in Chapter 3 were based on the SOM version of self-organizing maps, where the response is based on Euclidean distance similarity measure, and the weights are changed to minimize Euclidean difference between the input and the weight vector (Equation 3.15).

The default parameters, used in the basic self-organization simulation of Figure 3.6, are listed in Table E.1. The receptor surface was fully connected to each map unit. The connections were initialized to uniformly random values within $[0..1]$. The input consisted of single unoriented Gaussians whose centers were chosen from a uniformly random distribution so that they were evenly scattered over the receptor surface.

The magnification simulation (Figure 3.7) was run with the same network and learning parameters; the only difference was the distribution of the input Gaussians. The network was trained up to 10,000 iterations as before, at which point the input distribution was changed. Instead of a uniformly random distribution, in Figure 3.7a a Gaussian high-density area with center $(0.5, 0.5)$ and radius 0.01 was added to a uniformly random distribution of range $[0..0.2]$. In Figure 3.7b, the distribution consisted of a uniform distribution and two Gaussian high-density areas, one with center $(0.5, 0.25)$ and the other $(0.5, 0.75)$, and the major and minor axes of 0.05 and 0.0125 for both areas.

The three-dimensional model of ocular dominance (Figure 3.10) was abstracted further by representing the input location as a two-dimensional variable (x, y), instead of an activation pattern on a two-dimensional surface as in the two previous simulations. These values were drawn from a uniformly random distribution within $[0..1]$. The third variable, representing ocular dominance, was also uniformly randomly distributed, but within $[-0.15..0.15]$. The inputs were thus three-dimensional vectors of (x, y, z) values, fully connected to the map units. The network was self-organized in $t_f = 120,000$ input presentations, with the adaptation rate $= 0.1 \exp(-4.0\,t/t_f)$ and neighborhood width $= \max[13.3 \exp(-5.0\,t/t_f), 0.8]$.

The specifications for the SOM in handwritten digit recognition simulations will be described in Appendix F.2.

Parameter	Value
N	40
L	24
σ_u	0.1
t_f	40000

Parameter	Value	Used in	Description
α_S	$0.42\exp(-6.0\,t/t_\mathrm{f})$	Section 3.4.2	SOM learning rate
σ_h	$\max[13.3\exp(-5.0\,t/t_\mathrm{f}),0.5]$	Section 3.4.2	Neighborhood width

Table E.1. Defaults for SOM simulations. The α_S (i.e. α in Chapter 3) and σ_h parameters were reduced exponentially at every iteration t until they reached their minimal values as indicated. Some of these defaults are overridden in the individual SOM simulations, as described in the text.

F

Visual Coding Simulation Specifications

The simulations in Chapter 14 focused on demonstrating how LISSOM representations form a sparse, redundancy-reduced code that retains the salient information in the input and serves as an efficient foundation for later stages of the visual system. Except as indicated below, the simulations were based on the reduced LISSOM specifications of Appendix B and SOM specifications of Appendix E. In addition, backpropagation and perceptron networks were used for reconstruction and recognition, as described in the following sections.

F.1 Sparse Coding and Reconstruction

The sparse coding and input reconstruction experiments in Section 14.2 were based on the default reduced LISSOM parameters, except the training Gaussians were very long ($\sigma_a = 30$), only a single pattern per iteration was used ($s_d = 1$) to avoid overlapping such long patterns, the afferent strength was decreased ($\gamma_A = 0.7$) to compensate for the larger inputs, the lateral inhibitory radius was increased ($r_I = N_d/1.2$) to allow long-range interactions, the lateral inhibitory strength was increased ($\gamma_I = 4.0$) to compensate for the fixed weight of 1.0 being spread over more inhibitory connections, and the cortical density was reduced ($N_d = 48$) to make it practical to simulate these long connections.

The SoG network was based on the same parameters as the network with self-organized lateral connections, except the lateral excitatory and inhibitory strengths were adjusted in order to perform sparse coding. In the first experiment in section 14.2.3, γ_I was increased to 40 while γ_E remained at 0.9. In the second experiment, γ_I was reduced to 11.4 without changing γ_E. In the third experiment, $\gamma_I = 14$ and $\gamma_I = 0.46$.

The reconstruction input consisted of pairs of contours, each with three Gaussian segments with axis lengths $\sigma_a = 1.9$ and $\sigma_b = 1.2$, and with centers of each segment separated by 9.5 units along the contour. The center of each contour was chosen randomly from a uniform distribution in the central two-thirds of the retina, with a minimum separation of 14.3 units between the centers. The orientation of each contour was determined uniformly randomly as well. These patterns were chosen to show how the representation of visual input in the isotropic networks is degraded, i.e. how the interactions between unrelated contour elements cause the response to one or more of the elements to disappear.

The sparseness of the response was measured with population kurtosis K (Field 1994; Willmore and Tolhurst 2001):

$$K = \frac{1}{N^2} \sum_{ij} \left(\frac{\eta_{ij} - \bar{\eta}}{\sigma_\eta} \right)^4 - 3, \qquad \text{(F.1)}$$

where η_{ij} is the response of neuron (i, j) in the $N \times N$ network and $\bar{\eta}$ and σ_η are the mean and the standard deviation of the responses.

For input reconstruction, a fully connected feedforward backpropagation network (Chauvin and Rumelhart 1995; Hecht-Nielsen 1989; Parker 1982; Rumelhart et al. 1986; Werbos 1974) with one hidden layer and sigmoidal units was trained to map the V1 activity patterns to the corresponding input activity patterns in the retina. Because only a few cortical neurons receive input from the retinal receptors near the edges (Figure A.1; Section 6.2.1), only the central 24×24 area of the retina was included in the target pattern. Three different networks were trained, one for reconstruction from the initial V1 response, another from the settled LISSOM response, and a third from the settled response of the sum-of-Gaussians network.

An extensive search for appropriate backpropagation parameters was first done for each network with a set of 10,000 randomly generated input patterns and the corresponding V1 responses. A feedforward network with 500 hidden units and a learning rate of 0.25 for on-line backpropagation (where weights are changed after each input presentation, as opposed to after each pass through the training set) was found to perform consistently the best. The results are robust to relatively wide variations of these parameters: Doubling or halving these values led to only slightly weaker results.

A different dataset of 10,000 randomly generated patterns was then used to compare how well the input could be reconstructed from each of the three types of activity patterns. A 10-fold cross-validation experiment was run where 9000 patterns from this dataset were used for training, 500 for deciding when to stop training (based on the RMS error), and 500 for testing. The validation and testing sets had no overlap between the 10 runs; the training time varied between 37 and 40 epochs. The final performance was measured by counting how many of the reconstructed patterns were closest in Euclidean distance similarity measure to the actual input pattern in the test set.

F.2 Handwritten Digit Recognition

In the handwritten digit recognition simulations in Section 14.3, performance based on internal representations on SOM and LISSOM maps were compared. These representations were formed with the default parameter except as indicated below.

The initial SOM map was formed in eight epochs over the training set, with a learning rate of 0.01, linearly reducing the neighborhood width from 20 to 8. The LISSOM training then continued for another 30 epochs, with the inhibition radius of 20 and the excitation radius linearly decreasing from eight to one. The initial SOM was also trained for another 30 epochs, linearly decreasing the neighborhood width from eight to one.

In the LISSOM simulations, an adaptive version of the sigmoid activation function was used. As the map learns and changes its responses, a histogram of the unit's recent activity values is maintained and used to construct an activation function that approximates the cumulative probability of activation at a given level (Choe 1995). As a result, all units respond at different levels equally often, allowing competition and self-organization to occur robustly. Such adaptation was not necessary with the input distributions and network structures used in this book, which result in even distributions of activity already. However, adaptive sigmoids are useful in general in making the self-organizing process robust under various input and network conditions.

Appropriate learning parameters were found after some experimentation to be $\alpha_A = 0.005$, $\alpha_E = 0.003$, $\alpha_I = 0.003$, $\alpha_S = 0.006$, $\gamma_E = 1.05$, $\gamma_I = 1.35$, $\theta_l = 1.0$, and $\theta_u = 3.5$. In addition, the adaptation rate of the sigmoid $\alpha_\sigma = 0.1$ and four histogram bins were used (Choe 1995).

Twelve different splits into training and testing data were generated by randomly ordering the dataset and taking the first 2000 inputs for training and the last 992 for testing. On average, each split had 600 inputs in the training set that did not appear in other training sets.

The perceptrons were trained with the on-line version of the delta rule (i.e. adapting weights after each input presentation), with a learning rate $\alpha_P = 0.2$, for up to 500 epochs. These settings were found to be appropriate experimentally. Among the 2000 inputs used for LISSOM training, 1700 were used to train the perceptrons, and the remaining 300 were used as the validation set to determine when to stop training. After a good learning schedule and parameters were found in this way, the whole 2000 patterns were used to train the perceptron again, in order to utilize the small training set as well as possible. The final recognition performance of the total system was measured on the remaining 992 patterns, which neither the maps (LISSOM and SOM) nor the perceptrons had seen during training.

G

Calculating Feature Maps

Feature maps, such as orientation, ocular dominance, and direction maps, summarize the preferences of a large set of neurons at once. Each pixel in a feature map plot represents the preferred stimulus of that unit. The feature preferences can be measured using a number of algorithms, but the results from each algorithm are similar, as long as the neurons are strongly selective for that feature. For instance, most map measurement methods result in the same preferred orientation for units that are highly selective for orientation. They often differ slightly for unselective units, where the preference is not as clearly defined. The typical techniques used for measuring maps are first surveyed in this appendix. The details of the weighted average method are then presented, including how it was used to compute each type of feature map plot in this book.

G.1 Preference Map Algorithms

Preference maps can be calculated directly from the weight values of each neuron, or indirectly by presenting a set of input patterns and analyzing the responses of each neuron. Direct methods are more efficient and indirect methods more accurate, as will be described in the following subsections.

G.1.1 Estimating Maps from Weights

Some feature maps can be calculated directly from weight values. For instance, a map of preferred position can be estimated by computing the center of gravity of each neuron's afferent weights. Due to Hebbian learning, the afferent weights tend to reflect the response properties of the neuron, so the center of gravity is a good measure of what position in the input sheet the neuron prefers.

More generally, preference maps can be computed from a neuron's weights by fitting a parametric function to the afferent weights of each neuron; the parameters of the best fit constitute an estimate of the preferences of that neuron. For instance, an ellipsoidal Gaussian can be numerically optimized to fit the afferent weights of a neuron in a reduced LISSOM network, and the orientation of the resulting Gaussian provides an estimate of the neuron's orientation preference (Sirosh 1995). Unfortunately, it is difficult to ensure that any single parametric function will be a good match to all of the RF types that may be found in a network, particularly with natural images or random noise inputs. Thus, the parametric fitting method is difficult to use with the range of LISSOM simulations presented in this book.

Lateral interactions within the network can affect the feature preferences significantly under certain circumstances, such as during cortical reorganization (Chapter 6). In general, it is not feasible to extend methods based on direct weight analysis to include such interactions. In addition, these methods rely on internal information in the model that would not be available in animal experiments, and thus the results are not directly comparable to animal data. In such cases, a method based on neuronal responses must be used instead.

G.1.2 Discrete Pattern Method

Most map measurement methods, both for animals and for models, involve presenting a series of input patterns with varying parameter values and keeping track of the responses of each neuron. For example, in the discrete pattern method, an orientation map can be computed by presenting stimuli with several different orientations. The estimated orientation preference of the neuron is the orientation of the stimulus that led to the greatest response. Lateral interactions can be taken into account by measuring the responses after the network has settled.

In practice, more than one pattern is needed for each orientation, e.g. at different retinal positions and spatial frequencies, because a neuron will only respond to its preferred orientation if it is at the correct position. That is, even though responses will be collected only for the different values of the map parameter (such as orientation), the other parameters (such as location, spatial frequency, and eye of origin) must be varied to ensure that at least one appropriate pattern has been presented for each map parameter value. For each value of the map parameter, the peak response obtained using any combination of the other parameter values is stored. The map parameter value producing the peak response for any pattern tested is then taken as the preference of this neuron.

Any input pattern capable of eliciting a neural response can be used in this procedure, including oriented Gaussians and sine gratings. Sine gratings are more practical because fewer input patterns are needed to cover the space: they vary in only one spatial dimension (i.e. phase), whereas Gaussians vary over both x and y positions. With sine gratings, the map measurement procedure can also be seen as an approximation of discrete Fourier analysis.

Although the discrete pattern method is effective and allows taking lateral interactions into account, a large number of test patterns is necessary to achieve good resolution. For instance, to obtain orientation resolution of $1°$ would require sine gratings with a complete set of phases, typically 24, to be presented at each of 180 orientations. Entire self-organization simulations only require 10,000 input presentations, so calculating just two orientation maps takes nearly as many presentations

$(180 \times 24 \times 2 = 8640)$ as the entire simulation. Thus, in practice the discrete pattern method is either prohibitively expensive or can provide only low-resolution maps.

G.1.3 Weighted Average Method

The feature maps in this book are based on the weighted average (also known as the vector sum[1]) method, introduced by Blasdel and Salama (1986). This technique generalizes the discrete pattern method by providing a continuous estimate of preferences between the discrete patterns.

As in the discrete pattern method, inputs that cover the whole range of parameter values (e.g. combinations of orientations, frequencies, and phases) are presented, and for each value of the map parameter, the peak response of the neuron is recorded. The crucial difference is that the preference is not just the map parameter value that led to the peak response, but the weighted average of the peak responses to all map parameter values. For a periodic parameter like orientation, the averaging must be done in the vector domain, so that orientations just above and below zero (e.g. $10°$ and $170°$) average to $0°$ (e.g. instead of $85°$). For non-periodic parameters such as ocular dominance, retinotopy, or spatial frequency, the arithmetic weighted average is used instead.

In computing the preferred orientation, for each test orientation ϕ, other pattern parameters such as spatial frequency and phase are varied systematically, and the peak response $\hat{\eta}_\phi$ is recorded. A vector is then formed for each orientation ϕ with $\hat{\eta}_\phi$ as its length and 2ϕ as its orientation (because orientation is π-periodic, not 2π-periodic), and these vectors are summed together to form vector $\mathbf{V} = (V_x, V_y)$:

$$V_x = \sum_\phi \hat{\eta}_\phi \cos 2\phi,$$
$$V_y = \sum_\phi \hat{\eta}_\phi \sin 2\phi. \tag{G.1}$$

The preferred orientation of the neuron, θ, is estimated as half the orientation of $\mathbf{V}$:

$$\theta = \frac{1}{2}\text{atan}_2(V_y, V_x), \tag{G.2}$$

where $\text{atan}_2(x, y)$ is a function that returns $\tan^{-1}(x/y)$ with the quadrant of the result chosen based on the signs of both arguments. The magnitude of $\mathbf{V}$ can be taken as an estimate for orientation selectivity; its variance can be reduced by dividing with the sum of component vector magnitudes, resulting in normalized selectivity S:

$$S = \frac{\sqrt{V_x^2 + V_y^2}}{\sum_\phi \hat{\eta}_\phi}, \tag{G.3}$$

---

[1] This method should not be confused with the vector sum method for measuring perceived orientation of cortical responses (Section 7.2.1).

The neuron is highly selective if much of the response is in the direction of the preferred orientation, and unselective if the response is distributed widely across all orientations.

For example, assume that patterns were presented at orientations $0°$, $60°$, and $120°$, and phases 0, $\frac{\pi}{8}, \ldots, \frac{7\pi}{8}$, for a total of 24 patterns. For a given neuron, assume that the peak responses across all eight phases were 0.1 for $0°$, 0.4 for $60°$, and 0.8 for $120°$. The preferred orientation and selectivity of this neuron are

$$V_x = 0.1 \cos 0 + 0.4 \cos \frac{2\pi}{3} + 0.8 \cos \frac{4\pi}{3} = -0.50,$$
$$V_y = 0.1 \sin 0 + 0.4 \sin \frac{2\pi}{3} + 0.8 \sin \frac{4\pi}{3} = -0.35,$$

(G.4)

$$\theta = \frac{1}{2} \mathrm{atan}_2(V_y, V_x) = 107°,$$

(G.5)

$$S = \frac{\sqrt{V_x^2 + V_y^2}}{0.1 + 0.4 + 0.8} = 0.47.$$

(G.6)

Thus, this neuron is estimated to prefer an orientation that is intermediate between two test patterns, with a relatively low selectivity because it had a significant response to two of the patterns.

The weighted average method results in highly accurate, continuously valued estimates while requiring fewer input presentations than the discrete pattern method. For instance, in informal tests using the reduced LISSOM model, maps computed with as few as 24 input presentations (three orientations, each with eight phases) using the weighted average method had higher orientation resolution than those computed with 864 input presentations (36 orientations, each with 24 phases) using the discrete pattern method. The resulting maps were similar, but the weighted average map more accurately represented fine differences in the preferences of nearby units (verified by comparing the afferent weights). The minimum number of patterns needed depends on how broadly the neurons are tuned; in general, the method is effective as long as neurons have a significant response to at least two of the patterns.

Lateral interactions can be included in the weighted average method the same way as in the discrete pattern method, by recording the peak responses after settling. In many cases, however, responses to afferent stimulation alone (Equation 4.5) provide a sufficient approximation. The maps computed in this manner are consistent with those computed from the settled responses; typically, differences are seen only in unselective neurons, for which the preferences are less clearly defined. To save computation time, nearly all maps in this book were computed based on the afferent responses. The plasticity experiments in Chapter 6 are an important exception: With lesions, the dynamic equilibrium between afferent and lateral inputs is disturbed, and it is necessary to observe the actual settled responses.

The same algorithm can be applied to any input feature that can be varied systematically, such as ocularity or direction. In each case, peak responses are collected for each value of the map parameter, and a weighted average is computed to estimate the preferred value. If the parameter is periodic, the average is a vector sum, and the selectivity is based on its magnitude. Otherwise, an arithmetic mean is used for the average, and the selectivity is based on the highest response. Sections G.2–G.5 provide details for how these techniques were applied to measure each of the different types of feature maps.

G.2 Retinotopic Maps

The retinotopic maps in Chapter 6 were computed from the settled response, and therefore included the effect of lateral interactions. The input patterns were single Gaussians that varied in x and y position, each taking the values 0, 6, 12, 18, and 24 for a retina of size $R=36$. The preference in each dimension was computed as the arithmetic mean of each of the positions tested, weighted by the peak response to that position.

Elsewhere in the book the retinotopic maps were computed by finding the center of gravity of the afferent weights. For the self-organized maps in those chapters, this method is roughly equivalent to computing the position preference based on afferent stimulation, but more efficient.

G.3 Orientation Maps

The orientation maps were computed from the settled response in Chapter 6, and from afferent stimulation elsewhere. For most networks, they were measured using the weighted average method based on four orientations, i.e. $0°$, $45°$, $90°$, and $135°$, and 18 phases. Similar maps are obtained as long as at least three orientations and at least eight phases are included. For historical reasons, maps in Chapter 10 were based on 36 orientations and those in Chapter 11 on 18 orientations; the maps in Chapter 15 were calculated with the discrete pattern method with 36 orientations.

Because each simulation focused on a single-size LGN RF, the same spatial frequency was used for all test patterns (1.0 units on the 42×42 retina in Chapter 11 and 0.76 units on the 36×36 retina elsewhere). Orientation selectivity was calculated as in equation G.3, multiplied by 16 to highlight areas of low selectivity such as fractures and pinwheels.

G.4 Ocular Dominance Maps

For networks that included two eyes, ocular dominance and orientation preference were computed at the same time, both using the weighted average method. The various sine gratings were presented in only one eye at a time, with twice as many test patterns in total. The ocular dominance value was obtained as the weighted average of the peak response to any pattern in the left eye and the peak response to any pattern in the right eye, divided by their sum. Selectivity was computed by dividing the peak response to the dominant eye by the sum of the peak responses for the two eyes.

G.5 Direction Maps

Direction maps were computed like orientation maps, but using six different directions, 12 phases, and four speeds (ranging from 0.0 to 1.0 retinal pixels per step). For each direction, the sine grating orientation was chosen to be perpendicular to the direction of motion. Because direction is 2π-periodic (unlike orientation), the vectors in the sum represented the actual direction, rather than twice the orientation. The vector sum was computed just as for orientation, but without dividing the result by two. The direction selectivity was calculated as in equation G.3 and multiplied by 96 to highlight areas of low selectivity such as fractures and pinwheels.

G.6 Orientation Gradients

For any of the feature maps described above, a gradient plot can be calculated. For example, orientation gradient plots (such as those in Figures 5.1b and 5.10b) represent how abruptly the orientation preferences change across a given point in the map. The gradient is high at fractures, and low (and nearly constant) across linear zones.

To measure and visualize the orientation gradient, first the differences $D_{x,ij}$ and $D_{y,ij}$ in orientation preference of each unit (i, j) in the map and its preceding neighbor in the x and the y directions were calculated:

$$
\begin{aligned}
D_{x,ij} &= \Omega_{ij} - \Omega_{(i-1)j}, \\
D_{y,ij} &= \Omega_{ij} - \Omega_{i(j-1)},
\end{aligned}
\tag{G.7}
$$

where Ω_{ij} is the orientation that unit (i, j) prefers. Negative differences and differences larger than $90°$ were converted to the equivalent angles within $[0°..90°]$ (for example, $110°$ and $-70°$ are both equivalent to a $70°$ difference). The gradient magnitude D_{ij} is then given by

$$
D_{ij} = \sqrt{D_{x,ij}^2 + D_{y,ij}^2}.
\tag{G.8}
$$

These values were computed for each unit in the array (except those at the top and the left edge), and together they represent the gradient over the orientation map.

References

Abbott, L. F., and Marder, E. (1995). Activity-dependent regulation of neuronal conductances. In Arbib, M. A., editor, *The Handbook of Brain Theory and Neural Networks*, 63–65. Cambridge, MA: MIT Press. First edition.

Abeles, M. (1982). *Local Cortical Circuits: An Electrophysiological Study,* vol. 6 of *Studies of Brain Function*. Berlin: Springer.

Abeles, M. (1991). *Corticonics: Neuronal Circuits of the Cerebral Cortex*. Cambridge, UK: Cambridge University Press.

Abeles, M., Bergman, H., Gat, I., Meilijson, I., Seidemann, E., Tishby, N., and Vaadia, E. (1995). Cortical activity flips among quasi-stationary states. *Proceedings of the National Academy of Sciences, USA*, 92:8616–8620.

Abeles, M., Bergman, H., Margalit, E., and Vaadia, E. (1993). Spatiotemporal firing patterns in the frontal cortex of behaving monkeys. *Journal of Neurophysiology*, 70:1629–1638.

Abramov, I., Gordon, J., Hendrickson, A., Hainline, L., Dobson, V., and LaBossiere, E. (1982). The retina of the newborn human infant. *Science*, 217:265–267.

Acerra, F., Burnod, Y., and de Schonen, S. (2002). Modelling aspects of face processing in early infancy. *Developmental Science*, 5:98–117.

Achermann, B. (1995). Full-faces database. Copyright 1995, University of Bern, all rights reserved. http://iamwww.unibe.ch/ fkiwww/Personen/achermann.html.

Adorján, P., Levitt, J. B., Lund, J. S., and Obermayer, K. (1999). A model for the intra-cortical origin of orientation preference and tuning in macaque striate cortex. *Visual Neuroscience*, 16:303–318.

Adrian, E. D. (1926). The impulses produced by sensory nerve endings. *The Journal of Physiology*, 61:49–72.

Agüera y Arcas, B., and Fairhall, A. L. (2003). What causes a neuron to spike? *Neural Computation*, 15:1789–1807.

Ahmed, R., Anderson, J. C., Martin, K. A. C., and Charmaine, N. J. (1997). Map of the synapses onto layer 4 basket cells of the primary visual cortex of the cat. *Journal of Computational Neuroscience*, 380:230–242.

Albrecht, D. G., Farrar, S. B., and Hamilton, D. B. (1984). Spatial contrast adaptation characteristics of neurones recorded in the cat's visual cortex. *The Journal of Physiology*, 347:713–739.

Albus, K., and Wolf, W. (1984). Early postnatal development of neuronal function in the kitten's visual cortex: A laminar analysis. *The Journal of Physiology*, 348:153–185.

Alexander, D. M., Bourke, P. D., Sheridan, P., Konstandatos, O., and Wright, J. J. (2004). Intrinsic connections in tree shrew V1 imply a global to local mapping. *Vision Research*, 44:857–876.

Aloimonos, J., Weiss, I., and Bandyopadhyay, A. (1988). Active vision. *International Journal of Computer Vision*, 1:333–356.

Alvarez, P., and Squire, L. R. (1994). Memory consolidation and the medial temporal lobe: A simple network model. *Proceedings of the National Academy of Sciences, USA*, 91:7041–7045.

Amari, S. (1980). Topographic organization of nerve fields. *Bulletin of Mathematical Biology*, 42:339–364.

Amit, D. J. (1994). The Hebbian paradigm reintegrated: Local reverberations as internal representations. *Behavioral and Brain Sciences*, 18:617–626.

Anderson, J. A., and Rosenfeld, E., editors (1988). *Neurocomputing: Foundations of Research*. Cambridge, MA: MIT Press.

Andrade, M. A., Muro, E. M., and Morán, F. (2001). Simulation of plasticity in the adult visual cortex. *Biological Cybernetics*, 84:445–451.

Angelucci, A., Levitt, J. B., and Lund, J. S. (2002). Anatomical origins of the classical receptive field and modulatory surround field of single neurons in macaque visual cortical area V1. *Progress in Brain Research*, 136:373–388.

Arbib, M. A., Érdi, P., and Szentágothai, J. (1997). *Neural Organization: Structure, Function, and Dynamics*. Cambridge, MA: MIT Press.

Arbib, M. A., and Grethe, J. S., editors (2001). *Computing the Brain: A Guide to Neuroinformatics*. San Diego, CA: Academic Press.

Ascoli, G. A., Krichmar, J. L., Nasuto, S. J., and Senft, S. L. (2001). Generation, description and storage of dendritic morphology data. *Philosophical Transactions: Biological Sciences*, 356:1131–1145.

Atick, J. J. (1992). Could information theory provide an ecological theory of sensory processing? *Network: Computation in Neural Systems*, 3:213–251.

Atick, J. J., and Redlich, A. N. (1990). Towards a theory of early visual processing. *Neural Computation*, 2:308–320.

Azouz, R., and Gray, C. M. (2000). Dynamic spike threshold reveals a mechanism for synaptic coincidence detection in cortical neurons in vivo. *Proceedings of the National Academy of Sciences, USA*, 97:8110–8115.

Bach y Rita, P. (1972). *Brain Mechanisms in Sensory Substitution*. San Diego, CA: Academic Press.

Bach y Rita, P. (2004). Tactile sensory substitution studies. *Annals of the New York Academy of Sciences*, 1013:83–91.

Bailey, D., Feldman, J. A., Narayanan, S., and Lakoff, G. (1997). Modeling embodied lexical development. In Shafto, M. G., and Langley, P., editors, *Proceedings of the 19th Annual Conference of the Cognitive Science Society*, 19–24. Hillsdale, NJ: Erlbaum.

Bair, W., Zohary, E., and Newsome, W. T. (2001). Correlated firing in macaque visual area MT: Time scales and relationship to behavior. *The Journal of Neuroscience*, 21:1676–1697.

Bajcsy, R. (1988). Active perception. *Proceedings of the IEEE*, 78:996–1005.

Baldi, P. (1998). Probabilistic models of neuronal spike trains. In Giles, C. L., and Gori, M., editors, *Adaptive Processing of Sequences and Data Structures: International Summer School on Neural Networks, "E. R. Caianiello" — Tutorial Lectures*, Lecture Notes in Artificial Intelligence 1387, 198–228. Berlin: Springer.

Baldi, P., and Meir, R. (1990). Computing with arrays of coupled oscillators: An application to preattentive texture discrimination. *Neural Computation*, 2:458–471.

Ballard, D. H. (1991). Animate vision. *Artificial Intelligence*, 48:57–86.

Ballard, D. H., Hayhoe, M. M., Pook, P. K., and Rao, R. P. N. (1997). Deictic codes for the embodiment of cognition. *Behavioral and Brain Sciences*, 20:723–767.

Banks, M. S., and Salapatek, P. (1981). Infant pattern vision: A new approach based on the contrast sensitivity function. *Journal of Experimental Child Psychology*, 31:1–45.

Barker, A. T., Jalinous, R., and Freeston, I. L. (1985). Non-invasive magnetic stimulation of human motor cortex. *Lancet*, 1:1106–1107.

Barlow, H. B. (1972). Single units and sensation: A neuron doctrine for perceptual psychology? *Perception*, 1:371–394.

Barlow, H. B. (1985). The twelfth Bartlett memorial lecture: The role of single neurons in the psychology of perception. *The Quarterly Journal of Experimental Psychology*, 37A:121–145.

Barlow, H. B. (1989). Unsupervised learning. *Neural Computation*, 1:295–311.

Barlow, H. B. (1990). A theory about the functional role and synaptic mechanism of visual after-effects. In Blakemore, C., editor, *Vision: Coding and Efficiency*, 363–375. Cambridge, UK: Cambridge University Press.

Barlow, H. B. (1994). What is the computational goal of the neocortex? In Koch, C., and Davis, J. L., editors, *Large Scale Neuronal Theories of the Brain*, 1–22. Cambridge, MA: MIT Press.

Barlow, H. B., and Földiák, P. (1989). Adaptation and decorrelation in the cortex. In Durbin, R., Miall, C., and Mitchison, G., editors, *The Computing Neuron*, 54–72. Reading, MA: Addison-Wesley.

Barnard, K., Cardei, V., and Funt, B. (2002). A comparison of computational color constancy algorithms—part I: Methodology and experiments with synthesized data. *IEEE Transactions on Image Processing*, 11:972–984.

Barrow, H. G., and Bray, A. J. (1992). An adaptive neural model of early visual processing. In Aleksander, I., and Taylor, J. G., editors, *Artificial Neural Networks, 2: Proceedings of the 1992 International Conference on Artificial Neural Networks*, 881–884. Amsterdam: North-Holland.

Bartlett, M. S., Movellan, J. R., and Sejnowski, T. J. (2002). Face recognition by independent component analysis. *IEEE Transactions on Neural Networks*, 13:1450–1464.

Bartlett, M. S., and Sejnowski, T. J. (1997). Viewpoint invariant face recognition using independent component analysis and attractor networks. In Mozer, M. C., Jordan, M. I., and Petsche, T., editors, *Advances in Neural Information Processing Systems 9*, 817–823. Cambridge, MA: MIT Press.

Bartlett, M. S., and Sejnowski, T. J. (1998). Learning viewpoint-invariant face representations from visual experience in an attractor network. *Network: Computation in Neural Systems*, 9:399–417.

Bartlett, P. L., and Maass, W. (2003). Vapnik–Chervonenkis dimension of neural nets. In Arbib, M. A., editor, *The Handbook of Brain Theory and Neural Networks*, 1188–1192. Cambridge, MA: MIT Press. Second edition.

Bartrip, J., Morton, J., and de Schonen, S. (2001). Responses to mother's face in 3-week to 5-month-old infants. *British Journal of Developmental Psychology*, 19:219–232.

Bartsch, A. P., and van Hemmen, J. L. (2001). Combined Hebbian development of geniculo-cortical and lateral connectivity in a model of primary visual cortex. *Biological Cybernetics*, 84:41–55.

Basole, A., White, L. E., and Fitzpatrick, D. (2003). Mapping multiple features in the population response of visual cortex. *Nature*, 424:986–990.

Bauer, H.-U., Brockmann, D., and Geisel, T. (1997). Analysis of ocular dominance pattern formation in a high-dimensional self-organizing-map model. *Network: Computation in Neural Systems*, 8:17–33.

Bauer, H.-U., and Villman, T. (1997). Growing a hypercubical output space in a self-organizing feature map. *IEEE Transactions on Neural Networks*, 218–226.

Bauman, L. A., and Bonds, A. B. (1991). Inhibitory refinement of spatial frequency selectivity in single cells of the cat striate cortex. *Vision Research*, 31:933–944.

Beaudot, W. H. A. (2002). Role of onset synchrony in contour integration. *Vision Research*, 42:1–9.

Bechtel, W., and Abrahamsen, A. (2002). *Connectionism and the Mind: Parallel Processing, Dynamics, and Evolution in Networks*. Oxford, UK: Blackwell. Second edition.

Becker, S. (1992). An information-theoretic unsupervised learning algorithm for neural networks. Doctoral dissertation, Department of Computer Science, University of Toronto, Toronto, Canada.

Bednar, J. A. (1997). Tilt aftereffects in a self-organizing model of the primary visual cortex. Master's thesis, Department of Computer Sciences, The University of Texas at Austin, Austin, TX. Technical Report AI-TR-97-259.

Bednar, J. A. (2000). Internally generated activity, non-episodic memory, and emotional salience in sleep. *Behavioral and Brain Sciences*, 23:908–909. Commentary on the *Sleep and Dreaming* issue.

Bednar, J. A. (2002). Learning to see: Genetic and environmental influences on visual development. Doctoral dissertation, Department of Computer Sciences, The University of Texas at Austin, Austin, TX. Technical Report AI-TR-02-294.

Bednar, J. A., Kelkar, A., and Miikkulainen, R. (2004). Scaling self-organizing maps to model large cortical networks. *Neuroinformatics*, 2:275–302.

Bednar, J. A., and Miikkulainen, R. (1998). Pattern-generator-driven development in self-organizing models. In Bower, J. M., editor, *Computational Neuroscience: Trends in Research, 1998*, 317–323. New York: Plenum Press.

Bednar, J. A., and Miikkulainen, R. (2000a). Self-organization of innate face preferences: Could genetics be expressed through learning? In *Proceedings of the 17th National Conference on Artificial Intelligence and the 12th Annual Conference on Innovative Applications of Artificial Intelligence*, 117–122. Menlo Park, CA: AAAI Press.

Bednar, J. A., and Miikkulainen, R. (2000b). Tilt aftereffects in a self-organizing model of the primary visual cortex. *Neural Computation*, 12:1721–1740.

Bednar, J. A., and Miikkulainen, R. (2003a). Learning innate face preferences. *Neural Computation*, 15:1525–1557.

Bednar, J. A., and Miikkulainen, R. (2003b). Self-organization of spatiotemporal receptive fields and laterally connected direction and orientation maps. *Neurocomputing*, 52–54:473–480.

Bednar, J. A., and Miikkulainen, R. (2005). Constructing visual function through prenatal and postnatal learning. In Mareschal, D., Johnson, M. H., Sirois, S., Spratling, M., Thomas, M. S. C., and Westermann, G., editors, *Neuroconstructivism, Vol. 2: Perspectives and Prospects*. Oxford, UK: Oxford University Press. In press.

Beer, R. D. (2000). Dynamical approaches to cognitive science. *Trends in Cognitive Sciences*, 4:91–99.

Bell, A. J., and Sejnowski, T. J. (1997). The "independent components" of natural scenes are edge filters. *Vision Research*, 37:3327.

Ben-Hur, A., Horn, D., Siegelmann, H. T., and Vapnik, V. (2001). Support vector clustering. *Journal of Machine Learning Research*, 2:125–137.

Ben-Shahar, O., and Zucker, S. W. (2004). Geometrical computations explain projection patterns of long-range horizontal connections in visual cortex. *Neural Computation*, 16:445–476.

Ben-Yishai, R., Bar-Or, R. L., and Sompolinsky, H. (1995). Theory of orientation tuning in visual cortex. *Proceedings of the National Academy of Sciences, USA*, 92:3844–3848.

Berkley, M. A., Debruyn, B., and Orban, G. (1993). Illusory, motion, and luminance-defined contours interact in the human visual system. *Vision Research*, 34:209–216.

Berns, G. S., Dayan, P., and Sejnowski, T. J. (1993). A correlational model for the development of disparity selectivity in visual cortex that depends on prenatal and postnatal phases. *Proceedings of the National Academy of Sciences, USA*, 90:8277–81.

Beyer, H.-G., and Schwefel, H.-P. (2002). Evolution strategies: A comprehensive introduction. *Natural Computing*, 1:3–52.

Bienenstock, E. L., Cooper, L. N., and Munro, P. W. (1982). Theory for the development of neuron selectivity: Orientation specificity and binocular interaction in visual cortex. *The Journal of Neuroscience*, 2:32–48.

Binzegger, T., Douglas, R. J., and Martin, K. A. C. (2004). A quantitative map of the circuit of cat primary visual cortex. *The Journal of Neuroscience*, 24:8441–8453.

Bishop, C. M., Svensén, M., and Williams, C. K. I. (1998). GTM: The generative topographic mapping. *Neural Computation*, 10:215–234.

Bisley, J. W., and Goldberg, M. E. (2003). The role of the parietal cortex in the neural processing of saccadic eye movements. *Advances in Neurology*, 93:141–157.

Blackmore, J., and Miikkulainen, R. (1995). Visualizing high-dimensional structure with the incremental grid growing neural network. In Prieditis, A., and Russell, S., editors, *Machine Learning: Proceedings of the 12th Annual Conference*, 55–63. San Francisco: Kaufmann.

Blais, B. S., Cooper, L. N., and Shouval, H. Z. (2000). Formation of direction selectivity in natural scene environments. *Neural Computation*, 12:1057–1066.

Blake, A., and Yuille, A. L., editors (1992). *Active Vision*. Cambridge, MA: MIT Press.

Blakemore, C., and Carpenter, R. H. S. (1971). Lateral thinking about lateral inhibition. *Nature*, 234:418–419.

Blakemore, C., Carpenter, R. H. S., and Georgeson, M. A. (1970). Lateral inhibition between orientation detectors in the human visual system. *Nature*, 228:37–39.

Blakemore, C., and Cooper, G. F. (1970). Development of the brain depends on the visual environment. *Nature*, 228:477–478.

Blakemore, C., and van Sluyters, R. C. (1975). Innate and environmental factors in the development of the kitten's visual cortex. *The Journal of Physiology*, 248:663–716.

Blasdel, G. G. (1992a). Differential imaging of ocular dominance columns and orientation selectivity in monkey striate cortex. *The Journal of Neuroscience*, 12:3115–3138.

Blasdel, G. G. (1992b). Orientation selectivity, preference, and continuity in monkey striate cortex. *The Journal of Neuroscience*, 12:3139–3161.

Blasdel, G. G., Obermayer, K., and Kiorpes, L. (1995). Organization of ocular dominance and orientation columns in the striate cortex of neonatal macaque monkeys. *Visual Neuroscience*, 12:589–603.

Blasdel, G. G., and Salama, G. (1986). Voltage-sensitive dyes reveal a modular organization in monkey striate cortex. *Nature*, 321:579–585.

Bohte, S. M., and Mozer, M. C. (2005). Reducing spike train variability: A computational theory of spike-timing dependent plasticity. In *Advances in Neural Information Processing Systems 17*. Cambridge, MA: MIT Press. In press.

Bolhuis, J. J. (1999). Early learning and the development of filial preferences in the chick. *Behavioural Brain Research*, 98:245–252.

Bolhuis, J. J., and Honey, R. C. (1998). Imprinting, learning and development: From behaviour to brain and back. *Trends in Neurosciences*, 21:306–311.

Bolz, J., and Gilbert, C. D. (1986). Generation of end-inhibition in the visual cortex via interlaminar connections. *Nature*, 320:362–364.

Bonds, A. B. (1979). Development of orientation tuning in the visual cortex of kittens. In Freeman, R. D., editor, *Developmental Neurobiology of Vision*, 31–41. New York: Plenum Press.

Bosking, W. H., Zhang, Y., Schofield, B. R., and Fitzpatrick, D. (1997). Orientation selectivity and the arrangement of horizontal connections in tree shrew striate cortex. *The Journal of Neuroscience*, 17:2112–2127.

Bourgeois, J. P., Jastreboff, P. J., and Rakic, P. (1989). Synaptogenesis in visual cortex of normal and preterm monkeys: Evidence for intrinsic regulation of synaptic overproduction. *Proceedings of the National Academy of Sciences, USA*, 86:4297–4301.

Bower, J. M., and Beeman, D. (1998). *The Book of GENESIS: Exploring Realistic Neural Models with the GEneral NEural SImulation System*. Santa Clara, CA: Telos. Second edition.

Braastad, B. O., and Heggelund, P. (1985). Development of spatial receptive-field organization and orientation selectivity in kitten striate cortex. *Journal of Neurophysiology*, 53:1158–1178.

Brainard, D. H. (2004). Color constancy. In Chalupa, L. M., and Werner, J. S., editors, *The Visual Neurosciences*, 948–961. Cambridge, MA: MIT Press.

Bray, A. J., and Barrow, H. G. (1996). Simple cell adaptation in visual cortex: A computational model of processing in the early visual pathway. Technical Report CSRP 331, School of Cognitive and Computing Sciences, University of Sussex, Brighton, UK.

Britten, K. H., Shalden, M. N., Newsome, W. T., and Movshon, J. A. (1992). The analysis of visual motion: A comparison of neuronal and psychophysical performance. *The Journal of Neuroscience*, 12:4745–4765.

Bronson, G. W. (1974). The postnatal growth of visual capacity. *Child Development*, 45:873–890.

Brooks, R. A., Breazeal (Ferrell), C., Irie, R., Kemp, C. C., Marjanović, M., Scassellati, B., and Williamson, M. M. (1998). Alternative essences of intelligence. In *Proceedings of the 15th National Conference on Artificial Intelligence and the 10th Annual Conference on Innovative Applications of Artificial Intelligence*, 961–976. Menlo Park, CA: AAAI Press.

Bruns, A., Eckhorn, R., Jokeit, H., and Ebner, A. (2000). Amplitude envelope correlation detects coupling among incoherent brain signals. *Neuroreport*, 11:1509–1514.

Buonomano, D. V., and Merzenich, M. M. (1998). Cortical plasticity: From synapses to maps. *Annual Review of Neuroscience*, 21:149–186.

Burger, D., and Goodman, J. R. (1997). Billion-transistor architectures. *IEEE Computer*, 30:46–48.

Burger, T., and Lang, E. W. (1999). An incremental Hebbian learning model of the primary visual cortex with lateral plasticity and real input patterns. *Zeitschrift für Naturforschung C — A Journal of Biosciences*, 54:128–140.

Burger, T., and Lang, E. W. (2001). Self-organization of local cortical circuits and cortical orientation maps: A nonlinear Hebbian model of the visual cortex with adaptive lateral couplings. *Zeitschrift für Naturforschung C—A Journal of Biosciences*, 56:464–478.

Burkhalter, A., and Bernardo, K. L. (1989). Organization of corticocortical connections in human visual cortex. *Proceedings of the National Academy of Sciences, USA*, 86:1071–1075.

Burkhalter, A., Bernardo, K. L., and Charles, V. (1993). Development of local circuits in human visual cortex. *The Journal of Neuroscience*, 13:1916–1931.

Burton, A. M., Bruce, V., and Hancock, P. J. B. (1999). From pixels to people: A model of familiar face recognition. *Cognitive Science*, 23:1–31.

Bushnell, I. W. R. (1998). The origins of face perception. In Simion, F., and Butterworth, G., editors, *The Development of Sensory, Motor and Cognitive Capacities in Early Infancy: From Perception to Cognition*, 69–86. East Sussex, UK: Psychology Press.

Bushnell, I. W. R. (2001). Mother's face recognition in newborn infants: Learning and memory. *Infant and Child Development*, 10:67–74.

Bushnell, I. W. R., Sai, F., and Mullin, J. T. (1989). Neonatal recognition of the mother's face. *British Journal of Developmental Psychology*, 7:3–15.

Butts, D. A., Feller, M. B., Shatz, C. J., and Rokhsar, D. S. (1999). Retinal waves are governed by collective network properties. *The Journal of Neuroscience*, 19:3580–3593.

Buzsáki, G., and Draguhn, A. (2004). Neuronal oscillations in cortical networks. *Science*, 304:1926–1929.

Cai, D., DeAngelis, G. C., and Freeman, R. D. (1997). Spatiotemporal receptive field organization in the lateral geniculate nucleus of cats and kittens. *Journal of Neurophysiology*, 78:1045–1061.

Calford, M. B., Schmid, L. M., and Rosa, M. G. P. (1999). Monocular focal retinal lesions induce short-term topographic plasticity in adult visual cortex. *Proceedings: Biological Sciences*, 266:499–507.

Calford, M. B., Wang, C., Taglianetti, V., Waleszczyk, W. J., Burke, W., and Dreher, B. (2000). Plasticity in adult cat visual cortex (area 17) following circumscribed monocular lesions of all retinal layers. *The Journal of Physiology*, 524:587–602.

Calford, M. B., Wright, L. L., Metha, A. B., and Taglianetti, V. (2003). Topographic plasticity in primary visual cortex is mediated by local corticocortical connections. *The Journal of Neuroscience*, 23:6434–6442.

Callaway, C. W., Lydic, R., Baghdoyan, H. A., and Hobson, J. A. (1987). Pontogeniculooccipital waves: Spontaneous visual system activity during rapid eye movement sleep. *Cellular and Molecular Neurobiology*, 7:105–149.

Callaway, E. M., and Katz, L. C. (1990). Emergence and refinement of clustered horizontal connections in cat striate cortex. *The Journal of Neuroscience*, 10:1134–1153.

Callaway, E. M., and Katz, L. C. (1991). Effects of binocular deprivation on the development of clustered horizontal connections in cat striate cortex. *Proceedings of the National Academy of Sciences, USA*, 88:745–749.

Callaway, E. M., and Wiser, A. K. (1996). Contributions of individual layer 2–5 spiny neurons to local circuits in macaque primary visual cortex. *Visual Neuroscience*, 13:907–922.

Calvert, G. A. (2001). Crossmodal processing in the human brain: Insights from functional neuroimaging studies. *Cerebral Cortex*, 11:1110–1123.

Calvert, J. E., and Harris, J. P. (1988). Spatial frequency and duration effects on the tilt illusion and orientation acuity. *Vision Research*, 28:1051–1059.

Campbell, F. W., and Maffei, L. (1971). The tilt aftereffect: A fresh look. *Vision Research*, 11:833–840.

Campbell, S. R., Wang, D., and Jayaprakash, C. (1999). Synchrony and desynchrony in integrate-and-fire oscillators. *Neural Computation*, 11:1595–1619.

Campos, M. M., and Carpenter, G. A. (2000). Building adaptive basis functions with a continuous self-organizing map. *Neural Processing Letters*, 11:59–78.

Carney, T. (1982). Directional specificity in tilt aftereffect induced with moving contours: A reexamination. *Vision Research*, 22:1273–1275.

Carpenter, G. A. (2001). Neural network models of learning and memory: Leading questions and an emerging framework. *Trends in Cognitive Sciences*, 5:114–118.

Carpenter, R. H. S., and Blakemore, C. (1973). Interactions between orientations in human vision. *Experimental Brain Research*, 18:287–303.

Casagrande, V. A., and Norton, T. T. (1989). Lateral geniculate nucleus: A review of its physiology and function. In Leventhal, A. G., editor, *The Neural Basis of Visual Function*, vol. 4 of *Vision and Visual Dysfunction*, 41–84. Boca Raton, FL: CRC Press.

Catania, K. C., Lyon, D. C., Mock, O. B., and Kaas, J. H. (1999). Cortical organization in shrews: Evidence from five species. *The Journal of Comparative Neurology*, 410:55–72.

Catsicas, M., and Mobbs, P. (1995). Waves are swell. *Current Biology*, 5:977–979.

Celebrini, S., and Newsome, W. T. (1994). Neuronal and psychophysical sensitivity to motion signals in extrastriate MST of the macaque monkey. *The Journal of Neuroscience*, 14:4109–4124.

Chakravarthy, S. V., and Ghosh, J. (1996). A complex-valued associative memory for storing patterns as oscillatory states. *Biological Cybernetics*, 75:229–238.

Chance, F. S., Nelson, S. B., and Abbott, L. F. (1999). Complex cells as cortically amplified simple cells. *Nature Neuroscience*, 2:277–282.

Chang, L.-C., and Chang, F.-J. (2002). An efficient parallel algorithm for LISSOM neural network. *Parallel Computing*, 28:1611–1633.

Chapman, B. (2000). Necessity for afferent activity to maintain eye-specific segregation in ferret lateral geniculate nucleus. *Science*, 287:2479–2482.

Chapman, B., and Bonhoeffer, T. (1998). Overrepresentation of horizontal and vertical orientation preferences in developing ferret area 17. *Proceedings of the National Academy of Sciences, USA*, 95:2609–2614.

Chapman, B., Gödecke, I., and Bonhoeffer, T. (1999). Development of orientation preference in the mammalian visual cortex. *Journal of Neurobiology*, 41:18–24.

Chapman, B., and Stryker, M. P. (1993). Development of orientation selectivity in ferret primary visual cortex and effects of deprivation. *The Journal of Neuroscience*, 13:5251–5262.

Chapman, B., Stryker, M. P., and Bonhoeffer, T. (1996). Development of orientation preference maps in ferret primary visual cortex. *The Journal of Neuroscience*, 16:6443–6453.

Chauvin, Y., and Rumelhart, D. E., editors (1995). *Backpropagation: Theory, Architectures, and Applications*. Hillsdale, NJ: Erlbaum.

Chino, Y. M., Kaas, J. H., Smith, E. L., Langston, A. L., and Cheng, H. (1992). Rapid reorganization of cortical maps in adult cats following restricted deafferentation in retina. *Vision Research*, 32:789–796.

Chino, Y. M., Smith, E. L., Kaas, J. H., and Cheng, H. (1995). Receptive-field properties of deafferentated visual cortical neurons after topographic map reorganization in adult cats. *The Journal of Neuroscience*, 15:2417–2433.

Chklovskii, D. B., Mel, B. W., and Svoboda, K. (2004). Cortical rewiring and information storage. *Nature*, 431:782–788.

Chklovskii, D. B., Schikorski, T., and Stevens, C. F. (2002). Wiring optimization in cortical circuits. *Neuron*, 34:341–347.

Cho, S.-B. (1997). Self-organizing map with dynamical node splitting: Application to hand-written digit recognition. *Neural Computation*, 9:1345–1355.

Choe, Y. (1995). Laterally interconnected self-organizing feature map in handwritten digit recognition. Master's thesis, Department of Computer Sciences, The University of Texas at Austin, Austin, TX. Technical Report AI95-236.

Choe, Y. (2001). Perceptual grouping in a self-organizing map of spiking neurons. Doctoral dissertation, Department of Computer Sciences, The University of Texas at Austin, Austin, TX. Technical Report AI01-292.

Choe, Y. (2002). Second order isomorphism: A reinterpretation and its implications in brain and cognitive sciences. In Gray, W. D., and Schunn, C. D., editors, *Proceedings of the 24th Annual Conference of the Cognitive Science Society*, 190–195. Hillsdale, NJ: Erlbaum.

Choe, Y. (2003a). Analogical cascade: A theory on the role of the thalamo-cortical loop in brain function. *Neurocomputing*, 52–54:713–719.

Choe, Y. (2003b). Processing of analogy in the thalamocortical circuit. In *Proceedings of the International Joint Conference on Neural Networks*, 1480–1485. Piscataway, NJ: IEEE.

Choe, Y. (2004). The role of temporal parameters in a thalamocortical model of analogy. *IEEE Transactions on Neural Networks*, 15:1071–1082.

Choe, Y., and Bhamidipati, S. K. (2004). Autonomous acquisition of the meaning of sensory states through sensory-invariance driven action. In Ijspeert, A. J., Murata, M., and Wakamiya, N., editors, *Biologically Inspired Approaches to Advanced Information Technology*, Lecture Notes in Computer Science 3141, 176–188. Berlin: Springer.

Choe, Y., and Miikkulainen, R. (1997). Self-organization and segmentation with laterally connected spiking neurons. In *Proceedings of the 15th International Joint Conference on Artificial Intelligence*, 1120–1125. San Francisco: Kaufmann.

Choe, Y., and Miikkulainen, R. (1998). Self-organization and segmentation in a laterally connected orientation map of spiking neurons. *Neurocomputing*, 21:139–157.

Choe, Y., and Miikkulainen, R. (2000). A self-organizing neural network for contour integration through synchronized firing. In *Proceedings of the 17th National Conference on Artificial Intelligence and the 12th Annual Conference on Innovative Applications of Artificial Intelligence*, 123–128. Menlo Park, CA: AAAI Press.

Choe, Y., and Miikkulainen, R. (2004). Contour integration and segmentation in a self-organizing map of spiking neurons. *Biological Cybernetics*, 90:75–88.

Choe, Y., Miikkulainen, R., and Cormack, L. K. (2000). Effects of presynaptic and postsynaptic resource redistribution in Hebbian weight adaptation. *Neurocomputing*, 32–33:77–82.

Chouvet, G., Blois, R., Debilly, G., and Jouvet, M. (1983). La structure d'occurrence des mouvements oculaires rapides du sommeil paradoxal est similaire chez les jumeaux homozygotes [The structure of the occurrence of rapid eye movements in paradoxical sleep is similar in homozygotic twins]. *Comptes Rendus des Seances de l'Academie des Sciences – Serie III, Sciences de la Vie*, 296:1063–1068.

Churchland, P. S., Ramachandran, V. S., and Sejnowski, T. J. (1994). A critique of pure vision. In Koch, C., and Davis, J. L., editors, *Large Scale Neuronal Theories of the Brain*, 23–60. Cambridge, MA: MIT Press.

Churchland, P. S., and Sejnowski, T. J. (1992). *The Computational Brain*. Cambridge, MA: MIT Press.

Clark, A. (1999). An embodied cognitive science. *Trends in Cognitive Sciences*, 3:345–351.

Clifford, C. W., and Ibbotson, M. R. (2002). Fundamental mechanisms of visual motion detection: Models, cells and functions. *Progress in Neurobiology*, 68:409–437.

Cohen, L. B. (1998). An information-processing approach to infant perception and cognition. In Simion, F., and Butterworth, G., editors, *The Development of Sensory, Motor and Cognitive Capacities in Early Infancy: From Perception to Cognition*, 277–300. East Sussex, UK: Psychology Press.

Cohen, L. B., and Cashon, C. H. (2003). Infant perception and cognition. In Easterbrooks, M. A., Lerner, R. M., and Mistry, J., editors, *Handbook of Psychology, Vol. VI: Developmental Psychology*, 65–89. Hoboken, NJ: Wiley.

Cohen, P. R., and Beal, C. R. (2000). Natural semantics for a mobile robot. Technical Report 00-59, Department of Computer Science, University of Massachusettes, Amherst, MA.

Coltheart, M. (1971). Visual feature-analyzers and aftereffects of tilt and curvature. *Psychological Review*, 78:114–121.

Constantine-Paton, M., Cline, H. T., and Debski, E. (1990). Patterned activity, synaptic convergence, and the NMDA receptor in developing visual pathways. *Annual Review of Neuroscience*, 13:129–154.

Constantine-Paton, M., and Law, M. I. (1978). Eye-specific termination bands in tecta of three-eyed frogs. *Science*, 202:639–641.

Conway, B. R. (2003). Colour vision: A clue to hue in V2. *Current Biology*, 13:308–310.

Cooper, L. N., Intrator, N., Blais, B. S., and Shouval, H. Z. (2004). *Theory of Cortical Plasticity*. Singapore: World Scientific.

Coppola, D. M., White, L. E., Fitzpatrick, D., and Purves, D. (1998). Unequal representation of cardinal and oblique contours in ferret visual cortex. *Proceedings of the National Academy of Sciences, USA*, 95:2621–2623.

Cormack, L. K., and Riddle, R. B. (1996). Binocular correlation detection with oriented dynamic random-line stereograms. *Vision Research*, 36:2303–2310.

Cottrell, M., de Bodt, E., and Verleysen, M. (2001). A statistical tool to assess the reliability of self-organizing maps. In Allinson, N. M., Yin, H., Allinson, L. J., and Slack, J., editors, *Advances in Self-Organizing Maps*, 7–14. Berlin: Springer.

Cover, T. M., and Thomas, J. (1991). *Elements of Information Theory*. Hoboken, NJ: Wiley.

Crabtree, J. W., Collingridge, G. L., and Isaac, J. T. R. (1998). A new intrathalamic pathway linking modality-related nuclei in the dorsal thalamus. *Nature Neuroscience*, 1:389–394.

Crabtree, J. W., and Isaac, J. T. R. (2002). Intrathalamic pathways allowing modality-related and cross-modality switching in the dorsal thalamus. *The Journal of Neuroscience*, 22:8754–8761.

Crair, M. C. (1999). Neuronal activity during development: Permissive or instructive? *Current Opinion in Neurobiology*, 9:88–93.

Crair, M. C., Gillespie, D. C., and Stryker, M. P. (1998). The role of visual experience in the development of columns in cat visual cortex. *Science*, 279:566–570.

Crair, M. C., Horton, J. C., Antonini, A., and Stryker, M. P. (2001). Emergence of ocular dominance columns in cat visual cortex by 2 weeks of age. *The Journal of Comparative Neurology*, 430:235–249.

Crair, M. C., and Malenka, R. C. (1995). A critical period for long-term potentiation at thalamocortical synapses. *Nature*, 375:325–328.

Crick, F. (1984). Function of the thalamic reticular complex: The searchlight hypothesis. *Proceedings of the National Academy of Sciences, USA*, 81:4586–4950.

Crowley, J. C., and Katz, L. C. (1999). Development of ocular dominance columns in the absence of retinal input. *Nature Neuroscience*, 2:1125–1130.

Crowley, J. C., and Katz, L. C. (2000). Early development of ocular dominance columns. *Science*, 290:1321–1324.

Çürüklü, B., and Lansner, A. (2003). Quantitative assessment of the local and long-range horizontal connections within the striate cortex. In *Proceedings of the Second International Conference on Computational Intelligence, Robotics, and Autonomous Systems*. Piscataway, NJ: IEEE.

Dailey, M. N., and Cottrell, G. W. (1999). Organization of face and object recognition in modular neural network models. *Neural Networks*, 12:1053–1074.

Dalva, M. B., and Katz, L. C. (1994). Rearrangements of synaptic connections in visual cortex revealed by laser photostimulation. *Science*, 265:255–258.

Darian-Smith, C., and Gilbert, C. D. (1995). Topographic reorganization in the striate cortex of the adult cat and monkey is cortically mediated. *The Journal of Neuroscience*, 15:1631–1647.

Das, A., and Gilbert, C. D. (1997). Distortions of visuotopic map match orientation singularities in primary visual cortex. *Nature*, 387:594–598.

Datta, S. (1997). Cellular basis of pontine ponto-geniculo-occipital wave generation and modulation. *Cellular and Molecular Neurobiology*, 17:341–365.

Daugman, J. G. (1980). Two-dimensional spectral analysis of cortical receptive field profiles. *Vision Research*, 20:847–856.

Daw, N. (1995). *Visual Development*. New York: Plenum Press.

Dayan, P. (1993). Arbitrary elastic topologies and ocular dominance. *Neural Computation*, 5:392–401.

Dayan, P., and Abbott, L. F. (2001). *Theoretical Neuroscience: Computational and Mathematical Modeling of Neural Systems*. Cambridge, MA: MIT Press.

Dayan, P., Hinton, G. E., Neal, R. M., and Zemel, R. S. (1995). The Helmholtz machine. *Neural Computation*, 7:889–904.

de Gelder, B., and Rouw, R. (2000). Configural face processes in acquired and developmental prosopagnosia: Evidence for two separate face systems. *Neuroreport*, 11:3145–3150.

de Gelder, B., and Rouw, R. (2001). Beyond localisation: A dynamical dual route account of face recognition. *Acta Psychologica*, 107:183–207.

de Haan, M. (2001). The neuropsychology of face processing during infancy and childhood. In Nelson, C. A., and Luciana, M., editors, *Handbook of Developmental Cognitive Neuroscience*, 381–398. Cambridge, MA: MIT Press.

de Sa, V. R. (1994). Unsupervised classification learning from cross-modal environmental structure. Doctoral dissertation, Department of Computer Science, University of Rochester, Rochester, NY.

de Sa, V. R., and Ballard, D. H. (1997). Perceptual learning from cross-modal feedback. In Goldstone, R. L., Schyns, P. G., and Medin, D. L., editors, *Perceptual Learning*, vol. 36 of *Psychology of Learning and Motivation*, 309–351. San Diego, CA: Academic Press.

de Schonen, S., Mancini, J., and Liegeois, F. (1998). About functional cortical specialization: The development of face recognition. In Simion, F., and Butterworth, G., editors, *The Development of Sensory, Motor and Cognitive Capacities in Early Infancy: From Perception to Cognition*, 103–120. East Sussex, UK: Psychology Press.

De Schutter, E., and Bower, J. M. (1994a). An active membrane model of the cerebellar Purkinje cell. I. *Journal of Neurophysiology*, 71:375–400.

De Schutter, E., and Bower, J. M. (1994b). An active membrane model of the cerebellar Purkinje cell. II. *Journal of Neurophysiology*, 71:401–419.

De Valois, K. K., and Tootell, R. B. H. (1983). Spatial-frequency-specific inhibition in cat striate cortex cells. *The Journal of Physiology*, 336:359–376.

DeAngelis, G. C., Ghose, G. M., Ohzawa, I., and Freeman, R. D. (1999). Functional micro-organization of primary visual cortex: Receptive field analysis of nearby neurons. *The Journal of Neuroscience*, 19:4046–4064.

DeAngelis, G. C., Ohzawa, I., and Freeman, R. D. (1993). Spatiotemporal organization of simple-cell receptive fields in the cat's striate cortex. I. General characteristics and post-natal development. *Journal of Neurophysiology*, 69:1091–1117.

DeAngelis, G. C., Ohzawa, I., and Freeman, R. D. (1995). Receptive-field dynamics in the central visual pathways. *Trends in Neurosciences*, 18:451–458.

Desai, N. S., Rutherford, L. C., and Turrigiano, G. G. (1999). Plasticity in the intrinsic excitability of cortical pyramidal neurons. *Nature Neuroscience*, 2:515–520.

Diamond, S. (1974). Four hundred years of instinct controversy. *Behavior Genetics*, 4:237–252.

Doi, E., Inui, T., Lee, T.-W., Wachtler, T., and Sejnowski, T. J. (2003). Spatio-chromatic receptive field properties derived from information-theoretic analyses of cone mosaic responses to natural scenes. *Neural Computation*, 15:397–417.

Dong, D. W. (1995). Associative decorrelation dynamics: A theory of self-organization and optimization in feedback networks. In Tesauro, G., Touretzky, D. S., and Leen, T. K., editors, *Advances in Neural Information Processing Systems 7*, 925–932. Cambridge, MA: MIT Press.

Dong, D. W. (1996). Associative decorrelation dynamics in visual cortex. In Sirosh, J., Miikkulainen, R., and Choe, Y., editors, *Lateral Interactions in the Cortex: Structure and Function*. Austin, TX: The UTCS Neural Networks Research Group. Electronic book, ISBN 0-9647060-0-8, http://nn.cs.utexas.edu/web-pubs/htmlbook96.

Dong, D. W., and Hopfield, J. J. (1992). Dynamic properties of neural networks with adapting synapses. *Network*, 3:267–283.

Douglas, R. J., Koch, C., Mahowald, M., Martin, K. A. C., and Suarez, H. H. (1995). Recurrent excitation in neocortical circuits. *Science*, 269:981–985.

Douglas, R. J., and Martin, K. A. C. (2004). Neuronal circuits of the neocortex. *Annual Review of Neuroscience*, 27:419–451.

Doya, K., Selverston, A. I., and Rowat, P. F. (1995). A Hodgkin-Huxley type neuron model that learns slow non-spike oscillation. In Tesauro, G., Touretzky, D. S., and Leen, T. K., editors, *Advances in Neural Information Processing Systems 7*, 566–573. Cambridge, MA: MIT Press.

Dragoi, V., Sharma, J., Miller, E. K., and Sur, M. (2002). Dynamics of neuronal sensitivity in visual cortex and local feature discrimination. *Nature Neuroscience*, 5:883–891.

Dragoi, V., Sharma, J., and Sur, M. (2000). Adaptation-induced plasticity of orientation tuning in adult visual cortex. *Neuron*, 28:287–298.

Durbin, R., and Mitchison, G. (1990). A dimension reduction framework for understanding cortical maps. *Nature*, 343:644–647.

Easterbrook, M. A., Kisilevsky, B. S., Hains, S. M. J., and Muir, D. W. (1999). Faceness or complexity: Evidence from newborn visual tracking of facelike stimuli. *Infant Behavior and Development*, 22:17–35.

Eckhorn, R. (1999). Neural mechanisms of scene segmentation: Recordings from the visual cortex suggest basic circuits for linking field models. *IEEE Transactions on Neural Networks*, 10:464–479.

Eckhorn, R., Bauer, R., Jordan, W., Kruse, M., Munk, W., and Reitboeck, H. J. (1988). Coherent oscillations: A mechanism of feature linking in the visual cortex? *Biological Cybernetics*, 60:121–130.

Eckhorn, R., Gail, A. M., Bruns, A., Gabriel, A., Al-Shaikhli, B., and Saam, M. (2004). Different types of signal coupling in the visual cortex related to neural mechanisms of associative processing and perception. *IEEE Transactions on Neural Networks*, 15:1039–1052.

Eckhorn, R., Reitboeck, H. J., Arndt, M., and Dicke, P. (1990). Feature linking via synchronization among distributed assemblies: Simulations of results from cat visual cortex. *Neural Computation*, 2:293–307.

Edelman, S. (1996). Why have lateral connections in the visual cortex? In Sirosh, J., Miikkulainen, R., and Choe, Y., editors, *Lateral Interactions in the Cortex: Structure and Function*. Austin, TX: The UTCS Neural Networks Research Group. Electronic book, ISBN 0-9647060-0-8, http://nn.cs.utexas.edu/web-pubs/htmlbook96.

Eeckman, F. H., and Freeman, W. J. (1990). Correlations between unit firing and EEG in the rat olfactory system. *Brain Research*, 528:238–244.

Egan, J. P. (1975). *Signal Detection Theory and ROC Analysis*. San Diego, CA: Academic Press.

Eglen, S. J. (1997). Modeling the development of the retinogeniculate pathway. Doctoral dissertation, School of Cognitive and Computing Sciences, University of Sussex, Brighton, UK. Technical Report CSRP 467.

Ehrenstein, W. (1941). Über Abwandlungen der L. Hermannschen Helligkeitserscheinung. *Zeitschrift für Psychologie*, 150:83–91. Modifications of Brightness Phenomenon of L. Hermann; translated by A. Hogg. In Petry, S., and Meyer, G. E., editors (1987). *The Perception of Illusory Contours*, 35–39. Berlin: Springer.

Einhauser, W., Kayser, C., Konig, P., and Kording, K. P. (2002). Learning the invariance properties of complex cells from their responses to natural stimuli. *European Journal of Neuroscience*, 15:475–486.

Elder, J. H., and Goldberg, R. M. (2002). Ecological statistics for the Gestalt laws of perceptual organization of contours. *Journal of Vision*, 2:324–353.

Elliott, T., Howarth, C. I., and Shadbolt, N. R. (1996). Axonal processes and neural plasticity. I: Ocular dominance columns. *Cerebral Cortex*, 6:781–788.

Elliott, T., and Shadbolt, N. R. (1999). A neurotrophic model of the development of the retinogeniculocortical pathway induced by spontaneous retinal waves. *The Journal of Neuroscience*, 19:7951–7970.

Elliott, T., and Shadbolt, N. R. (2002). Multiplicative synaptic normalization and a nonlinear Hebb rule underlie a neurotrophic model of competitive synaptic plasticity. *Neural Computation*, 14:1311–1322.

Elliott, T., and Shadbolt, N. R. (2003). Developmental robotics: Manifesto and application. *Philosophical Transactions: Mathematical, Physical and Engineering Sciences*, 361:2187–2206.

Elman, J. L., Bates, E. A., Johnson, M. H., Karmiloff-Smith, A., Parisi, D., and Plunkett, K. (1996). *Rethinking Innateness: A Connectionist Perspective on Development*. Cambridge, MA: MIT Press.

Elston, G. N., and Rosa, M. G. P. (1998). Morphological variation of layer III pyramidal neurons in the occipitotemporal pathway of the macaque monkey visual cortex. *Cerebral Cortex*, 8:278–294.

Emery, D. L., Royo, N. C., Fischer, I., Saatman, K. E., and McIntosh, T. K. (2003). Plasticity following injury to the adult central nervous system: Is recapitulation of a developmental state worth promoting? *Journal of Neurotrauma*, 20:1271–1292.

Engel, A. K., König, P., Kreiter, A. K., and Singer, W. (1991a). Interhemispheric synchronization of oscillatory neuronal responses in cat visual cortex. *Science*, 252:1177–1179.

Engel, A. K., Kreiter, A. K., König, P., and Singer, W. (1991b). Synchronization of oscillatory neuronal responses between striate and extrastriate visual cortical areas of the cat. *Proceedings of the National Academy of Sciences, USA*, 88:6048–6052.

Ernst, U., Pawelzik, K., Sahar-Pikielny, C., and Tsodyks, M. (2001). Intracortical origin of visual maps. *Nature Neuroscience*, 4:431–436.

Erwin, E., and Miller, K. D. (1998). Correlation-based development of ocularly matched orientation and ocular dominance maps: Determination of required input activities. *The Journal of Neuroscience*, 18:9870–9895.

Erwin, E., Obermayer, K., and Schulten, K. J. (1992a). Self-organizing maps: Ordering, convergence properties and energy functions. *Biological Cybernetics*, 67:47–55.

Erwin, E., Obermayer, K., and Schulten, K. J. (1992b). Self-organizing maps: Stationary states, metastability and convergence rate. *Biological Cybernetics*, 67:35–45.

Erwin, E., Obermayer, K., and Schulten, K. J. (1995). Models of orientation and ocular dominance columns in the visual cortex: A critical comparison. *Neural Computation*, 7:425–468.

Eurich, C. W., Pawelzik, K., Ernst, U., Cowan, J. D., and Milton, J. G. (1999). Dynamics of self-organized delay adaptation. *Physical Review Letters*, 82:1594–1597.

Eurich, C. W., Pawelzik, K., Ernst, U., Thiel, A., Cowan, J. D., and Milton, J. G. (2000). Delay adaptation in the nervous system. *Neurocomputing*, 32–33:741–748.

Eysel, U. T., and Schweigart, G. (1999). Increased receptive field size in the surround of chronic lesions in the adult cat visual cortex. *Cerebral Cortex*, 9:101–109.

Fahle, M. (1993). Figure-ground discrimination from temporal information. *Proceedings: Biological Sciences*, 254:199–203.

Fahle, M., Edelman, S., and Poggio, T. (1995). Fast perceptual learning in hyperacuity. *Vision Research*, 35:3003–3013.

Farah, M. J., Wilson, K. D., Drain, M., and Tanaka, J. N. (1998). What is "special" about face perception? *Psychological Review*, 105:482–498.

Farkas, I., and Miikkulainen, R. (1999). Modeling the self-organization of directional selectivity in the primary visual cortex. In *Proceedings of the Ninth International Conference on Artificial Neural Networks*, 251–256. Berlin: Springer.

Felleman, D. J., and Van Essen, D. C. (1991). Distributed hierarchical processing in primate cerebral cortex. *Cerebral Cortex*, 1:1–47.

Fellenz, W. A., and Taylor, J. G. (2002). Establishing retinotopy by lateral-inhibition type homogeneous neural fields. *Neurocomputing*, 48:313–322.

Feller, M. B. (1999). Spontaneous correlated activity in developing neural circuits. *Neuron*, 22:653–656.

Feller, M. B., Butts, D. A., Aaron, H. L., Rokhsar, D. S., and Shatz, C. J. (1997). Dynamic processes shape spatiotemporal properties of retinal waves. *Neuron*, 19:293–306.

Feller, M. B., Wellis, D. P., Stellwagen, D., Werblin, F. S., and Shatz, C. J. (1996). Requirement for cholinergic synaptic transmission in the propagation of spontaneous retinal waves. *Science*, 272:1182–1187.

Ferrari, F., Manzotti, R., Nalin, A., Benatti, A., Cavallo, R., Torricelli, A., and Cavazzutti, G. (1986). Visual orientation to the human face in the premature and fullterm newborn. *The Italian Journal of Neurological Sciences*, 5:53–60.

Ferster, D. (1994). Linearity of synaptic interactions in the assembly of receptive fields in cat visual cortex. *Current Opinion in Neurobiology*, 4:563–568.

Ferster, D., and Lindström, S. (1985). Synaptic excitation of neurons in area 17 of the cat by intracortical axon collaterals of cortico-geniculate cells. *The Journal of Physiology*, 367:233–252.

Field, D. J. (1994). What is the goal of sensory coding? *Neural Computation*, 6:559–601.

Field, D. J., Hayes, A., and Hess, R. F. (1993). Contour integration by the human visual system: Evidence for a local association field. *Vision Research*, 33:173–193.

Field, T. M., Cohen, D., Garcia, R., and Greenberg, R. (1984). Mother–stranger face discrimination by the newborn. *Infant Behavior and Development*, 7:19–25.

Findlay, J. M. (1998). Active vision: Visual activity in everyday life. *Current Biology*, 8:R640–R642.

Findlay, J. M., and Gilchrist, I. D. (2003). *Active Vision: The Psychology of Looking and Seeing*. Oxford, UK: Oxford University Press.

Finkel, L. H., and Edelman, G. M. (1989). Integration of distributed cortical systems by reentry: A computer simulation of interactive functionally segregated visual areas. *The Journal of Neuroscience*, 9:3188–3208.

Fisher, S. A., Fisher, T. M., and Carew, T. J. (1997). Multiple overlapping processes underlying short-term synaptic enhancement. *Trends in Neurosciences*, 20:170–177.

Fisken, R. A., Garey, L. J., and Powell, T. P. S. (1975). The intrinsic, association and commissural connections of area 17 of the visual cortex. *Philosophical Transactions of the Royal Society of London. Series B, Biological Sciences*, 272:487–536.

FitzHugh, R. (1961). Impulses and physiological states in models of nerve membrane. *Biophysics Journal*, 1:445–466.

Fitzpatrick, D., Schofield, B. R., and Strote, J. (1994). Spatial organization and connections of iso-orientation domains in the tree shrew striate cortex. In *Society for Neuroscience Abstracts*, vol. 20, 837. Washington, DC: Society for Neuroscience.

Fogel, D. B. (1999). *Evolutionary Computation: Toward a New Philosophy of Machine Intelligence*. Piscataway, NJ: IEEE. Second edition.

Földiák, P. (1990). Forming sparse representations by local anti-Hebbian learning. *Biological Cybernetics*, 64:165–170.

Földiák, P. (1991a). Learning invariance from transformation sequences. *Neural Computation*, 3:194–200.

Földiák, P. (1991b). Models of sensory coding. Doctoral dissertation, Churchill College, University of Cambridge, Cambridge, UK. Department of Engineering Technical Report CUED/F-INFENG/TR 91.

Freeman, E., Driver, J., Sagi, D., and Zhaoping, L. (2003). Top-down modulation of lateral interactions in early vision: Does attention affect integration of the whole or just perception of the parts? *Current Biology*, 13:985–989.

Freeman, W. J., and Burke, B. C. (2003). A neurobiological theory of meaning in perception. Part IV: Multicortical patterns of amplitude modulation in gamma EEG. *International Journal of Bifurcation and Chaos*, 13:2857–2866.

Friedrich, R. W., and Laurent, G. (2001). Dynamic optimization of odor representations by slow temporal patterning of mitral cell activity. *Science*, 291:889–894.

Fritzke, B. (1994). Growing cell structures—a self-organizing network for unsupervised and supervised learning. *Neural Networks*, 7:1441–1460.

Fritzke, B. (1995). Growing grid: A self-organizing network with constant neighborhood range and adaptation strength. *Neural Processing Letters*, 2:9–13.

Fu, Y. X., Djupsund, K., Gao, H., Hayden, B., Shen, K., and Dan, Y. (2002). Temporal specificity in the cortical plasticity of visual space representation. *Science*, 296:1999–2003.

Fuhs, M. C., Redish, A. D., and Touretzky, D. S. (1998). A visually driven hippocampal place cell model. In Bower, J. M., editor, *Computational Neuroscience: Trends in Research*, 379–384. New York: Plenum Press.

Fukushima, K., and Miyake, S. (1982). Neocognitron: A self-organizing neural network model for a mechanism of visual pattern recognition. In Amari, S., and Arbib, M. A., editors, *Competition and Cooperation in Neural Nets*, Lecture Notes in Biomathematics 45, 267–285. Berlin: Springer.

Gabbiani, F., and Koch, C. (1998). Principles of spike train analysis. In Koch, C., and Segev, I., editors, *Methods in Neuronal Modeling: From Ions to Networks*, 313–360. Cambridge, MA: MIT Press. Second edition.

Gandhi, S. P., Heeger, D. J., and Boynton, G. M. (1999). Spatial attention affects brain activity in human primary visual cortex. *Proceedings of the National Academy of Sciences, USA*, 96:3314–3319.

Gardner, D. (2004). Neurodatabase.org: Networking the microelectrode. *Nature Neuroscience*, 7:486–487.

Garris, M. D. (1992). Design and collection of a handwriting sample image database. *Social Science Computer Review*, 10:196–214.

Gauthier, I., and Logothetis, N. K. (2000). Is face recognition not so unique, after all? *Cognitive Neuropsychology*, 17:125–142.

Gauthier, I., and Nelson, C. A. (2001). The development of face expertise. *Current Opinion in Neurobiology*, 11:219–224.

Geisler, W. S., and Albrecht, D. G. (1995). Bayesian analysis of identification performance in the primary visual cortex: Nonlinear mechanisms and stimulus certainty. *Vision Research*, 35:2723–2730.

Geisler, W. S., and Albrecht, D. G. (1997). Visual cortex neurons in monkeys and cats: Detection, discrimination, and identification. *Visual Neuroscience*, 14:897–919.

Geisler, W. S., Perry, J. S., Super, B. J., and Gallogly, D. P. (2001). Edge co-occurrence in natural images predicts contour grouping performance. *Vision Research*, 41:711–724.

Geisler, W. S., and Super, B. J. (2000). Perceptual organization of two-dimensional patterns. *Psychological Review*, 107:677–708.

Geisler, W. S., Thornton, T., Gallogly, D. P., and Perry, J. S. (2000). Image structure models of texture and contour visibility. In *Search and Target Acquisition (Recherche et acquisition d'objectifs)*, RTO Meeting Proceedings 45, 15/1–15/8. Hull, Québec: Canada Communication Group.

Gelbtuch, M. H., Calvert, J. E., Harris, J. P., and Phillipson, O. T. (1986). Modification of visual orientation illusions by drugs which influence dopamine and GABA neurones: Differential effects on simultaneous and successive illusions. *Psychopharmacology*, 90:379–383.

Geman, S., Bienenstock, E. L., and Doursat, R. (1992). Neural networks and the bias/variance dilemma. *Neural Computation*, 4:1–58.

George, M. S., Wassermann, E. M., Williams, W. A., Steppel, J., Pascual-Leone, A., Basser, P., Hallett, M., and Post, R. M. (1996). Changes in mood and hormone levels after rapid-rate transcranial magnetic stimulation (rTMS) of the prefrontal cortex. *The Journal of Neuropsychiatry and Clinical Neurosciences*, 8:172–180.

Gerstner, W. (1998a). Hebbian learning of pulse timing in the barn owl auditory system. In Maass, W., and Bishop, C. M., editors, *Pulsed Neural Networks*, 353–377. Cambridge, MA: MIT Press.

Gerstner, W. (1998b). Spiking neurons. In Maass, W., and Bishop, C. M., editors, *Pulsed Neural Networks*, 3–54. Cambridge, MA: MIT Press.

Gerstner, W., and Kistler, W. M. (2002). *Spiking Neuron Models: Single Neurons, Populations, Plasticity*. Cambridge, UK: Cambridge University Press.

Gerstner, W., and van Hemmen, J. L. (1992). Associative memory in a network of spiking neurons. *Network*, 3:139–164.

Ghahramani, Z., and Hinton, G. E. (1998). Hierarchical non-linear factor analysis and topographic maps. In Jordan, M. I., Kearns, M. J., and Solla, S. A., editors, *Advances in Neural Information Processing Systems 10*, 486–492. Cambridge, MA: MIT Press.

Ghose, G. M., and Ts'o, D. Y. (1997). Form processing modules in primate area V4. *Journal of Neurophysiology*, 77:2191–2196.

Gibson, J. J. (1950). *The Perception of the Visual World*. Boston: Houghton Mifflin.

Gibson, J. J. (1979). *The Ecological Approach to Visual Perception*. Boston: Houghton Mifflin.

Gibson, J. J., and Radner, M. (1937). Adaptation, after-effect and contrast in the perception of tilted lines. *Journal of Experimental Psychology*, 20:453–467.

Giese, M. A. (1998). *Dynamic Neural Field Theory for Motion Perception*. Berlin: Springer.

Gilbert, C. D. (1992). Horizontal integration and cortical dynamics. *Neuron*, 9:1–13.

Gilbert, C. D. (1994). Circuitry, architecture and functional dynamics of visual cortex. In Bock, G. R., and Goode, J. A., editors, *Higher-Order Processing in the Visual System*, Ciba Foundation Symposium 184, 35–62. Hoboken, NJ: Wiley.

Gilbert, C. D. (1998). Adult cortical dynamics. *Physiological Reviews*, 78:467–485.

Gilbert, C. D., Das, A., Ito, M., Kapadia, M. K., and Westheimer, G. (1996). Spatial integration and cortical dynamics. *Proceedings of the National Academy of Sciences, USA*, 93:615–622.

Gilbert, C. D., Hirsch, J. A., and Wiesel, T. N. (1990). Lateral interactions in visual cortex. In *The Brain*, vol. LV of *Cold Spring Harbor Symposia on Quantitative Biology*, 663–677. Cold Spring Harbor, NY: Cold Spring Harbor Laboratory Press.

Gilbert, C. D., and Wiesel, T. N. (1979). Morphology and intracortical projections of functionally identified neurons in cat visual cortex. *Nature*, 280:120–125.

Gilbert, C. D., and Wiesel, T. N. (1983). Clustered intrinsic connections in cat visual cortex. *The Journal of Neuroscience*, 3:1116–1133.

Gilbert, C. D., and Wiesel, T. N. (1989). Columnar specificity of intrinsic horizontal and corticocortical connections in cat visual cortex. *The Journal of Neuroscience*, 9:2432–2442.

Gilbert, C. D., and Wiesel, T. N. (1990). The influence of contextual stimuli on the orientation selectivity of cells in primary visual cortex of the cat. *Vision Research*, 30:1689–1701.

Gilbert, C. D., and Wiesel, T. N. (1992). Receptive field dynamics in adult primary visual cortex. *Nature*, 356:150–152.

Glover, M., Hamilton, A., and Smith, L. S. (2002). Analogue VLSI leaky integrate-and-fire neurons and their use in a sound analysis system. *Analog Integrated Circuits and Signal Processing*, 30:91–100.

Goddard, N. H., Hucka, M., Howell, F., Cornelis, H., Shankar, K., and Beeman, D. (2001). Towards NeuroML: Model description methods for collaborative modelling in neuroscience. *Philosophical Transactions: Biological Sciences*, 356:1209–1228.

Gödecke, I., and Bonhoeffer, T. (1996). Development of identical orientation maps for two eyes without common visual experience. *Nature*, 379:251–254.

Gödecke, I., Kim, D. S., Bonhoeffer, T., and Singer, W. (1997). Development of orientation preference maps in area 18 of kitten visual cortex. *European Journal of Neuroscience*, 9:1754–1762.

Goldberg, D. E. (1989). *Genetic Algorithms in Search, Optimization and Machine Learning*. Reading, MA: Addison-Wesley.

Goldman-Rakic, P. S. (1980). Morphological consequences of prenatal injury to the primate brain. In McConnell, P. S., Boer, G. J., Romijn, H. J., van de Poll, N. E., and Corner, M. A., editors, *Adaptive Capabilities of the Nervous System: Proceedings of the 11th International Summer School of Brain Research*, vol. 53 of *Progress in Brain Research*, 3–19. Amsterdam: Elsevier.

Goldstone, R. L. (2003). Learning to perceive while perceiving to learn. In Kimchi, R., Behrmann, M., and Olson, C., editors, *Perceptual Organization in Vision: Behavioral and Neural Perspectives*, 233–278. Hillsdale, NJ: Erlbaum.

Gomez, F., and Miikkulainen, R. (1997). Incremental evolution of complex general behavior. *Adaptive Behavior*, 5:317–342.

Goodale, M. A., and Milner, A. D. (1992). Separate visual pathways for perception and action. *Trends in Neurosciences*, 15:20–25.

Goodall, S., Reggia, J. A., Chen, Y., Ruppin, E., and Whitney, C. (1997). A computational model of acute focal cortical lesions. *Stroke*, 28:101–109.

Goodhill, G. J. (1993). Topography and ocular dominance: A model exploring positive correlations. *Biological Cybernetics*, 69:109–118.

Goodhill, G. J., and Cimponeriu, A. (2000). Analysis of the elastic net model applied to the formation of ocular dominance and orientation columns. *Network: Computation in Neural Systems*, 11:153–168.

Goodhill, G. J., and Löwel, S. (1995). Theory meets experiment: Correlated neural activity helps determine ocular dominance column periodicity. *Trends in Neurosciences*, 18:437–439.

Goodhill, G. J., and Willshaw, D. J. (1990). Application of the elastic net algorithm to the formation of ocular dominance stripes. *Network: Computation in Neural Systems*, 1:41–59.

Goodhill, G. J., and Willshaw, D. J. (1994). Elastic net model of ocular dominance: Overall stripe pattern and monocular deprivation. *Neural Computation*, 6:615–621.

Göppert, J., and Rosenstiel, W. (1997). The continuous interpolating self-organizing map. *Neural Processing Letters*, 5:185–192.

Gorchetchnikov, A. (2000). Introduction of threshold self-adjustment improves the convergence in feature-detective neural nets. *Neurocomputing*, 32–33:385–390.

Goren, C. C., Sarty, M., and Wu, P. Y. (1975). Visual following and pattern discrimination of face-like stimuli by newborn infants. *Pediatrics*, 56:544–549.

Gould, E., Reeves, A. J., Graziano, M. S. A., and Gross, C. G. (1999). Neurogenesis in the neocortex of adult primates. *Science*, 286:548–552.

Gove, A., Grossberg, S., and Mingolla, E. (1993). Brightness perception, illusory contours and corticogeniculate feedback. In *World Congress on Neural Networks*, vol. I, 25–28. Hillside, NJ: Erlbaum.

Govindan, V. K., and Shivaprasad, A. P. (1990). Character recognition — a review. *Pattern Recognition*, 23:671–683.

Graham, D. B., and Allinson, N. M. (1998). Automatic face representation and classification. In Nixon, M. S., and Carter, J. N., editors, *Proceedings of the Ninth British Machine Vision Conference*, 64–73. Malvern, UK: BMVA Press.

Grajski, K. A., and Merzenich, M. M. (1990). Hebb-type dynamics is sufficient to account for the inverse magnification rule in cortical somatotopy. *Neural Computation*, 2:71–84.

Gray, C. M. (1999). The temporal correlation hypothesis of visual feature integration: Still alive and well. *Neuron*, 24:31–47.

Gray, C. M., Konig, P., Engel, A. K., and Singer, W. (1989). Oscillatory responses in cat visual cortex exhibit inter-columnar synchronization which reflects global stimulus properties. *Nature*, 338:334–337.

Gray, C. M., and McCormick, D. A. (1996). Chattering cells: Superficial pyramidal neurons contributing to the generation of synchronous oscillations in the visual cortex. *Science*, 274:109–113.

Gray, C. M., and Singer, W. (1987). Stimulus-specific neuronal oscillations in the cat visual cortex: A cortical functional unit. In *Society for Neuroscience Abstracts*, vol. 13, 404.3. Washington, DC: Society for Neuroscience.

Gray, M. S., Lawrence, D. T., Golomb, B. A., and Sejnowski, T. J. (1995). A perceptron reveals the face of sex. *Neural Computation*, 7:1160–1164.

Greenlee, M. W., and Magnussen, S. (1987). Saturation of the tilt aftereffect. *Vision Research*, 27:1041–1043.

Grieve, K. L., and Sillito, A. M. (1995). Non-length-tuned cells in layer II/III and IV of the visual cortex: The effect of blockade of layer VI on responses to stimuli of different lengths. *Experimental Brain Research*, 104:12–20.

Grinvald, A., Lieke, E. E., Frostig, R. D., and Hildesheim, R. (1994). Cortical point-spread function and long-range lateral interactions revealed by real-time optical imaging of macaque monkey primary visual cortex. *The Journal of Neuroscience*, 14:2545–2568.

Gross, C. G., Rocha-Miranda, C. E., and Bender, D. B. (1972). Visual properties of neurons in inferotemporal cortex of the macaque. *Journal of Neurophysiology*, 35:96–111.

Grossberg, S. (1976). On the development of feature detectors in the visual cortex with applications to learning and reaction-diffusion systems. *Biological Cybernetics*, 21:145–159.

Grossberg, S. (1999). How does the cerebral cortex work? Learning, attention, and grouping by the laminar circuits of visual cortex. *Spatial Vision*, 12:125–254.

Grossberg, S., and Mingolla, E. (1985). Neural dynamics of form perception: Boundary completion, illusory figures, and neon color spreading. *Psychological Review*, 92:173–211.

Grossberg, S., Mingolla, E., and Ross, W. D. (1997). Visual brain and visual perception: How does the cortex do perceptual grouping? *Trends in Neurosciences*, 20:106–111.

Grossberg, S., and Olson, S. J. (1994). Rules for the cortical map of ocular dominance and orientation columns. *Neural Networks*, 7:883–894.

Grossberg, S., and Seitz, A. (2003). Laminar development of receptive fields, maps and columns in visual cortex: The coordinating role of the subplate. *Cerebral Cortex*, 13:852–863.

Grossberg, S., and Williamson, J. R. (2001). A neural model of how horizontal and interlaminar connections of visual cortex develop into adult circuits that carry out perceptual grouping and learning. *Cerebral Cortex*, 11:37–58.

Grubb, M. S., Rossi, F. M., Changeux, J.-P., and Thompson, I. (2003). Abnormal functional organization in the dorsal lateral geniculate nucleus of mice lacking the $\beta2$ subunit of the nicotinic acetylcholine receptor. *Neuron*, 40:1161–1172.

Gustafsson, B., and Wigström, H. (1988). Physiological mechanisms underlying long-term potentiation. *Trends in Neurosciences*, 11:156–162.

Hadjikhani, N., and Roland, P. E. (1998). Cross-modal transfer of information between the tactile and the visual representations in the human brain: A positron emission tomographic study. *The Journal of Neuroscience*, 18:1072–1084.

Haessly, A., Sirosh, J., and Miikkulainen, R. (1995). A model of visually guided plasticity of the auditory spatial map in the barn owl. In *Proceedings of the 17th Annual Conference of the Cognitive Science Society*, 154–158. Hillsdale, NJ: Erlbaum.

Haith, G. L. (1998). Modeling activity-dependent development in the retinogeniculate projection. Doctoral dissertation, Department of Psychology, Stanford University, Palo Alto, CA.

Halgren, E., Dale, A. M., Sereno, M. I., Tootell, R. B. H., Marinkovic, K., and Rosen, B. R. (1999). Location of human face-selective cortex with respect to retinotopic areas. *Human Brain Mapping*, 7:29–37.

Hallett, M. (2000). Transcranial magnetic stimulation and the human brain. *Nature*, 406:147–150.

Han, S. K., Kim, W. S., and Kook, H. (1998). Temporal segmentation of the stochastic oscillator neural network. *Physical Review E*, 58:2325–2334.

Hanson, S. J., Matsuka, T., and Haxby, J. V. (2004). Combinatorial codes in ventral temporal lobe for object recognition: Haxby (2001) revisited: Is there a "face" area? *Neuroimage*, 23:156–166.

Harnad, S., Pace-Schott, E., Blagrove, M., and Solms, M., editors (2003). *Sleep and Dreaming: Scientific Advances and Reconsiderations*. Cambridge, UK: Cambridge University Press.

Harris, L., and Jenkin, M., editors (1998). *Vision and Action*. Cambridge, UK: Cambridge University Press.

Hasselmo, M. E., Bodelón, C., and Wyble, B. P. (2002). A proposed function for hippocampal theta rhythm: Separate phases of encoding and retrieval enhance reversal of prior learning. *Neural Computation*, 14:793–817.

Hasselmo, M. E., Rolls, E. T., and Baylis, G. C. (1989). The role of expression and identity in the face-selective responses of neurons in the temporal visual cortex of the monkey. *Behavioural Brain Research*, 32:203–218.

Hastie, T., and Stuetzle, W. (1989). Principal curves. *Journal of the American Statistical Association*, 84:502–516.

Hata, Y., Tsumoto, T., Sato, H., Hagihara, K., and Tamura, H. (1993). Development of local horizontal interactions in cat visual cortex studied by cross-correlation analysis. *Journal of Neurophysiology*, 69:40–56.

Haussler, D. (1988). Quantifying inductive bias: AI learning algorithms and Valiant's learning framework. *Artificial Intelligence*, 36:177–221.

Haxby, J. V., Gobbini, M. I., Furey, M. L., Ishai, A., Schouten, J. L., and Pietrini, P. (2001). Distributed and overlapping representations of faces and objects in ventral temporal cortex. *Science*, 293:2425–2430.

Haxby, J. V., Horwitz, B., Ungerleider, L. G., Maisog, J. M., Pietrini, P., and Grady, C. L. (1994). The functional organization of human extrastriate cortex: A PET-rCBF study of selective attention to faces and locations. *The Journal of Neuroscience*, 14:6336–6353.

Hayes, W. P., and Meyer, R. L. (1988a). Optic synapse number but not density is constrained during regeneration onto surgically halved tectum in goldfish: HRP-EM evidence that optic fibers compete for fixed numbers of postsynaptic sites on the tectum. *Journal of Computational Neurology*, 274:539–559.

Hayes, W. P., and Meyer, R. L. (1988b). Retinotopically inappropriate synapses of subnormal density formed by misdirected optic fibers in goldfish tectum. *Developmental Brain Research*, 38:304–312.

Haykin, S. (1994). *Neural Networks: A Comprehensive Foundation*. New York: Macmillan.

Hebb, D. O. (1949). *The Organization of Behavior: A Neuropsychological Theory*. Hoboken, NJ: Wiley.

Hecht-Nielsen, R. (1989). Theory of the backpropagation neural network. In *Proceedings of the International Joint Conference on Neural Networks*, vol. I, 593–605. Piscataway, NJ: IEEE.

Hecht-Nielsen, R. (2002). A theory of thalamocortex. In Hecht-Nielsen, R., and McKenna, T., editors, *Computational Models for Neuroscience: Human Cortical Information Processing*, 85–124. Berlin: Springer.

Heeger, D. J., Boynton, G. M., Demb, J. B., Seidemann, E., and Newsome, W. T. (1999). Motion opponency in visual cortex. *The Journal of Neuroscience*, 19:7162–7174.

Hempel, C. M., Hartman, K. H., Wang, X.-J., Turrigiano, G. G., and Nelson, S. B. (2000). Multiple forms of short-term plasticity at excitatory synapses in rat medial prefrontal cortex. *Journal of Neurophysiology*, 83:3031–3041.

Henry, G. H. (1989). Afferent inputs, receptive field properties and morphological cell types in different laminae of the striate cortex. In Leventhal, A. G., editor, *The Neural Basis of Visual Function*, vol. 4 of *Vision and Visual Dysfunction*, 223–245. Boca Raton, FL: CRC Press.

Hensch, T. K., Fagiolini, M., Mataga, N., Stryker, M. P., Baekkeskov, S., and Kash, S. F. (1998). Local GABA circuit control of experience-dependent plasticity in developing visual cortex. *Science*, 282:1604–1608.

Hensch, T. K., and Stryker, M. P. (2004). Columnar architecture sculpted by GABA circuits in developing cat visual cortex. *Science*, 303:1678–1681.

Hershenson, M., Kessen, W., and Munsinger, H. (1967). Pattern perception in the human newborn: A close look at some positive and negative results. In Wathen-Dunn, W., editor, *Models for the Perception of Speech and Visual Form: Proceedings of a Symposium*, 282–290. Cambridge, MA: MIT Press.

Hess, R. F., and Dakin, S. C. (1997). Absence of contour linking in peripheral vision. *Nature*, 390:602–604.

Hess, R. F., Hayes, A., and Field, D. J. (2004). Contour integration and cortical processing. *Journal of Physiology - Paris*, 97:105–119.

Hines, M. L., and Carnevale, N. T. (1997). The NEURON simulation environment. *Neural Computation*, 9:1179–1209.

Hirsch, H. V. B. (1985). The role of visual experience in the development of cat striate cortex. *Cellular and Molecular Neurobiology*, 5:103–121.

Hirsch, H. V. B., and Spinelli, D. (1970). Visual experience modifies distribution of horizontally and vertically oriented receptive fields in cats. *Science*, 168:869–871.

Hirsch, J. A., Alonso, J. M., Reid, R. C., and Martinez, L. M. (1998a). Synaptic integration in striate cortical simple cells. *The Journal of Neuroscience*, 18:9517–9528.

Hirsch, J. A., Gallagher, C. A., Alonso, J. M., and Martinez, L. M. (1998b). Ascending projections of simple and complex cells in layer 6 of the cat striate cortex. *The Journal of Neuroscience*, 18:8086–8094.

Hirsch, J. A., and Gilbert, C. D. (1991). Synaptic physiology of horizontal connections in the cat's visual cortex. *The Journal of Neuroscience*, 11:1800–1809.

Hochreiter, S., and Schmidhuber, J. (1999). Source separation as a by-product of regularization. In Kearns, M. S., Solla, S. A., and Cohn, D. A., editors, *Advances in Neural Information Processing Systems 11*, 459–465. Cambridge, MA: MIT Press.

Hodgkin, A. L., and Huxley, A. F. (1952). A quantitative description of membrane current and its application to conduction and excitation in nerve. *The Journal of Physiology*, 117:500–544.

Hoffman, D. D. (1998). *Visual Intelligence: How We Create What We See*. New York: Norton.

Holland, J. H. (1975). *Adaptation in Natural and Artificial Systems: An Introductory Analysis with Applications to Biology, Control and Artificial Intelligence*. Ann Arbor, MI: University of Michigan Press.

Hopkins, R. O., Myers, C. E., Shohamy, D., Grossman, S., and Gluck, M. (2003). Impaired probabilistic category learning in hypoxic subjects with hippocampal damage. *Neuropsychologia*, 41:1919–1928.

Hoppensteadt, F. C., and Izhikevich, E. M. (1997). *Weakly Connected Neural Networks*. Berlin: Springer.

Horn, D., Levy, N., and Ruppin, E. (1998). Memory maintenance via neuronal regulation. *Neural Computation*, 10:1–18.

Horn, D., and Opher, I. (1998). Collective excitation phenomena and their applications. In Maass, W., and Bishop, C. M., editors, *Pulsed Neural Networks*, 297–320. Cambridge, MA: MIT Press.

Horn, D., and Usher, M. (1989). Neural networks with dynamical thresholds. *Physical Review A*, 40:1036–1044.

Horn, D., and Usher, M. (1992). Oscillatory model of short term memory. In Moody, J. E., Hanson, S. J., and Lippmann, R. P., editors, *Advances in Neural Information Processing Systems, 4*, 125–132. San Francisco: Kaufmann.

Horn, G. (1985). *Memory, Imprinting, and the Brain: An Inquiry Into Mechanisms*. Oxford, UK: Clarendon Press.

Horne, J. A. (1988). *Why We Sleep: The Functions of Sleep in Humans and Other Mammals*. Oxford, UK: Oxford University Press.

Horton, J. C., and Hocking, D. R. (1996). An adult-like pattern of ocular dominance columns in striate cortex of newborn monkeys prior to visual experience. *The Journal of Neuroscience*, 16:1791–1807.

Howard, I. P., and Templeton, W. B. (1966). *Human Spatial Orientation*. Hoboken, NJ: Wiley.

Hubel, D. H., and Wiesel, T. N. (1959). Receptive fields of single neurons in the cat's striate cortex. *The Journal of Physiology*, 148:574–591.

Hubel, D. H., and Wiesel, T. N. (1962). Receptive fields, binocular interaction and functional architecture in the cat's visual cortex. *The Journal of Physiology*, 160:106–154.

Hubel, D. H., and Wiesel, T. N. (1965). Receptive fields and functional architecture in two nonstriate visual areas (18 and 19) of the cat. *Journal of Neurophysiology*, 28:229–289.

Hubel, D. H., and Wiesel, T. N. (1967). Cortical and callosal connections concerned with the vertical meridian of visual fields of the cat. *Journal of Neurophysiology*, 30:1561–1573.

Hubel, D. H., and Wiesel, T. N. (1968). Receptive fields and functional architecture of monkey striate cortex. *The Journal of Physiology*, 195:215–243.

Hubel, D. H., and Wiesel, T. N. (1974). Sequence regularity and geometry of orientation columns in the monkey striate cortex. *The Journal of Comparative Neurology*, 158:267–294.

Hubel, D. H., Wiesel, T. N., and LeVay, S. (1977). Plasticity of ocular dominance columns in monkey striate cortex. *Philosophical Transactions of the Royal Society of London. Series B, Biological Sciences*, 278:377–409.

Hubener, M., Shoham, D., Grinvald, A., and Bonhoeffer, T. (1997). Spatial relationships among three columnar systems in cat area 17. *The Journal of Neuroscience*, 17:9270–9284.

Hugh, G. S., Laubach, M., Nicolelis, M. A. L., and Henriquez, C. S. (2002). A simulator for the analysis of neuronal ensemble activity: Application to reaching tasks. *Neurocomputing*, 44–46:847–854.

Hugues, E., Guilleux, F., and Rochel, O. (2002). Contour detection by synchronization of integrate-and-fire neurons. In Bülthoff, H. H., Lee, S.-W., Poggio, T., and Wallraven, C., editors, *Biologically Motivated Computer Vision: Second International Workshop*, Lecture Notes in Computer Science 2525, 60–69. Berlin: Springer.

Humphrey, A. L., Saul, A. B., and Feidler, J. C. (1998). Strobe rearing prevents the convergence of inputs with different response timings onto area 17 simple cells. *Journal of Neurophysiology*, 80:3005–3020.

Hurri, J., and Hyvarinen, A. (2003). Temporal and spatiotemporal coherence in simple-cell responses: A generative model of natural image sequences. *Network: Computation in Neural Systems*, 14:527–551.

Hyvärinen, A., and Hoyer, P. O. (2001). A two-layer sparse coding model learns simple and complex cell receptive fields and topography from natural images. *Vision Research*, 41:2413–2423.

Ilmoniemi, R. J., Virtanen, J., Ruohonen, J., Karhu, J., Aronen, H. J., Näätänen, R., and Katila, T. (1997). Neuronal responses to magnetic stimulation reveal cortical reactivity and connectivity. *Neuroreport*, 8:3537–3540.

Issa, N. P., Trachtenberg, J. T., Chapman, B., Zahs, K. R., and Stryker, M. P. (1999). The critical period for ocular dominance plasticity in the ferret's visual cortex. *The Journal of Neuroscience*, 19:6965–6978.

Issa, N. P., Trepel, C., and Stryker, M. P. (2001). Spatial frequency maps in cat visual cortex. *The Journal of Neuroscience*, 20:8504–8514.

Izhikevich, E. M. (2001). Resonate-and-fire neurons. *Neural Networks*, 14:883–894.

Izhikevich, E. M. (2003). Simple model of spiking neurons. *IEEE Transactions on Neural Networks*, 14:1569–1572.

Jani, N. G., and Levine, D. S. (2000). A neural network theory of proportional analogy-making. *Neural Networks*, 13:149–183.

Jefferys, J. G. R., Traub, R. D., and Whittington, M. A. (1996). Neuronal networks for induced '40 Hz' rhythms. *Trends in Neurosciences*, 19:202–208.

Jensen, K., and Mody, I. (2001). L-type Ca^{2+} channel-mediated short-term plasticity of GABAergic synapses. *Nature Neuroscience*, 4:975–976.

Jockusch, S. (1990). A neural network which adapts its structure to a given set of patterns. In Eckmiller, R., Hartmann, G., and Hauske, G., editors, *Parallel Processing in Neural Systems and Computers*, 169 172. Amsterdam: North-Holland.

Johnson, M. H., Dziurawiec, S., Ellis, H., and Morton, J. (1991). Newborns' preferential tracking of face-like stimuli and its subsequent decline. *Cognition*, 40:1–19.

Johnson, M. H., and Mareschal, D. (2001). Cognitive and perceptual development during infancy. *Current Opinion in Neurobiology*, 11:213–218.

Johnson, M. H., and Morton, J. (1991). *Biology and Cognitive Development: The Case of Face Recognition*. Oxford, UK: Blackwell.

Joliot, M., Ribary, U., and Llinás, R. (1994). Human oscillatory brain activity near 40 Hz coexists with cognitive temporal binding. *Proceedings of the National Academy of Sciences, USA*, 91:11748–11751.

Jolliffe, I. T. (1986). *Principal Component Analysis*. Berlin: Springer.

Jones, J. P., and Palmer, L. A. (1987). The two-dimensional spatial structure of simple receptive fields in cat striate cortex. *Journal of Neurophysiology*, 58:1187–1211.

Jouvet, M. (1980). Paradoxical sleep and the nature-nurture controversy. In McConnell, P. S., Boer, G. J., Romijn, H. J., van de Poll, N. E., and Corner, M. A., editors, *Adaptive Capabilities of the Nervous System: Proceedings of the 11th International Summer School of Brain Research*, vol. 53 of *Progress in Brain Research*, 331–346. Amsterdam: Elsevier.

Jouvet, M. (1998). Paradoxical sleep as a programming system. *Journal of Sleep Research*, 7:1–5.

Jouvet, M. (1999). *The Paradox of Sleep: The Story of Dreaming*. Cambridge, MA: MIT Press.

Kaas, J. H. (1991). Plasticity of sensory and motor maps in adult animals. *Annual Review of Neuroscience*, 14:137–167.

Kaas, J. H. (2000). Why is brain size so important: Design problems and solutions as neocortex gets bigger or smaller. *Brain and Mind*, 1:7–23.

Kaas, J. H. (2001a). The mutability of sensory representations after injury in adult mammals. In Shaw, C. A., and McEachern, J. C., editors, *Toward a Theory of Neuroplasticity*, 323–334. East Sussex, UK: Psychology Press.

Kaas, J. H. (2001b). Reorganization of sensory and motor systems in adult mammals after injury. In Kaas, J. H., editor, *The Mutable Brain: Dynamic and Plastic Features of the Developing and Mature Brain*, 165–242. Chur, Switzerland: Harwood.

Kaas, J. H., Krubitzer, L. A., Chino, Y. M., Langston, A. L., Polley, E. H., and Blair, N. (1990). Reorganization of retinotopic cortical maps in adult mammals after lesions of the retina. *Science*, 248:229–231.

Kalarickal, G. J., and Marshall, J. A. (2002). Rearrangement of receptive field topography after intracortical and peripheral stimulation: The role of plasticity in inhibitory pathways. *Network: Computation in Neural Systems*, 13:1–40.

Kambhatla, N., and Leen, T. K. (1997). Dimension reduction by local principal component analysis. *Neural Computation*, 9:1493–1516.

Kammen, D. M., Holmes, P. J., and Koch, C. (1989). Origin of oscillations in visual cortex: Feedback versus local coupling. In Cotterill, R. M. J., editor, *Models of Brain Functions*, 273–284. Cambridge, UK: Cambridge University Press.

Kandel, E. R., Schwartz, J. H., and Jessell, T. M. (1991). *Principles of Neural Science*. Amsterdam: Elsevier. Third edition.

Kandel, E. R., Schwartz, J. H., and Jessell, T. M. (2000). *Principles of Neural Science*. New York: McGraw-Hill. Fourth edition.

Kanerva, P. (1998). Dual role of analogy in the design of a cognitive computer. In Holyoak, K., Gentner, D., and Kokinov, B., editors, *Advances in Analogy Research: Integration of Theory and Data from the Cognitive, Computational, and Neural Sciences*, 164–170. Sofia, Bulgaria: NBU Press.

Kanizsa, G. (1955). Margini quasi-Percettivi in Campi con Stimolazione Omogenea. *Rivista di Psicologia*, 49:7–30. Quasiperceptual Margins in Homogeneously Stimulated Fields; translated by W. Gerbino. In Petry, S., and Meyer, G. E., editors (1987). *The Perception of Illusory Contours*, 40–49. Berlin: Springer.

Kanizsa, G. (1976). Subjective contours. *Scientific American*, 234:48–52.

Kanwisher, N. (2000). Domain specificity in face perception. *Nature Neuroscience*, 3:759–763.

Kanwisher, N., McDermott, J., and Chun, M. M. (1997). The fusiform face area: A module in human extrastriate cortex specialized for face perception. *The Journal of Neuroscience*, 17:4302–4311.

Kapadia, M. K., Gilbert, C. D., and Westheimer, G. (1994). A quantitative measure for short-term cortical plasticity in human vision. *The Journal of Neuroscience*, 14:451–457.

Kapadia, M. K., Ito, M., Gilbert, C. D., and Westheimer, G. (1995). Improvement in visual sensitivity by changes in local context: Parallel studies in human observers and in V1 of alert monkeys. *Neuron*, 15:843–856.

Karni, A., and Bertini, G. (1997). Learning perceptual skills: Behavioral probes into adult cortical plasticity. *Current Opinion in Neurobiology*, 7:530–535.

Kasamatsu, T., Kitano, M., Sutter, E. E., and Norcia, A. M. (1998). Lack of lateral inhibitory interactions in visual cortex of monocularly deprived cats. *Vision Research*, 38:1–12.

Kaski, S., Kangas, J., and Kohonen, T. (1998). Bibliography of self-organizing map (SOM) papers: 1981–1997. *Neural Computing Surveys*, 1:102–350.

Katz, L. C., and Callaway, E. M. (1992). Development of local circuits in mammalian visual cortex. *Annual Review of Neuroscience*, 15:31–56.

Katz, L. C., and Shatz, C. J. (1996). Synaptic activity and the construction of cortical circuits. *Science*, 274:1133–1138.

Keeler, J., and Rumelhart, D. E. (1992). A self-organizing integrated segmentation and recognition neural network. In Moody, J. E., Hanson, S. J., and Lippmann, R. P., editors, *Advances in Neural Information Processing Systems 4*, 496–504. San Francisco: Kaufmann.

Keesing, R., Stork, D. G., and Shatz, C. J. (1992). Retinogeniculate development: The role of competition and correlated retinal activity. In Moody, J. E., Hanson, S. J., and Lippmann, R. P., editors, *Advances in Neural Information Processing Systems 4*, 91–97. San Francisco: Kaufmann.

Kellman, P. J., Yin, C., and Shapley, T. F. (1998). A common mechanism for illusory and occluded object completion. *Journal of Experimental Psychology: Human Perception and Performance*, 24:859–869.

Kim, D. (2004). A spiking neuron model for synchronous flashing of fireflies. *Biosystems*, 76:7–20.

Kim, D. S., and Bonhoeffer, T. (1994). Reverse occlusion leads to a precise restoration of orientation preference maps in visual cortex. *Nature*, 370:370–372.

Kiorpes, L., and Kiper, D. C. (1996). Development of contrast sensitivity across the visual field in macaque monkeys (*Macaca nemestrina*). *Vision Research*, 36:239–247.

Kirillov, A. B., and Woodward, D. J. (1993). Synchronization of spiking neurons: Transmission delays, noise and NMDA receptors. In *World Congress on Neural Networks*, 594–597. Hillsdale, New Jersey: Erlbaum.

Kisvárday, Z. F., and Eysel, U. T. (1992). Cellular organization of reciprocal patchy networks in layer III of cat visual cortex (area 17). *Neuroscience*, 46:275–286.

Kisvárday, Z. F., Kim, D. S., Eysel, U. T., and Bonhoeffer, T. (1994). Relationship between lateral inhibitory connections and the topography of the orientation map in the cat visual cortex. *European Journal of Neuroscience*, 6:1619–1632.

Kleiner, K. A. (1987). Amplitude and phase spectra as indices of infants' pattern preferences. *Infant Behavior and Development*, 10:49–59.

Kleiner, K. A. (1993). Specific vs. non-specific face-recognition device. In de Boysson-Bardies, B., editor, *Developmental Neurocognition: Speech and Face Processing in the First Year of Life*, 103–108. Dordrecht, The Netherlands: Kluwer.

Knoblauch, A., and Palm, G. (2003). Synchronization of neuronal assemblies in reciprocally connected cortical areas. *Theory in Biosciences*, 122:37–54.

Knudsen, E. I., and Knudsen, P. F. (1985). Vision calibrates sound localization in developing barn owls. *The Journal of Neuroscience*, 9:3306–3313.

Ko, J., and Byun, H. (2003). N-division output coding method applied to face recognition. *Pattern Recognition Letters*, 24:3115–3123.

Köhler, W., and Wallach, H. (1944). Figural after-effects: An investigation of visual processes. *Proceedings of the American Philosophical Society*, 88:269–357.

Kohn, A., and Movshon, J. A. (2003). Neuronal adaptation to visual motion in area MT of the macaque. *Neuron*, 39:681–691.

Kohonen, T. (1982a). Analysis of a simple self-organizing process. *Biological Cybernetics*, 44:135–140.

Kohonen, T. (1982b). Self-organized formation of topologically correct feature maps. *Biological Cybernetics*, 43:59–69.

Kohonen, T. (1989). *Self-Organization and Associative Memory*. Berlin: Springer. Third edition.

Kohonen, T. (1990). The self-organizing map. *Proceedings of the IEEE*, 78:1464–1480.

Kohonen, T. (1993). Physiological interpretation of the self-organizing map algorithm. *Neural Networks*, 6:895–905.

Kohonen, T. (2001). *Self-Organizing Maps*. Berlin: Springer. Third edition.

Kohonen, T., Kaski, S., Lagus, K., Salojärvi, J., Honkela, J., Paatero, V., and Saarela, A. (2000). Self-organization of a massive document collection. *IEEE Transactions on Neural Networks*, 11:574–585.

Kolen, J. F., and Pollack, J. B. (1990). Scenes from exclusive-or: Back propagation is sensitive to initial conditions. In *Proceedings of the 12th Annual Conference of the Cognitive Science Society*, 868–875. Hillsdale, NJ: Erlbaum.

Kosslyn, S. M., and Sussman, A. L. (1995). Roles of imagery in perception: Or, there is no such thing as immaculate perception. In Gazzaniga, M. S., editor, *The Cognitive Neurosciences*, 1035–1041. Cambridge, MA: MIT Press.

Kötter, R. (2004). Online retrieval, processing, and visualization of primate connectivity data from the CoCoMac database. *Neuroinformatics*, 2:127–144.

Koulakov, A. A., and Chklovskii, D. B. (2001). Orientation preference patterns in mammalian visual cortex: A wire length minimization approach. *Neuron*, 29:519–527.

Kovacs, I., and Julesz, B. (1993). A closed curve is much more than an incomplete one: Effect of closure in figure-ground segmentation. *Proceedings of the National Academy of Sciences, USA*, 90:7495–7497.

Koza, J. R. (1992). *Genetic Programming: On the Programming of Computers by Means of Natural Selection*. Cambridge, MA: MIT Press.

Kozma, R., Alvarado, M., Rogers, L., Lau, B., and Freeman, W. J. (2001). Emergence of un-correlated common-mode oscillations in the sensory cortex. *Neurocomputing*, 38–40:747–755.

Krüger, N., and Wörgötter, F. (2002). Multi modal estimation of collinearity and parallelism in natural image sequences. *Network: Computation in Neural Systems*, 13:553–576.

Kuhlmann, L., Burkitt, A. N., Paolini, A., and Clark, G. M. (2002). Summation of spatiotemporal input patterns in leaky integrate-and-fire neurons: Application to neurons in the cochlear nucleus receiving converging auditory nerve fiber input. *Journal of Computational Neuroscience*, 12:55–73.

LaBerge, D. (1995). *Attentional Processing: The Brain's Art of Mindfulness*. Cambridge, MA: Harvard University Press.

LaBerge, D., and Buchsbaum, M. S. (1990). Positron emission tomographic measurements of pulvinar activity during an attention task. *The Journal of Neuroscience*, 10:613–619.

Lamme, V. A., Super, H., and Spekreijse, H. (1998). Feedforward, horizontal, and feedback processing in the visual cortex. *Current Opinion in Neurobiology*, 8:529–535.

Lancaster, J. L., Narayana, S., Wenzel, D., Luckemeyer, J., Roby, J., and Fox, P. (2004). Evaluation of an image-guided, robotically positioned transcranial magnetic stimulation system. *Human Brain Mapping*, 22:329–340.

Lander, E. S., et al. (2001). Initial sequencing and analysis of the human genome. *Nature*, 409:860–921.

Landisman, C. E., and Ts'o, D. Y. (2002a). Color processing in macaque striate cortex: Electrophysiological properties. *Journal of Neurophysiology*, 87:3138–3151.

Landisman, C. E., and Ts'o, D. Y. (2002b). Color processing in macaque striate cortex: Relationships to ocular dominance, cytochrome oxidase, and orientation. *Journal of Neurophysiology*, 87:3126–3137.

Landy, M. S., Maloney, L. T., and Pavel, M., editors (1995). *Exploratory Vision: The Active Eye*. Berlin: Springer.

Langley, P., Choi, D., and Shapiro, D. (2004). A cognitive architecture for physical agents. Technical report, Institute for the Study of Learning and Expertise, Palo Alto, CA.

Lapicque, M. L. (1907). Recherches quantitatives sur l'excitation électrique des nerfs traitée comme une polarisation [Quantitative studies on electric excitation of nerves treated as polarization]. *Journal de Physiologie et Pathologie General*, 9:620–635.

Lawrence, S., Giles, C. L., Tsoi, A. C., and Back, A. D. (1997). Face recognition: A convolutional neural network approach. *IEEE Transactions on Neural Networks*, 8:98–113.

LeCun, Y., Jackel, L. D., Bottou, L., Cortes, C., Denker, J. S., Drucker, H., Guyon, I., Muller, U. A., Sackinger, E., Simard, P., and Vapnik, V. (1995). Learning algorithms for classification: A comparison on handwritten digit recognition. In Oh, J. H., Kwon, C., and Cho, S., editors, *Neural Networks: The Statistical Mechanics Perspective. Proceedings of the CTP-PBSRI Joint Workshop on Theoretical Physics*, 261–276. Singapore: World Scientific.

Lee, C. W., Eglen, S. J., and Wong, R. O. L. (2002a). Segregation of ON and OFF retinogeniculate connectivity directed by patterned spontaneous activity. *Journal of Neurophysiology*, 88:2311–2321.

Lee, K., and Lee, Y. (2000a). A framework of two-stage combination of multiple recognizers for handwritten numerals. In Mizoguchi, R., and Slaney, J., editors, *Topics in Artificial Intelligence: 6th Pacific Rim International Conference on Artificial Intelligence*, Lecture Notes in Artificial Intelligence 1886, 617–626. Berlin: Springer.

Lee, S.-H., and Blake, R. (1999). Visual form created solely from temporal structure. *Science*, 284:1165–1168.

Lee, S.-H., and Blake, R. (2001). Neural synergy in visual grouping: When good continuation meets common fate. *Vision Research*, 41:2057–2064.

Lee, S.-I., and Lee, S.-Y. (2000b). Top-down attention control at feature space for robust pattern recognition. In Lee, S.-W., Bülthoff, H. H., and Poggio, T., editors, *Biologically Motivated Computer Vision: First IEEE International Workshop*, Lecture Notes in Computer Science 1811, 129–138. Berlin: Springer.

Lee, S.-W. (1996). Off-line recognition of totally unconstrained handwritten numerals using multilayer cluster neural network. *IEEE Transactions on Pattern Analysis and Machine Intelligence*, 18:648–652.

Lee, T. S., and Nguyen, M. (2001). Dynamics of subjective contour formation in early visual cortex. *Proceedings of the National Academy of Sciences, USA*, 98:1907–1911.

Lee, T.-W., Wachtler, T., and Sejnowski, T. J. (2002b). Color opponency is an efficient representation of spectral properties in natural scenes. *Vision Research*, 42:2095–2103.

Lennie, P. (2003). The cost of cortical computation. *Current Biology*, 13:493–497.

Leonards, U., and Singer, W. (1998). Two segmentation mechanisms with differential sensitivity for colour and luminance contrast. *Vision Research*, 38:101–109.

Leonards, U., Singer, W., and Fahle, M. (1996). The influence of temporal phase difference on texture segmentation. *Vision Research*, 36:2689–2697.

Leopold, D. A., O'Toole, A. J., Vetter, T., and Blanz, V. (2001). Prototype-referenced shape encoding revealed by high-level aftereffects. *Nature Neuroscience*, 4:89–94.

Leow, W. K. (1994). VISOR: Learning visual schemas in neural networks for object recognition and scene analysis. Doctoral dissertation, Department of Computer Sciences, The University of Texas at Austin, Austin, TX. Technical Report AI94-219.

Leow, W. K., and Miikkulainen, R. (1997). Visual schemas in neural networks for object recognition and scene analysis. *Connection Science*, 9:161–200.

Lesher, G. W., and Mingolla, E. (1995). Illusory contour formation. In Arbib, M. A., editor, *The Handbook of Brain Theory and Neural Networks*, 481–483. Cambridge, MA: MIT Press. First edition.

Levine, D. S., and Grossberg, S. (1976). Visual illusions in neural networks: Line neutralization, tilt after effect, and angle expansion. *Journal of Theoretical Biology*, 61:477–504.

Levy, I., Hasson, U., Avidan, G., Hendler, T., and Malach, R. (2001). Center-periphery organization of human object areas. *Nature Neuroscience*, 4:533–539.

Lewis, J. W., and Van Essen, D. C. (2000). Corticocortical connections of visual, sensorimotor, and multimodal processing areas in the parietal lobe of the macaque monkey. *The Journal of Comparative Neurology*, 428:112–137.

Li, P., Farkas, I., and MacWhinney, B. (2004). Early lexical development in a self-organizing neural network. *Neural Networks*, 17:1345–1362.

Li, Z. (1998). A neural model of contour integration in the primary visual cortex. *Neural Computation*, 10:903–940.

Li, Z. (1999). Visual segmentation by contextual influences via inter-cortical interactions in the primary visual cortex. *Network: Computation in Neural Systems*, 10:187–212.

Linsker, R. (1986a). From basic network principles to neural architecture: Emergence of orientation columns. *Proceedings of the National Academy of Sciences, USA*, 83:8779–8783.

Linsker, R. (1986b). From basic network principles to neural architecture: Emergence of orientation-selective cells. *Proceedings of the National Academy of Sciences, USA*, 83:8390–8394.

Linsker, R. (1986c). From basic network principles to neural architecture: Emergence of spatial-opponent cells. *Proceedings of the National Academy of Sciences, USA*, 83:7508–7512.

Lippe, W. R. (1994). Rhythmic spontaneous activity in the developing avian auditory system. *The Journal of Neuroscience*, 14:1486–1495.

Lisman, J. (1998). What makes the brain's ticker tock. *Nature*, 394:132–133.

Liu, X., and Wang, D. (1999). Range image segmentation using an oscillatory network. *IEEE Transactions of Neural Networks*, 10:564–573.

Livingstone, M. S., and Hubel, D. H. (1984a). Anatomy and physiology of a color system in the primate visual cortex. *The Journal of Neuroscience*, 4:309–356.

Livingstone, M. S., and Hubel, D. H. (1984b). Specificity of intrinsic connections in primate primary visual cortex. *The Journal of Neuroscience*, 4:2830–2835.

Löwel, S. (1994). Ocular dominance column development: Strabismus changes the spacing of adjacent columns in cat visual cortex. *The Journal of Neuroscience*, 14:7451–7468.

Löwel, S., Bischof, H. J., Leutenecker, B., and Singer, W. (1988). Topographic relations between ocular dominance and orientation columns in the cat striate cortex. *Experimental Brain Research*, 71:33–46.

Löwel, S., and Singer, W. (1992). Selection of intrinsic horizontal connections in the visual cortex by correlated neuronal activity. *Science*, 255:209–212.

Luhmann, H. J., Martínez Millán, L., and Singer, W. (1986). Development of horizontal intrinsic connections in cat striate cortex. *Experimental Brain Research*, 63:443–448.

Lund, J. S., Yoshioka, T., and Levitt, J. B. (1993). Comparison of intrinsic connectivity in different areas of macaque monkey cerebral cortex. *Cerebral Cortex*, 3:148–162.

Lytton, W. W. (2002). *From Computer to Brain: Foundations of Computational Neuroscience*. Berlin: Springer.

Lytton, W. W., and Sejnowski, T. J. (1991). Simulations of cortical pyramidal neurons synchronized by inhibitory interneurons. *Journal of Neurophysiology*, 66:1059–1079.

Maass, W. (1997). Networks of spiking neurons: The third generation of neural network models. *Neural Networks*, 10:1659–1671.

Maass, W. (1998). Computing with spiking neurons. In Maass, W., and Bishop, C. M., editors, *Pulsed Neural Networks*, 55–85. Cambridge, MA: MIT Press.

Maffei, L., and Galli-Resta, L. (1990). Correlation in the discharges of neighboring rat retinal ganglion cells during prenatal life. *Proceedings of the National Academy of Sciences, USA*, 87:2861–2864.

Magnussen, S., and Johnsen, T. (1986). Temporal aspects of spatial adaptation: A study of the tilt aftereffect. *Vision Research*, 26:661–672.

Magnussen, S., and Kurtenbach, W. (1980). Adapting to two orientations: Disinhibition in a visual aftereffect. *Science*, 207:908–909.

Mainen, Z. F., and Sejnowski, T. J. (1998). Modeling active dendritic processes in pyramidal neurons. In Koch, C., and Segev, I., editors, *Methods in Neuronal Modeling: From Ions to Networks*, 170–209. Cambridge, MA: MIT Press. Second edition.

Mäkelä, P., Näsänen, R., Rovamo, J., and Melmoth, D. (2001). Identification of facial images in peripheral vision. *Vision Research*, 41:599–610.

Malach, R., Amir, Y., Harel, M., and Grinvald, A. (1993). Relationship between intrinsic connections and functional architecture revealed by optical imaging and in vivo targeted biocytin injections in the primate striate cortex. *Proceedings of the National Academy of Sciences, USA*, 90:10469–10473.

Maldonado, P. E., Gödecke, I., Gray, C. M., and Bonhoeffer, T. (1997). Selectivity in pinwheel centers in cat striate cortex. *Science*, 276:1551–1555.

Maquet, P., and Phillips, S. C. (1998). Functional brain imaging of human sleep. *Journal of Sleep Research*, 7:42–47.

Marcus, G. F. (2003). *The Algebraic Mind: Integrating Connectionism and Cognitive Science*. Cambridge, MA: MIT Press.

Marder, E., and Calabrese, R. L. (1996). Principles of rhythmic motor pattern generation. *Physiological Reviews*, 76:687–717.

Mareschal, D., Johnson, M. H., Sirois, S., Spratling, M., Thomas, M. S. C., and Westermann, G., editors (2005a). *Neuroconstructivism, Vol. 1: How the Brain Constructs Cognition*. Oxford, UK: Oxford University Press. In press.

Mareschal, D., Johnson, M. H., Sirois, S., Spratling, M., Thomas, M. S. C., and Westermann, G., editors (2005b). *Neuroconstructivism, Vol. 2: Perspectives and Prospects*. Oxford, UK: Oxford University Press. In press.

Markman, A. B., and Dietrich, E. (2000). Extending the classical view of representation. *Trends in Cognitive Sciences*, 4:470–475.

Markram, H., Lübke, J., Frotscher, M., and Sakmann, B. (1997). Regulation of synaptic efficacy by coincidence of postsynaptic APs and EPSPs. *Science*, 275:213–215.

Marks, G. A., Shaffery, J. P., Oksenberg, A., Speciale, S. G., and Roffwarg, H. P. (1995). A functional role for REM sleep in brain maturation. *Behavioural Brain Research*, 69:1–11.

480   References

Marr, D. (1982). *Vision*. New York: Freeman.

Marsalek, P., Koch, C., and Maunsell, J. H. R. (1997). On the relationship between sublinear input and spike output jitter in individual neurons. *Proceedings of the National Academy of Sciences, USA*, 94:735–740.

Marshall, J. A. (1990). Self-organizing neural networks for perception of visual motion. *Neural Networks*, 3:45–74.

Marshall, J. A., and Alley, R. (1996). A self-organizing neural network that learns to detect and represent visual depth from occlusion events. In Sirosh, J., Miikkulainen, R., and Choe, Y., editors, *Lateral Interactions in the Cortex: Structure and Function*. Austin, TX: The UTCS Neural Networks Research Group. Electronic book, ISBN 0-9647060-0-8, http://nn.cs.utexas.edu/web-pubs/htmlbook96.

Martin, G. L., Rashid, M., Chapman, D., and Pittman, J. A. (1993). Learning to see where and what: Training a net to make saccades and recognize handwritten characters. In Giles, C. L., Hanson, S. J., and Cowan, J. D., editors, *Advances in Neural Information Processing Systems 5*, 441–447. San Francisco: Kaufmann.

Martinez-Conde, S., Macknik, S. L., and Hubel, D. H. (2000). Microsaccadic eye movements and firing of single cells in the striate cortex of macaque monkeys. *Nature Neuroscience*, 3:251–258.

Masini, R., Antonietti, A., and Moja, E. A. (1990). An increase in strength of tilt aftereffect associated with tryptophan depletion. *Perceptual and Motor Skills*, 70:531–539.

Mastronarde, D. N., Humphrey, A. L., and Saul, A. B. (1991). Lagged Y cells in the cat lateral geniculate nucleus. *Visual Neuroscience*, 7:191–200.

Maurer, D., and Barrera, M. (1981). Infants' perception of natural and distorted arrangements of a schematic face. *Child Development*, 52:196–202.

Mayer, N., Herrmann, J. M., and Geisel, T. (2001). Signatures of natural image statistics in cortical simple cell receptive fields. *Neurocomputing*, 38:279–284.

Mazziotta, J., Toga, A., Evans, A., Fox, P., Lancaster, J. L., Zilles, K., Woods, R., Paus, T., Simpson, G., Pike, B., Holmes, C., Collins, L., Thompson, P., MacDonald, D., Iacoboni, M., Schormann, T., Amunts, K., Palomero-Gallagher, N., Geyer, S., Parsons, L., Narr, K., Kabani, N., Le Goualher, G., Feidler, J. C., Smith, K., Boomsma, D., Hulshoff Pol, H., Cannon, T., Kawashima, R., and Mazoyer, B. (2001). A four-dimensional probabilistic atlas of the human brain. *Journal of the American Medical Informatics Association*, 8:401–430.

McCasland, J. S., Bernardo, K. L., Probst, K. L., and Woolsey, T. A. (1992). Cortical local circuit axons do not mature after early deafferentation. *Proceedings of the National Academy of Sciences, USA*, 89:1832–1836.

McClelland, J. L., and Rogers, T. T. (2003). The parallel distributed processing approach to semantic cognition. *Nature Reviews Neuroscience*, 4:1–14.

McCormick, B. H., Choe, Y., Koh, W., Abbott, L. C., Keyser, J., Melek, Z., Doddapaneni, P., and Mayerich, D. M. (2004a). Construction of anatomically correct models of mouse brain networks. *Neurocomputing*, 58–60:379–386.

McCormick, B. H., Mayerich, D. M., Abbott, L. C., Gutierrez-Osuna, R., Keyser, J., Choe, Y., Koh, W., and Busse, B. L. (2004b). Whole mouse brain mapped at submicron resolution using knife-edge scanning microscope. In *Society for Neuroscience Abstracts*, Program No. 1033.4. Washington, DC: Society for Neuroscience.

McDonald, C. T., and Burkhalter, A. (1993). Organization of long-range inhibitory connections within rat visual cortex. *The Journal of Neuroscience*, 13:768–781.

McGraw, P. V., Walsh, V., and Barrett, B. T. (2004). Motion-sensitive neurones in V5/MT modulate perceived spatial position. *Current Biology*, 14:1090–1093.

McGuire, B. A., Gilbert, C. D., Rivlin, P. K., and Wiesel, T. N. (1991). Targets of horizontal connections in macaque primary visual cortex. *The Journal of Comparative Neurology*, 305:370–392.

McGurk, H., and MacDonald, J. (1976). Hearing lips and seeing voices. *Nature*, 264:746–748.

McIlhagga, W. H., and Mullen, K. T. (1996). Contour integration with colour and luminance contrast. *Vision Research*, 36:1265–1279.

McLaughlin, T., Torborg, C. L., Feller, M. B., and O'Leary, D. D. (2003). Retinotopic map refinement requires spontaneous retinal waves during a brief critical period of development. *Neuron*, 40:1147–1160.

Meister, M., Wong, R. O. L., Baylor, D. A., and Shatz, C. J. (1991). Synchronous bursts of action-potentials in the ganglion cells of the developing mammalian retina. *Science*, 252:939–943.

Meltzoff, A. N., and Moore, A. K. (1993). Why faces are special to infants — On connecting the attraction of faces and infants' ability for imitation and cross-modal processing. In de Boysson-Bardies, B., editor, *Developmental Neurocognition: Speech and Face Processing in the First Year of Life*, 211–226. Dordrecht, The Netherlands: Kluwer.

Menon, V. (1990). Dynamic aspects of signaling in distributed neural systems. Doctoral dissertation, Department of Computer Sciences, The University of Texas at Austin, Austin, TX. Technical Report TR-90-36.

Menon, V. (1991). Population oscillations in neuronal groups. *International Journal of Neural Systems*, 2:237–262.

Meredith, M. A., and Stein, B. E. (1986). Visual, auditory, and somatosensory convergence on cells in superior colliculus results in multisensory integration. *Journal of Neurophysiology*, 56:640–662.

Merigan, W. H., and Maunsell, J. H. R. (1993). How parallel are the primate visual pathways? *Annual Review of Neuroscience*, 16:369–402.

Merzenich, M. M., Nelson, R. J., Stryker, M. P., Cynader, M. S., Schoppmann, A., and Zook, J. M. (1984). Somatosensory cortical map changes following digit amputation in adult monkeys. *The Journal of Comparative Neurology*, 224:591–605.

Merzenich, M. M., Recanzone, G. H., Jenkins, W. M., and Grajski, K. A. (1990). Adaptive mechanisms in cortical networks underlying cortical contributions to learning and nondeclarative memory. In *The Brain*, vol. LV of *Cold Spring Harbor Symposia on Quantitative Biology*, 873–887. Cold Spring Harbor, NY: Cold Spring Harbor Laboratory Press.

Meunier, C., and Segev, I. (2002). Playing the devil's advocate: Is the Hodgkin–Huxley model useful? *Trends in Neurosciences*, 25:558–563.

Meyerson, R. G., and Palmer, S. E. (2004). Change blindness in synchrony grouping. *Journal of Vision*, 4:496a.

Miikkulainen, R. (1991). Self-organizing process based on lateral inhibition and synaptic resource redistribution. In Kohonen, T., Mäkisara, K., Simula, O., and Kangas, J., editors, *Proceedings of the 1991 International Conference on Artificial Neural Networks*, 415–420. Amsterdam: North-Holland.

Miikkulainen, R. (1992). Trace feature map: A model of episodic associative memory. *Biological Cybernetics*, 66:273–282.

Miikkulainen, R. (1993). *Subsymbolic Natural Language Processing: An Integrated Model of Scripts, Lexicon, and Memory*. Cambridge, MA: MIT Press.

Miikkulainen, R., Bednar, J. A., Choe, Y., and Sirosh, J. (1997). Self-organization, plasticity, and low-level visual phenomena in a laterally connected map model of the primary visual cortex. In Goldstone, R. L., Schyns, P. G., and Medin, D. L., editors, *Perceptual*

Learning, vol. 36 of *Psychology of Learning and Motivation*, 257–308. San Diego, CA: Academic Press.

Miller, G. A. (1956). The magical number seven, plus or minus two: Some limits on our capacity of processing information. *Psychological Review*, 63:81–97.

Miller, K. D. (1994). A model for the development of simple cell receptive fields and the ordered arrangement of orientation columns through activity-dependent competition between ON- and OFF-center inputs. *The Journal of Neuroscience*, 14:409–441.

Miller, K. D., Erwin, E., and Kayser, A. (1999). Is the development of orientation selectivity instructed by activity? *Journal of Neurobiology*, 41:44–57.

Miller, K. D., Keller, J. B., and Stryker, M. P. (1989). Ocular dominance column development: Analysis and simulation. *Science*, 245:605–615.

Miller, K. D., and MacKay, D. J. C. (1994). The role of constraints in Hebbian learning. *Neural Computation*, 6:100–126.

Milner, A. D., and Goodale, M. A. (1993). Visual pathways to perception and action. *Progress in Brain Research*, 95:317–337.

Mirollo, R. E., and Strogatz, S. H. (1990). Synchronization of pulse-coupled biological oscillators. *SIAM Journal of Applied Mathematics*, 50:1645–1662.

Mirsky, J. S., Nadkarni, P. M., Healy, M. D., Miller, P. L., and Shepherd, G. M. (1998). Database tools for integrating and searching membrane property data correlated with neuronal morphology. *Journal of Neuroscience Methods*, 82:105–121.

Mitchell, D. E., and Muir, D. W. (1976). Does the tilt aftereffect occur in the oblique meridian? *Vision Research*, 16:609–613.

Mitchell, M. (1996). *An Introduction to Genetic Algorithms*. Cambridge, MA: MIT Press.

Miyashita, M., Kim, D. S., and Tanaka, S. (1997). Cortical direction selectivity without directional experience. *Neuroreport*, 8:1187–1191.

Miyashita, M., and Tanaka, S. (1992). A mathematical model for the self-organization of orientation columns in visual cortex. *Neuroreport*, 3:69–72.

Miyashita-Lin, E. M., Hevner, R., Wassarman, K. M., Martinez, S., and Rubenstein, J. L. (1999). Early neocortical regionalization in the absence of thalamic innervation. *Science*, 285:906–909.

Molnár, Z., Higashi, S., and López-Bendito, G. (2003). Choreography of early thalamocortical development. *Cerebral Cortex*, 13:661–669.

Mondloch, C. J., Lewis, T. L., Budreau, D. R., Maurer, D., Dannemiller, J. L., Stephens, B. R., and Kleiner-Gathercoal, K. A. (1999). Face perception during early infancy. *Psychological Science*, 10:419–422.

Moody, J., and Darken, C. (1990). Fast learning in networks of locally-tuned processing units. *Neural Computation*, 1:281–294.

Moorcroft, W. H. (1995). [The function of sleep] Comments on the symposium and an attempt at synthesis. *Behavioural Brain Research*, 69:207–210.

Mori, S., Kaufmann, W. E., Davatzikos, C., Stieltjes, B., Amodei, L., Fredericksen, K., Pearlson, G. D., Melhem, E. R., Solaiyappan, M., Raymond, G. V., Moser, H. W., and van Zijl, P. C. M. (2002). Imaging cortical association tracts in the human brain using diffusion-tensor-based axonal tracking. *Magnetic Resonance in Medicine*, 215–223.

Morris, J. S., Ohman, A., and Dolan, R. J. (1999). A subcortical pathway to the right amygdala mediating "unseen" fear. *Proceedings of the National Academy of Sciences, USA*, 96:1680–1685.

Moscovitch, M., and Nadel, L. (1998). Consolidation and the hippocampal complex revisited: In defense of the multiple-trace model. *Current Opinion in Neurobiology*, 8:297–300.

Movshon, J. A., and van Sluyters, R. C. (1981). Visual neural development. *Annual Review of Psychology*, 32:477–522.

Muir, D. W., and Over, R. (1970). Tilt aftereffects in central and peripheral vision. *Journal of Experimental Psychology*, 85:165–170.

Müller, T., Stetter, M., Hubener, M., Sengpiel, F., Bonhoeffer, T., Gödecke, I., Chapman, B., Löwel, S., and Obermayer, K. (2000). An analysis of orientation and ocular dominance patterns in the visual cortex of cats and ferrets. *Neural Computation*, 12:2573–2595.

Mundel, T., Dimitrov, A., and Cowan, J. D. (1997). Visual cortex circuitry and orientation tuning. In Mozer, M. C., Jordan, M. I., and Petsche, T., editors, *Advances in Neural Information Processing Systems 9*, 887–893. Cambridge, MA: MIT Press.

Murray, M., Sharma, S., and Edwards, M. A. (1982). Target regulation of synaptic number in the compressed retinotectal projection of goldfish. *Journal of Computational Neurology*, 209:374–385.

Murray, S. O., Schrater, P. R., and Kersten, D. (2004). Perceptual grouping and the interactions between visual cortical areas. *Neural Networks*, 17:695–705.

Myhr, K. L., Lukasiewicz, P. D., and Wong, R. O. L. (2001). Mechanisms underlying developmental changes in the firing patterns of ON and OFF retinal ganglion cells during refinement of their central projections. *The Journal of Neuroscience*, 21:8664–8671.

Nachson, I. (1995). On the modularity of face recognition: The riddle of domain specificity. *Journal of Clinical and Experimental Neuropsychology*, 17:256–275.

Nadel, L., and Moscovitch, M. (1997). Memory consolidation, retrograde amnesia and the hippocampal complex. *Current Opinion in Neurobiology*, 7:217–227.

Nagumo, J. S., Arimato, S., and Yoshizawa, S. (1962). An active pulse transmission line simulating a nerve axon. *Proceedings of the IRE*, 50:2061–2070.

Nakayama, K., and Shimojo, S. (1992). Experiencing and perceiving visual surfaces. *Science*, 257:1357–1363.

Nass, M. M., and Cooper, L. N. (1975). A theory for the development of feature detecting cells in visual cortex. *Biological Cybernetics*, 19:1–18.

National Park Service (1995). Image database. www.freestockphotos.com/NPS.

Nelson, J. I. (1995). Visual scene perception: Neurophysiology. In Arbib, M. A., editor, *The Handbook of Brain Theory and Neural Networks*, 1024–1028. Cambridge, MA: MIT Press. First edition.

Niebur, E., and Wörgötter, F. (1993). Orientation columns from first principles classical visual receptive field. *Biological Cybernetics*, 70:1–13.

Nischwitz, A., and Glünder, H. (1995). Local lateral inhibition: A key to spike synchronization? *Biological Cybernetics*, 73:389–400.

Nolfi, S., and Parisi, D. (1994). Desired answers do not correspond to good teaching inputs in ecological neural networks. *Neural Processing Letters*, 1:1–4.

Nowak, L. G., and Bullier, J. (1997). The timing of information transfer in the visual system. In Rockland, K. S., Kaas, J. H., and Peters, A., editors, *Extrastriate Cortex in Primates*, vol. 12 of *Cerebral Cortex*, 205–241. New York: Plenum Press.

Nudo, R. J., Wise, B. M., Fuentes, F., and Milliken, G. W. (1996). Neural substrates for the effects of rehabilitative training on motor recovery after ischemic infarct. *Science*, 272:1791–1794.

Nugent, A. K., Keswani, R. N., Woods, R. L., and Peli, E. (2003). Contour integration in peripheral vision reduces gradually with eccentricity. *Vision Research*, 43:2427–2437.

Obermayer, K., and Blasdel, G. G. (1993). Geometry of orientation and ocular dominance columns in the monkey striate cortex. *The Journal of Neuroscience*, 13:4114–4129.

Obermayer, K., Blasdel, G. G., and Schulten, K. J. (1992). Statistical–mechanical analysis of self-organization and pattern formation during the development of visual maps. *Physical Review A*, 45:7568–7589.

Obermayer, K., Ritter, H., and Schulten, K. J. (1990a). Large-scale simulation of a self-organizing neural network: Formation of a somatotopic map. In Eckmiller, R., Hartmann, G., and Hauske, G., editors, *Parallel Processing in Neural Systems and Computers*, 71–74. Amsterdam: North-Holland.

Obermayer, K., Ritter, H., and Schulten, K. J. (1990b). Large-scale simulations of self-organizing neural networks on parallel computers: Application to biological modelling. *Parallel Computing*, 14:381–404.

Obermayer, K., Ritter, H., and Schulten, K. J. (1990c). A neural network model for the formation of topographic maps in the CNS: Development of receptive fields. In *International Joint Conference on Neural Networks* (San Diego, CA), vol. II, 423–429. Piscataway, NJ: IEEE.

Obermayer, K., Ritter, H., and Schulten, K. J. (1990d). A principle for the formation of the spatial structure of cortical feature maps. *Proceedings of the National Academy of Sciences, USA*, 87:8345–8349.

Obermayer, K., Sejnowski, T. J., and Blasdel, G. G. (1995). Neural pattern formation via a competitive Hebbian mechanism. *Behavioural Brain Research*, 66:161–167.

O'Donovan, M. J. (1999). The origin of spontaneous activity in developing networks of the vertebrate nervous system. *Current Opinion in Neurobiology*, 9:94–104.

Oja, E. (1982). A simplified neuron model as a principal component analyzer. *Journal of Mathematical Biology*, 15:267–273.

Oja, E. (1989). Neural networks, principal components, and subspaces. *International Journal of Neural Systems*, 1:61–68.

Oja, E., and Kaski, S., editors (1999). *Kohonen Maps*. Amsterdam: Elsevier.

Oja, M., Kaski, S., and Kohonen, T. (2003). Bibliography of self-organizing map (SOM) papers: 1998-2001 addendum. *Neural Computing Surveys*, 3:1–156.

O'Keefe, J., and Burgess, N. (1996). Geometric determinants of the place fields of hippocampal neurones. *Nature*, 381:425–428.

O'Keefe, J., and Reece, M. (1993). Phase relationship between hippocampal place units and the hippocampal theta rhythm. *Hippocampus*, 3:317–330.

Oksenberg, A., Shaffery, J. P., Marks, G. A., Speciale, S. G., Mihailoff, G., and Roffwarg, H. P. (1996). Rapid eye movement sleep deprivation in kittens amplifies LGN cell-size disparity induced by monocular deprivation. *Developmental Brain Research*, 97:51–61.

Olshausen, B. A. (2003). Principles of image representation in visual cortex. In Chalupa, L. M., and Werner, J. S., editors, *The Visual Neurosciences*, 1603–1615. Cambridge, MA: MIT Press.

Olshausen, B. A., Anderson, C. H., and Van Essen, D. C. (1995). A multiscale dynamic routing circuit for forming size- and position-invariant object representations. *Journal of Computational Neuroscience*, 2:45–62.

Olshausen, B. A., Anderson, C. H., and Van Essen, D. C. (1996). A neurobiological model of visual attention and invariant pattern recognition based on dynamic routing of information. *The Journal of Neuroscience*, 13:4700–4719.

Olshausen, B. A., and Field, D. J. (1997). Sparse coding with an overcomplete basis set: A strategy employed by V1? *Vision Research*, 37:3311–3325.

Olson, S. J., and Grossberg, S. (1998). A neural network model for the development of simple and complex cell receptive fields within cortical maps of orientation and ocular dominance. *Neural Networks*, 11:189–208.

Oram, M. W., Wiener, M. C., Lestienne, R., and Richmond, B. J. (1999). Stochastic nature of precisely timed spike patterns in visual system neuronal responses. *Journal of Neurophysiology*, 81:3021–3033.

O'Regan, J. K., and Noë, A. (2001). A sensorimotor account of vision and visual consciousness. *Behavioral and Brain Sciences*, 24:939–973.

O'Reilly, R. C., and Munakata, Y. (2000). *Computational Explorations in Cognitive Neuroscience: Understanding the Mind by Simulating the Brain*. Cambridge, MA: MIT Press.

Osan, R., and Ermentrout, B. (2002). Development of joint ocular dominance and orientation selectivity maps in a correlation-based neural network model. *Neurocomputing*, 44–46:561–566.

O'Toole, A. J., Millward, R. B., and Anderson, J. A. (1988). A physical system approach to recognition memory for spatially transformed faces. *Neural Networks*, 1:179–199.

O'Toole, B. I. (1979). Exposure-time and spatial-frequency effects in the tilt illusion. *Perception*, 8:557–564.

Pallas, S. L., and Finlay, B. L. (1991). Compensation for population-size mismatches in the hamster retinotectal system: Alterations in the organization of retinal projections. *Visual Neuroscience*, 6:271–281.

Palmer, S. E. (1999). *Vision Science: Photons to Phenomenology*. Cambridge, MA: MIT Press.

Panchev, C., and Wermter, S. (2001). Hebbian spike-timing dependent self-organization in pulsed neural networks. In Rattay, F., editor, *World Congress on Neuroinformatics: Part II, Proceedings*, 378–385. Vienna: ARGESIM/ASIM-Verlag.

Paradiso, M. A., Shimojo, S., and Nakayama, K. (1989). Subjective contours, tilt aftereffects, and visual cortical organization. *Vision Research*, 29:1205–1213.

Parker, A. J., and Newsome, W. T. (1998). Sense and the single neuron: Probing the physiology of perception. *Annual Review of Neuroscience*, 21:227–277.

Parker, D. B. (1982). Learning-logic. Invention Report S81-64, File 1, Office of Technology Licensing, Stanford University, Palo Alto, CA.

Parks, T. E. (1980). Letter to the editor. *Perception*, 9:723.

Pascalis, O., de Schonen, S., Morton, J., Deruelle, C., and Fabre-Grenet, M. (1995). Mother's face recognition by neonates: A replication and an extension. *Infant Behavior and Development*, 18:79–85.

Pearson, J. C., Finkel, L. H., and Edelman, G. M. (1987). Plasticity in the organization of adult cortical maps: A computer simulation based on neuronal group selection. *The Journal of Neuroscience*, 7:4209–4223.

Pei, X., Vidyasagar, T. R., Volgushev, M., and Creutzfeldt, O. D. (1994). Receptive field analysis and orientation selectivity of postsynaptic potentials of simple cells in cat visual cortex. *The Journal of Neuroscience*, 14:7130–7140.

Peinado, A., Yuste, R., and Katz, L. C. (1993). Extensive dye-coupling between rat neocortical neurons during the period of circuit formation. *Neuron*, 14:103–114.

Penn, A. A., and Shatz, C. J. (1999). Brain waves and brain wiring: The role of endogenous and sensory-driven neural activity in development. *Pediatric Research*, 45:447–458.

Perrett, D. I. (1992). Organization and functions of cells responsive to faces in the temporal cortex. *Philosophical Transactions: Biological Sciences*, 335:23–30.

Peterhans, E., von der Heydt, R., and Baumgartner, G. (1986). Neuronal responses to illusory contour stimuli reveal stages of visual cortical processing. In Pettigrew, J. D., Sanderson, K. J., and Levick, W. R., editors, *Visual Neuroscience*, 343–351. Cambridge, UK: Cambridge University Press.

Petrov, Y. (2002). Disparity capture by flanking stimuli: A measure for the cooperative mechanism of stereopsis. *Vision Research*, 42:809–813.

Petry, S., and Meyer, G. E., editors (1987). *The Perception of Illusory Contours*. Berlin: Springer.

Pettet, M. W., and Gilbert, C. D. (1992). Dynamic changes in receptive-field size in cat primary visual cortex. *Proceedings of the National Academy of Sciences, USA*, 89:8366–8370.

Pettet, M. W., McKee, S. P., and Grzywacz, N. M. (1998). Constraints on long range interactions mediating contour detection. *Vision Research*, 38:865–879.

Pfeifer, R., and Scheier, C. (1997). Sensory-motor coordination: The metaphor and beyond. *Robotics and Autonomous Systems*, 20:157–178.

Pfeifer, R., and Scheier, C. (1998). Representation in natural and artificial agents: An embodied cognitive science perspective. *Zeitschrift für Naturforschung C — A Journal of Biosciences*, 53:480–503.

Pfleger, B., and Bonds, A. B. (1995). Dynamic differentiation of $GABA_A$-sensitive influences on orientation selectivity of complex cells in the cat striate cortex. *Experimental Brain Research*, 104:81–88.

Philipona, D., O'Regan, J. K., and Nadal, J.-P. (2003). Is there something out there? Inferring space from sensorimotor dependencies. *Neural Computation*, 15:2029–2050.

Phillips, P. J., Wechsler, H., Huang, J., and Rauss, P. (1998). The FERET database and evaluation procedure for face recognition algorithms. *Image and Vision Computing*, 16:295–306.

Piepenbrock, C., and Obermayer, K. (2002). Cortical orientation map development from natural images: The role of cortical response amplification in V1. In Backhaus, W., editor, *Neuronal Coding of Perceptual Systems: Proceedings of the International School of Biophysics*, 161–168. Singapore: World Scientific.

Piepenbrock, C., Ritter, H., and Obermayer, K. (1996). Cortical map development driven by spontaneous retinal activity waves. In von der Malsburg, C., von Seelen, W., Vorbrüggen, J. C., and Sendhoff, B., editors, *Proceedings of the Sixth International Conference on Artificial Neural Networks*, Lecture Notes in Computer Science 1112, 427–432. Berlin: Springer.

Pinsk, M. A., Doniger, G. M., and Kastner, S. (2004). Push–pull mechanism of selective attention in human extrastriate cortex. *Journal of Neurophysiology*, 92:622–629.

Polat, U., Mizobe, K., Pettet, M. W., Kasamatsu, T., and Norcia, A. M. (1998). Collinear stimuli regulate visual responses depending on cell's contrast threshold. *Nature*, 391:580–584.

Polat, U., Norcia, A. M., and Sagi, D. (1996). The pattern and functional significance of long-range interactions in human visual cortex. In Sirosh, J., Miikkulainen, R., and Choe, Y., editors, *Lateral Interactions in the Cortex: Structure and Function*. Austin, TX: The UTCS Neural Networks Research Group. Electronic book, ISBN 0-9647060-0-8, http://nn.cs.utexas.edu/web-pubs/htmlbook96.

Pollen, D. A. (1999). On the neural correlates of visual perception. *Cerebral Cortex*, 9:4–19.

Pompeiano, O., Pompeiano, M., and Corvaja, N. (1995). Effects of sleep deprivation on the postnatal development of visual-deprived cells in the cat's lateral geniculate nucleus. *Archives Italiennes de Biologie*, 134:121–140.

Prazdny, K. (1983). Illusory contours are not caused by simultaneous brightness contrast. *Perception and Psychophysics*, 34:403–404.

Previc, F. H. (1990). Functional specialization in the lower and upper visual fields in humans: Its ecological origins and neurophysiological implications. *Behavioral and Brain Sciences*, 13:519–575.

Prince, D. A., and Huguenard, J. R. (1988). Functional properties of neocortical neurons. In Rakic, P., and Singer, W., editors, *Neurobiology of Neocortex*, 153–176. Hoboken, NJ: Wiley.

Prodöhl, C., Würtz, R. P., and von der Malsburg, C. (2003). Learning the gestalt rule of collinearity from object motion. *Neural Computation*, 15:1865–1896.

Prut, Y., Vaadia, E., Bergman, H., Haalman, I., Slovin, H., and Abeles, M. (1998). Spatiotemporal structure of cortical activity: Properties and behavioral relevance. *Journal of Neurophysiology*, 79:2857–2874.

Puce, A., Allison, T., Gore, J. C., and McCarthy, G. (1995). Face-sensitive regions in human extrastriate cortex studied by functional MRI. *Journal of Neurophysiology*, 74:1192–1199.

Purves, D. (1988). *Body and Brain: A Trophic Theory of Neural Connections*. Cambridge, MA: Harvard University Press.

Purves, D., and Lichtman, J. W. (1985). *Principles of Neural Development*. Sunderland, MA: Sinauer.

Pylyshyn, Z. W. (2000). Situating vision in the world. *Trends in Cognitive Sciences*, 4:197–207.

Qin, Y.-L., McNaughton, B. L., Skaggs, W. E., and Barnes, C. A. (1997). Memory reprocessing in corticocortical and hippocampocortical neuronal ensembles. *Philosophical Transactions: Biological Sciences*, 352:1525–1533.

Raizada, R. D. S., and Grossberg, S. (2001). Context-sensitive binding by the laminar circuits of V1 and V2: A unified model of perceptual grouping, attention, and orientation contrast. *Visual Cognition*, 8:431–466.

Rakic, P. (1988). Specification of cerebral cortical areas. *Science*, 241:170–176.

Rall, W. (1962). Theory of physiological properties of dendrites. *Annals of the New York Academy of Sciences*, 96:1071–1092.

Rall, W. (1977). Core conductor theory and cable properties of neurons. In Kandel, E. R., Brookhart, J. M., and Mountcastle, V. B., editors, *The Handbook of Physiology, Section 1: The Nervous System, Vol. 1: Cellular Biology of Neurons*, 39–97. Bethesda, MD: American Physiological Society.

Rall, W., and Agmon-Snir, H. (1998). Cable theory for dendritic neurons. In Koch, C., and Segev, I., editors, *Methods in Neuronal Modeling: From Ions to Networks*, 27–92. Cambridge, MA: MIT Press. Second edition.

Ramoa, A. S., Mower, A. F., Liao, D., and Jafri, S. I. (2001). Suppression of cortical NMDA receptor function prevents development of orientation selectivity in the primary visual cortex. *The Journal of Neuroscience*, 21:4299–4309.

Rao, R. P. N., and Ballard, D. H. (1995). Natural basis functions and topographic memory for face recognition. In *Proceedings of the 14th International Joint Conference on Artificial Intelligence*, 10–17. San Francisco: Kaufmann.

Rao, R. P. N., and Ballard, D. H. (1997). Efficient encoding of natural time varying images produces oriented space-time receptive fields. Technical Report 97.4, Department of Computer Science, University of Rochester, Rochester, New York.

Rao, R. P. N., Olshausen, B. A., and Lewicki, M. S., editors (2002). *Probabilistic Models of the Brain: Perception and Neural Function*. Cambridge, MA: MIT Press.

Rao, S. C., Toth, L. J., and Sur, M. (1997). Optically imaged maps of orientation preference in primary visual cortex of cats and ferrets. *The Journal of Comparative Neurology*, 387:358–370.

Rechtschaffen, A. (1998). Current perspectives on the function of sleep. *Perspectives in Biology and Medicine*, 41:359–390.

Rector, D. M., Poe, G. R., Redgrave, P., and Harper, R. M. (1997). A miniature CCD video camera for high-sensitivity light measurements in freely behaving animals. *Journal of Neuroscience Methods*, 78:85–91.

Redies, C., Crook, J. M., and Creutzfeldt, O. D. (1986). Neuronal responses to borders with and without luminance gradients in cat visual cortex and dorsal lateral geniculate nucleus. *Experimental Brain Research*, 61:469–481.

Regehr, W. G., Delaney, K. R., and Tank, D. W. (1994). The role of presynaptic calcium in short-term enhancement at the hippocampal mossy fiber synapse. *The Journal of Neuroscience*, 14:523–537.

Regier, T. (1996). *The Human Semantic Potential: Spatial Language and Constrained Connectionism*. Cambridge, MA: MIT Press.

Rehn, M., and Lansner, A. (2004). Sequence memory with dynamical synapses. *Neurocomputing*, 58–60:271–278.

Reinagel, P., and Zador, A. M. (1999). Natural scene statistics at the center of gaze. *Network: Computation in Neural Systems*, 10:1–10.

Reitboeck, H. J., Stoecker, M., and Hahn, C. (1993). Object separation in dynamic neural networks. In *Proceedings of the IEEE International Conference on Neural Networks* (San Francisco, CA), 638–641. Piscataway, NJ: IEEE.

Rensink, R. A., and Enns, J. T. (1998). Early completion of occluded objects. *Vision Research*, 38:2489–2505.

Repp, B. H., and Penel, A. (2002). Auditory dominance in temporal processing: New evidence from synchronization with simultaneous visual and auditory sequences. *Journal of Experimental Psychology: Human Perception and Performance*, 28:1085–1099.

Revow, M., Williams, C. K. I., and Hinton, G. E. (1995). Using generative models for handwritten digit recognition. *IEEE Transactions on Pattern Analysis and Machine Intelligence*, 18:592–606.

Rieke, F., Warland, D., de Ruyter van Steveninck, R., and Bialek, W. (1997). *Spikes: Exploring the Neural Code*. Cambridge, MA: MIT Press.

Riesenhuber, M., Bauer, H.-U., Brockmann, D., and Geisel, T. (1998). Breaking rotational symmetry in a self-organizing map model for orientation map development. *Neural Computation*, 10:717–730.

Ringach, D. L. (2004). Mapping receptive fields in primary visual cortex. *The Journal of Physiology*, 558:717–728.

Rinzel, J., and Ermentrout, B. (1998). Analysis of neural excitability and oscillations. In Koch, C., and Segev, I., editors, *Methods in Neuronal Modeling: From Ions to Networks*, 251–291. Cambridge, MA: MIT Press. Second edition.

Ritter, H. (1991). Asymptotic level density for a class of vector quantization processes. *IEEE Transactions on Neural Networks*, 2:173–175.

Ritter, H., Martinetz, T., and Schulten, K. J. (1992). *Neural Computation and Self-Organizing Maps: An Introduction*. Reading, MA: Addison-Wesley.

Ritter, H., Obermayer, K., Schulten, K. J., and Rubner, J. (1991). Self-organizing maps and adaptive filters. In *Models of Neural Networks*, 281–306. Berlin: Springer.

Robert, A. (1999). Lamination and within-area integration in the neocortex. Doctoral dissertation, Department of Cognitive Science, University of California at San Diego, San Diego, CA.

Rochester, N., Holland, J. H., Haibt, L. H., and Duda, W. L. (1956). Tests on a cell assembly theory of the action of the brain, using a large digital computer. *IRE Transactions on Information Theory*, 2:80–93. Reprinted in Anderson and Rosenfeld (1988), 68–79.

Rockel, A. J., Hiorns, R. W., and Powell, T. P. S. (1980). The basic uniformity in structure of the neocortex. *Brain*, 103:221–244.

Rockland, K. S. (1985). Anatomical organization of primary visual cortex (area 17) in the ferret. *The Journal of Comparative Neurology*, 241:225–236.

Rockland, K. S., Lund, J. S., and Humphrey, A. L. (1982). Anatomical binding of intrinsic connections in striate cortex of tree shrews (*Tupaia glis*). *The Journal of Comparative Neurology*, 209:41–58.

Rodieck, R. W. (1965). Quantitative analysis of cat retinal ganglion cell response to visual stimuli. *Vision Research*, 5:583–601.

Rodman, H. R. (1994). Development of inferior temporal cortex in the monkey. *Cerebral Cortex*, 4:484–498.

Rodman, H. R., Skelly, J. P., and Gross, C. G. (1991). Stimulus selectivity and state dependence of activity in inferior temporal cortex of infant monkeys. *Proceedings of the National Academy of Sciences, USA*, 88:7572–7575.

Rodriques, J. S., and Almeida, L. B. (1990). Improving the learning speed in topological maps of patterns. In *Proceedings of the International Neural Networks Conference*, 813–816. Dordrecht, The Netherlands: Kluwer.

Roffwarg, H. P., Muzio, J. N., and Dement, W. C. (1966). Ontogenetic development of the human sleep-dream cycle. *Science*, 152:604–619.

Rojer, A. S., and Schwartz, E. L. (1990). Cat and monkey cortical columnar patterns modeled by bandpass-filtered 2D white noise. *Biological Cybernetics*, 62:381–391.

Rolls, E. T. (1990). The representation of information in the temporal lobe visual cortical areas of macaques. In Eckmiller, R., editor, *Advanced Neural Computers*, 69–78. Amsterdam: Elsevier.

Rolls, E. T. (1992). Neurophysiological mechanisms underlying face processing within and beyond the temporal cortical visual areas. *Philosophical Transactions: Biological Sciences*, 335:11–21.

Rolls, E. T. (2000). Functions of the primate temporal lobe cortical visual areas in invariant visual object and face recognition. *Neuron*, 27:205–218.

Rolls, E. T., Baylis, G. C., Hasselmo, M. E., and Nalwa, V. (1989). The effect of learning on the face selective responses of neurons in the cortex in the superior temporal sulcus of the monkey. *Experimental Brain Research*, 76:153–164.

Rolls, E. T., and Milward, T. (2000). A model of invariant object recognition in the visual system: Learning rules, activation functions, lateral inhibition, and information-based performance measures. *Neural Computation*, 12:2547–2572.

Roque Da Silva Filho, A. C. (1992). Investigation of a generalized version of Amari's continuous model for neural networks. Doctoral dissertation, School of Cognitive and Computing Sciences, University of Sussex, Brighton, UK.

Rosen, D. J., Rumelhart, D. E., and Knudsen, E. I. (1995). A connectionist model of the owl's sound localization system. In Tesauro, G., Touretzky, D. S., and Leen, T. K., editors, *Advances in Neural Information Processing Systems 7*, 606–613. Cambridge, MA: MIT Press.

Ross, W. D., Grossberg, S., and Mingolla, E. (2000). Visual cortical mechanisms of perceptual grouping: Interacting layers, networks, columns, and maps. *Neural Networks*, 13:571–588.

Rotenberg, V. S. (1992). Sleep and memory. I: The influence of different sleep stages on memory. *Neuroscience & Biobehavioral Reviews*, 16:497–502.

Roweis, S. T., and Saul, L. K. (2000). Nonlinear dimensionality reduction by locally linear embedding. *Science*, 290:2323–2326.

Rowley, H. A., Baluja, S., and Kanade, T. (1998). Neural network-based face detection. *IEEE Transactions on Pattern Analysis and Machine Intelligence*, 20:23–38.

Rubin, N., Nakayama, K., and Shapley, R. (1996). Enhanced perception of illusory contours in the lower versus upper visual hemifields. *Science*, 271:651–653.

Ruf, B., and Schmitt, M. (1998). Self-organization of spiking neurons using action potential timing. *IEEE Transactions on Neural Networks*, 9:575–578.

Rumelhart, D. E., Hinton, G. E., and Williams, R. J. (1986). Learning internal representations by error propagation. In Rumelhart, D. E., and McClelland, J. L., editors, *Parallel Distributed Processing: Explorations in the Microstructure of Cognition, Vol. 1: Foundations*, 318–362. Cambridge, MA: MIT Press.

Ruthazer, E. S., and Stryker, M. P. (1996). The role of activity in the development of long-range horizontal connections in area 17 of the ferret. *The Journal of Neuroscience*, 16:7253–7269.

Sabatini, S. P. (1996). Recurrent inhibition and clustered connectivity as a basis for Gabor-like receptive fields in the visual cortex. In Sirosh, J., Miikkulainen, R., and Choe, Y., editors, *Lateral Interactions in the Cortex: Structure and Function*. Austin, TX: The UTCS Neural Networks Research Group. Electronic book, ISBN 0-9647060-0-8, http://nn.cs.utexas.edu/web-pubs/htmlbook96.

Sabatini, S. P., Solari, F., and Secchi, L. (2004). A continuum-field model of visual cortex stimulus-driven behaviour: Emergent oscillations and coherence fields. *Neurocomputing*, 57:411–433.

Sackett, G. P. (1966). Monkeys reared in isolation with pictures as visual input: Evidence for an innate releasing mechanism. *Science*, 154:1468–1473.

Sackett, G. P. (1970). Unlearned responses, differential rearing, experiences, and the development of social attachments by rhesus monkeys. In Rosenblum, L. A., editor, *Primate Behavior: Developments in Field and Laboratory Research*, vol. 1, 111–140. San Diego, CA: Academic Press.

Saito, D. N., Okada, T., Morita, Y., Yonekura, Y., and Sadato, N. (2003). Tactile-visual cross-modal shape matching: A functional MRI study. *Cognitive Brain Research*, 17:14–25.

Sajda, P., and Finkel, L. H. (1992). A neural network model of object segmentation and feature binding in visual cortex. In *International Joint Conference on Neural Networks*, 43–48. Piscataway, NJ: IEEE.

Sakamoto, S. (2004). Synaptic weight normalization effects for topographic mapping formation. *Neural Networks*, 17:1109–1120.

Salzman, C. D., Britten, K. H., and Newsome, W. T. (1990). Cortical microstimulation influences perceptual judgements of motion direction. *Nature*, 346:174–177, Erratum 346:589.

Sanger, T. D. (1989). Optimal unsupervised learning in a single-layer linear feedforward neural network. *Neural Networks*, 2:459–473.

Saudargiene, A., Porr, B., and Wörgötter, F. (2004). How the shape of pre-and postsynaptic signals can influence STDP: A biophysical model. *Neural Computation*, 16:595–625.

Saul, A. B., and Humphrey, A. L. (1992). Evidence of input from lagged cells in the lateral geniculate nucleus to simple cells in cortical area 17 of the cat. *Journal of Neurophysiology*, 68:1190–1208.

Sceniak, M. P., Hawken, M. J., and Shapley, R. (2001). Visual spatial characterization of macaque V1 neurons. *Journal of Neurophysiology*, 85:1873–1887.

Schaffer, J. D., Whitley, D., and Eshelman, L. J. (1992). Combinations of genetic algorithms and neural networks: A survey of the state of the art. In Whitley, D., and Schaffer, J., editors, *Proceedings of the International Workshop on Combinations of Genetic Algorithms and Neural Networks*, 1–37. Los Alamitos, CA: IEEE Computer Society Press.

Schmid, L. M., Rosa, M. G. P., and Calford, M. B. (1995). Retinal detachment induces massive immediate reorganization in visual cortex. *Neuroreport*, 6:1349–1353.

Schmidt, K. E., Kim, D. S., Singer, W., Bonhoeffer, T., and Löwel, S. (1997). Functional specificity of long-range intrinsic and interhemispheric connections in the visual cortex of strabismic cats. *The Journal of Neuroscience*, 17:5480–5492.

Schrater, P. R., Knill, D. C., and Simoncelli, E. P. (2001). Perceiving visual expansion without optic flow. *Nature*, 410:816–819.

Schumann, F. (1900). Beiträge zur Analyse der Gesichtswahrnehmungen. Erste Abhandlung: Einige Beobachtungen über die Zusammenfassung von Gesichtseindrücken zu Einheiten [Contributions to the analysis of visual perceptions. First paper: Some observations on the grouping of visual impressions into wholes]. *Zeitschrift für Psychologie und Physiologie der Sinnesorgane*, 23:1–32.

Schwark, H. D., and Jones, E. G. (1989). The distribution of intrinsic cortical axons in area 3b of cat primary somatosensory cortex. *Experimental Brain Research*, 78:501–513.

Schyns, P. G., Goldstone, R. L., and Thibaut, J.-P. (1998). The development of features in object concepts. *Behavioral and Brain Sciences*, 21:1–54.

Sclar, G., and Freeman, R. D. (1982). Orientation selectivity in the cat's striate cortex is invariant with stimulus contrast. *Experimental Brain Research*, 46:457–461.

Sejnowski, T. J. (1995). Neural networks: Sleep and memory. *Current Biology*, 5:832–834.

Sejnowski, T. J., and Rosenberg, C. R. (1987). Parallel networks that learn to pronounce English text. *Complex Systems*, 1:145–168.

Sengpiel, F., and Kind, P. C. (2002). The role of activity in development of the visual system. *Current Biology*, 12:R818–R826.

Sengpiel, F., Stawinski, P., and Bonhoeffer, T. (1999). Influence of experience on orientation maps in cat visual cortex. *Nature Neuroscience*, 2:727–732.

Senn, W., Segev, I., and Tsodyks, M. (1998). Reading neuronal synchrony with depressing synapses. *Neural Computation*, 10:815–819.

Senseman, D. M. (1996). High-speed optical imaging of afferent flow through rat olfactory bulb slices: Voltage-sensitive dye signals reveal periglomerular cell activity. *The Journal of Neuroscience*, 16:313–324.

Sergent, J. (1989). Structural processing of faces. In Young, A. W., and Ellis, H. D., editors, *Handbook of Research on Face Processing*, 57–91. Amsterdam: Elsevier.

Seung, H. S., and Lee, D. D. (2000). The manifold ways of perception. *Science*, 290:2268–2269.

Seung, H. S., Lee, D. D., Reis, B. Y., and Tank, D. W. (2000). The autapse: A simple illustration of short-term analog memory storage by tuned synaptic feedback. *Journal of Computational Neuroscience*, 9:171–185.

Sharma, J., Angelucci, A., and Sur, M. (2000). Induction of visual orientation modules in auditory cortex. *Nature*, 404:841–847.

Shastri, L. (2002). Episodic memory and cortico-hippocampal interactions. *Trends in Cognitive Sciences*, 6:162–168.

Shatz, C. J. (1990). Impulse activity and the patterning of connections during CNS development. *Neuron*, 5:745–756.

Shatz, C. J. (1992). The developing brain. *Scientific American*, 267:61–67.

Shatz, C. J. (1996). Emergence of order in visual system development. *Proceedings of the National Academy of Sciences, USA*, 93:602–608.

Shatz, C. J., and Stryker, M. P. (1978). Ocular dominance in layer IV of the cat's visual cortex and the effects of monocular deprivation. *The Journal of Physiology*, 281:267–283.

Shepherd, G. M. (2003). *The Synaptic Organization of the Brain*. Oxford, UK: Oxford University Press. Fifth edition.

Sherman, S. M., and Guillery, R. W. (2001). *Exploring the Thalamus*. San Diego, CA: Academic Press.

Sheth, B. R., Sharma, J., Rao, S. C., and Sur, M. (1996). Orientation maps of subjective contours in visual cortex. *Science*, 274:2110–2115.

Shimojo, S., Kamitani, Y., and Nishida, S. (2001). Afterimage of perceptually filled-in surface. *Science*, 293:1677–1680.

Shipley, T. F., and Kellman, P. J. (1992). Strength of visual interpolation depends on the ratio of physically specified to total edge length. *Perception and Psychophysics*, 52:97–106.

Shipp, S., Blanton, M., and Zeki, S. (1998). A visuo-somatomotor pathway through superior parietal cortex in the macaque monkey: Cortical connections of areas V6 and V6A. *European Journal of Neuroscience*, 10:3171–3193.

Shiu, L.-P., and Pashler, H. (1992). Improvement in line orientation discrimination is retinally local but dependent on cognitive set. *Perception and Psychophysics*, 52:582–588.

Shmuel, A., and Grinvald, A. (1996). Functional organization for direction of motion and its relationship to orientation maps in cat area 18. *The Journal of Neuroscience*, 16:6945–6964.

Shouno, H., and Kurata, K. (2001). Formation of a direction map by projection learning using Kohonen's self-organization map. *Biological Cybernetics*, 85:241–246.

Shouval, H. Z. (1995). Formation and organization of receptive fields, with an input environment composed of natural scenes. Doctoral dissertation, Department of Physics, Brown University, Providence, RI.

Shouval, H. Z., Goldberg, D. H., Jones, J. P., Beckerman, M., and Cooper, L. N. (2000). Structured long-range connections can provide a scaffold for orientation maps. *The Journal of Neuroscience*, 20:1119–1128.

Shouval, H. Z., Intrator, N., and Cooper, L. N. (1997). BCM network develops orientation selectivity and ocular dominance in natural scene environment. *Vision Research*, 37:3339–3342.

Shouval, H. Z., Intrator, N., Law, C. C., and Cooper, L. N. (1996). Effect of binocular cortical misalignment on ocular dominance and orientation selectivity. *Neural Computation*, 8:1021–1040.

Siegel, J. M. (1999). The evolution of REM sleep. In Lydic, R., and Baghdoyan, H. A., editors, *Handbook of Behavioral State Control: Cellular and Molecular Mechanisms*, 87–100. Boca Raton, FL: CRC Press.

Sigman, M., Cecchi, G. A., Gilbert, C. D., and Magnasco, M. O. (2001). On a common circle: Natural scenes and gestalt rules. *Proceedings of the National Academy of Sciences, USA*, 98:1935–1940.

Sillito, A. M. (1979). Inhibitory mechanisms influencing complex cell orientation selectivity and their modification at high resting discharge. *The Journal of Physiology*, 289:33–53.

Sillito, A. M., Jones, H. E., Gerstein, G. L., and West, D. C. (1994). Feature-linked synchronization of thalamic relay cell firing induced by feedback from the visual cortex. *Nature*, 369:479–482.

Simion, F., Cassia, V. M., Turati, C., and Valenza, E. (2001). The origins of face perception: Specific versus non-specific mechanisms. *Infant and Child Development*, 10:59–66.

Simion, F., Valenza, E., and Umiltà, C. (1998a). Mechanisms underlying face preference at birth. In Simion, F., and Butterworth, G., editors, *The Development of Sensory, Motor and Cognitive Capacities in Early Infancy: From Perception to Cognition*, 87–102. East Sussex, UK: Psychology Press.

Simion, F., Valenza, E., Umiltà, C., and Dalla Barba, B. (1998b). Preferential orienting to faces in newborns: A temporal-nasal asymmetry. *Journal of Experimental Psychology: Human Perception and Performance*, 24:1399–1405.

Simoncelli, E. P., and Olshausen, B. A. (2001). Natural image statistics and neural representation. *Annual Review of Neuroscience*, 24:1193–1216.

Sincich, L. C., and Blasdel, G. G. (2001). Oriented axon projections in primary visual cortex of the monkey. *The Journal of Neuroscience*, 21:4416–4426.

Singer, W. (1993). Synchronization of cortical activity and its putative role in information processing and learning. *Annual Review of Physiology*, 55:349–374.

Singer, W. (1999). Neuronal synchrony: A versatile code for the definition of relations? *Neuron*, 24:49–65.

Singer, W., and Gray, C. M. (1995). Visual feature integration and the temporal correlation hypothesis. *Annual Review of Neuroscience*, 18:555–586.

Singer, W., Gray, C. M., Engel, A. K., König, P., Artola, A., and Bröcher, S. (1990). Formation of cortical cell assemblies. In *The Brain*, vol. LV of *Cold Spring Harbor Symposia on Quantitative Biology*, 939–952. Cold Spring Harbor, NY: Cold Spring Harbor Laboratory Press.

Sirosh, J. (1995). A self-organizing neural network model of the primary visual cortex. Doctoral dissertation, Department of Computer Sciences, The University of Texas at Austin, Austin, TX. Technical Report AI95-237.

Sirosh, J., and Miikkulainen, R. (1993). How lateral interaction develops in a self-organizing feature map. In *Proceedings of the IEEE International Conference on Neural Networks* (San Francisco, CA), 1360–1365. Piscataway, NJ: IEEE.

Sirosh, J., and Miikkulainen, R. (1994a). Cooperative self-organization of afferent and lateral connections in cortical maps. *Biological Cybernetics*, 71:66–78.

Sirosh, J., and Miikkulainen, R. (1994b). Modeling cortical plasticity based on adapting lateral interaction. In Bower, J. M., editor, *The Neurobiology of Computation: The Proceedings of the Third Annual Computation and Neural Systems Conference*, 305–310. Dordrecht, The Netherlands: Kluwer.

Sirosh, J., and Miikkulainen, R. (1996a). A neural network model of topographic reorganization following cortical lesions. In *Computational Medicine, Public Health and Biotechnology: Building a Man in the Machine. Proceedings of the First World Congress*, vol. 5 of *Mathematical Biology and Medicine*. Singapore: World Scientific.

Sirosh, J., and Miikkulainen, R. (1996b). Self-organization and functional role of lateral connections and multisize receptive fields in the primary visual cortex. *Neural Processing Letters*, 3:39–48.

Sirosh, J., and Miikkulainen, R. (1997). Topographic receptive fields and patterned lateral interaction in a self-organizing model of the primary visual cortex. *Neural Computation*, 9:577–594.

Sirosh, J., Miikkulainen, R., and Bednar, J. A. (1996a). Self-organization of orientation maps, lateral connections, and dynamic receptive fields in the primary visual cortex. In Sirosh, J., Miikkulainen, R., and Choe, Y., editors, *Lateral Interactions in the Cortex: Structure and Function*. Austin, TX: The UTCS Neural Networks Research Group. Electronic book, ISBN 0-9647060-0-8, http://nn.cs.utexas.edu/web-pubs/htmlbook96.

Sirosh, J., Miikkulainen, R., and Choe, Y., editors (1996b). *Lateral Interactions in the Cortex: Structure and Function*. Austin, TX: The UTCS Neural Networks Research Group. Electronic book, ISBN 0-9647060-0-8, http://nn.cs.utexas.edu/web-pubs/htmlbook96.

Slater, A. (1993). Visual perceptual abilities at birth: Implications for face perception. In de Boysson-Bardies, B., editor, *Developmental Neurocognition: Speech and Face Processing in the First Year of Life*, 125–134. Dordrecht, The Netherlands: Kluwer.

Slater, A., and Johnson, S. P. (1998). Visual sensory and perceptual abilities of the newborn: Beyond the blooming, buzzing confusion. In Simion, F., and Butterworth, G., editors, *The Development of Sensory, Motor and Cognitive Capacities in Early Infancy: From Perception to Cognition*, 121–142. East Sussex, UK: Psychology Press.

Slater, A., and Kirby, R. (1998). Innate and learned perceptual abilities in the newborn infant. *Experimental Brain Research*, 123:90–94.

Slater, A., Morison, V., and Somers, M. (1988). Orientation discrimination and cortical function in the human newborn. *Perception*, 17:597–602.

Smith, A. T., and Over, R. (1977). Orientation masking and the tilt illusion with subjective contours. *Perception*, 6:441–447.

Smith, C. (1996). Sleep states, memory processes and synaptic plasticity. *Behavioural Brain Research*, 78:49–56.

Snippe, H. P. (1996). Parameter extraction from population codes: A critical assessment. *Neural Computation*, 8:511–529.

Sober, S. J., Stark, J. M., Yamasaki, D. S., and Lytton, W. W. (1997). Receptive field changes after strokelike cortical ablation: A role for activation dynamics. *Journal of Neurophysiology*, 78:3438–3443.

Sohn, J.-W., Zhang, B.-T., and Kaang, B.-K. (1999). Temporal pattern recognition using a spiking neural network with delays. In *Proceedings of the International Joint Conference on Neural Networks*, 2590–2593. Piscataway, NJ: IEEE.

Somers, D. C., Toth, L. J., Todorov, E., Rao, S. C., Kim, D.-S., Nelson, S. B., Siapas, A. G., and Sur, M. (1996). Variable gain control in local cortical circuitry supports context-dependent modulation by long-range connections. In Sirosh, J., Miikkulainen, R., and Choe, Y., editors, *Lateral Interactions in the Cortex: Structure and Function*. Austin, TX: The UTCS Neural Networks Research Group. Electronic book, ISBN 0-9647060-0-8, http://nn.cs.utexas.edu/web-pubs/htmlbook96.

Song, S., Miller, K. D., and Abbott, L. F. (2000). Competitive Hebbian learning through spike-timing-dependent synaptic plasticity. *Nature Neuroscience*, 3:919–926.

Sporns, O., Tononi, G., and Edelman, G. E. (1991). Modeling perceptual grouping and figure-ground segregation by means of active reentrant connections. *Proceedings of the National Academy of Sciences, USA*, 88:129–33.

Stanley, K. O., and Miikkulainen, R. (2002). Evolving neural networks through augmenting topologies. *Evolutionary Computation*, 10:99–127.

Stein, B. E., and Meredith, M. A. (1993). *The Merging of the Senses*. Cambridge, MA: MIT Press.

Stein, B. E., Meredith, M. A., Huneycutt, W. S., and McDade, L. (1989). Behavioral indices of multisensory integration: Orientation to visual cues is affected by auditory stimuli. *The Journal of Cognitive Neuroscience*, 1:12–24.

Stellwagen, D., and Shatz, C. J. (2002). An instructive role for retinal waves in the development of retinogeniculate connectivity. *Neuron*, 33:357–367.

Stemmler, M., Usher, M., and Niebur, E. (1995). Lateral interactions in primary visual cortex: A model bridging physiology and psychophysics. *Science*, 269:1877–1880.

Steriade, M., Paré, D., Bouhassira, D., Deschênes, M., and Oakson, G. (1989). Phasic activation of lateral geniculate and perigeniculate thalamic neurons during sleep with ponto-geniculo-occipital waves. *The Journal of Neuroscience*, 9:2215–2229.

Stetter, M., Müller, A., and Lang, E. W. (1994). Neural network model for the coordinated formation of orientation preference and orientation selectivity maps. *Physical Review E*, 50:4167–4181.

Stettler, D. D., Das, A., Bennett, J., and Gilbert, C. D. (2002). Lateral connectivity and contextual interactions in macaque primary visual cortex. *Neuron*, 36:739–750.

Stevens, B., Tanner, S., and Fields, R. D. (1998). Control of myelination by specific patterns of neural impulses. *The Journal of Neuroscience*, 18:9303–9311.

Strasburger, H., and Rentschler, I. (1996). Contrast-dependent dissociation of visual recognition and detection fields. *European Journal of Neuroscience*, 8:1787–1791.

Stringer, S. M., and Rolls, E. T. (2002). Invariant object recognition in the visual system with novel views of 3D objects. *Neural Computation*, 14:2585–2596.

Suenaga, H., and Ishikawa, M. (2000). Self-organizing map with a variable-size competitive layer. In Lee, S.-Y., editor, *Proceedings of the Seventh International Conference on Neural Information Processing*, 727–731.

Sulston, J. E., and Horvitz, H. R. (1977). Post-embryonic cell lineages of the nematode, *Caenorhabditis elegans*. *Developmental Biology*, 56:110–156.

Sur, M., Angelucci, A., and Sharma, J. (1999). Rewiring cortex: The role of patterned activity in development and plasticity of neocortical circuits. *Journal of Neurobiology*, 41:33–43.

Sur, M., Garraghty, P. E., and Roe, A. W. (1988). Experimentally induced visual projections in auditory thalamus and cortex. *Science*, 242:1437–1441.

Sur, M., and Leamey, C. A. (2001). Development and plasticity of cortical areas and networks. *Nature Reviews Neuroscience*, 2:251–262.

Sutherland, N. S. (1961). Figural after-effects and apparent size. *Quarterly Journal of Psychology*, 13:222–228.

Sutor, B., and Luhmann, H. J. (1995). Development of excitatory and inhibitory postsynaptic potentials in the rat neocortex. *Perspectives on Developmental Neurobiology*, 2:409–419.

Sutton, G. G., Reggia, J. A., Armentrout, S. L., and D'Autrechy, C. L. (1994). Cortical map reorganization as a competitive process. *Neural Computation*, 6:1–13.

Swindale, N. V. (1980). A model for the formation of ocular dominance stripes. *Proceedings of the Royal Society of London. Series B, Biological Sciences*, 215:243–264.

Swindale, N. V. (1992). A model for the coordinated development of columnar systems in primate striate cortex. *Biological Cybernetics*, 66:217–230.

Swindale, N. V. (1996). The development of topography in the visual cortex: A review of models. *Network: Computation in Neural Systems*, 7:161–247.

Switkes, E., Mayer, M. J., and Sloan, J. A. (1978). Spatial frequency analysis of the visual environment: Anisotropy and the carpentered environment hypothesis. *Vision Research*, 18:1393–1399.

Tanaka, S. (1990). Theory of self-organization of cortical maps: Mathematical framework. *Neural Networks*, 3:625–640.

Tarr, M. J., and Gauthier, I. (2000). FFA: A flexible fusiform area for subordinate-level visual processing automatized by expertise. *Nature Neuroscience*, 3:764–769.

Tavazoie, S. F., and Reid, R. C. (2000). Diverse receptive fields in the lateral geniculate nucleus during thalamocortical development. *Nature Neuroscience*, 3:608–616.

Taylor, J. G., and Alavi, F. N. (1996). A basis for long-range inhibition across cortex. In Sirosh, J., Miikkulainen, R., and Choe, Y., editors, *Lateral Interactions in the Cortex: Structure and Function*. Austin, TX: The UTCS Neural Networks Research Group. Electronic book, ISBN 0-9647060-0-8, http://nn.cs.utexas.edu/web-pubs/htmlbook96.

Tenenbaum, J. B., de Silva, V., and Langford, J. C. (2000). A global geometric framework for nonlinear dimensionality reduction. *Science*, 290:2319–2323.

Terman, D., and Wang, D. (1995). Global competition and local cooperation in a network of neural oscillators. *Physica D*, 81:148–176.

Thomas, H. (1965). Visual-fixation responses of infants to stimuli of varying complexity. *Child Development*, 36:629–638.

Thompson, E., and Varela, F. J. (2001). Radical embodiment: Neural dynamics and consciousness. *Trends in Cognitive Sciences*, 5:418–425.

Thompson, I. (1997). Cortical development: A role for spontaneous activity? *Current Biology*, 7:R324–R326.

Thomson, A. M., and Deuchars, J. (1994). Temporal and spatial properties of local circuits in neocortex. *Trends in Neurosciences*, 17:119–126.

Tiňo, P., and Nabney, I. (2002). Hierarchical GTM: Constructing localized non-linear projection manifolds in a principled way. *IEEE Transactions on Pattern Analysis and Machine Intelligence*, 24:639–659.

Tolhurst, D. J., and Thompson, P. G. (1975). Orientation illusions and aftereffects: Inhibition between channels. *Vision Research*, 15:967–972.

Tonkes, B., Blair, A. D., and Wiles, J. (2000). Evolving learnable languages. In Solla, S. A., Leen, T. K., and Muller, K.-R., editors, *Advances in Neural Information Processing Systems 12*, 66–72. Cambridge, MA: MIT Press.

Touretzky, D. S. (2002). The rodent navigation circuit. In Sharp, P. E., editor, *The Neural Basis of Navigation: Evidence from Single Cell Recording*, 217–233. Dordrecht, The Netherlands: Kluwer.

Tovée, M. J. (1998). Face processing: Getting by with a little help from its friends. *Current Biology*, 8:R317–R320.

Trappenberg, T. P. (2002). *Fundamentals of Computational Neuroscience*. Oxford, UK: Oxford University Press.

Treves, A. (1997). On the perceptual structure of face space. *Biosystems*, 40:189–196.

Troyer, T. W., Krukowski, A. E., Priebe, N. J., and Miller, K. D. (1998). Contrast-invariant orientation tuning in cat visual cortex: Thalamocortical input tuning and correlation-based intracortical connectivity. *The Journal of Neuroscience*, 18:5908–5927.

Tsien, J. Z. (2000). Linking Hebb's coincidence-detection to memory formation. *Current Opinion in Neurobiology*, 10:266–273.

Ts'o, D. Y., Frostig, R. D., Lieke, E. E., and Grinvald, A. (1990). Functional organization of primate visual cortex revealed by high resolution optical imaging. *Science*, 249:417–420.

Ts'o, D. Y., Roe, A. W., and Gilbert, C. D. (2001). A hierarchy of the functional organization for color, form and disparity in primate visual area V2. *Vision Research*, 41:1333–1349.

Tsodyks, M., Pawelzik, K., and Markram, H. (1998). Neural networks with dynamic synapses. *Neural Computation*, 10:821–835.

Turney, P. D. (1996). How to shift bias: Lessons from the Baldwin effect. *Evolutionary Computation*, 4:271–295.

Turrigiano, G. G. (1999). Homeostatic plasticity in neuronal networks: The more things change, the more they stay the same. *Trends in Neurosciences*, 22:221–227.

Turrigiano, G. G., Leslie, K. R., Desai, N. S., Rutherford, L. C., and Nelson, S. B. (1998). Activity-dependent scaling of quantal amplitude in neocortical neurons. *Nature*, 391:892–896.

Tversky, T., Geisler, W. S., and Perry, J. S. (2004). Contour grouping: Closure effects are explained by good continuation and proximity. *Vision Research*, 44:2769–2777.

Tversky, T., and Miikkulainen, R. (2002). Modeling directional selectivity using self-organized delay-adaptation maps. *Neurocomputing*, 44–46:679–684.

Ullman, S. (1976). Filling-in the gaps: The shape of subjective contours and a model for their generation. *Biological Cybernetics*, 25:1–6.

Usher, M., and Donnelly, N. (1998). Visual synchrony affects binding and segmentation in perception. *Nature*, 394:179–182.

Usher, M., Stemmler, M., and Niebur, E. (1996). The role of lateral connections in visual cortex: Dynamics and information processing. In Sirosh, J., Miikkulainen, R., and Choe, Y., editors, *Lateral Interactions in the Cortex: Structure and Function*. Austin, TX: The UTCS Neural Networks Research Group. Electronic book, ISBN 0-9647060-0-8, http://nn.cs.utexas.edu/web-pubs/htmlbook96.

Utgoff, P., and Mitchell, T. (1982). Acquisition of appropriate bias for inductive concept learning. In *Proceedings of the Second National Conference on Artificial Intelligence*, 414–417. Menlo Park, CA: AAAI Press.

Vaadia, E., Haalman, I., Abeles, M., Bergman, H., Prut, Y., Slovin, H., and Aertsen, A. (1995). Dynamics of neuronal interactions in monkey cortex in relation to behavioral events. *Nature*, 373:515–518.

Vaitkevicius, H., Karalius, M., Meskauskas, A., Sinius, J., and Sokolov, E. (1983). A model for the monocular line orientation analyzer. *Biological Cybernetics*, 48:139–147.

Valentin, D., Abdi, H., O'Toole, A. J., and Cottrell, G. W. (1994). Connectionist models of face processing: A survey. *Pattern Recognition*, 27:1209–1230.

Valenza, E., Simion, F., Cassia, V. M., and Umiltà, C. (1996). Face preference at birth. *Journal of Experimental Psychology: Human Perception and Performance*, 22:892–903.

Valiant, L. G. (1994). *Circuits of the Mind*. Oxford, UK: Oxford University Press.

van der Zwan, R., and Wenderoth, P. (1994). Psychophysical evidence for area V2 involvement in the reduction of subjective contour tilt aftereffects by binocular rivalry. *Visual Neuroscience*, 11:823–830.

van der Zwan, R., and Wenderoth, P. (1995). Mechanisms of purely subjective contour tilt aftereffects. *Vision Research*, 35:2547–2557.

Van Essen, D. C. (2003). Organization of visual areas in macaque and human cerebral cortex. In Chalupa, L. M., and Werner, J. S., editors, *The Visual Neurosciences*, 507–521. Cambridge, MA: MIT Press.

Van Essen, D. C. (2004). Surface-based approaches to spatial localization and registration in primate cerebral cortex. *Neuroimage*, 23(Suppl.):S97–S107.

Van Essen, D. C., Anderson, C. H., and Felleman, D. J. (1992). Information processing in the primate visual system: An integrated systems perspective. *Science*, 255:419–423.

Van Horn, J. D., Grafton, S. T., Rockmore, D., and Gazzaniga, M. S. (2004). Sharing neuroimaging studies of human cognition. *Nature Neuroscience*, 7:473–481.

van Vreeswijk, C., and Abbott, L. F. (1994). When inhibition not excitation synchronizes neural firing. *Journal of Computational Neuroscience*, 1:313–321.

VanRullen, R., Delorme, A., and Thorpe, S. J. (2001). Feed-forward contour integration in primary visual cortex based on asynchronous spike propagation. *Neurocomputing*, 38–40:1003–1009.

Vapnik, V., and Chervonenkis, A. (1971). On the uniform convergence of relative frequencies of events to their probabilities. *Theory of Probability and its Applications*, 16:264–280.

Venter, J. C., et al. (2001). The sequence of the human genome. *Science*, 291:1304–1351.

Vidyasagar, T. R. (1990). Pattern adaptation in cat visual cortex is a co-operative phenomenon. *Neuroscience*, 36:175–179.

Vidyasagar, T. R., and Mueller, A. (1994). Function of GABA inhibition in specifying spatial frequency and orientation selectivities in cat striate cortex. *Experimental Brain Research*, 98:31–38.

Viola, P., and Jones, M. (2004). Robust real-time object detection. *International Journal of Computer Vision*, 57:137–154.

von der Heydt, R., and Peterhans, E. (1989). Mechanisms of contour perception in monkey visual cortex. I. Lines of pattern discontinuity. *The Journal of Neuroscience*, 9:1731–1748.

von der Malsburg, C. (1973). Self-organization of orientation-sensitive cells in the striate cortex. *Kybernetik*, 15:85–100. Reprinted in Anderson and Rosenfeld (1988), 212–227.

von der Malsburg, C. (1981). The correlation theory of brain function. Internal Report 81-2, Department of Neurobiology, Max-Planck-Institute for Biophysical Chemistry, Göttingen, Germany.

von der Malsburg, C. (1986a). Am I thinking assemblies? In Palm, G., and Aertsen, A., editors, *Brain Theory: Proceedings of the First Trieste Meeting on Brain Theory*, 161–176. Berlin: Springer.

von der Malsburg, C. (1986b). A neural cocktail-party processor. *Biological Cybernetics*, 54:29–40.

von der Malsburg, C. (1987). Synaptic plasticity as basis of brain organization. In Changeux, J.-P., and Konishi, M., editors, *The Neural and Molecular Bases of Learning*, 411–432. Hoboken, NJ: Wiley.

von der Malsburg, C. (1999). The what and why of binding: The modeler's perspective. *Neuron*, 24:95–104.

von der Malsburg, C. (2003). Dynamic link architecture. In Arbib, M. A., editor, *The Handbook of Brain Theory and Neural Networks*, 365–368. Cambridge, MA: MIT Press. Second edition.

von der Malsburg, C., and Buhmann, J. (1992). Sensory segmentation with coupled neural oscillators. *Biological Cybernetics*, 67:233–242.

von der Malsburg, C., and Singer, W. (1988). Principles of cortical network organization. In Rakic, P., and Singer, W., editors, *Neurobiology of Neocortex*, 69–99. Hoboken, NJ: Wiley.

von der Malsburg, C., and Willshaw, D. J. (1977). How to label nerve cells so that they can interconnect in an ordered fashion. *Proceedings of the National Academy of Sciences, USA*, 74:5176–5178.

von Melchner, L., Pallas, S. L., and Sur, M. (2000). Visual behaviour mediated by retinal projections directed to the auditory pathway. *Nature*, 404:871–876.

Waleszczyk, W. J., Wang, C., Young, J. M., Burke, W., Calford, M. B., and Dreher, B. (2003). Laminar differences in plasticity in area 17 following retinal lesions in kittens or adult cats. *European Journal of Neuroscience*, 17:2351–2368.

Wallace, M. T., McHaffie, J. G., and Stein, B. E. (1997). Visual response properties and visuotopic representation in the newborn monkey superior colliculus. *Journal of Neurophysiology*, 78:2732–2741.

Wallis, G. M. (1994). Neural mechanisms underlying processing in the visual areas of the occipital and temporal lobes. Doctoral dissertation, Corpus Christi College, Oxford University, Oxford, UK.

Wallis, G. M., and Rolls, E. T. (1997). Invariant face and object recognition in the visual system. *Progress in Neurobiology*, 51:167–194.

Walsh, V., and Cowey, A. (2000). Transcranial magnetic stimulation and cognitive neuroscience. *Nature Reviews Neuroscience*, 1:73–79.

Walton, G. E., Armstrong, E. S., and Bower, T. G. R. (1997). Faces as forms in the world of the newborn. *Infant Behavior and Development*, 20:537–543.

Walton, G. E., and Bower, T. G. R. (1993). Newborns form "prototypes" in less than 1 minute. *Psychological Science*, 4:203–205.

Wandell, B. A. (1995). *Foundations of Vision*. Sunderland, MA: Sinauer.

Wang, D. (1995). Emergent synchrony in locally coupled neural oscillators. *IEEE Transactions on Neural Networks*, 6:941–948.

Wang, D. (1996). Synchronous oscillations based on lateral connections. In Sirosh, J., Miikkulainen, R., and Choe, Y., editors, *Lateral Interactions in the Cortex: Structure and Function*. Austin, TX: The UTCS Neural Networks Research Group. Electronic book, ISBN 0-9647060-0-8, http://nn.cs.utexas.edu/web-pubs/htmlbook96.

Wang, D. (1999). Relaxation oscillators and networks. In Webster, J. G., editor, *Wiley Encyclopedia of Electrical and Electronics Engineering*, 396–405. Hoboken, NJ: Wiley.

Wang, D. (2000). On connectedness: A solution based on oscillatory correlation. *Neural Computation*, 12:131–139.

Wang, D., and Brown, G. J. (1999). Separation of speech from interfering sounds based on oscillatory correlation. *IEEE Transactions on Neural Networks*, 10:684–697.

Wang, G., Tanaka, K., and Tanifuji, M. (1996). Optical imaging of functional organization in the monkey inferotemporal cortex. *Science*, 272:1665–1668.

Wang, X.-J. (2001). Synaptic reverberation underlying mnemonic persistent activity. *Trends in Neurosciences*, 24:455–463.

Ware, C., and Mitchell, D. E. (1974). The spatial selectivity of the tilt aftereffect. *Vision Research*, 14:735–737.

Watt, R. J., and Phillips, W. A. (2000). The function of dynamic grouping in vision. *Trends in Cognitive Sciences*, 4:447–454.

Weber, C. (2001). Self-organization of orientation maps, lateral connections, and dynamic receptive fields in the primary visual cortex. In *Proceedings of the International Conference on Artificial Neural Networks*, Lecture Notes in Computer Science 2130, 1147–1152. Berlin: Springer.

Webster, M. A., and MacLin, O. H. (1999). Figural aftereffects in the perception of faces. *Psychonomic Bulletin and Review*, 6:647–653.

Wehrhahn, C., and Westheimer, G. (1993). Temporal asynchrony interferes with vernier acuity. *Visual Neuroscience*, 10:13–19.

Weiss, Y., Edelman, S., and Fahle, M. (1993). Models of perceptual learning in vernier hyperacuity. *Neural Computation*, 5:695–718.

Weitzel, L., Kopecz, K., Spengler, C., Eckhorn, R., and Reitboeck, H. J. (1997). Contour segmentation with recurrent neural networks of pulse-coding neurons. In *Proceedings of the 7th International Conference on Computer Analysis of Images and Patterns*, 337–344. Berlin: Springer.

Weliky, M., Bosking, W. H., and Fitzpatrick, D. (1996). A systematic map of direction preference in primary visual cortex. *Nature*, 379:725–728.

Weliky, M., Kandler, K., Fitzpatrick, D., and Katz, L. C. (1995). Patterns of excitation and inhibition evoked by horizontal connections in visual cortex share a common relationship to orientation columns. *Neuron*, 15:541–552.

Weliky, M., and Katz, L. C. (1997). Disruption of orientation tuning in visual cortex by artificially correlated neuronal activity. *Nature*, 386:680–685.

Wenderoth, P., and Johnstone, S. (1988). The different mechanisms of the direct and indirect tilt illusions. *Vision Research*, 28:301–312.

Weng, J., McClelland, J. L., Pentland, A., Sporns, O., Stockman, I., Sur, M., and Thelen, E. (2001). Autonomous mental development by robots and animals. *Science*, 291:599–600.

Werbos, P. J. (1974). Beyond regression: New tools for prediction and analysis in the behavioral sciences. Doctoral dissertation, Department of Applied Mathematics, Harvard University, Cambridge, MA.

Wersing, H., Steil, J. J., and Ritter, H. (2001). A competitive layer model for feature binding and segmentation. *Neural Computation*, 13:357–387.

Westheimer, G. (1990). Simultaneous orientation contrast for lines in the human fovea. *Vision Research*, 30:1913–1921.

White, E. L. (1989). *Cortical Circuits: Synaptic Organization of the Cerebral Cortex — Structure, Function, and Theory*. Basel, Switzerland: Birkhäuser.

White, L. E., Bosking, W. H., Weliky, M., and Fitzpatrick, D. (1996). Direction selectivity and horizontal connections in layers 2/3 of ferret primary visual cortex (V1). In *Society for Neuroscience Abstracts*, vol. 22, 1610. Washington, DC: Society for Neuroscience.

White, L. E., Bosking, W. H., Williams, S. M., and Fitzpatrick, D. (1999). Maps of central visual space in ferret V1 and V2 lack matching inputs from the two eyes. *The Journal of Neuroscience*, 19:7089–7099.

White, L. E., Coppola, D. M., and Fitzpatrick, D. (2001). The contribution of sensory experience to the maturation of orientation selectivity in ferret visual cortex. *Nature*, 411:1049–1052.

White, L. E., and Fitzpatrick, D. (2003). Dark-rearing prevents the development of direction selectivity in ferret visual cortex. In *Society for Neuroscience Abstracts*, Program No. 567.12. Washington, DC: Society for Neuroscience.

Widrow, B., and Hoff, M. E. (1960). Adaptive switching circuits. In *1960 IRE WESCON Convention Record*, Part 4, 96–104. New York: IRE. Reprinted in Anderson and Rosenfeld (1988), 126–134.

Wiesel, T. N. (1982). Postnatal development of the visual cortex and the influence of the environment. *Nature*, 299:583–591.

Wilkinson, R. A., Garris, M. D., and Geist, J. (1993). Machine-assisted human classification of segmented characters for OCR testing and training. In D'Amato, D. P., editor, *Character Recognition Technologies, Proceedings of SPIE 1906*, 208–217. Bellingham, WA: SPIE.

Willmore, B., and Tolhurst, D. J. (2001). Characterizing the sparseness of neural codes. *Network: Computation in Neural Systems*, 12:255–270.

Willshaw, D. J., and von der Malsburg, C. (1976). How patterned neural connections can be set up by self-organization. *Proceedings of the Royal Society of London. Series B, Biological Sciences*, 194:431–445.

Willshaw, D. J., and von der Malsburg, C. (1979). A marker induction mechanism for the establishment of ordered neural mappings: Its application to the retinotectal problem. *Philosophical Transactions of the Royal Society of London. Series B, Biological Sciences*, 287:203–243.

Wilson, H. R., and Cowan, J. D. (1972). Excitatory and inhibitory interactions in localized populations of model neurons. *Biophysical Journal*, 12:1–24.

Wilson, H. R., and Humanski, R. (1993). Spatial frequency adaptation and contrast gain control. *Vision Research*, 33:1133–1149.

Wilson, M. A., and McNaughton, B. L. (1994). Reactivation of hippocampal ensemble memories during sleep. *Science*, 265:676–679.

Wimbauer, S., Wenisch, O. G., Miller, K. D., and van Hemmen, J. L. (1997a). Development of spatiotemporal receptive fields of simple cells: I. Model formulation. *Biological Cybernetics*, 77:453–461.

Wimbauer, S., Wenisch, O. G., van Hemmen, J. L., and Miller, K. D. (1997b). Development of spatiotemporal receptive fields of simple cells: II. Simulation and analysis. *Biological Cybernetics*, 77:463–477.

Wiskott, L., and Sejnowski, T. J. (1998). Constrained optimization for neural map formation: A unifying framework for weight growth and normalization. *Neural Computation*, 10:671–716.

Wiskott, L., and von der Malsburg, C. (1996). Face recognition by dynamic link matching. In Sirosh, J., Miikkulainen, R., and Choe, Y., editors, *Lateral Interactions in the Cortex: Structure and Function*. Austin, TX: The UTCS Neural Networks Research Group. Electronic book, ISBN 0-9647060-0-8, http://nn.cs.utexas.edu/web-pubs/htmlbook96.

Wolfe, J., and Palmer, L. A. (1998). Temporal diversity in the lateral geniculate nucleus of cat. *Visual Neuroscience*, 15:653–675.

Wolfe, J. M. (1984). Short test flashes produce large tilt aftereffects. *Vision Research*, 24:1959–1964.

Wong, R. O. L. (1999). Retinal waves and visual system development. *Annual Review of Neuroscience*, 22:29–47.

Wong, R. O. L., Meister, M., and Shatz, C. J. (1993). Transient period of correlated bursting activity during development of the mammalian retina. *Neuron*, 11:923–938.

Wong, S. T., and Koslow, S. H. (2001). Human brain program research progress in bioinformatics/neuroinformatics. *Journal of the American Medical Informatics Association*, 8:103–104.

Wu, S., Amari, S., and Nakahara, H. (2002). Population coding and decoding in a neural field: A computational study. *Neural Computation*, 14:999–1026.

Wurtz, R. H., Yamasaki, D. S., Duffy, C. J., and Roy, J. P. (1990). Functional specialization for visual motion processing in primate cerebral cortex. In *The Brain*, vol. LV of *Cold Spring Harbor Symposia on Quantitative Biology*, 717–727. Cold Spring Harbor, NY: Cold Spring Harbor Laboratory Press.

Xiao, Y., Wang, Y., and Felleman, D. J. (2003). A spatially organized representation of color in macaque cortical area V2. *Nature*, 421:535–539.

Xu, L., and Oja, E. (1990). Adding top-down expectation into the learning procedure of self-organizing maps. In *International Joint Conference on Neural Networks* (Washington, DC), vol. II, 531–534. Hillsdale, NJ: Erlbaum.

Yaeger, L. S., Webb, B. J., and Lyon, R. F. (1998). Combining neural networks and context-driven search for online, printed handwriting recognition in the NEWTON. *AI Magazine*, 19:73–89.

Yang, M.-H., Kriegman, D., and Ahuja, N. (2002). Detecting faces in images: A survey. *IEEE Transactions on Pattern Analysis and Machine Intelligence*, 24:34–58.

Yao, X. (1999). Evolving artificial neural networks. *Proceedings of the IEEE*, 87:1423–1447.

Yen, S.-C., and Finkel, L. H. (1997). Identification of salient contours in cluttered images. In *Proceedings of IEEE Computer Society Conference on Computer Vision and Pattern Recognition*, 273–279. Los Alamitos, CA: IEEE Computer Society Press.

Yen, S.-C., and Finkel, L. H. (1998). Extraction of perceptually salient contours by striate cortical networks. *Vision Research*, 38:719–741.

Yilmaz, A., and Shah, M. (2002). Automatic feature detection and pose recovery for faces. In *Proceedings of the Fifth Asian Conference on Computer Vision*, 284–289.

Yu, Y., and Choe, Y. (2004). Angular disinhibition effect in a modified Poggendorff illusion. In Forbus, K. D., Gentner, D., and Regier, T., editors, *Proceedings of the 26th Annual Conference of the Cognitive Science Society*, 1500–1505. Hillsdale, NJ: Erlbaum.

Yu, Y., Yamauchi, T., and Choe, Y. (2004). Explaining low-level brightness-contrast illusions using disinhibition. In *Biologically Inspired Approaches to Advanced Information Technology*, Lecture Notes in Computer Science 3141, 166–175. Berlin: Springer.

Yuille, A. L., Kammen, D. M., and Cohen, D. S. (1989). Quadrature and the development of orientation selective cortical cells by Hebb rules. *Biological Cybernetics*, 61:183–194.

Yuille, A. L., Kolodny, J. A., and Lee, C. W. (1996). Dimension reduction, generalized deformable models and the development of ocularity and orientation. *Neural Networks*, 9:309–319.

Yuste, R., Nelson, D. A., Rubin, W. W., and Katz, L. C. (1995). Neuronal domains in developing neocortex: Mechanisms of coactivation. *Neuron*, 14:7–17.

Zador, A. M., and Pearlmutter, B. A. (1996). VC dimension of an integrate-and-fire neuron model. *Neural Computation*, 8:611–624.

Zemel, R. S., Dayan, P., and Pouget, A. (1998). Probabilistic interpretation of population codes. *Neural Computation*, 10:403–430.

Zepeda, A., Sengpiel, F., Guagnelli, M. A., Vaca, L., and Arias, C. (2004). Functional reorganization of visual cortex maps after ischemic lesions is accompanied by changes in expression of cytoskeletal proteins and NMDA and $GABA_A$ receptor subunits. *The Journal of Neuroscience*, 24:1812–1821.

Zhang, L. I., Tao, H. W., Holt, C. E., Harris, W. A., and Poo, M.-M. (1998). A critical window for cooperation and competition among developing retinotectal synapses. *Nature*, 395:37–44.

Zhao, L., and Chubb, C. (2001). The size-tuning of the face-distortion after-effect. *Vision Research*, 41:2979–2994.

Zhou, Y.-D., and Fuster, J. M. (2000). Visuo-tactile cross-modal associations in cortical somatosensory cells. *Proceedings of the National Academy of Sciences, USA*, 97:9777–9782.

Ziemke, T. (1999). Rethinking grounding. In Riegler, A., Peschl, M., and von Stein, A., editors, *Understanding Representation in the Cognitive Sciences: Does Representation Need Reality?*, 177–199. Dordrecht, The Netherlands: Kluwer.

Zucker, R. S. (1989). Short-term synaptic plasticity. *Annual Review of Neuroscience*, 12:13–31.

Zucker, S. W. (1995). Perceptual grouping. In Arbib, M. A., editor, *The Handbook of Brain Theory and Neural Networks*, 725–727. Cambridge, MA: MIT Press. First edition.

Author Index

Subject Index

Page numbers in **bold** refer to main discussion.